AF788824

MECHANICS OF FRACTURE INITIATION AND PROPAGATION

ENGINEERING APPLICATION OF FRACTURE MECHANICS

Editor-in-Chief: George C. Sih

1. G.C. Sih and L. Faria (eds.): *Fracture Mechanics Methodology*. Evaluation of Structural Components Integrity. 1984 ISBN 90-247-2941-6
2. E.E. Gdoutos: *Problems of Mixed Mode Crack Propagation*. 1984 ISBN 90-247-3055-4
3. A Carpinteri and A.R. Ingraffea (eds.): *Fracture Mechanics of Concrete*. Material Characterization and Testing. 1984 ISBN 90-247-2959-9
4. G.C. Sih and A. DiTommaso (eds): *Fracture Mechanics of Concrete*. Structural Application and Numerical Calculation. 1985 ISBN 90-247-2960-2
5. A. Carpinteri: *Mechanical Damage and Crack Growth in Concrete*. Plastic Collapse to Brittle Fracture. 1986 ISBN 90-247-3233-6
6. J.W. Provan (ed.): *Probabilistic Fracture Mechanics and Reliability*. 1987 ISBN 90-247-3334-0
7. A.A. Baker and R. Jones (eds.): *Bonded Repair of Aircraft Structures*. 1987 ISBN 90-247-3606-4
8. J.T. Pindera and M.-J. Pindera: *Isodyne Stress Analysis*. 1989 ISBN 0-7923-0269-9
9. G.C. Sih and E.E. Gdoutos (eds.): *Mechanics and Physics of Energy Density*. Characterization of Material – Structive Behaviour with and without Damage. 1991 ISBN 0-7923-0604-X
10. E.E. Gdoutos: *Fracture Mechanics Criteria and Applications*. 1990 ISBN 0-7923-0605-8
11. G.C. Sih: *Mechanics of Fracture Initiation and Propagation*. 1991 ISBN 0-7923-0877-8

Volume 11

Mechanics of Fracture Initiation and Propagation

Surface and volume energy density applied as failure criterion

G. C. Sih

Institute of Fracture and Solid Mechanics
Lehigh University, Bethlehem, Pennsylvania

Kluwer Academic Publishers

Dordrecht / Boston / London

Library of Congress Cataloging-in-Publication Data

```
Sih, G. C. (George C.)
   Mechanics of fracture initiation and propagation : surface and
 volume energy density applied as failure criterion / G.C. Sih.
      p.   cm. -- (Engineering application of fracture mechanics ;
 11)
   Includes bibliographical references and index.
   ISBN 0-7923-0877-8
   1. Fracture mechanics.   I. Title.   II. Series.
 TA409.S55  1990
 620.1'126--dc20                                          90-44394
```

ISBN 0-7923-0877-8

Published by Kluwer Academic Publishers,
P.O. Box 17, 3300 AA Dordrecht, The Netherlands.

Kluwer Academic Publishers incorporates the publishing programmes of
D. Reidel, Martinus Nijhoff, Dr W. Junk and MTP Press.

Sold and distributed in the U.S.A. and Canada
by Kluwer Academic Publishers,
101 Philip Drive, Norwell, MA 02061, U.S.A.

In all other countries, sold and distributed
by Kluwer Academic Publishers Group,
P.O. Box 322, 3300 AH Dordrecht, The Netherlands.

Printed on acid-free paper

All Rights Reserved
© 1991 Kluwer Academic Publishers
No part of the material protected by this copyright notice may be reproduced or
utilized in any form or by any means, electronic or mechanical,
including photocopying, recording or by any information storage and
retrieval system, without written permission from the copyright owner.

Printed in the Netherlands

Contents

Series on engineering application of fracture mechanics

Fracture mechanics technology has received considerable attention in recent years and has advanced to the stage where it can be employed in engineering design to prevent against the brittle fracture of high-strength materials and highly constrained structures. While research continued in an attempt to extend the basic concept to the lower strength and higher toughness materials, the technology advanced rapidly to establish material specifications, design rules, quality control and inspection standards, code requirements, and regulations for safe operation. Among these are the fracture toughness testing procedures of the American Society of Testing Materials (ASTM), the American Society of Mechanical Engineers (ASME) Boiler and Pressure Vessel Codes for the design of nuclear reactor components, etc. Step-by step fracture detection and prevention procedures are also being developed by the industry, government and university to guide and regulate the design of engineering products. This involves the interaction of individuals from the different sectors of the society that often presents a problem in communication. The transfer of new research findings to the users is now becoming a slow, tedious and costly process.

One of the practical objectives of this series on *Engineering Application of Fracture Mechanics* is to provide a vehicle for presenting the experience of real situations by those who have been involved in applying the basic knowledge of fracture mechanics in practice. It is time that the subject should be presented in a systematic way to the practising engineers as well as to the students in universities at least to all those who are likely to bear a responsibility for safe and economic design. Even though the current theory of linear elastic fracture mechanics (LEFM) is limited to brittle fracture behavior, it has already provided a remarkable improvement over the conventional methods not accounting for initial defects that are inevitably present in all materials and structures. The potential of the fracture mechanics technology, however, has not been fully recognized. There remains much to be done in constructing a quantitative theory of material damage that can reliably translate small specimen data to the design of large size structural components. The work of the physical metallurgists and the fracture mechanicians should also be brought together by reconciling the details of the material microstructure with the assumed continua of the computational methods. It is with the the aim of developing a wider appreciation of the fracture mechanics technology applied to the design of engineering structures such as aircrafts, ships, bridges, pavements, pressure vessels, off-shore structures, pipelines, etc. that this series is being developed.

Undoubtedly, the successful application of any technology must rely on the soundness of the underlying basic concepts and mathematical models and how they reconcile with each other. This goal has been accomplished to a large extent by the book series on *Mechanics of Fracture* started in 1972. The seven published volumes offer a wealth of information on the effects of defects or cracks in cylindrical bars, thin and thick plates, shells, composites and solids in three dimensions. Both static and dynamic loads are considered. Each volume contains an introductory chapter that illustrates how the strain energy criterion can be used to analyze the combined influence of defect size, component geometry and size, loading, material properties, etc. The criterion is particularly effective for treating mixed mode fracture where the crack propagates in a non-self similar fashion. One of the major difficulties that continuously perplex the practitioners in fracture mechanics is the selection of an appropriate fracture criterion without which no reliable prediction of failure could be made. This requires much discernment, judgement and experience. General conclusions based on the agreement of theory and experiment for a limited number of physical phenomena should be avoided.

Looking into the future the rapid advancement of modern technology will require more sophisticated concepts in design. The microchips used widely in electronics and advanced composites developed for aerospace applications are just some of the more well-known examples. The more efficient use of materials in previously unexperienced environments is no doubt needed. Fracture mechanics should be extended beyond the range of LEFM. To be better understood is the entire process of material damage that includes crack initiation, slow growth and eventual termination by fast crack propagation. Material behavior characterized from the uniaxial tensile tests must be related to more complicated stress states. These difficulties should be overcome by unifying metallurgical and fracture mechanics studies, particularly in assessing the results with consistency.

This series is therefore offered to emphasize the applications of fracture mechanics technology that could be employed to assure the safe behavior of engineering products and structures. Unexpected failures may or may not be critical in themselves but they can often be annoying, time-wasting and discrediting of the technical community.

Bethlehem, Pennsylvania
1987

G. C. SIH
Editor-in-Chief

Preface

The assessment of crack initiation and/or propagation has been the subject of many past discussions on fracture mechanics. Depending on how the chosen failure criterion is combined with the solution of a particular theory of continuum mechanics, the outcome could vary over a wide range. Modelling of the material damage process could be elusive if the scale level of observation is left undefined. The specification of physical dimension alone is not sufficient because time and temperature also play an intimate role. It is only when the latter two variables are fixed that failure predictions can be simplified. The *sudden* fracture of material with a pre-existing crack is a case in point. Barring changes in the local temperature,* the energy released to create a unit surface area of an existing crack can be obtained by considering the change in elastic energy of the system before and after crack extension. Such a quantity has been referred to as the critical energy release rate, G_c, or stress intensity factor, K_{1c}. Other parameters, such as the crack opening displacement (COD), path-independent J-integral, etc., have been proposed; their relation to the fracture process is also based on the energy release concept. These one-parameter approaches, however, are unable simultaneously to account for the failure process of crack initiation, propagation and onset of rapid fracture. A review on the use of G, K_1, COD, J, etc., has been made by Sih [1, 2].

More than two decades ago, Sih [3] proposed the use of the volume energy density as a fracture criterion in contrast to the idea of specific surface energy or surface energy density. This provided an alternative approach to failure prediction for the same stress solution. The distinctions were emphasized in the series *Mechanics of Fracture* [4] where failure analyses were carefully separated from stress analyses and addressed only in the introduc-

* Temperature change is inherently related to energy dissipation that could take place over a region atomic, microscopic or macroscopic in size.

tory chapters. As the volume energy density criterion offered new directions of research, opinions were diversified with reference to the reasons [5–10] why different criteria led to different predictions. The new criterion gained momentum and credibility as the aircraft [11–17] and pipeline [18] industries recognized its usefulness in analyzing failure. Other past applications have been summarized in [19] and two recent books [20, 21]. It is with the intention of updating the development of the volume energy density (dW/dV) criterion that this book is devoted to the mechanics of fracture initiation and propagation. No effort will be made to review the general field of fracture mechanics, coverage of which can be found in most textbooks dealing with failure analysis, but rather the emphases will be placed on failure as characterized by the combined effects of dilatation and distortion. To this end, much of the material in this volume is extracted from the series *Mechanics of Fracture* [4]. Thermoelastic and hygrothermoelastic behavior of cracks [22] and isoenergy density theory [23, 24] applied to analyze crack initiation and growth are additional topics normally not covered elsewhere. Not to be underestimated is the unique association of dW/dV with the surface energy density dW/dA [23, 24]. In retrospect, it was the neglect of the rate change of volume with surface dV/dA that led to the decoupling of dW/dA and dW/dV; that is, the limit of $dV/dA \to 0$ contributed to the elimination of size effect in the development of continuum mechanics theories. The inclusion of size/time/temperature interaction effects requires retainment of finiteness of element size or dV/dA in the analysis. Achieved in the isoenergy density theory formulation [23, 24] is the adjoinment of stress and failure analysis such that they are no longer considered as separate entities but unified as one operation. A unique characterization of the material damage process can thus be made, free from possible ambiguities arising from specifying material properties at the atomic, microscopic and macroscopic scale level. Synchronization of the deformation and temperature fluctuation leads to failure predictions that can differ quantitatively and qualitatively from those found in the classical theories of continuum mechanics. These situations are discussed in [25–28] and can be found from the references in this book.

Chapter 1 begins with a discussion of mixed mode fracture where the applied load is directed at an angle with respect to the crack plane. The additional presence of a Mode II stress field leads to nonself-similar crack growth, a condition that is excluded from the definition of the energy release rate quantity G or the Mode I stress intensity factor k_1. As a departure from the classical approach of Griffith [29], failure is assumed to initiate in a unit volume of material at a finite distance ahead of the main crack. A strain energy density factor S being the coefficient of the $1/r$ singular behavior of the volume energy density dW/dV can thus be defined. The distance r is measured from the crack tip. The stationary values of S with reference to the coordinates of the element under consideration are assumed to coincide with the location of crack initiation. The elastic crack tip stress field involving both k_1 and k_2 is employed to illustrate how the direction of crack initiation can be determined from the relative minimum of S or $S_{\min}$ with reference to the angle θ at a fixed value of r. The onset of crack initiation is assumed to occur when $S_{\min}$ reaches a critical value S_{cr} which can be related to k_{1c} even though the stress field involves both k_1 and k_2. Crack growth directions for an angle crack subjected to uniaxial tension and compression are obtained for the same material or S_{cr} value. The results agreed well with available experimental data for plexiglass plates in tension and glass plates in compression. They also explained why the tensile and compressive strength of relatively brittle materials could differ by a wide margin if the orientation of the initial defects in the specimens is not accounted for in the analysis.

Crack growth in three dimensions is much more complicated than that in two dimensions. This is discussed in Chapter 2. Instead of considering the failure of one element ahead of the crack tip, there would be a series of elements next to a crack border that may fail simultaneously. In addition to Mode I and II stress fields, Mode III can also prevail. The ways in which Mode I, II and III are combined can change along the crack front as in the case of an embedded elliptical crack subjected to general loading conditions. Intuition does not always apply in three dimensions. At a critical stress lower than Mode I, crack extension may occur under a mixed mode stress field. This corrresponds to a through crack whose plane is tilted with reference to the plate surface giving rise to Mode III in addition to Mode I. Nonalignment of the load and crack plane is quite common for specimens made of high-strength metal alloys. The fracture surface would be warped as in cup-and-cone failure. Prediction of a crack growth profile is made in segments by assuming that S_j/r_j for the j-th growth increment remains constant on the crack front. This assumption is applied to obtain the segment of crack growth initiating from a three-dimensional elliptical crack under tension and compression at an angle. Exhibited are the highly nonself-similar features of the new crack surface.

Presented in Chapter 3 are crack extensions in thin plates and shells. Simplifications on plate and shell thickness and boundary conditions can affect the stress solution near the crack tip. The conventional subdivision of thin and thick plates or shells does not hold in regions where the stress or strain gradients are influenced by geometric discontinuities. The requirement $(h/R_0)_{\max} \leq \frac{1}{20}$ is not sufficient for satisfying the thin shell assumptions. Here, h is the shell thickness and R_0 the smallest radius of curvature. At small distances (as compared with h) from the crack, the shell would appear to be thick and the assumptions* of the classical thin shell theory would lead to an inaccurate description of the local stress state. The replacement of boundary conditions on the twisting moment and shearing force by an equivalent of Kirchhoff shear results in the satisfaction of only four boundary conditions instead of five, as required on the crack surface. As an *a priori*, the qualitative character of the crack front stresses in a plate or shell should be the same as that derived from the three-dimensional theory of elasticity. By including the transverse shear deformation, Sih and Hagendorf [30] have developed a tenth-order thin shell theory such that satisfaction of the five natural boundary conditions on the crack surface is made possible. This is an improvement over the classical theory which considers only four conditions. Solutions showing the influence of shell curvature, shell thickness, crack length, etc., on the stress intensity factor are given and they can differ appreciably from those obtained by the classical theory corresponding to an eighth-order system of equations. In general, the state of plane strain should be preserved near the crack front. Except in a thin layer next to a free surface, the normal stresses are related as $\sigma_z = \nu(\sigma_x + \sigma_y)$, in which σ_z and σ_x, σ_y are the transverse and in-plane normal stresses while ν stands for the Poisson ratio. The stress solution in the surface layer can be constructed separately by adopting the concept of a boundary layer within which the stress intensity factors decrease sharply and tend to zero** on the free surface. This desirable feature of the solution can serve as a useful guideline in selecting the appropriate plate or shell theory.

As pointed out in Chapter 4, the stresses and displacements due to dynamic loading can be greatly different from those associated with corresponding static loading. This difference is attributable to inertia effects in the physical system. Inertia alters the load transmission characteristics and the physical properties of the material depending on the time-dependent nature of loading. Irreversible damage such as microcracks can be inflicted into the material. In a continuum theory, this is reflected macroscopically through differences in parameters identified with dynamic properties from those for static properties. Both the yield and ultimate strength tend to increase with the rate of loading for most metals. When a body is loaded quickly, there is less time for yielding or microscopic damage to take place and, hence, energy is released within a short time interval by rapid macro-

---

* Points near the crack edge in a plate or shell, which lie on the normal to the middle surface before deformation, cannot be assumed to remain on the normal after deformation. This in effect implies that shear forces cannot be disregarded.

** The three-dimensional theory of elasticity yields a stress solution different from $1/\sqrt{r}$ on the free surface where r is a radial distance. This implies the vanishing of stress intensity factors. Since a hypothetical surface of zero thickness cannot be conceived physically, the necessity of admitting a boundary layer in engineering applications is apparent.

crack propagation leading to material separation or fragmentation. Fracture mechanics is concerned with the prediction of this instability point through a parameter generally known as the fracture toughness. The question that often arises is whether or not this parameter should be different for a given material when loaded dynamically and when loaded statically. Conceptually speaking, if, indeed, fracture toughness is to be measured by the load at the point of macrocrack instability, then attention should be focused on the sudden creation of a small segment of new macrocrack surface. Any damage to the material which occurred at the microscopic level, such as yielding, prior to load instability does no more than reduce the available energy for driving the macrocrack and certainly should not be considered as part of the energy used to create a new macrocrack surface. There is reason to believe that local yielding (microscopic damage) and the path of brittle fracture (macroscopic damage) do not coincide. In fact, the macrocrack always tends to propagate around or bypass the yielded portion of the material [31].

Much has been discussed with reference to crack behavior subsequent to the point of incipient fracture. Whether or not a crack will arrest itself depends directly on the load transmission characteristics of the system. If energy is constantly supplied to the crack tip region, crack motion will prevail and continue to do so. Such a situation arises in a constant load specimen where the moving crack is engulfed in a uniform tensile stress field. A fixed head displacement specimen will eventually lead to crack arrest since the energy input to the crack tip region decreases in time provided, of course, that the specimen size is sufficiently large so that complete separation does not occur. Another feature of self-arrest appears in situations where the distance between the energy source and the crack tip increases in time as in the splitting of a long double-cantilever specimen. The load transmission characteristics of a specimen or structure can directly affect the velocity and acceleration of a crack; their association with the resistance of a material to dynamic fracture has not been made clear in the literature. Scatter in dynamic fracture data is specimen specific because of overload, an effect that deserves more attention.

One of the prerequisites for applying the energy release rate concept is that the solid must contain an initial line crack and that the same functional form of the energy expression applies before and after the crack extends by a small segment. These conditions are not valid for blunt notches because they do not fail in a self-similar manner. Since the volume energy density criterion focuses attention on a local element in the material, its application is independent of the existence and/or shape of the defect. Chapter 5 shows how failure initiation sites can be identified along the notch front with relative minima of the strain energy density factor S. Elasticity solutions for an internal elliptical notch and two external hyperbolic notches are used in conjunction with the S-criterion to determine the failure load.

The results are shown to approach those of the crack solutions for narrow ellipses or hyperbolas, and to display satisfactory agreement with published experimental data under both tensile and compressive loading conditions for the case of an elliptical notch. Crack trajectories emanating from notch boundaries are also found to give reasonable predictions if the interim period between fracture initiation and termination is sufficiently small.

The presence of moisture and/or temperature can affect the crack initiation and growth condition. Sih, Michopoulos and Chou [22] have derived the volume energy density function for diffusion and deformation coupled problems. A brief summary of their work is presented in Chapter 6 for the T300/5208 composite. Specific results are discussed for a crack subjected to moisture or temperature boundary conditions. Moisture diffusion can elevate the intensity of the local stresses ahead of the crack in addition to that caused by mechanical loads. A coupled thermoelastic solution is obtained if moisture changes are neglected. Further reduction to the classical linear thermoelasticity theory can be made when the coupling effects become vanishingly small. Slow crack growth owing to thermal changes can thus be determined without considering the interdependence of crack motion and temperature field.

Chapter 7 ponders on the failure of composites where the usual assumption of material isotropy and homogeneity may no longer hold in general. Failure can correspond to fracture of the different constituents within a composite and cannot always be modelled by a dominant crack as in metal alloys. The fracture toughness concept for the homogeneous material would not apply if the critical energy released at instability changed with the load angles and sizes of the cracks. Material anisotropy and inhomogeneity imply a nonuniqueness in failure characterization and defy standardization. Specimen tests would only serve to control the quality of composites in fabrication.

The failure prediction of composite structures from a small-scale laboratory specimen is not reliable as many of the processing variables cannot be accounted for theoretically, and agreement between analysis and experiments can only be made in special instances. Therefore, no attempt will be made to characterize composites. Discussed will be those cases where fracture mechanics can be applied. More recent developments on the application of energy density to analyze composite failure have been made by Sih [32, 33].

Two analytical models are proposed for predicting the failure of unidirectional fiber-reinforced composites subjected to a load that is oriented at an arbitrary angle with the crack plane. The first model assumes that the composite is an ideal homogeneous anisotropic body and the second assumes that failure takes place in a layer of matrix material. Brittle fracture is assumed to occur when the strain energy stored in an element ahead of the crack reaches a critical value, S_{cr}. Results are presented for E-glass fiber-reinforced plastics, stainless steel fiber-reinforced aluminum and graphite fiber epoxy materials and compared with available experimental data. The degree of material nonhomogeneity – which depends on the fiber volume fraction and the difference of elastic properties of the fiber and matrix – affects the state of affairs in the region ahead of the crack tip. This nonhomogeneity is found to be more significant in composites with a plastic resin matrix than with metal. The matrix-cracking model agrees well with the experimental results on precracked Scotchply 1002 sheets provided that the load direction is not too closely aligned with the fiber orientation.

The strain energy density criterion applies equally well to the dynamic fracture of unidirectional fibrous composites. Under impact, waves are reflected and refracted from the fibers and create a complex pattern. By assuming that matrix cracking is the dominant mode of failure, the maximum intensification of the local strain energy density is found to occur very quickly when the waves traveling from one end of the crack tip reach the other. The critical dynamic stress level is determined as a function of the geometry and material parameters of the composite. When S_{min} exceeds S_{cr}, the crack begins to propagate with a velocity that depends on the degree of nonhomogeneity in the composite system. This dependence is discussed for a crack running in the matrix material.

Two failure modes that commonly arise in angle-ply laminates made of layered unidirectional composites are presented. They are known as thru-lamina and interlaminar cracking. The latter has also been referred to as delamination. Results are given for a balanced four-layered laminate. Possible delamination is assumed to coincide with the locations of minimum strain energy density. Such a location indeed occurs in a small region confined to the free edges.

Emphasized in Chapter 8 is the importance to support experimental data by theory. Empiricisms provide little understanding on the fundamentals of the material damage process. Quantitative assessment of ductile fracture, fatigue, etc., is still not being carried out on a consistent basis. There is no reason why the different types of failure modes could not be characterized by a single theory in mechanics.

Material separation is inherently an irreversible path-dependent process. It is not sufficient to model the degree of irreversibility simply as nonlinear behavior on a load–deformation diagram. Unless the nonlinear behavior is correlated with material damage, no physical meaning can be attached to the data and no new experience can be gained to advance theory. Typically, in metal fracture, a crack first spreads in a stable fashion before rapid growth. During this stage, the crack can change shape and size depending on the local rate of energy dissipation. The complete process should be determinable from the test specimens with or without considering the presence of initial defects. The validity of a theory is judged from its ability to correlate data collected from material and/or fracture testings.

The stage is thus set for Chapter 9 to discuss how the material damage process can be uniquely characterized by applying the isoenergy density theory [23, 24]. Analyses on stress and failure are considered to be mutually interacting. Surface energy density dW/dA is assumed to depend on the volume energy density dW/dV. The former dominates near the surface where load is initially applied, while the latter increases with time as the disturbances spread into the volume. No longer neglected is the rate change of volume with surface area, dV/dA. Thermal and mechanical effects are synchronized for describing the process of nonequilibrium material damage. Stress and strain is not presumed but determined according to the combined influence of load/geometry/material for all material elements. Exhibited is the highly nonhomogeneous character of the state of affairs near a crack tip, even though the initial material may be homogeneous. Another major difference is the dependence of surface tractions on the local inertia forces which is significant near the crack free surface. Results obtained from the isoenergy density theory differed significantly from those found from plasticity theory for a stationary and slowly growing crack.

Finally, the size/time/temperature interaction is emphasized for relating observations made at the atomic, microscopic and macroscopic scale. Sih and Chao [34] have shown that, at one scale level, failure modes belonging to the same material damage process may have a different appearance at another scale level. Depending on the loading type and rate, dilatation/distortion and cooling/heating tend to flip-flop as the scale level of observation is altered; this also depends, of course, on the selected time interval within which the events are observed. This alternating behavior is intrinsic to the nonequilibrium and nonhomogeneous character of material damage.

The contents in the nine chapters of this volume give a summary of the progress made by those associated with the Institute of Fracture and Solid Mechanics at Lehigh University over the past two decades. Many of the ideas and analyses will become more fully understood as more results are made available. I would specifically like to thank Mrs Barbara DeLazaro and Mrs Connie Weaver for their assistance not only in typing the manuscripts but also in proofreading.

March 1990 G. C. SIH
Bethlehem, Pennsylvania

References

[1] Sih, G. C., Fracture mechanics of engineering structural components, ***Fracture Mechanics Methodology: Evaluation of Structural Components Integrity***, edited by G. C. Sih and L. L. Faria, Martinus Nijhoff Publishers, The Netherlands, pp. 35–101 (1984).

[2] Sih, G. C., Fracture mechanics in two decades, ***Fracture Mechanics: Perspectives and Directions*** (Twentieth Symposium), ASTM STP 1020, R. P. Wei and R. P. Gangloff, eds., American Society for Testing and Materials, Phila., pp. 9–28 (1989).

[3] Sih, G. C., Some basic problems in fracture mechanics and new concepts, ***Journal of Engineering Fracture Mechanics*** **5**, No. 2, pp. 365–377 (1973).

[4] ***Mechanics of Fracture***, Vol. I to Vol. VII, edited by G. C. Sih, Martinus Nijhoff Publishers, The Netherlands (1973–1982).

[5] Sih, G. C., Reply to discussion on some observations on Sih's strain energy density approach for fracture prediction by I. Finnie and H. O. Weiss, ***Journal of International Fracture*** **10**, pp. 279–283 (1974).

[6] Sih, G. C. and Gdoutos, E. E., Discussion on a modified strain energy density criterion applied to crack propagation by P. S. Theocaris and N. P. Andrianopoulos, ***Journal of Applied Mechanics*** **49**, pp. 678–679 (1982).

[7] Sih, G. C., Moyer, E. T., Jr. and Gdoutos, E. E., Discussion on the Mises elastic–plastic boundary as the core region in fracture criteria by P. S. Theocaris and N. P. Andrianopoulos, ***Journal of Engineering Fracture Mechanics*** **18**, No. 3, pp. 731–734 (1983).

[8] Sih, G. C. and Madenci, E., Discussion on a further examination on the application of the strain energy density theory to the angled crack problem by K. J. Chang, ***Trans. ASME, Journal of Applied Mechanics*** **50**, pp. 234–236 (1983).

[9] Sih, G. C. and Tzou, D. Y., Discussion on criteria for brittle fracture in biaxial tension by S. K. Maiti and R. A. Smith, ***Journal of Engineering Fracture Mechanics*** **21**, No. 5, pp. 977–981 (1985).

[10] Sih, G. C. and Tzou, D. Y., Discussion on subsurface crack propagation due to surface traction in sliding wear by H. C. Sin and N. P. Suh, ***Journal of Applied Mechanics*** **52**, pp. 239–240 (1985) and Tzou, D. Y. and Sih, G. C., Crack growth prediction of subsurface crack in yielded material, ***Journal of Applied Mechanics*** **52**, pp. 237–238 (1985).

[11] Au, N. N. and Lin, S. N., Mixed mode fractures in space launch pressure vessels, ***Mixed Mode Crack Propagation***, edited by G. C. Sih and P. S. Theocaris, Sijthoff and Noordhoff Publishers, The Netherlands, pp. 55–76 (1981).

[12] Au, N. N., Safe-life analysis of structures subjected to general in-plane loadings, ***Application of Fracture Mechanics to Materials and Structures***, G. C. Sih, E. Sommer and W. Dahl, eds., Martinus Nijhoff Publishers, The Netherlands, pp. 317–336 (1984).

[13] Au, N. N., Application of the S_c-theory to flightworthiness evaluation of space launch vehicles, ***Role of Fracture Mechanics in Modern Technology***, edited by G. C. Sih and H. Nisitani, VNU Science Publishers, The Netherlands, pp. 69–88 (1987).

[14] Badaliance, R., Application of strain energy density factor of fatigue crack growth analysis, ***Journal of Engineering Fracture Mechanics*** **13**, No. 3, pp. 657–666 (1980).

[15] Badaliance, R., Fatigue life prediction: metal and composites, ***Fracture Mechanics Technology***, edited by G. C. Sih and L. Faria, Martinus Nijhoff Publishers, The Netherlands, pp. 1–34 (1984).

[16] Janardhana, M. N., Brown, K. C. and Jones, R., Designing for tolerance to impact damage at fastener holes in graphite/epoxy laminates under compression, ***Journal of Theoretical and Applied Fracture Mechanics*** **5**, No. 1, pp. 51–55 (1986).

[17] Jones, R., Molent, L., Baker, A. A. and Davis, M. J., Bonded repair of metallic components: thick sections, ***Journal of Theoretical and Applied Fracture Mechanics*** **9**, No. 1, pp. 61–70 (1988).

[18] Civallero, M., Mirabile, M. and Sih, G. C., Fracture mechanics in pipeline technology, in ***Analytical and Experimental Fracture Mechanics***, edited by G. C. Sih and M.

Mirabile, Sijthoff and Noordhoff Publishers, The Netherlands, pp.157–174 (1981).

[19] *Absorbed Specific Energy and/or Energy Density Criterion*, G. C. Sih, E. Czoboly and F. Gillemot, eds., Martinus Nijhoff Publishers, The Netherlands (1982).

[20] Gdoutos, E. E., *Problems of Mixed Mode Crack Propagation*, Vol. II in series on *Engineering Application of Fracture Mechanics*, Martinus Nijhoff Publishers, The Netherlands (1984).

[21] Carpinteri, A., *Crack Growth and Material Damage in Concrete: Plastic Collapse and Brittle Fracture*, Vol. V in series on *Engineering Application of Fracture Mechanics*, Martinus Nijhoff Publishers, The Netherlands (1986).

[22] Sih, G. C., Michopoulos, J. G. and Chou, S. C., *Hygrothermoelasticity*, Martinus Nijhoff Publishers (1986).

[23] Sih, G. C., Mechanics and physics of energy density and rate of change of volume with surface, *Journal of Theoretical and Applied Fracture Mechanics* **4**, No. 3, pp. 157–173 (1985).

[24] Sih, G. C., Thermomechanics of solids: nonequilibrium and irreversibility, *Journal of Theoretical and Applied Fracture Mechanics* **9**, No. 3, pp. 175–198 (1988).

[25] Sih, G. C. and Tzou, D. Y., Heating preceded by cooling ahead of crack: macrodamage free zone, *Journal of Theoretical and Applied Fracture Mechanics* **6**, No. 2, pp. 103–111 (1986).

[26] Sih, G. C., Tzou, D. Y. and Michopoulos, J. G., Secondary temperature fluctuation in cracked 1020 steel specimen loaded monotonically, *Journal of Theoretical and Applied Fracture Mechanics* **7**, No. 2, pp. 79–89 (1987).

[27] Sih, G. C., Lieu, F. L. and Chao, C. K., Thermal/mechanical damage of 6061-T6 aluminum tensile specimen, *Journal of Theoretical and Applied Fracture Mechanics* **7**, No. 2, pp. 67–78 (1987).

[28] Sih, G. C. and Tzou, D. Y., Irreversibility and damage of SAFC-40R steel specimen in uniaxial tension, *Journal of Theoretical and Applied Fracture Mechanics* **7**, No. 1, pp. 23–30 (1987).

[29] Griffith, A. A., The phenomena of rupture and flow in solids, *Philosophical Transactions of the Royal Society* **A221**, pp. 163–198 (1921).

[30] Sih, G. C. and Hagendorf, H. C., A new theory of spherical shells with cracks, thin-shell structures: *Theory, Experiment and Design*, Y. C. Fung and E. E. Sechler, eds., Prentice Hall, New Jersey, pp. 519–545 (1974).

[31] Sih, G. C., Fracture toughness concept, *Properties Related to Fracture Toughness*, ASTM STP 605, American Society for Testing and Materials, Phila., pp. 3–15 (1976).

[32] Mechanics of composite materials and structures, in *Advanced Composite Materials and Structures*, G. C. Sih and S. E. Hsu, eds., VNU Science Publishers, The Netherlands, pp. 3–24 (1987).

[33] Sih, G. C., Microstructure and damage dependence of advanced composite material behavior, in *Composite Material Response: Constitutive Relations and Damage Mechanisms*, G. C. Sih, G. F. Smith, I. H. Marshall and J. J. Wu, eds., Elsevier Applied Science, London, pp. 1–23 (1987).

[34] Sih, G. C. and Chao, C. K., Scaling of size time/temperature associated with damage of uniaxial tensile specimens – Part 1: Progressive damage in uniaxial tensile specimen, pp. 93–108 and Part 2: Progressive damage in uniaxial compressive specimen, pp. 109–119, *Journal of Theoretical and Applied Fracture Mechanics* **12**, No. 2 (1989).

1

A special theory of crack propagation

1.1. Historical remarks

The Griffith concept [1, 2] of imperfection instability in a solid was the first step toward predicting the fracture strength of solids. The basic idea behind his theory is that a crack will begin to propagate if the elastic energy released by its growth is greater than the energy required to create the fractured surfaces. As a model, Griffith considered the problem of a crack of length $2a$ in a plate under tension σ as in Figure 1.1(a). He then found that the critical stress σ_{cr} required for crack growth is

$$\sigma_{\mathrm{cr}} a^{1/2} = \left(\frac{2E\gamma}{\pi}\right)^{1/2} \tag{1.1}$$

where E is Young's modulus and γ the specific surface energy. Since the quantity $(2E\gamma/\pi)^{1/2}$ contains only material constants, the factor $\sigma_{\mathrm{cr}} a^{1/2}$ should be an intrinsic material parameter. Twice the specific surface energy γ is equal to the critical elastic energy release rate $G_{1\mathrm{c}}$, i.e., $G_{1\mathrm{c}} = 2\gamma$. The experiments Griffith performed on glass show that the values of $\sigma_{\mathrm{cr}} a^{1/2}$ were indeed the same over a wide range of crack lengths.

The concept of crack energy release leads to serious drawbacks in carrying out the mathematical details for cracks in a combined stress field. The energy release concept assumes the direction of crack propagation to be known *a priori*. Hence, the Griffith theory can only treat problems with the crack lying in a plane normal to the applied stress as in Figure 1.1(a). A simple question, such as what will be the direction of crack propagation if the crack was inclined at an angle β to the loading axis, has not yet been answered satisfactorily. In such a case, equation (1.1) is obviously no longer valid. Moreover, a reversal of loading, as shown in Figure 1.1(b), will produce crack propagation along a different path. A realistic theory of fracture mechanics should be able to explain the fracture phenomena for both types of loading on the inclined crack in Figures 1.1(a) and 1.1(b).

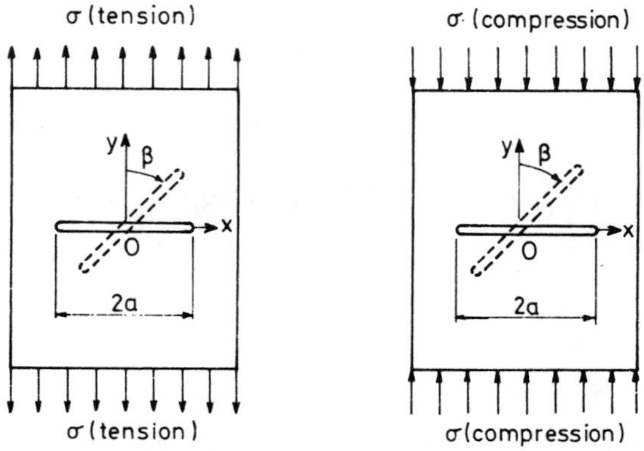

(a) Crack growth (b) No crack growth

Figure 1.1. **A central crack in a plate.**

Irwin [3] in applying the Griffith's concept to solve fracture problems recognized the importance of the intensity of the local stress field. He proposed three modes of crack extension which are identified by their respective stress intensity factors k_1, k_2 and k_3. The Mode I intensity factor k_1 is in fact related to the Griffith energy release rate G_1 as

$$G_1 = \frac{\pi k_1^2}{E} \quad \text{(generalized plane stress).} \tag{1.2}$$

Similar relations for $G_2 = \pi k_2^2/E$ and $G_3 = k_3^2/2\mu$ with μ being the shear modulus of elasticity were formally computed by assuming that the crack extends in a plane collinear with the original crack as in Mode I. In fact, the physical meaning of Mode II and Mode III crack extension was never questioned and only a few attempts [4, 5] have been made to study the direction of Mode II and Mode III crack propagation. To the author's best knowledge, experimental values of G_{2c} and G_{3c} were never recorded and hence the present day understanding of crack propagation is restricted to Mode I problems.

The crack model of Barenblatt [6] should also be mentioned. In his theory, Barenblatt chooses to remove the $r^{-1/2}$ stress singularity by considering a cohesive zone along a line ahead of the crack. The criterion of fracture is based on the concept of a critical modulus of cohesion. Subsequently, Dugdale [7] used a mathematically similar but physically different model to study the crack tip plastic zone sizes in thin sheets. A generalization of some of these physical concepts was later made by Rice [8] through the application of a path-independent integral which is identical in form to a component of the energy momentum tensor introduced by Eshelby [9] to characterize generalized forces on dislocations in elastic fields. The above crack models which put more emphasis on the physical aspects of the material, are all inherently restricted to cracks that extend along the line of load symme-

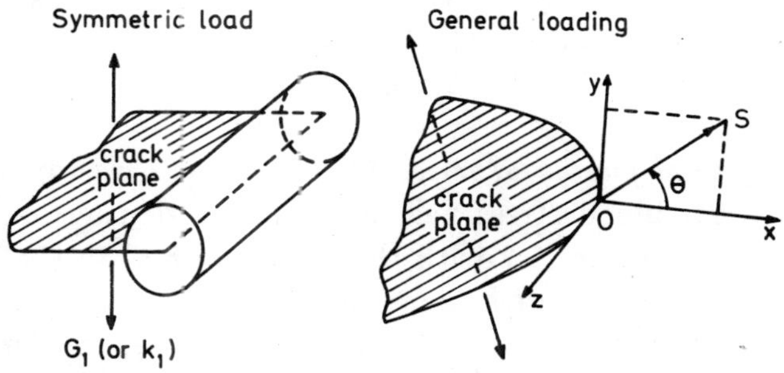

(a) Scalar theory (b) Director theory

Figure 1.2. **Comparison of scalar and director theory.**

try. Although this limitation is not essential to the experimental studies of fracture toughness, it presents a major set-back to the prediction of applied stress for crack initiation in structural members. In the design of bridges and aircraft, the stress state around the crack tip is in most cases of the mixed type where the assumption of Mode I fracture would be unrealistic.

1.2. The strain energy density concept

In a series of recent papers, Sih [10–12] has proposed a theory of fracture based on the field strength of the local strain energy density which marks a fundamental departure from the classical and current concepts. The theory requires no calculation on the energy release rate and thus possesses the inherent advantage of being able to treat all mixed mode crack extension problems for the first time. Unlike the conventional theory of G and k which measures only the amplitude of the local stresses, the fundamental parameter in the new theory, the "strain energy density" factor S, is also direction sensitive. The difference between k (or G) and S is analogous to the difference between a scalar and a vector.

Referring to Figures 1.2(a) and 1.2(b), the Griffith–Irwin theory can be viewed as a scalar theory in that it specifies only the critical value of a scalar G_{1c} (or k_{1c}) at incipient fracture. The direction of crack propagation is always preassumed to be normal to the load. Moreover, the crack front must be straight such that G or k does not vary along the leading edge of the crack. In addition, a scalar theory cannot yield the correct material parameter if two or more stress intensity factors are present along the crack border. The S-factor in the Sih theory behaves like a director. It senses the direction of least resistance by attaining a stationary value with respect to the angle θ as indicated in Figure 1.2(b). As it will be shown, the stationary value of $S_{\min}$ can be used as an intrinsic material parameter whose value at the point of crack instability S_{cr} is independent of the crack geometry and loading. In the general context, the Griffith–Irwin theory is the special case when $\theta = 0$ and the director S coincides with the x-axis.

Special form of local energy density. Consider the three-dimensional case of a crack in a combined stress field and focus attention on a coordinate system (x, y, z) in Figure 1.2 with the x-axis normal to the crack, the y-axis perpendicular to the crack plane and the z-axis tangent to the crack border. While the origin 0 traces the crack periphery, the functional form of the stress components in r, θ [13] remains unchanged, i.e.,

$$\sigma_x = \frac{k_1}{(2r)^{1/2}} \cos(\theta/2)[1 - \sin(\theta/2)\sin(3\theta/2)] - \\ - \frac{k_2}{(2r)^{1/2}} \sin(\theta/2)[2 + \cos(\theta/2)\cos(3\theta/2)] + \cdots \quad (1.3a)$$

$$\sigma_y = \frac{k_1}{(2r)^{1/2}} \cos(\theta/2)[1 + \sin(\theta/2)\sin(3\theta/2)] + \\ + \frac{k_2}{(2r)^{1/2}} \sin(\theta/2)\cos(\theta/2)\cos(3\theta/2) + \cdots \quad (1.3b)$$

$$\tau_{xy} = \frac{k_1}{(2r)^{1/2}} \cos(\theta/2)\sin(\theta/2)\cos(3\theta/2)] + \\ + \frac{k_2}{(2r)^{1/2}} \cos(\theta/2)[1 - \sin(\theta/2)\sin(3\theta/2)] + \cdots \quad (1.3c)$$

$$\sigma_z = 2\nu \frac{k_1}{(2r)^{1/2}} \cos(\theta/2) - 2\nu \frac{k_2}{(2r)^{1/2}} \sin(\theta/2) + \cdots \quad (1.3d)$$

$$\tau_{xz} = -\frac{k_3}{(2r)^{1/2}} \sin(\theta/2) + \cdots \quad (1.3e)$$

$$\tau_{yz} = \frac{k_3}{(2r)^{1/2}} \cos(\theta/2) + \cdots \quad (1.3f)$$

where the nonsingular terms have been dropped and r, θ are the polar components in the yz-plane (Figure 1.3). For an elastic material, the strain energy stored in the element $dV = dx\,dy\,dz$ under a general three-dimensional stress system is

$$dW = \left[\frac{1}{2E}(\sigma_x^2 + \sigma_y^2 + \sigma_z^2) - \frac{\nu}{E}(\sigma_x\sigma_y + \sigma_y\sigma_z + \sigma_z\sigma_x) + \\ + \frac{1}{2\mu}(\tau_{xy}^2 + \tau_{xz}^2 + \tau_{yz}^2)\right] dV. \quad (1.4)$$

Substituting equations (1.3) into (1.4) yields the quadratic form for the

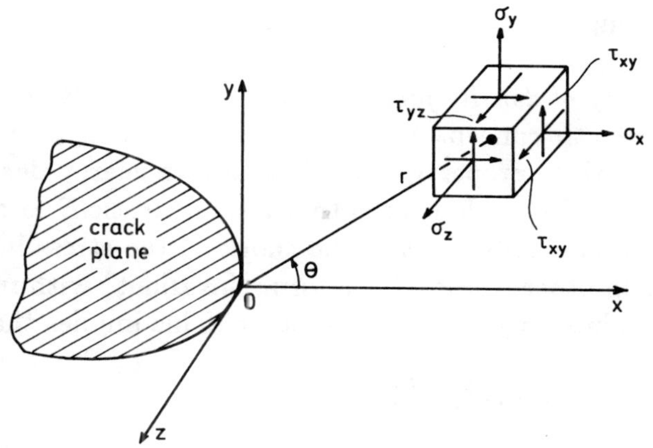

Figure 1.3. Stress components near crack border.

strain energy density function

$$\frac{dW}{dV} = \frac{1}{r}(a_{11}k_1^2 + 2a_{12}k_1k_2 + a_{22}k_2^2 + a_{33}k_3^2) + \cdots \tag{1.5}$$

Note that the higher order terms in r have been neglected and that the strain energy density function near the crack possesses a $1/r$ energy singularity. Hence, the quadratic

$$S = a_{11}k_1^2 + 2a_{12}k_1k_2 + a_{22}k_2^2 + a_{33}k_3^2 \tag{1.6}$$

represents the amplitude or the intensity of the strain energy density field and it varies with the polar angle θ in Figure 1.3. The coefficients a_{ij} $(i, j = 1,\ 2,\ 3)$ are given by

$$a_{11} = \frac{1}{16\mu}[(3 - 4\nu - \cos\theta)(1 + \cos\theta)] \tag{1.7a}$$

$$a_{12} = \frac{1}{16\mu}2\sin\theta[\cos\theta - (1 - 2\nu)] \tag{1.7b}$$

$$a_{22} = \frac{1}{16\mu}[4(1-\nu)(1-\cos\theta) + (1+\cos\theta)(3\cos\theta - 1)] \tag{1.7c}$$

$$a_{33} = \frac{1}{4\mu} \tag{1.7d}$$

where ν is Poisson's ratio.

Equations (1.3) represent the general form of the crack border stress field involving the three stress intensity factors k_1, k_2, and k_3. For two-dimensional problems where the crack extends in the xy-plane, the stress intensity factors do not vary along the crack front and S depends only on one variable, namely the angle θ. In three dimensions, k_1, k_2 and k_3 may occur simultaneously and they can also vary from point to point on the crack border. An example of this is given in [12].

Invariant property of density factor. Before proceeding with the application of the field strength concept to the fracture problem, it is worthwhile to examine the invariant property of the local strain energy density field. The important point to be made here is that the quadratic S in equations (1.6) has a value that is independent of the choice of the stress intensity factors k_1, k_2 and k_3. In other words, S is a form "invariant" with reference to one system of stress intensity factors to another. This implies that

$$S(k_1, k_2, k_3) = S(k_1', k_2', k_3') \tag{1.8}$$

where k_i' and k_i are the stress intensity factors of two different systems and S possesses the inherent property of being a constant. Equation (1.8) defines a quadric surface for each state of stress intensity factors and the discussion of this quadric parallels closely that of the strain or stress quadric of Cauchy in the theory of elasticity. Briefly, equation (1.6) may be solved as an eigenvalue problem. For a nontrivial solution, the three eigenvalues are found to be

$$\lambda^2 - (a_{11} + a_{22})\lambda + (a_{11}a_{22} - a_{12}^2) = 0 \tag{1.9a}$$

$$\lambda_3 = a_{33}. \tag{1.9b}$$

Using relations of a_{ij} in equation (1.7), equation (1.9a) gives

$$\lambda_1 = \frac{1}{16\mu}[4(1-2\nu) + (1-\cos\theta)^2] \tag{1.10a}$$

$$\lambda_2 = \frac{1}{16\mu}(1+\cos\theta)^2. \tag{1.10b}$$

Hence, the quadric surface in eigenform is

$$\lambda_1 K_1^2 + \lambda_2 K_2^2 + \lambda_3 K_3^2 = \text{constant} \tag{1.11}$$

and the associated transformation

$$\begin{bmatrix} \cos(\theta/2) & \sin(\theta/2) & 0 \\ -\sin(\theta/2) & \cos(\theta/2) & 0 \\ 0 & 0 & 1 \end{bmatrix} \begin{bmatrix} k_1 \\ k_2 \\ k_3 \end{bmatrix} = \begin{bmatrix} K_1 \\ K_2 \\ K_3 \end{bmatrix}. \tag{1.12}$$

The solution determines the points on a central quadric surface for which the distance from the origin is stationary relative to neighboring points. With this background, the property of the quadric S will be shown to play an important role in the theory of crack propagation.

1.3. Fundamental hypotheses on crack initiation and direction

Since the strain energy density factor S in equation (1.6) has some attributes of the intensity of a force field associated with some type of potential, it is natural to inquire on the relationship between S and the potential energy Σ in the system. If the cracked body is subjected to tractions only, then the potential energy is equal to the negative of the strain energy.* Now let P stand for the potential energy per unit volume of the element located at a distance r from the crack border as shown in Figure 1.3, i.e., $P = \mathrm{d}\Sigma/\mathrm{d}V$, and similarly let U be the strain energy per unit volume given by $U = \mathrm{d}W/\mathrm{d}V$. Making use of the relation $P = -U$ and $U = S/r$, the potential energy per unit volume becomes

$$P = -\frac{S}{r}. \tag{1.13}$$

Basic assumptions. Two fundamental hypotheses** of crack extension will now be laid down:

Hypothesis (1): ***The crack will spread in the direction of maximum potential energy density.***

Hypothesis (2): ***The critical intensity S_{cr} of this potential field governs the onset of crack propagation.***

Note that for crack propagation to take place in the xy-plane the direction of maximum potential energy density must be found. In two-dimensional problems, the direction of crack propagation can be determined by a single variable θ and hence hypothesis (1) can be satisfied by the application of the calculus of variations. A necessary condition for the potential energy density P to have a stationary value is that

$$\frac{\partial P}{\partial \theta} = 0 \quad \text{at } \theta = \theta_0. \tag{1.14}$$

The value of θ_0, which makes P a maximum, determines the angle of the plane along which the crack spreads and can be found by further requiring that

$$\frac{\partial^2 P}{\partial \theta^2} < 0 \quad \text{at } \theta = \theta_0 \tag{1.15}$$

which is a position of unstable equilibrium. From the canonical form of S in equation (1.11), it can be concluded that S is positive definite and thus P is negative.

---

* The opposite holds for displacement loading conditions.

** These hypotheses will be further discussed in a general theory of crack propagation being developed by Sih [14].

Rewriting the conditions in equations (1.14) and (1.15) in terms of the strain energy density function renders

$$\frac{\partial S}{\partial \theta} = 0, \qquad \frac{\partial^2 S}{\partial \theta^2} > 0 \quad \text{at } \theta = \theta_0 \tag{1.16}$$

which are the necessary and sufficient conditions for S to be a minimum. Hypothesis (1) is thus equivalent to the assumption that

crack initiation will start in a radial direction along which the strain energy density is a minimum.

An essential point to be made here is that the above criterion is based on the local density of the energy field in the crack tip region and requires no special assumption on the direction in which the energy released by the separating crack surfaces is computed as in the Griffith theory [1, 2] and others. This removes the fundamental difficulties involved in the past for computing energy release rate in mixed mode problems. It is now clear that any fracture criterion based on a single stress parameter such as k_1 alone will not be sufficient to describe the problem of mixed mode fracture. Furthermore, the Mode II energy release rate commonly computed as $G_2 = \pi k_2^2/E$, where k_2 is the stress intensity factor for skew symmetric loadings, is not valid since in Mode II the crack does not run straight ahead as in Mode I. Hence the crack energy release for a mixed mode problem cannot be obtained by simply adding G_1 and G_2. It should also be emphasized that the so-called "crack closure" method though conceptually correct has misled a number of previous investigators to solve the branch crack or kinked crack problem, analytically as well as numerically. This type of boundary value problems is extremely difficult to solve and thus far no effective solutions have been found.

1.4. Prediction of crack growth direction

One of the least understood problems in fracture mechanics has been on the estimate of applied loads at crack instability in mixed mode situations where a crack can spread in any direction depending on the relative orientation of the load and crack position. Strangely enough, this problem has never been examined seriously. Admittedly, this is not an important problem in fracture toughness testing studies since all experiments could be carried out under Mode I crack extension. However, the omission of the mixed mode effect in predicting the applied stress to trigger fracture in structures could be dangerously optimistic. For this reason, the classical fracture mechanics approach is basically a theory for characterizing materials and has limited use in structural design.

A major shortcoming of the classical theory is that it cannot predict the direction of crack propagation. It is basically a theory for cracks that propagate in a self-similar manner, although Griffith did postulate that the crack will open up in the plane normal to the direction of maximum stress. At first sight, this statement appears to be plausible and has received a certain degree of acceptance in the literature. The validity and completeness of Griffith's postulate has never been clearly understood and is not a simple matter to settle. It will be shown that the predictions based on the maximum stress criterion do deviate from those of the strain energy density criterion and can be questioned simply on physical grounds. Additional discussions on this subject will be brought up in the general theory of crack propagation in a future communication.

At the present time, it is more pertinent to illustrate the method of determining the direction of crack propagation by considering a few simple examples. All of the crack problems treated by the energy density theory are taken to be in a state of plane strain.

Mode I crack extension. In his classical paper, Griffith examined the problem of an infinite body containing a central crack of length $2a$ subjected to applied stress σ normal to the crack plane as in Figure 1.1(a) or 1.4(a). Because of load symmetry, the direction of crack propagation never entered into the problem. Suppose now that this is an unknown in the problem, then it is necessary to apply equations (1.14) and (1.15) or (1.16).

For the Griffith crack

$$k_1 = \sigma a^{1/2}, \qquad k_2 = 0 \tag{1.17}$$

and hence equation (1.6) takes the simple form

$$S = \frac{\sigma^2 a}{16\mu}[(3 - 4\nu - \cos\theta)(1 + \cos\theta)]. \tag{1.18}$$

Differentiating S with respect to θ and setting $\partial S/\partial\theta = 0$, two possible solutions are found

$$\theta = 0 \quad \text{and} \quad \cos\theta = 1 - 2\nu. \tag{1.19}$$

The second derivative of S with θ gives

$$\frac{\partial^2 S}{\partial\theta^2} = \frac{\sigma^2 a}{8\mu}[\cos 2\theta - (1 - 2\nu)\cos\theta]. \tag{1.20}$$

Inserting the results in equation (1.19) into (1.20), it is found that the solution $\theta_0 = 0$ yields $\partial^2 S/\partial\theta^2 > 0$ and thus S is a minimum:

$$S_{\min} = \frac{(1 - 2\nu)\sigma^2 a}{4\mu}. \tag{1.21}$$

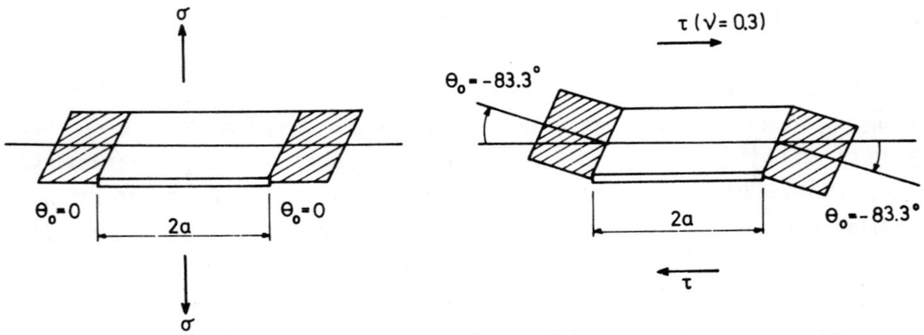

(a) Mode I (b) Mode II

Figure 1.4. Mode I and Mode II crack extension.

The plane $\theta_0 = 0$ corresponds to the direction of maximum potential energy, a position of unstable equilibrium. The critical applied stress to initiate crack growth becomes

$$\sigma_{cr} = [4\mu S_{cr}/(1-2\nu)a]^{1/2} \tag{1.22}$$

where the parameter $\sigma_{cr}a^{1/2}$ is a material constant as in the classical case. Thus for cracks propagating in a self-similar fashion S_{cr} can be related to γ in the Griffith theory and to k_{1c} or G_{1c} as follows:

$$\sigma_{cr}^2 a = [4\mu/(1-2\nu)]S_{cr} = [2E/\pi(1-\nu^2)]\gamma = k_{1c}^2. \tag{1.23}$$

Mode II crack extension. Let an infinite solid with a through crack of length $2a$ in Figure 1.4(b) be subjected to shear stresses τ such that the crack occupies a plane of skew symmetry. For this problem

$$k_1 = 0, \qquad k_2 = \tau a^{1/2} \tag{1.24}$$

and the direction of crack growth is no longer obvious. Applying the conditions stated in equation (1.16) to

$$S = \frac{\tau^2 a}{16\mu}[4(1-\nu)(1-\cos\theta) + (1+\cos\theta)(3\cos\theta - 1)] \tag{1.25}$$

it is found that

$$\cos\theta_0 = \frac{1-2\nu}{3} \tag{1.26}$$

which gives S a minimum or P a maximum. The other solution $\sin\theta = 0$ is not of interest in discussing crack propagation. The predicted fracture angle θ_0 in equation (1.26) depends on Poisson ratio ν as shown in Table 1.1.

A point to be recognized is that the energy density theory predicts the crack to initiate in a plane which is no longer collinear with the crack itself. Furthermore, this direction depends on the material property of the solid.

Table 1.1. Mode II fracture angle.

ν	0.0	0.1	0.2	0.3	0.4	0.5
θ_0	−70.5°	−75.6°	−79.3°	−83.3°	−87.2°	−90.0°

This marks a fundamental departure from the classical concept of fracture mechanics. The minimum value of S is found to be

$$S_{\min} = \frac{[8(1-\nu) - 4\nu^2]\tau^2 a}{48\mu}. \tag{1.27}$$

In passing, it should be mentioned that the maximum stress criterion in [4] predicts a fixed angle of $\theta_0 = -70.5°$ which corresponds to a material with zero Poisson's ratio in Table 1.1.

Mixed mode crack extension. The first study of the initial direction of crack growth in the presence of both k_1 and k_2 was made in [4] for the problem of a crack of length $2a$ inclined at an angle β with the loading axis as in Figure 1.1. The stress intensity factors k_1 and k_2 for this problem are [15]

$$k_1 = \sigma a^{1/2} \sin^2 \beta \tag{1.28a}$$

$$k_2 = \sigma a^{1/2} \sin \beta \cos \beta \tag{1.28b}$$

where σ is the applied stress. It was assumed that the crack will start to extend in the plane which is normal to the maximum circumferential stress σ_θ in accordance with the condition

$$k_1 \sin \theta_0 + k_2(3 \cos \theta_0 - 1) = 0 \tag{1.29}$$

for determining the initial angle of crack growth θ_0. Using equation (1.28), k_1 and k_2 may be eliminated. This renders

$$\sin \theta_0 + (3 \cos \theta_0 - 1) \cot \beta = 0, \quad \beta \neq 0 \tag{1.30}$$

which contains no elastic constants. This result implies that for the crack configuration and loading condition given in Figure 1.1 the initial angle of crack growth is independent of the material properties.

Turning now to the strain energy density theory which states that crack propagates in the direction of

$$S = \sigma^2 a(a_{11} \sin^2 \beta + 2a_{12} \sin \beta \cos \beta + a_{22} \cos^2 \beta) \sin^2 \beta \tag{1.31}$$

being a minimum. The coefficients a_{ij} in equation (1.31) as derived from equations (1.6) and (1.28) are those given in equations (1.7). Differentiating equation (1.31) with respect to θ and setting the result to zero, the fracture angle θ_0 for a given position of the crack specified by β can be calculated from

$$2(1-2\nu) \sin(\theta_0 - 2\beta) - 2 \sin[2(\theta_0 - \beta)] - \sin 2\theta_0 = 0, \quad \beta \neq 0. \tag{1.32}$$

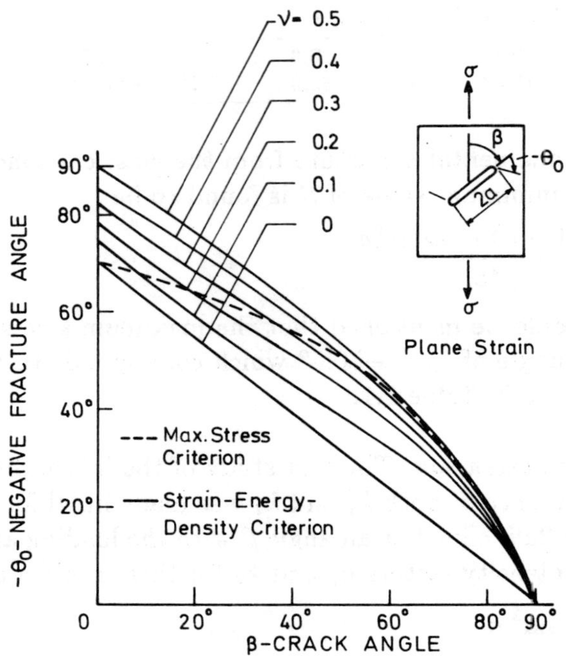

Figure 1.5. Crack angle versus fracture angle in tension.

(a) ***Uniaxial tension.*** The numerical results of equation (1.32) for negative values of θ_0 and positive σ are shown in Figure 1.5 which is a plot of the fracture angle θ_0 ***versus*** the crack angle β from 0° to 90°. The curve based on the maximum stress criterion which is dotted agrees well with equation (1.32) for large values of β and represents a lower bound for small values of β. In general, it can be taken as an average curve.

The validity of these predictions can be checked with the results of a series of experiments [4] performed on the specimen in Figure 1.5. Plexiglass sheets of approximately $9'' \times 18'' \times \frac{3}{16}''$ were used with a central crack of approximately $2''$ in length positioned at angles of β from 30° to 80° in increments of 10°. The initial fracture angles at both ends of the crack were measured. The experimental data for four sets of tests are given in Table 1.2 with $(\theta_0)_{avg}$ being the average fracture angle of all the measured values. The last two rows give the theoretical calculations of equation (1.29) for the maximum stress criterion and equation (1.32) based on the strain energy density criterion with $\nu = \frac{1}{3}$. As it can be seen the agreement between theory and experiment is good. Additional experiments are needed to verify the influence of Poisson's ratio on the fracture angle θ_0 as predicted by the solid curves in Figure 1.5.

(b) ***Uniaxial compression.*** There exists another set of solutions of positive θ_0 to equation (1.32) which will also yield a minimum on S. Physically,

Table 1.2. Measured and calculated values of the fracture angle.

β		30°	40°	50°	60°	70°	80°
	1	−64°	−55.5°	−50°	−40°	−29°	−17°
θ_0	2	−60°	−52°	−50°	−43.5°	−30.5°	−18°
@ right	3	−63°	−57°	−53°	−44.5°	—	−15.5°
	4	—	−57°	−52°	−43.5°	—	—
	1	−65°	−58°	−50.5°	−44°	−31.5°	−18.5°
θ_0	2	—	−53°	−52°	−40°	−31°	−17.5°
@ left	3	−60°	−55°	−51.5°	−46°	−31.5°	−17°
	4	—	−57°	−50°	−43°	—	—
$(\theta_0)_{avg}$		−62.4°	−55.6°	−51.1°	−43.1°	−30.7°	−17.3°
Eq.(1.29)		−60.2°	−55.7°	−50.2°	−43.2°	−33.2°	−19.3°
Eq.(1.32)		−63.5°	−56.7°	−49.5°	−41.5°	−31.8°	−18.5°

the positive values of θ_0 correspond to the problem of Figure 1.1b in which the inclined crack is now under uniaxial compression. Since S depends on σ^2, equation (1.31) contains both the solutions of uniaxial tension ($+\sigma$) and compression ($-\sigma$). A plot of the positive fracture angle θ_0 against β is given in Figure 1.6 for different values of Poisson's ratio. In contrast to tensile loading, where the crack tends to become horizontal, the crack path under uniaxial compression is towards the direction of loading. Such a phenomenon has indeed been observed by Hock and Bieniawski [17] who have tested a number of glass plates with an inclined crack under uniaxial compression. Unfortunately, they did not report the initial angle of crack extension so that a comparison of the theoretical results in Figure 1.6 with experiments cannot be made at this time. Nevertheless, they did publish the values of the applied stress σ_{cr} to initiate crack growth for different positions of the crack. It will be shown that their results on the variations of $\sigma_{cr}a^{1/2}$ with β are indeed predictable from the present theory.

1.5. Intrinsic property of strain energy density factor

One of the principal aims of fracture mechanics is to characterize the behavior of materials in the presence of flaws or cracks. This requires a clear distinction of the difference between S and S_{cr}. The strain energy density factor S is simply the leading coefficient of the series expansion of dW/dV about the crack point $r = 0$ in Figure 1.3 and it varies as a function of the polar angle θ. In the vicinity of the crack tip, S may be regarded as a crack resistance force with the interpretation that the crack tends to run in the direction of least resistance that corresponds to S reaching a minimum. Once S has attained a critical value S_{cr} at the point of incipient fracture it may be regarded as a crack extension force which should be independent of loading conditions and crack configuration. When verified experimentally,

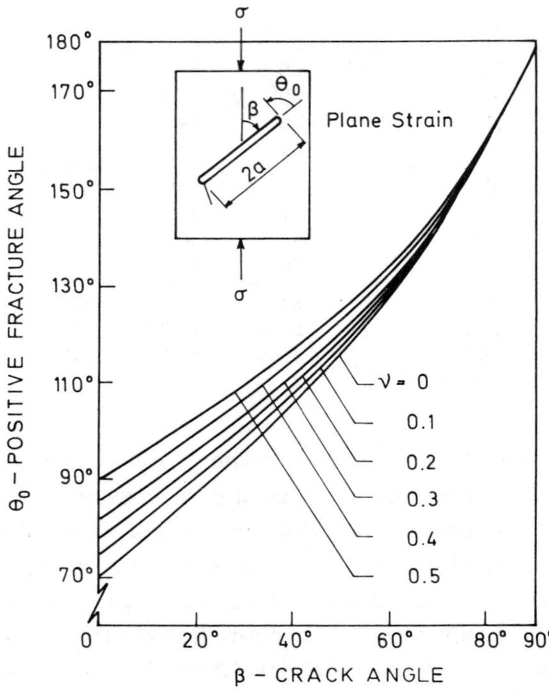

Figure 1.6. Crack angle versus fracture angle in compression.

S_{cr} can be used as a material constant that serves as an indication of the fracture toughness of the material.

In order to become familiar with the strain energy density factor S, which can be viewed as a vector-like quantity, further considerations will be given to the inclined crack problem. For the case of uniaxial tension, the crack will spread in the negative θ-direction in a plane for which the crack resistance force S is a minimum. Figure 1.7 shows a plot of $16\mu S_{\min}/\sigma^2 a$ against β for ν varying from 0 to 0.4 inclusive. In general, the quantity $16\mu S_{\min}/\sigma^2 a$ increases with the crack angle β reaching a maximum along the axis of Mode I crack extension. As $S_{\min}$ will be used as a material constant, the above statement implies that the lowest value of the applied stress σ_{cr} to initiate crack propagation occurs at $\beta = \pi/2$ for a material with low Poisson's ratio. A similar graph for uniaxial compression is shown in Figure 1.8. An interesting point to be observed here is that the quantity $16\mu S_{\min}/\sigma^2 a$ first increases with the crack angle β reaching a peak and then decreases in magnitude. The peak value is a function of Poisson's ratio. This suggests that given $16\mu S_{\min}/\sigma^2 a =$ constant there exists a critical angle β_0 at which the critical applied compressive stress is a minimum.

Having completed the preliminaries, the theoretical results will now be compared with the experimental data obtained on DTD 5050-5$\frac{1}{2}$% Zn alu-

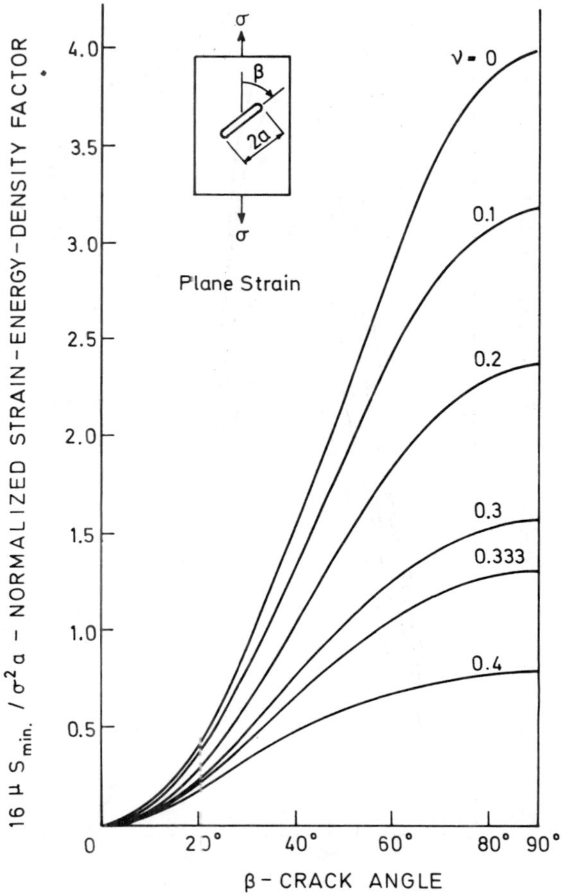

Figure 1.7. Variations of density factor with crack angle for tensile loading.

minum alloy in [18]. All the tests were carried out in specimens containing an inclined crack. The measured values of $\sigma_{cr}a^{1/2}$ for different crack sizes and failure loads are plotted against the crack angle β in Figure 1.9. The solid and dotted curves represent the predicted values for the aluminum alloy with k_{1c} equal to 28.2 ksi $\sqrt{\text{in.}}$ and 29.2 ksi $\sqrt{\text{in.}}$, respectively. The agreement is good. The same data is given in Figure 1.10 with the critical stress density factor S_{cr} normalized with respect to its value $(S_{cr})_{\pi/2}$ corresponding to Mode I crack extension and S_{cr} remained essentially constant. Although corrections for plasticity ahead of the crack can be made, it is not considered to be essential in this discussion.

The fracture mechanics of an inclined crack under compression is basically different from that of extension. For glass with a Poisson's ratio of $\nu =$ 0.25, the theoretical curve in Figure 1.11 predicts a critical angle of $\beta_0 \simeq$ 37° at which the applied stress to initiate fracture is a minimum. In [17],

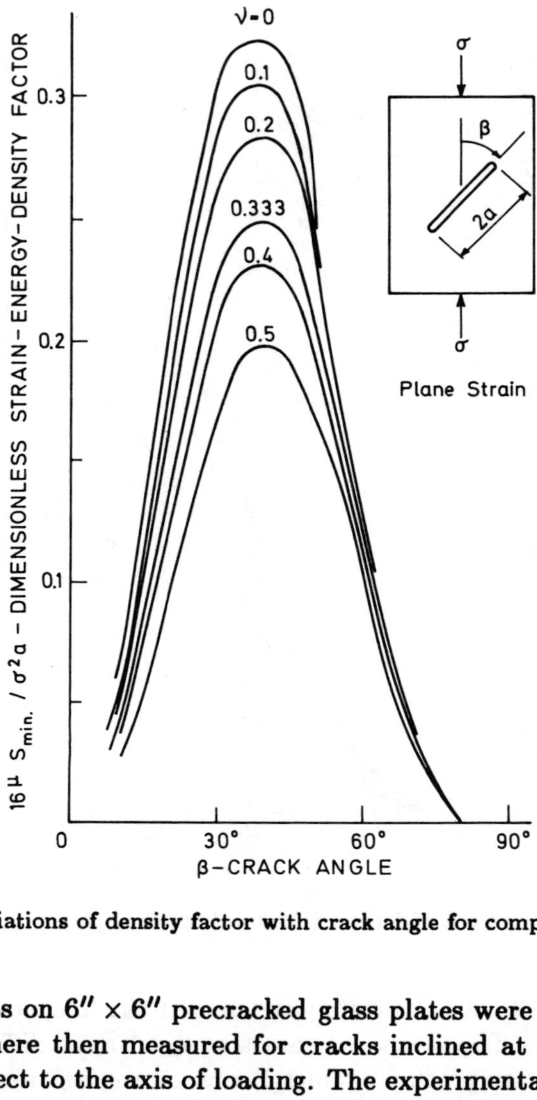

Figure 1.8. Variations of density factor with crack angle for compressive loading.

compression tests on 6″ × 6″ precracked glass plates were performed. The critical loads where then measured for cracks inclined at various different angles with respect to the axis of loading. The experimental curve is dotted and indicates a critical angle $\beta_0 \simeq 31°$. For cracks loaded under remote compression, there is a tendency for the crack surfaces to come into contact and to rub against one another. Thus, the important point to be made here may not be in the quantitative agreement of theory and experiment but in the trend of the failure stress variations with crack angle for fracture under compression which has been predicted by a theory based on the concept of a strain energy density factor.

In addition, the new theory can explain that the apparent compressive strength of brittle materials can be many times greater than their tensile strength depending upon the geometry and loading conditions. Griffith has attempted to use the maximum stress criterion to explain this apparent

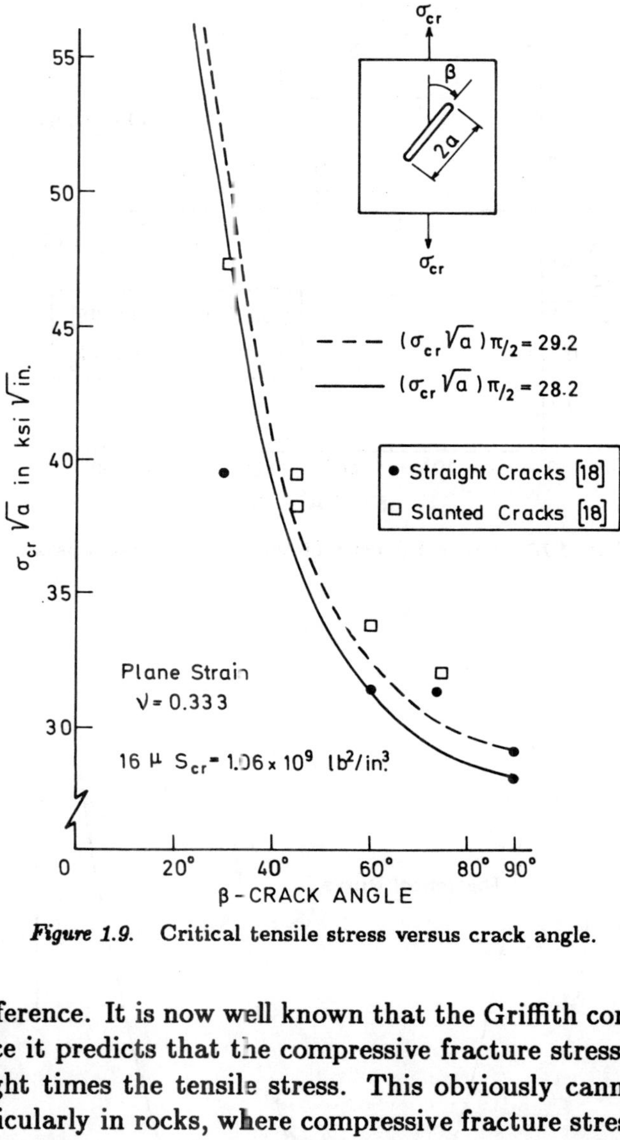

Figure 1.9. Critical tensile stress versus crack angle.

strength* difference. It is now well known that the Griffith concept must be defective since it predicts that the compressive fracture stress of a material is exactly eight times the tensile stress. This obviously cannot be true in general, particularly in rocks, where compressive fracture stress in excess of one hundred times the tensile stress have been reported. McClintock and Walsh [16] pointed out this defect in the Griffith conception and modified the Griffith model by assuming that the cracks close up under compression, developing friction on the sliding crack surfaces. However, as they point out, the coefficient of surface friction would have to be unrealistically high in order to explain compressive fracture stress in excess of ten times the

* The strength or the intrinsic property of a material and the maximum stress at failure are two different quantities. The failure to observe such a distinction has caused numerous misconceptions in the open literature.

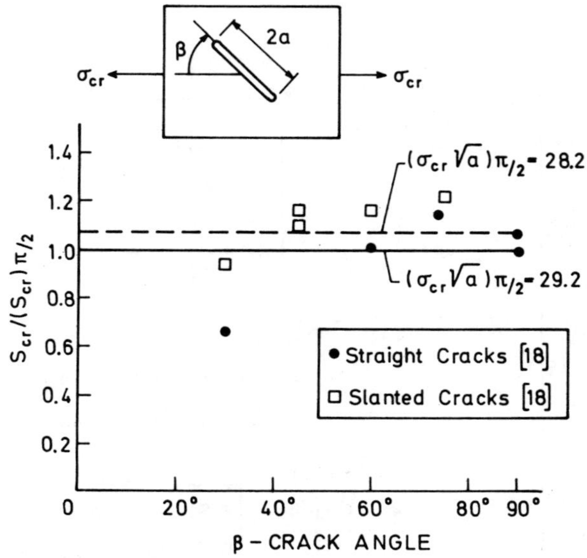

Figure 1.10. **Critical density factor as a material constant.**

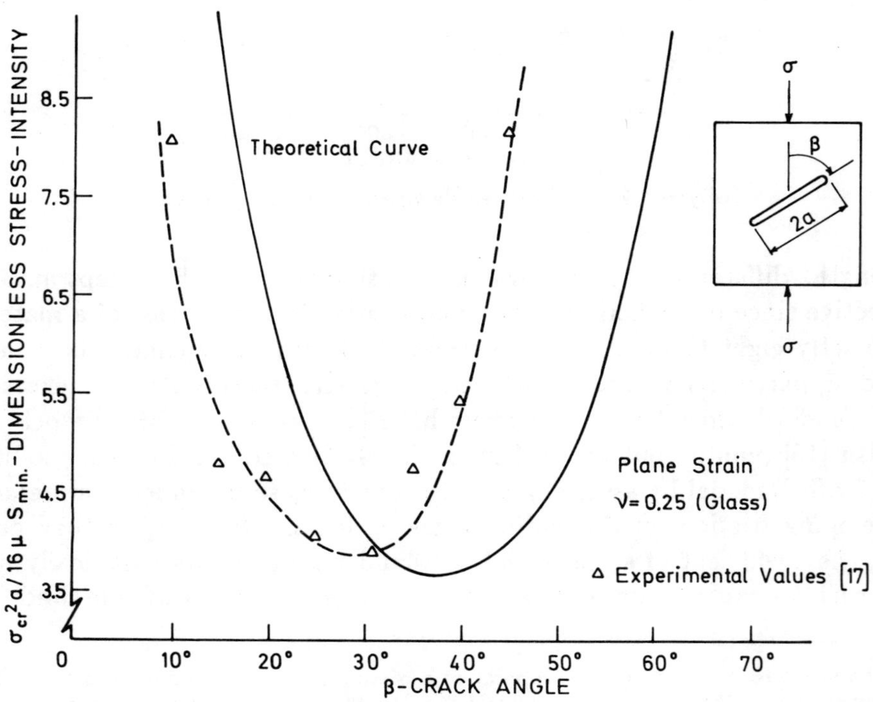

Figure 1.11. **Critical compressive stress versus crack angle.**

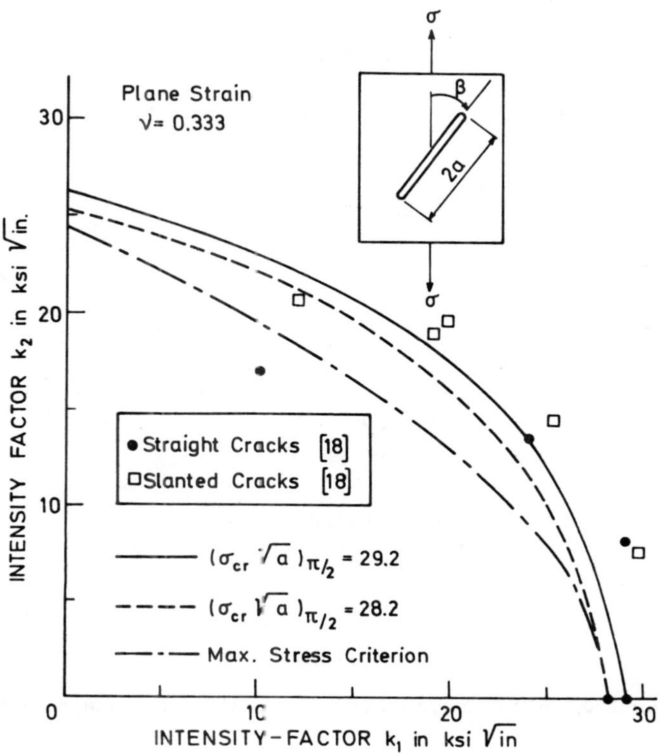

Figure 1.12. Mixed mode fracture criterion of k_{1c} versus k_{2c}.

tensile stress. The deficiency lies in using the maximum stress theory as the criterion of fracture and the failure to realize that the fracture stress does not represent the strength of the material.

The answer to the problem posed by Griffith is given by the curves in Figures 1.7 and 1.8 or Figures 1.9 and 1.11. On a qualitative basis, it is easily seen that for sufficiently large values of β, say 60° or 70°, the ratio of σ_{cr} (compression) to σ_{cr} (tension) can be very large. This ratio depends on the Poisson ratio and the position of the crack relative to the applied load. It should be emphasized that the material possesses only one strength characterized by the critical value of the strain energy density factor S_{cr} regardless of the nature of loading. The fact that Griffith chooses to distinguish between compressive strength and tensile strength of the same material is in itself a weakness of the theory.

1.6. Mixed mode fracture criterion

Having shown that S_{cr} can be used as a material constant, a mixed mode fracture criterion can be stated. The critical values of k_1 and k_2, i.e., k_{1c}

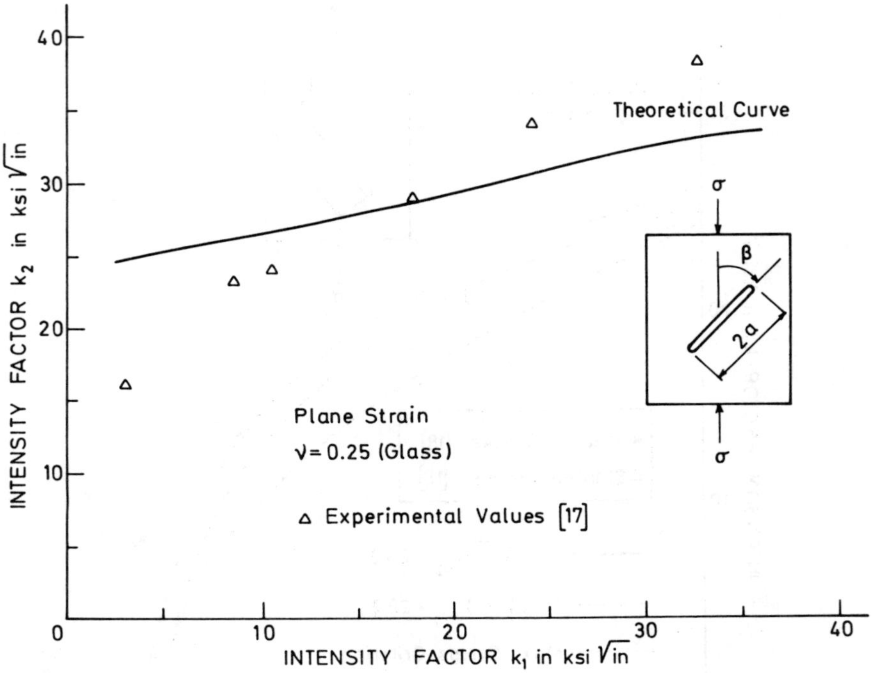

Figure 1.13. Mixed mode criterion for cracks under compression.

and k_{2c}, in a given problem will lie on a curve in the k_1, k_2-plane determined by the hypotheses stated earlier. The theoretical values of k_1 and k_2 may be determined from equations (1.31) and (1.32) for a given material, i.e., a specific value of S_{cr}. From the reported values of k_{1c} in [18] on the aluminum alloy, the values of $4.8(\mu S_{cr})^{1/2} = 28.2$ psi $\sqrt{\text{in.}}$ and 29.2 psi $\sqrt{\text{in.}}$ are used and the theoretical plots of k_2 *versus* k_1 are given in Figure 1.12. The third curve represents the prediction based on a criterion of maximum stress [4]. It is evident that the strain energy density theory is closer to the experimental results. The same observation has been made on plexiglass plates tested in [4]; i.e., the measured points of (k_{1c}, k_{2c}) lie outside of the k_1k_2-curve of the maximum stress criterion.

The k_1k_2-curve governing the mixed mode fracture of cracks under remote compression is basically different from that of tension. First, the curve does not intersect the k_1-axis which implies the obvious fact that Mode I crack extension does not exist in compression. This can be easily verified by solving equations (1.31) and (1.32) for k_1 and k_2 with the constraint that the crack angle and fracture angle satisfies the relations dictated by the curves given in Figure 1.6. For a glass with $\nu = 0.25$, the theoretical prediction gives a slanted curve in the k_1, k_2-plane as shown in Figure 1.13. Again the qualitative feature of the solution is in agreement.

1.7. Concluding remarks

A theory based on the concept of a strain energy density factor S has been presented. The stationary values of this density factor can predict the direction of crack growth under mixed mode conditions, whereas it is well known that the classical theory of Griffith lacks this basic feature of predicting the direction of crack propagation. In addition, the critical value S_{cr} has been shown to be independent of the crack geometry and loading and hence it can be used as a material parameter for measuring the resistance against fracture. Using the example of an inclined crack, the theory predicts two basic solutions, one for tensile loading and the other for compressive loading. The theoretical results for both of these loadings are in good agreement with the available experimental data on crack extension in combined stress fields where Mode I is mixed with Mode II.

Strain energy density factors S for a variety of mixed mode crack problems have already been developed and experimentations are underway to establish the quadric surface of failure for each of the following problem areas:

1. Cracks in a generally anisotropic body.
2. Cohesive and adhesive failure of layered composites.
3. Vibration and impact of cracked bodies.
4. Classical and higher order plate and shell theories.
5. Miscellaneous crack problems of fundamental nature.

The present theory has opened the door to a new and fruitful area of research in fracture mechanics. The future progress will depend largely on the willingness of the practitioners in the field to accept this new concept. There is no doubt that the classical theory will have to be replaced so that technology in fracture mechanics can advance and provide solutions to numerous previously unanswered questions, particularly in the area of applying fracture mechanics to structure problems. The negligence of the mixed mode effect in design can lead to drastic errors on the prediction of the applied stress to cause fracture.

This work represents a simplified version of a more general theory of fracture [14] which has already been developed far beyond the basic concept presented here. The special theory being the first deviation from the classical thought can be easily understood and immediately applied for resolving many practical problems in the field.

1.8. References

[1] Griffith, A. A., The phenomena of rupture and flow in solids, *Philosophical Transactions, Royal Society of London, Series* **A221**, pp. 163–198 (1921).

[2] Griffith, A. A., The theory of rupture, *Proceedings of First International Congress of Applied Mechanics, Delft*, pp. 55–93 (1924).

[3] Irwin, G. R., Fracture mechanics, in *Structural Mechanics*, Pergamon Press, New York, pp. 557–592 (1960).

[4] Erdogan, F. and Sih, G. C., On the crack extension of plates under plane loading and transverse shear, *Journal of Basic Engineering* **85**, No. 4, p. 519–527 (1963).

[5] Knauss, W. G., An observation of crack propagation in anti-plane shear, *International Journal of Fracture Mechanics* **6**, pp. 183–187 (1970).

[6] Barenblatt, G. I., On some basic ideas of the theory of equilibrium cracks forming during brittle fracture, *Problems of Continuum Mechanics*, Society for Industrial and Applied Mathematics, Philadelphia, Pa., pp. 22–38 (1961).

[7] Dugdale, D. S., Yielding of steel sheets containing slits, *Journal of Mechanics and Physics of Solids* **8**, pp. 100–104 (1960).

[8] Rice, J. R., A path independent integral and the approximate analysis of strain concentration by notches and cracks, *Journal of Applied Mechanics* **35**, pp. 379–386 (1986).

[9] Eshelby, J. D., The determination of the elastic field of an ellipsoidal inclusion and related problems, *Proceedings of Royal Society London, Series A* **241**, pp. 376–396 (1957).

[10] Sih, G. C., Strain energy density factor applied to mixed mode crack problems, *International Journal of Fracture Mechanics* **10**, pp. 305–321 (1974).

[11] Sih, G. C., Some basic problems in fracture mechanics and new concepts, *Journal of Engineering Fracture Mechanics* **5**, pp. 365–377 (1973).

[12] Sih, G. C., Application of the strain energy density theory to fundamental fracture problems, *Recent Advances in Engineering Science*, edited by T. S. Chang, Scientific Publishers, Inc., Boston, Vol. 6, pp. 221–234 (1975).

[13] Sih, G. C. and Liebowitz, H., Mathematical theories of brittle fracture, *Mathematical Fundamentals of Fracture*, edited by H. Liebowitz, Academic Press, New York, Vol. 2, pp. 67–190 (1968).

[14] Sih, G. C., Outlook on fracture mechanics, *The Mechanism of Fracture*, edited by V. S. Goel, American Society of Metal Publication, Salt Lake City, Utah, pp. 1–6, (1985).

[15] Sih, G. C., Paris, P. C. and Erdogan, F., Crack-tip stress-intensity factor for plane extension and plate bending problems, *Journal of Applied Mechanics* **29**, No. 2, pp. 306–312 (1962).

[16] McClintock, F. A. and Walsh, J. B., Friction on Griffith cracks in rocks under pressure, *Proceedings of the 4th U.S. National Congress of Applied Mechanics*, pp. 1015–1021 (1962).

[17] Hock, E. and Bieniawski, Z. T., *Fracture propagation mechanics in hard rock*, Technical Report – Rock Mech. Div., South African Council for Scientific and Industrial Research (1965).

[18] Pook, L. P., The effect of crack angle on fracture toughness, *Journal of Engineering Fracture Mechanics* **3**, pp. 205–218 (1966).

2

A three-dimensional strain energy density factor theory of crack propagation

2.1. Preliminary remarks

It is common knowledge that a material will fail under applied load either by separation or segmentation. This process may occur suddenly, or over a long time interval, depending on the temperature, rate of loading, and other variables present. A material that breaks after it has deformed appreciably may have lost its structural usefulness long before it actually separates. In this case, the initiation of yielding or plastic flow may represent failure, and a criterion of yielding such as maximum shear stress or maximum distortional energy may be used. When a material breaks with very little deformation the separation process usually takes place suddenly, and is known as brittle fracture. For materials under uniaxial loading, such as tensile specimens, the maximum principal stress has customarily been used in a brittle failure criterion.

The critical values of parameters used in the conventional failure criteria are usually measured from laboratory specimens under simple tension and torsion. The specimens are also assumed to be homogeneous, continuous materials, with no consideration given to mechanical defects which are inherent in the real material. For this reason, the conventional failure criteria are inadequate when applied to structural members containing initial flaws. This is basically because the material response and stress distribution around the flaws are very different from those in the tensile bar specimen. The purpose of fracture mechanics is to take the dimensions of the initial defects of the material into account and to determine the critical load or displacement applied remotely from the defect that will cause catastrophic failure of the whole structure. Therefore, it is essential to relate the resistance of the material to crack propagation to a parameter that can characterize the fracture toughness of the material. Aside from the influence of the thermodynamic variables, such a parameter should first be invariant with respect to the direction of crack propagation relative to the applied load. Hence, a realistic

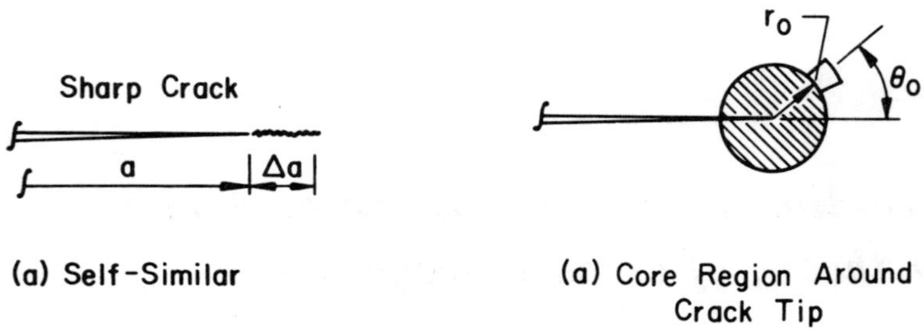

Figure 2.1. Crack tip region.

fracture criterion should have the ability to determine the path of failure upon reaching the critical load.

The most widely used fracture criterion for assessing the onset of rapid fracture in materials without substantial permanent deformation is that of Griffith [1]. Although the present-day interpretation of this theory is basically different from the original version,* the central feature of the criterion is retained by the equation

$$\sigma_{cr} a^{1/2} = \left[\frac{2E\gamma}{\pi(1-\nu^2)}\right]^{1/2} = \text{material constant} \tag{2.1}$$

in which σ_{cr} is the critical applied stress, a the half crack length, E the elastic modulus, v Poisson's ratio and γ the specific surface energy. One of the most disturbing features of equation (2.1) is that γ for a solid has been used liberally in the literature without a precise definition and frequently misinterpreted.** The now accepted explanation of the Griffith theory relies on the energy release concept. It is assumed that fracture occurs when the increase in free surface energy associated with the extension of a sharp crack is supplied by the elastic energy released by this crack growth, say Δa as shown in Figure 2.1(a). It is commonly taken for granted that a crack actually extends into the material ahead in a continuous fashion from its tip. On physical grounds, such an assumption obviously lacks proper justification. What is used as a mathematical convenience should not be confused with what actually happens in reality. A continuous segment of crack extension is no more than a necessary preliminary to the calculation of energy release rate.

It is known from experimental observation on real materials that some elements of the material at a finite distance ahead of the crack fail first prior to the onset of rapid crack growth. Recently, Sih [2, 3] has proposed

* A detailed discussion in the improper application of surface tension of a liquid to the fracture of a narrow ellipse as originally proposed by Griffith has been given by Sih [2].

** Sih [2] emphasized the necessity of scale shifting should γ be used to explain fracture at both the macroscopic and microscopic level.

the use of strain energy density factor S associated with an interior element ahead of the crack in a criterion of fracture as shown in Figure 2.1(b). The S-factor is equal to the core region radius r_0 in Figure 2.1(b) multiplied by dW/dV, the amount of strain energy per unit volume of the element. Local instability,* i.e., failure of the element of Figure 2.1(b), is assumed to occur when the local energy density factor, S, reaches a critical value, S_{cr}, which is characteristic of the material. The new flaw then links up with the crack tip. For a through crack of length $2a$ subjected to uniform applied stress σ, Sih [2, 3] has assumed that fracture occurs when

$$\sigma_{cr}a^{1/2} = \left[\frac{2ES_{cr}}{(1+\nu)(1-2\nu)}\right]^{1/2} = \text{material constant.} \tag{2.2}$$

Although equation (2.2) is similar in form to equation (2.1), it is based on an entirely different premise. The ease of computation of the strain energy density factor for an element situated in an arbitrary position with reference to the crack tip offers a simple method for analyzing the mixed mode crack problem** in which the applied load is not normal to the crack plane. If the tensile loading axis makes an angle β with a crack of length $2a$, equation (2.2) must be modified as

$$\sigma_{cr}a^{1/2} = \left[\frac{S_{cr}}{F(\beta,\theta_0)}\right]^{1/2} \tag{2.3}$$

where

$$F(\beta,\theta_0) = (a_{11}\sin^2\beta + 2a_{12}\sin\beta\cos\beta + a_{22}\cos^2\beta)\sin^2\beta. \tag{2.4}$$

The coefficients a_{ij} $(i,j = 1,2)$ are given by (Sih [3])

$$16\mu a_{11} = (3-4\nu-\cos\theta_0)(1+\cos\theta_0) \tag{2.5a}$$

$$8\mu a_{12} = \sin\theta_0[\cos\theta_0 - (1-2\nu)] \tag{2.5b}$$

$$16\mu a_{22} = 4(1-\nu)(1-\cos\theta_0) + (1+\cos\theta_0)(3\cos\theta_0 - 1). \tag{2.5c}$$

The direction of crack growth θ_0 for a given β, according to Sih [3,5], is toward the point near the crack tip with the minimum value of the strain

---

* Depending on the load and geometry, the local failure may lead to failure of the entire body.

** The importance of the mixed mode crack extension problem was also recognized by Griffith [4] who, in his second paper on fracture, attempted to predict the failure of a narrow ellipse situated at an arbitrary position in a biaxial stress field. To this end, he did not pursue the energy release rate concept introduced in his first paper of 1921, but postulated an independent criterion that fracture would occur when the maximum peripheral stress on the free surface of the ellipse would reach locally a critical value. In retrospect, it would have been extremely unprofitable, if not impossible, to apply the energy release rate concept rigorously to cracks that do not propagate in a self-similar manner.

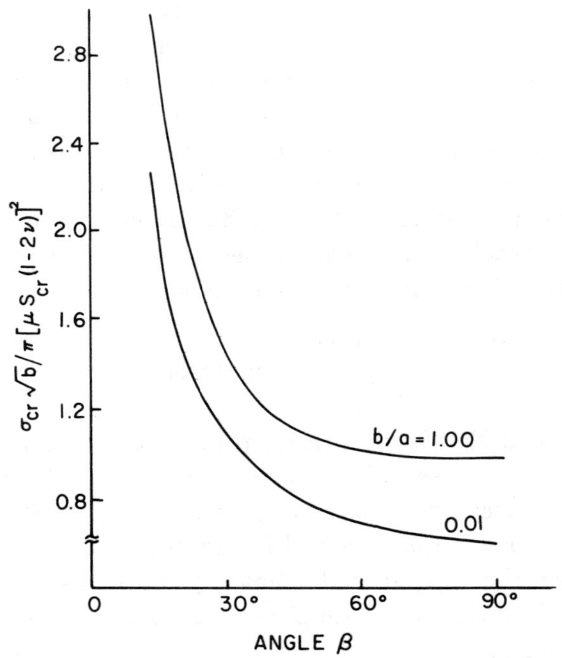

Figure 2.2. **Critical tensile stress versus angle β.**

energy density factor S. When the uniaxial tension is not too closely aligned along the crack plane, the minimum value of S occurs in a region in which the energy of volume change is greater than the energy of distortion. It should be emphasized that the quantity $\sigma_{cr}a^{1/2}$ in equation (2.3) is no longer a constant since the function $F(\beta, \theta_0)$ changes for every β or θ_0. Experimental data (Sih [5], Sih and Kipp [6]) on polymethylmethacrylate plates with angle cracks show that S_{cr} remains reasonably constant for all positions of the load. In comparing theory and experiment, Sih and Kipp [6] have noted the necessity of observing the distance r_0 at which the fracture angle θ_0 is measured.* A slight variation in r_0 could lead to large changes in θ_0. This is because the fracture path in the region of crack initiation can be severely distorted. In fact, Kipp and Sih [7] have shown that the curvature of the local fracture path changes direction when the angle β between the load and crack is varied.

Until now the criterion for fracture initiation has been applied only to cracks that extend completely through the material in one of the coordinate directions. The strain energy density concept has been extended by Sih and

* Sih [9] has suggested that r_0 be found from a notch sensitive strain energy density criterion or S-criterion by assuming that the crack possesses a very small but finite radius of curvature ρ. The core region radius r_0 is selected so that the predictions based on the notch sensitive S-criterion at $r = r_0$ match those for a perfectly sharp crack with $\rho = 0$.

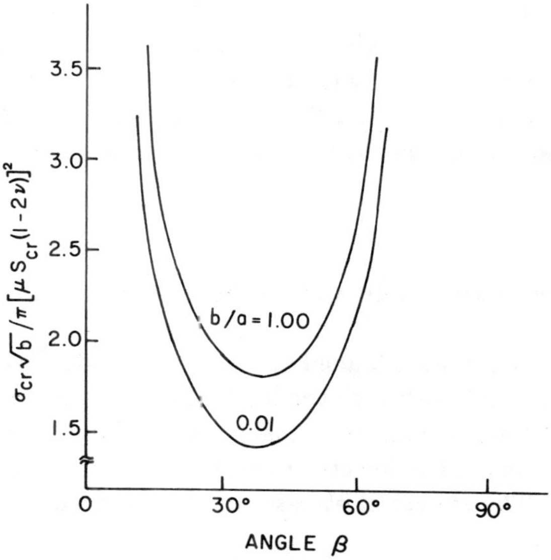

Figure 2.3. **Critical compressive stress versus angle β.**

Cha [8] to analyze cracks embedded in the solid. Figure (2.2) gives a plot of the critical applied stress as a function of the angle β between the crack plane and uniaxial tension σ by holding S_{cr} constant. Results are given for two different shapes of crack, namely, circular and elliptical or a rectangular with one side being infinite in length (i.e., $a = \infty$). Regardless of the crack shapes, the curves decrease monotonically as the angle β is increased. The minimum value of the applied stress required to initiate fracture of a given crack occurs when $\beta = 90°$ which corresponds to tension applied normal to the crack plane. It is interesting to note that the results of a two-dimensional through crack coincide with those of a needle-like embedded crack having an aspect ratio of $b/a = 0.01$, where b and a are, respectively, the semi-minor and semi-major axes of an ellipse. At a given angle β, the penny-shaped configuration, $b/a = 1$, yields the largest critical stress. This indicates that an irregularly-shaped crack might tend to grow into a plane circle.

The S-criterion applies equally well to crack initiation under compression.* Predictions of the critical compressive stresses σ_{cr} for two different crack shapes are presented in Figure 2.3. A minimum value of σ_{cr} prevails when β is approximately 40° nearly independent of any change in crack shapes. The magnitude of σ_{cr} required to initiate fracture, however, increases very rapidly for small and large angles of β. This is one of the reasons why

* The analysis assumes that the cracks will not close up under compression and is only intended to demonstrate the qualitative features of the variations of the critical compressive stress with the load angle β. McClintock and Walsh [10] modified the Griffith [4] solution and showed that the effect of normal and frictional forces on the crack surfaces caused by compression plays a minor role.

the compressive fracture stress can exceed* many times the tensile fracture stress if β is larger than the angle 40° which represents the orientation of the most vulnerable internal crack with the load. Recall that σ_{cr} in tension for large β is relatively low. Figures 2.2 and also 2.3 show that for $\beta < 40°$ both σ_{cr} in tension and compression are large and hence their ratio is relatively small.

2.2. Generalized stress field near crack border

Structural members are often loaded in ways that produce three-dimensional stress states. In general, although the loading may appear outwardly simple, a complex state of stress can exist inside the medium, particularly in the neighborhood of mechanical defects or cracks, where the stresses can undergo sharp variations. The effects of stresses of this kind can frequently lead to unexpected failure.

In the neighborhood of cracks of irregular shapes, the stress state is often triaxial in nature and the problem of predicting the surface of crack propagation is most difficult. Boundary value crack problems in three dimensions are rarely easy to solve. Applying results of the analysis is also difficult, for the crack border stress field will involve all three stress intensity factors, which will affect the initiation of material separation along certain planes. In addition, it is necessary to develop a failure criterion involving a suitable combination of the stress intensity factors reaching certain critical conditions. Previous analyses (Sih [11], Hartranft and Sih [12]) on three-dimensional crack border stresses were concerned with problems in which the crack front is a straight line as shown in Figure 2.4. The results were expressed explicitly in terms of a set of cylindrical polar coordinates (r, θ, z) and the stress intensity factors k_j $(j = 1, 2, 3)$ depended only on the variable z measured along the crack edge. The singular variation in r and the angular distribution in θ of the stresses were given in a plane normal to the crack edge. In such a plane, the local stress field, except for the k_j-factors, was the same as that found in two-dimensional crack problems. Kassir and Sih [13] solved the problem of an elliptical internal flat crack subjected to arbitrary loadings. It was shown that the stress intensity factors k_j $(j = 1, 2, 3)$ vary along the crack border, and the three-dimensional stresses near the crack border can be expressed explicitly in terms of a convenient set of coordinates

---

* Griffith [4] used a maximum stress criterion and predicted that the uniaxial compressive failure load of a brittle material is eight times its tensile failure load. This is in contrast to experimental observation where the ratio of the fracture load in compression and tension can be as high as 10^3. McClintock and Walsh [10] increased this ratio to 10 by assuming that the crack surfaces close up under compression and slide against one another with a coefficient of friction up to unity.

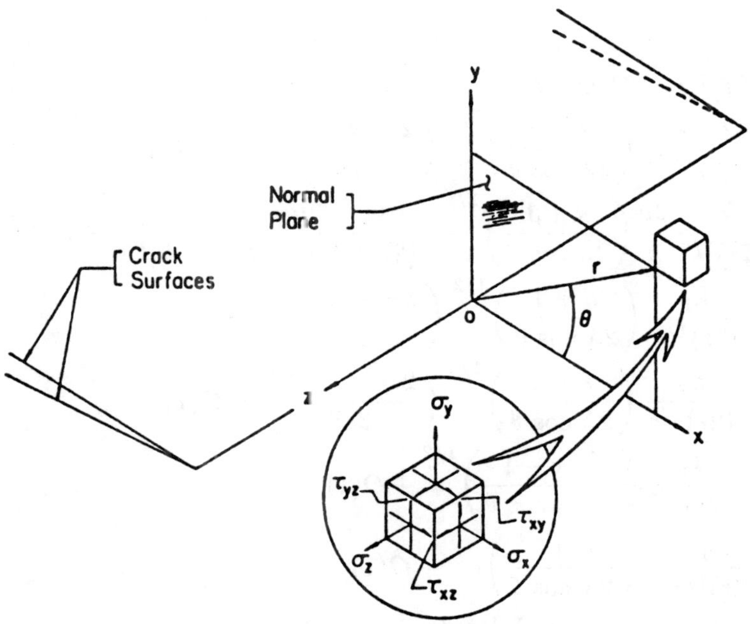

Figure 2.4. Element referred to a straight crack front.

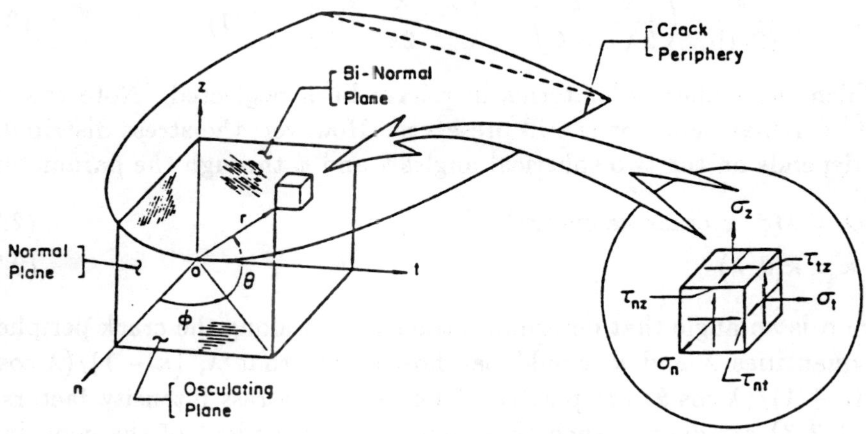

Figure 2.5. Crack border coordinate system.

r and θ in a plane normal to the edge of the crack.

To investigate failure, the stresses at points on small spheres centered at various points of the crack front are compared. This is accomplished by expressing the stresses in terms of the local spherical coordinates (r, θ, ϕ) shown in Figure 2.5. For an elliptical crack, these expressions can be written as

$$\sigma_n = \frac{k_1}{(2r)^{1/2}}\left(\frac{\kappa+1}{2\lambda\cos\theta}\right)^{1/2}\left(\frac{2-\kappa+\kappa^2}{2\kappa^3}\right) - $$
$$- \frac{k_2}{(2r)^{1/2}}\left(\frac{\kappa-1}{2\lambda\cos\theta}\right)^{1/2}\left(\frac{2+\kappa+3\kappa^2}{2\kappa^3}\right) + O(1) \qquad (2.6a)$$
$$\sigma_t = \frac{k_1}{(2r)^{1/2}}\frac{2\nu}{\kappa}\left(\frac{\kappa+1}{2\lambda\cos\theta}\right)^{1/2} - \frac{k_2}{(2r)^{1/2}}\frac{2\nu}{\kappa}\left(\frac{\kappa-1}{2\lambda\cos\theta}\right)^{1/2} + O(1) \qquad (2.6b)$$
$$\sigma_z = -\frac{k_1}{(2r)^{1/2}}\left(\frac{\kappa+1}{2\lambda\cos\theta}\right)^{1/2}\left(\frac{2-\kappa-3\kappa^2}{2\kappa^3}\right) + $$
$$+ \frac{k_2}{(2r)^{1/2}}\left(\frac{\kappa-1}{2\lambda\cos\theta}\right)^{1/2}\left(\frac{2+\kappa-\kappa^2}{2\kappa^3}\right) + O(1) \qquad (2.6c)$$
$$\tau_{nt} = -\frac{k_3}{(2r)^{1/2}}\frac{1}{\kappa}\left(\frac{\kappa-1}{2\lambda\cos\theta}\right)^{1/2} + O(1) \qquad (2.6d)$$
$$\tau_{tz} = \frac{k_3}{(2r)^{1/2}}\frac{1}{\kappa}\left(\frac{\kappa+1}{2\lambda\cos\theta}\right)^{1/2} + O(1) \qquad (2.6e)$$
$$\tau_{nz} = \frac{k_1}{(2r)^{1/2}}\left(\frac{\kappa-1}{2\lambda\cos\theta}\right)^{1/2}\left(\frac{2+\kappa-\kappa^2}{2\kappa^3}\right) + $$
$$+ \frac{k_2}{(2r)^{1/2}}\left(\frac{\kappa+1}{2\lambda\cos\theta}\right)^{1/2}\left(\frac{2-\kappa+\kappa^2}{2\kappa^3}\right) + O(1) \qquad (2.6f)$$

in which the higher order terms in r have been neglected. Note that the $1/r^{1/2}$ singular behavior is still preserved. However, the stress distribution now depends on the two spherical angles θ and ϕ through the parameters

$$\lambda = \lambda(\phi, \alpha, \text{crack geometry}) \qquad (2.7a)$$
$$\kappa = \kappa(\theta, \lambda) \qquad (2.7b)$$

where α is an angle that determines the location along the crack periphery. The quantities λ and κ should be chosen such that λ, $(\kappa-1)/(\lambda\cos\theta)$ and $(\kappa+1)/(\lambda\cos\theta)$ are positive definite. The stress intensity factors k_j $(j = 1, 2, 3)$ are defined such that they are independent of the coordinate variables (r, θ, ϕ) and depend only on the angle α, the nature of loading, and the crack geometry. When $\phi = 0$, equations (2.6a) to (2.6f) reduce to those in the normal plane nz reported in the open literature (Sih [11], Hartranft and Sih [12]).

2.3. Strain energy density factor

When all three stress intensity factors are present along the crack border the fracture criterion should express that fracture occurs when some combination of them reaches a critical value, i.e.,

$$f(k_1, k_2, k_3) = f_{cr}. \tag{2.8}$$

Such a criterion can be developed by referring to the strain energy stored in a volume element ahead of the crack as shown in Figure 2.5. With reference to the coordinate system (n, t, z) in Figure 2.5, the strain energy stored in a volume element $dV = dn\,dt\,dz$ is a quadratic form of the stresses $\sigma_n, \sigma_t, \ldots, \tau_{nz}$:

$$\begin{aligned}\frac{dW}{dV} &= \frac{1}{2E}(\sigma_n^2 + \sigma_t^2 + \sigma_z^2) - \frac{\nu}{E}(\sigma_n\sigma_t + \sigma_t\sigma_z + \sigma_z\sigma_n) + \\ &+ \frac{1+\nu}{E}(\tau_{nt}^2 + \tau_{tz}^2 + \tau_{nz}^2)\end{aligned} \tag{2.9}$$

where ν is Poisson's ratio and E is Young's modulus of the material. Substituting the stresses from equations (2.6) into (2.9), the strain energy per unit volume may be written as

$$\frac{dW}{dV} = \frac{S}{r} + \text{nonsingular terms} \tag{2.10}$$

which becomes singular in the limit as $r \to 0$. It is apparent that the factor S is defined only if $r \neq 0$ and hence the volume element is always kept at a finite distance away from the crack border.

The magnitude, S, of the $1/r$ energy field in equation (2.10) will be referred to as the strain energy density factor. It can be expressed as a quadratic form in the stress intensity factors k_1, k_2 and k_3:

$$S = a_{11}k_1^2 + 2a_{12}k_1k_2 + a_{22}k_2^2 + a_{33}k_3^2. \tag{2.11}$$

The factors k_j $(j = 1, 2, 3)$ will vary with the location along the crack border and the coefficients a_{11}, a_{12}, a_{22} and a_{33} with the spherical angles θ and ϕ as given by

$$a_{11} = \frac{\kappa + 1}{16\mu\lambda\kappa^2 \cos\theta}\left[2(1-2\nu) + \frac{\kappa - 1}{\kappa}\right] \tag{2.12a}$$

$$a_{12} = \frac{(\kappa^2 - 1)^{1/2}}{8\mu\lambda\kappa^2 \cos\theta}\left[\frac{1}{\kappa} - (1-2\nu)\right] \tag{2.12b}$$

$$a_{22} = \frac{1}{16\mu\lambda\kappa^2 \cos\theta}\left[4(1-\nu)(\kappa - 1) + \frac{1}{\kappa}(\kappa+1)(3-\kappa)\right] \tag{2.12c}$$

$$a_{33} = \frac{1}{4\mu\lambda\kappa \cos\theta}. \tag{2.12d}$$

In the special case of $\phi = 0$, the coefficients a_{ij} reduce to those for the two-dimensional crack problems (Sih [3, 5]).

There are two basic ways in which a volume element absorbs energy: an element can store energy by 'dilatation' and/or 'distortion'. The former is associated with change in volume and the latter is associated with change in shape. When an element of the material exceeds a certain limit in volume change, brittle fracture can occur. Hence, it is useful to separate S into two parts:

$$S = S_{\mathrm{v}} + S_{\mathrm{d}} \tag{2.13}$$

where S_{v} represents that portion of S due to the volume change and S_{d}, that due to shape change. The quantities S_{v} and S_{d} are given by

$$S_{\mathrm{v}} = b_{11}k_1^2 + 2b_{12}k_1k_2 + b_{22}k_2^2 + b_{33}k_3^2 \tag{2.14a}$$

$$S_{\mathrm{d}} = c_{11}k_1^2 + 2c_{12}k_1k_2 + c_{22}k_2^2 + c_{33}k_3^2 \tag{2.14b}$$

The coefficients by b_{ij} and c_{ij} stand for

$$b_{11} = \frac{(1+\nu)(1-2\nu)(\kappa+1)}{12\mu\lambda\kappa^2\cos\theta} \tag{2.15a}$$

$$b_{12} = -\frac{(1+\nu)(1-2\nu)(\kappa^2-1)^{1/2}}{12\mu\lambda\kappa^2\cos\theta} \tag{2.15b}$$

$$b_{22} = \frac{(1+\nu)(1-2\nu)(\kappa-1)}{12\mu\lambda\kappa^2\cos\theta} \tag{2.15c}$$

$$b_{33} = 0 \tag{2.15d}$$

and

$$c_{11} = \frac{\kappa+1}{16\mu\lambda\kappa^2\cos\theta}\left[\frac{2(1-2\nu)^2}{3} + \frac{\kappa-1}{\kappa}\right] \tag{2.16a}$$

$$c_{12} = \frac{(\kappa^2-1)^{1/2}}{8\mu\lambda\kappa^2\cos\theta}\left[\frac{1}{\kappa} - \frac{(1-2\nu)^2}{3}\right] \tag{2.16b}$$

$$c_{22} = \frac{\kappa-1}{16\mu\lambda\kappa^2\cos\theta}\left[\frac{(1+\kappa)(3-\kappa)}{\kappa(\kappa-1)} + \frac{8(1-\nu+\nu^2)}{3}\right] \tag{2.16c}$$

$$c_{33} = \frac{1}{4\mu\lambda\kappa\cos\theta}. \tag{2.16d}$$

2.4. Basic assumptions of the theory

Sih [3, 5] has proposed that crack propagation initiating from an inherent flaw in the material is determined by the energy density in the region of high stress elevation. The locations of maximum yielding* and of brittle fracture

* The factor S_{d} in equation (2.14b) is equivalent to the Huber–von Mises–Hencky yield criterion. However, the problem of yielding will not be discussed in this paper.

can be found from the stationary values of the strain energy density factor, S. The basic hypotheses of the strain energy density factor theory as applied to three-dimensional crack problems may be outlined as follows:

Hypothesis (1): ***The direction of crack propagation at any point along the crack border is toward the region with the minimum value of strain energy density factor S as compared with other regions on the same spherical surface surrounding the point.***

Hypothesis (2): ***Crack extension occurs when the strain energy density factor in the region determined by hypothesis (1), $S = S_{\min}$, reaches a critical value, say S_{cr}.***

Hypothesis (3): ***The length, r_0, of the initial crack extension is assumed to be proportional to $S_{\min}$ such that $S_{\min}/r_0$ remains constant along the new crack front.***

Hypothesis (1) and (2) are sufficient for determining where and when unstable crack propagation occurs in a two-dimensional problem in which all the elements at the same distance r_0 from the straight crack front are assumed to fail simultaneously.* The relation $S_{cr} = r_0(\mathrm{d}W/\mathrm{d}V)_{cr}$ implies that if r_0 is constant along the crack front then the value S_{cr} is uniquely related to $(\mathrm{d}W/\mathrm{d}V)_{cr}$. In three dimensions, the crack front is generally curved and the distance r_0 may vary from one point on the crack border to the next as shown in Figure 2.6. The location of the initial fracture points is described by a locus of elements with coordinates (r_0, θ_0, ϕ_0) determined by holding $(\mathrm{d}W/\mathrm{d}V)_{cr} = S_{\min}/r_0$ constant. Hypothesis (3) is, therefore, necessary for predicting the shape of the initial portion of the fracture surface. This surface generally will follow a path along which the elements experience more volume change than shape change.

The prediction of the initial crack growth based on the elasticity solution for the original crack geometry is justified if the time interval between the onset of rapid fracture and the final separation of a brittle material is extremely short. The material then does not have sufficient time for stress redistribution as the crack profile changes during propagation. Hence, the initial stress distribution around the stationary crack at incipient fracture determines the crack path reasonably well. This has been noted by Kipp and Sih [7] for a two-dimensional crack propagating under mixed mode conditions.

---

* This assumption is reasonable for a through crack in materials that do not distort or deform appreciably in the thickness direction prior to the onset of rapid crack extension. Such a condition can be satisfied for metal alloys if the plate specimens are sufficiently thick.

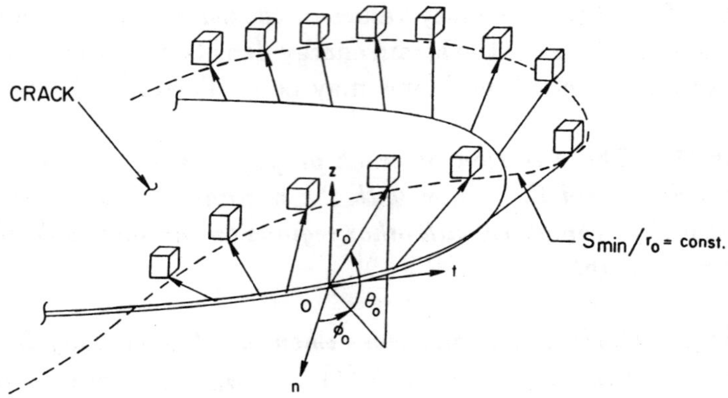

Figure 2.6. Line of constant S_{min}/r_0.

2.5. Minimum values of S

In order to apply the strain energy density theory to crack problems in three dimensions, the variations of the minimum values of S along the crack border will be studied. It is seen from equation (2.11) that S is a continuous function of θ and ϕ through the coefficients a_{ij}. The ranges of θ and ϕ are $[-\pi, \pi]$ and $[-\pi/2, \pi/2]$ respectively, and λ is positive definite in these ranges. The continuous function $S(\theta, \phi)$ is said to have a local minimum at (θ_0, ϕ_0) in the region $R(-\pi \leq 0 \leq \pi; -\pi/2 < \phi < \pi/2)$, if $S(\theta, \phi) \geq S(\theta_0, \phi_0)$ for all points (θ, ϕ) in the neighborhood of the point (θ_0, ϕ_0) in R.

If the function $S(\theta, \phi)$ is differentiable in the region R and a stationary value is attained at an interior point (θ_0, ϕ_0), then the first derivatives of $S(\theta, \phi)$ with respect to each variable vanish at (θ_0, ϕ_0), i.e.,

$$\frac{\partial S}{\partial \theta} = 0 \quad \text{and} \quad \frac{\partial S}{\partial \phi} = 0 \quad \text{at} \quad (\theta_0, \phi_0). \tag{2.17}$$

If S is a relative minimum at (θ_0, ϕ_0), then usually the second derivatives of S satisfy the conditions

$$\left(\frac{\partial^2 S}{\partial \theta \partial \phi}\right)^2 - \frac{\partial^2 S}{\partial \theta^2}\frac{\partial^2 S}{\partial \phi^2} < 0; \quad \frac{\partial^2 S}{\partial \theta^2} > 0 \quad \text{and} \quad \frac{\partial^2 S}{\partial \phi^2} > 0. \tag{2.18}$$

For crack problems, S possesses two local minima* on every plane, $\phi =$ constant. Depending on the curvature of the crack border, points on the crack front will in general join with a region in planes different from the normal plane, i.e., $\phi_0 \neq 0$. Making use of equations (2.13) and (2.14), the minimum values of $S(\theta_0, \phi_0)$ can be separated into two parts, i.e., energy absorbed by volume change $S_v(\theta_0, \phi_0)$ and by shape change $S_d(\theta_0, \phi_0)$. In

* The two local maxima in the plane ϕ = constant correspond to the directions of maximum yielding under remote tensile and compressive loads.

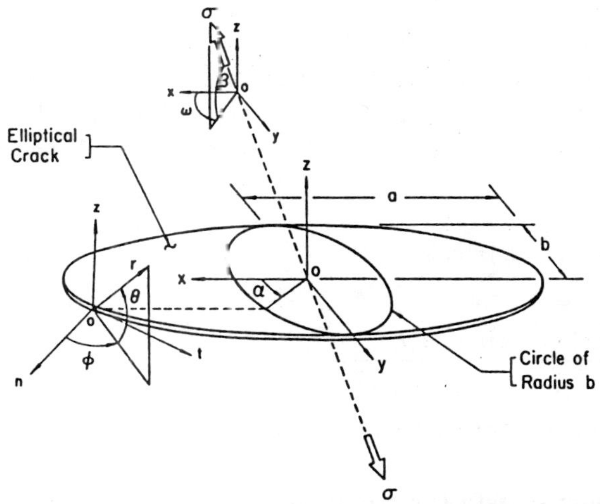

Figure 2.7. Elliptical crack subjected to inclined loading.

the direction of minimum S, S_v is greater than S_d when the applied normal stress may or may not be greater than the applied shear stress. Results of this type of calculation for an elliptical crack will now follow.

2.6. Application of the S-theory to the elliptical crack problem

The classical fracture criterion, based on the strain energy release rate or critical stress intensity factor, is limited to the problems involving symmetric loads since the analysis applies only to the case where the extended crack surface remains in the plane of the initial crack. If the crack is inclined to the axis of loading, the extended portion of the crack surface may no longer preserve the condition of self-similarity. Since the amount of energy used to initiate failure depends largely on the nature and position of the load relative to the location of the flaws or cracks, it is essential to be able to predict the location of failure within the solid.

As an example on the application of the strain energy density theory to three-dimensional crack problems, the geometry of a flat elliptical crack will be considered. Referring to Figure 2.7, an embedded elliptical crack with semi-major axis a and semi-minor axis b lies in the xy-plane and is described by the equation $x^2/a^2 + y^2/b^2 = 1$. The position of an element in Figure 2.7 referred to the coordinates, (x, y, z) may be written as

$$x = a \cos \alpha + r \cos \theta \frac{b \cos \alpha \cos \phi - a \sin \alpha \sin \phi}{(a^2 \sin^2 \alpha + b^2 \cos^2 \alpha)^{1/2}} \tag{2.19a}$$

$$y = b \sin \alpha + r \cos \theta \frac{b \cos \alpha \cos \phi + a \sin \alpha \cos \phi}{(a^2 \sin^2 \alpha + b^2 \cos^2 \alpha)^{1/2}} \tag{2.19b}$$

$$z = r \sin\theta. \tag{2.19c}$$

Note that on the crack front $r = 0$, equations (2.19) reduce to the well-known parametric equations of an ellipse $x = a\cos\alpha$ and $y = b\sin\alpha$. The local spherical coordinates (r, θ, ϕ) can be related to the ellipsoidal coordinates (ξ, η, ς) as

$$\xi = \frac{ab\lambda(1+\kappa)r\cos\theta}{(a^2\sin^2\alpha + b^2\cos^2\alpha)^{1/2}} \tag{2.20a}$$

$$\eta = \frac{ab\lambda(1-\kappa)r\cos\theta}{(a^2\sin^2\alpha + b^2\cos^2\alpha)^{1/2}} \tag{2.20b}$$

$$\varsigma = -(a^2\sin^2\alpha + b^2\cos^2\alpha) \tag{2.20c}$$

where ξ, η, ς must satisfy the conditions

$$-a^2 \le \varsigma \le -b^2 \le \eta \le 0 \le \xi < \infty. \tag{2.21}$$

For the present problem of an elliptical crack, the parameters λ and κ in equations (2.6) and (2.20) take the forms

$$\lambda = \cos\phi + \frac{b^2 - a^2}{ab}\sin\phi\sin\alpha\cos\alpha \tag{2.22a}$$

$$\kappa = \pm\left\{1 + \left[(\tan\theta)/\lambda\right]^2\right\}^{1/2} \tag{2.22b}$$

which depend on the spherical angles θ and ϕ as well as on the parametric angle α and the ratio b/a. Moreover, equation (2.21) requires that

$$\lambda > 0 \quad \text{and} \quad \kappa = \text{sign}(\cos\theta)\left\{1 + \left[(\tan\theta)/\lambda\right]^2\right\}^{1/2} \tag{2.23}$$

in the region $R(-\pi \le 0 \le \pi;\ -\pi/2 < \phi < \pi/2)$.

When a uniform load σ is applied at an angle β with reference to the crack plane and at an angle ω with the x-axis as shown in Figure 2.7, the stress intensity factors are given by (Kassir and Sih [13])

$$k_1 = \frac{\sigma\sin^2\beta}{E(k)}\left(\frac{b}{a}\right)^{1/2}(a^2\sin^2\alpha + b^2\cos^2\alpha)^{1/4} \tag{2.24a}$$

$$k_2 = -\frac{\sigma\sin\beta\cos\beta}{(a^2\sin^2\alpha + b^2\cos^2\alpha)^{1/4}}\left[1 - \left(\frac{b}{a}\right)^2\right]\left(\frac{b}{a}\right)^{1/2}$$
$$\times\left[\frac{a\sin\alpha\sin\omega}{(k^2 + \nu k'^2)E(k) - \nu k'^2K(k)} + \frac{b\cos\alpha\cos\omega}{(k^2 - \nu)E(k) + k'^2K(k)}\right] \tag{2.24b}$$

$$k_3 = \frac{(1-\nu)\sigma \sin\beta \cos\beta}{(a^2 \sin^2\alpha + b^2 \cos^2\alpha)^{1/4}} \left[1 - \left(\frac{b}{a}\right)^2\right] \left(\frac{b}{a}\right)^{1/2}$$
$$\times \left[\frac{a \sin\alpha \cos\omega}{(k^2-\nu)E(k) + k'^2 K(k)} - \frac{b \cos\alpha \sin\omega}{(k^2+\nu k'^2)E(k) - \nu k'^2 K(k)}\right] . \quad (2.24c)$$

The quantities $K(k)$ and $E(k)$ are complete elliptic integrals of the first and second kinds associated with the argument $k = [1 - (b/a)^2]^{1/2}$ where $k'^2 = 1 - k^2$. If the applied stress σ changes from tension to compression, then σ in equations (2.24) must be replaced by $-\sigma$. The directions of crack initiation for tensile and compressive loads as predicted by the strain energy density criterion will be discussed subsequently.

Tensile loading. Suppose that the cracked solid is stretched by an inclined uniform load σ. The directions of crack initiation (θ_0, ϕ_0) can be found by inserting the k_j-factors in equations (2.24) into (2.11) and applying the conditions given in equations (2.17) and (2.18). Based on the S-criterion, the planes ϕ_0 for various values of α around the crack periphery are obtained. It is first found that the crack will not run in the normal plane nz (i.e., $\phi_0 \neq 0°$) except at the intersections of the major and minor axes with the crack border. The locations of the ϕ_0-planes in which the crack does extend are given in Table 2.1 for different ratios of b/a. For a penny-shaped crack $b/a=1$, fracture will always occur in the normal plane $\phi_0 = 0$. The corresponding values of θ_0 in the ϕ_0-planes will depend on the crack geometry, load position and Poisson's ratio of the material. For the special case of $b/a = 0.5$, $\nu = \frac{1}{3}$, and $\beta = 60°$ (where the applied tension is in the xz-plane), the angles θ_0 together with the minimum values of $S(\theta_0, \phi_0)$ or S_{min} around the elliptical crack border are shown in Table 2.2. The minimum of S which occurs at a negative* angle θ_0 is the one which corresponds to tensile loading, for the crack tends to grow so that its periphery approaches that of a large circular crack normal to the applied load. According to the assumptions in the S-theory, the point at $\alpha = 90°$ will be one** where S_{min} first reaches the critical value S_{cr}. This corresponds to the ends of the minor axis, where both

* The positive angle at which the other local minimum occurs corresponds to the location of fracture initiated under compression where the crack tends to run into the direction of loading.

** If failure of the element at $\alpha = 90°$ (or elements at $\alpha = \pm 90°$ for $\omega = 0$) corresponds to unstable fracture of the entire solid, it then suffices to consider only S_{min} in the element at $\alpha = 90°$ reaching S_{cr} as failure. In general, the S-theory, based on the asymptotic solution of the stress field, assumes that a line of elements each at a different distance r_0 from the crack front fail simultaneously at the onset of unstable fracture. In any experiments, the loading, crack geometry and microstructure (or discrete nature) of the material will have an influence on r_0.

Table 2.1. Values of ϕ_0 for different α and b/a.

α \ b/a	1	0.9	0.5	0.1
0°	0°	0°	0	0
1°	0°	−0.211°	− 1.499°	− 9.801°
15°	0°	−3.021°	−20.556°	−67.999°
30°	0°	−5.223°	−33.004°	−76.869°
45°	0°	−6.025°	−36.869°	−78.578°
60°	0°	−5.223°	−33.004°	−76.869°
75°	0°	−3.021°	−20.556°	−67.999°
89°	0°	−0.211°	− 1.499°	− 9.801°
90°	0°	0°	0°	0°

Table 2.2. Negative angles θ_0 and S_{min} for $b/a = \frac{1}{2}$, $\nu = \frac{1}{3}$, $\omega = 0°$ and $\beta = 60°$.

α	θ_0	$16\mu S_{min}/\sigma^2 b$	$16\mu S_d/\sigma^2 b$	$16\mu S_v/\sigma^2 b$
0°	−47.54°	0.46128	0.12074	0.34053
15°	−44.72°	0.48518	0.15144	0.33374
30°	−35.79°	0.53590	0.21521	0.32069
45°	−24.45°	0.59957	0.27590	0.32366
60°	−16.39°	0.69799	0.33738	0.36060
75°	− 9.79°	0.83233	0.41270	0.41963
90°	0.00°	0.91176	0.45721	0.45454

S_v associated with volume change, and S_d associated with shape change, are greatest. Table 2.2 also shows that S_v is greater than S_d except at $\alpha = 90°$ where Mode III shear dominates.

For $\omega = 0$ and $\nu = \frac{1}{3}$, the variation of the negative fracture angle θ_0 with the crack angle β is displayed in Figure 2.8 for $b/a = 0.5$. The fracture angle $(-\theta_0)$ decreases monotonically with increasing β for small values of α corresponding to the region in which Mode I and II dominate. As α is increased into the region where Mode I and III loading dominates, the fracture angle $(-\theta_0)$ first increases and then decreases to zero at $\beta = 90°$. Additional curves showing the influence of b/a and ω on the fracture angle can be found in Appendix 2.1.

Having computed (θ_0, ϕ_0), the minima of S can be found from equation (2.11) since $S_{min} = S(\theta_0, \phi_0)$. A plot of the normalized strain energy density factor $16\mu S_{min}/\sigma^2 b$ versus the crack angle β is shown in Figure 2.9 for various positions (α) along the crack front and for a fixed $b/a = 0.1$. Note that S_{min} for a given β is the greatest at the end of the minor axis, $\alpha = 90°$, and the smallest at the end of the major axis, $\alpha = 0°$. Since the value of S_{min} at failure is postulated to be constant for a given material, a higher value of $16\mu S_{min}/\sigma^2 b$ corresponds to a lower value of failure stress σ at constant b. For a narrow ellipse $b/a = 0.1$, all the curves peak at $\beta \simeq 60°$. Refer to Appendix 2.1 for more information.

The lowest stress required to initiate fracture may be estimated by set-

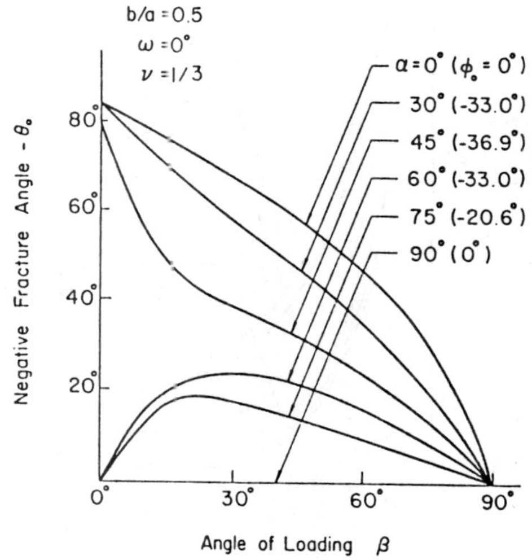

Figure 2.8. Negative fracture angles for $b/a = 0.5$ and $\omega = 0°$.

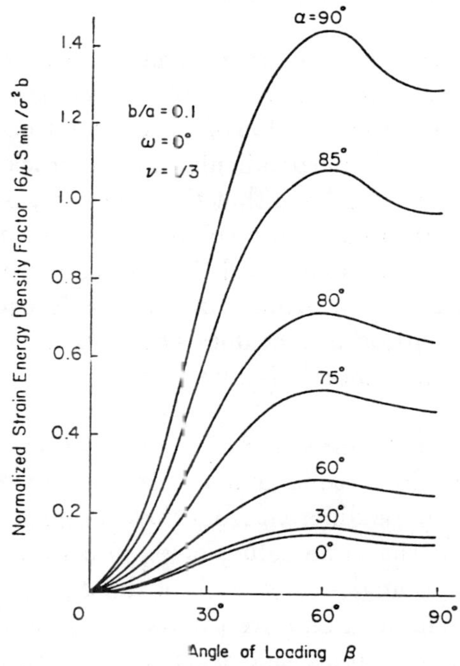

Figure 2.9. Variations of $S_{\min}$ with loading angle β for $b/a = 0.1$ and various α.

ting $S_{\min} = S_{\mathrm{cr}}$ at $\alpha = 90°$. The curves in Figure 2.10 represent a plot of the failure loads σ_{cr} versus β and show that σ_{cr} required to fracture the solid increases as b/a is increased. The two-dimensional plane strain crack

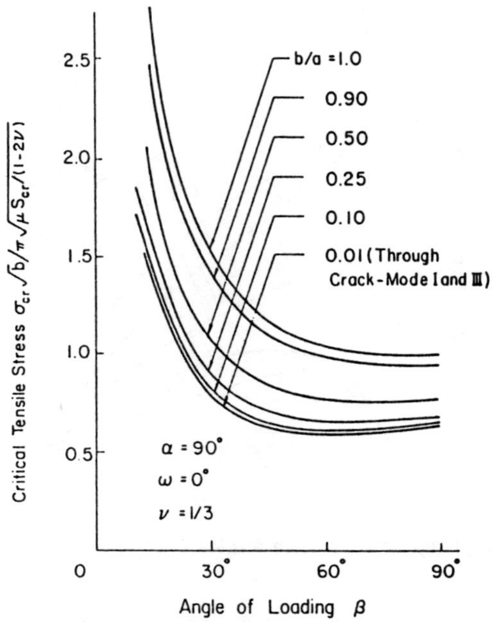

Figure 2.10. Critical tensile stress versus angle β for $\alpha = 90°$, $\omega = 0°$ and various b/a.

solution corresponds to the three-dimensional solution of a very narrow elliptical crack with $b/a = 0.01$ at $\alpha = 90°$. In contrast to one's physical intuition, the lowest permissible value of σ_{cr} does not occur at $\beta = 90°$, i.e., when the applied tension is perpendicular to the crack surface. It can be seen from Figure 2.10 for small b/a that the lowest σ_{cr} occurred at $\beta \simeq 60°$. This behavior can be attributed to a condition of mixed mode loading and exhibited independently by Mode I and III loading of a plane strain through crack.* The details are given in Appendix 2.2. The foregoing example shows that Mode I crack extension in fact does not always yield the conservative solution with the lowest allowable failure stress.

Compressive loading. If the direction of the applied load σ in Figure 2.7 is reversed such that the elastic solid is in a state of compression,** the surface of fracture initiation will be different from that of tensile loading. Equations (2.17) and (2.18) yield the same values of ϕ_0 in compression as in tension. Refer to Table 2.1. The angles θ_0 in each of the ϕ_0-plane around the crack border, however, are changed and are tabulated in Table 2.3 together with S_{min}. Positive values of θ_0 imply that fracture tends to initiate towards the

---

* A similar problem involving a crack slanted in the thickness direction has been discussed by Sih [9].

** Although there is a tendency of the crack surfaces to come in contact when compressed, this will not affect the results qualitatively. The two-dimensional problem of cracks propagating under compression has been discussed by Sih [3].

Table 2.3. Positive angles θ_0 and S_{min} for $b/a = \frac{1}{3}$, $\nu = \frac{1}{3}$, $\omega = 0°$ and $\beta = 60°$.

α	θ_0	$16\mu S_{min}/\sigma^2 b$	$16\mu S_d/\sigma^2 b$	$16\mu S_v/\sigma^2 b$
0°	124.78°	0.08839	0.07782	0.01056
15°	126.73°	0.11403	0.10662	0.00740
30°	132.95°	0.16961	0.16629	0.00332
45°	143.56°	0.22798	0.22660	0.00138
60°	156.182°	0.28923	0.28885	0.00038
75°	168.37°	0.36001	0.35999	0.00001
90°	180.00°	0.40039	0.40039	0.0000

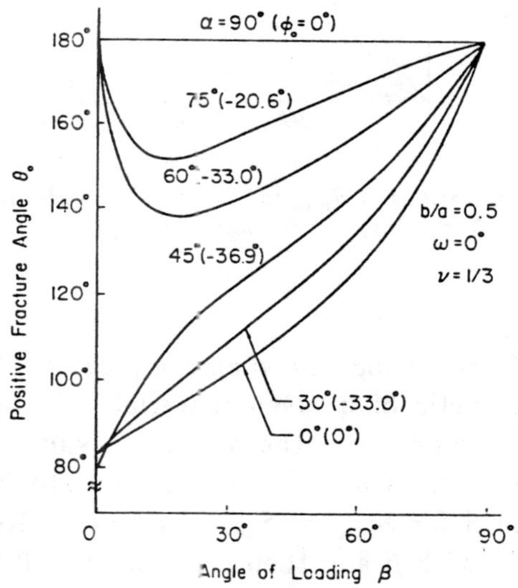

Figure 2.11. Positive fracture angles for $b/a = 0.5$ and $\omega = 0°$.

direction of the compressive load. The predicted fracture angles θ_0 become increasingly larger as α is increased from the end of the major axis, $\alpha = 0°$, to the end of the minor axis, $\alpha = 90°$. For a brittle material, which cannot distort appreciably, the S-criterion predicts that fracture initiates at the point, $\alpha \approx 90°$, where S_{min} is the largest.

Theoretical predictions of the positive fracture angle θ_0 as a function of β are presented graphically in Figure 2.11 for $b/a = 0.5$. The angle ω is taken as zero with $\nu = \frac{1}{3}$. The effects of mixed Modes I and II for curves initiating at $\theta_0 \simeq 80°$ and Modes I and III for curves starting at $\theta_0 = 180°$ are also observed in compression for various parametric angles α. Figure 2.12 gives the variations of S_{min} with β for different values of α with $b/a = 0.1$. The compressive load is directed in the plane of the major axis, $\omega = 0°$. The nondimensionalized strain energy density factor $16\mu S_{min}/\sigma^2 b$ increases as α is increased for a given b/a. Other curves are given in Appendix 2.3.

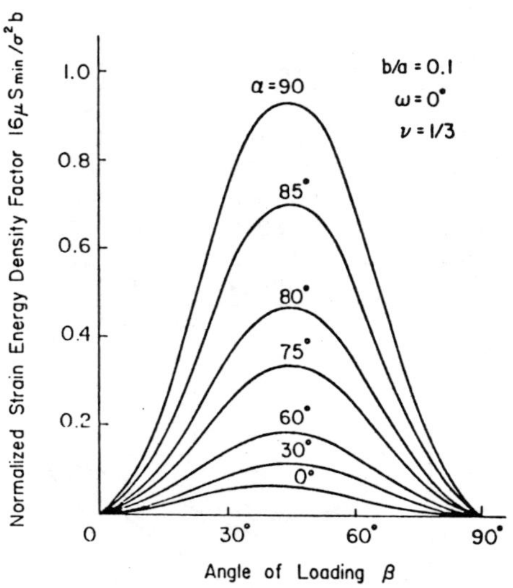

Figure 2.12. Variations of S_{min} in compression with angle β for $b/a = 0.1$, $\omega = 0°$ and various α.

It is assumed that fracture* occurs when the greatest S_{min} attains a critical value S_{cr} characteristic of a given material. For an elliptical crack subjected to uniaxial compression, the maximum value of S_{min} occurs at $\alpha = 90°$. The critical stress σ_{cr} required to initiate fracture is plotted against the load angle β for $S_{min} = S_{cr} =$ constant. Referring to Figure 2.13, σ_{cr} first decreases with β and then increases rapidly for large β. This behavior of σ_{cr} versus β is very different from that of tensile loading and implies that there exists an angle $0° < \beta < 90°$ at which the applied stress σ_{cr} required to initiate fracture is minimum. Figure 2.13 gives the results for $\omega = 0$ and different values of b/a. The minimum σ_{cr} is seen to occur** at $\beta \simeq 45°$ and is not sensitive to changes in b/a. The two-dimensional plane strain solution of Mode I and III loading prevails at $\alpha \simeq 90°$ when the minor axis of the ellipse is approximately one tenth of the major axis. More results can be found in Appendix 2.3.

---

* For materials such as rocks, secondary or minute cracks can grow slowly in a stable fashion prior to fragmentation of the solid. In the present model of a single crack, fracture corresponds with the largest fully developed macroflaw at the onset of unstable crack propagation.

** Sih [3] has predicted that for two-dimensional crack problems the theoretical angle β, at which σ_{cr} is a minimum, occurs at approximately 37° for $\nu = \frac{1}{4}$. This result agrees well with experimental observation for crack extension in glass.

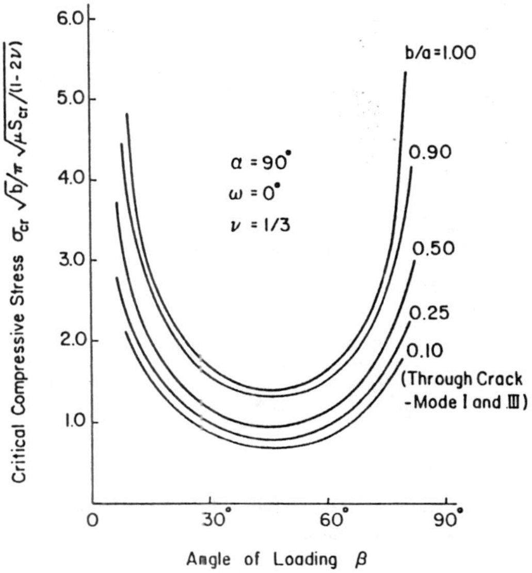

Figure 2.13. Critical compressive stress versus angle β for various b/a at $\alpha = 90°$ and $\omega = 0°$.

2.7. Shape of fracture increment

The third hypothesis of the theory presented in Section 2.4 is concerned with the amount as well as the direction of initial crack growth in a region near the crack border. The influence of this hypothesis on computed shapes of the surface of new crack growth will now be considered. For embedded cracks in a complex stress state, points of fracture do not occur at the same distance away from the crack front and crack propagation will not be confined to the original plane of the crack. The problem of finding the initial segment of the extended crack surface has not been treated adequately in the literature.

The conventional fracture mechanics concept of k_{1c} (Irwin [14]) can only be applied to the case of a flat crack subjected to normal tensile load. In this particular case, fracture initiation is assumed to occur at the point where the stress intensity factor k_1 is the greatest. In a triaxial stress field, a crack may not grow coplanarly. The experiments described by Kipp and Sih [7] have shown that a crack does not grow coplanarly when the load makes an angle with the direction normal to the crack plane. The calculation of strain energy release rate for a crack cannot be made by arbitrarily assuming various possible shapes of crack extensions. This is because the assumed shapes may not necessarily occur in nature, and the amount of energy released during crack extension depends directly on the new crack geometry and the direction of crack propagation.

The strain energy density theory does not make *a priori* assumptions

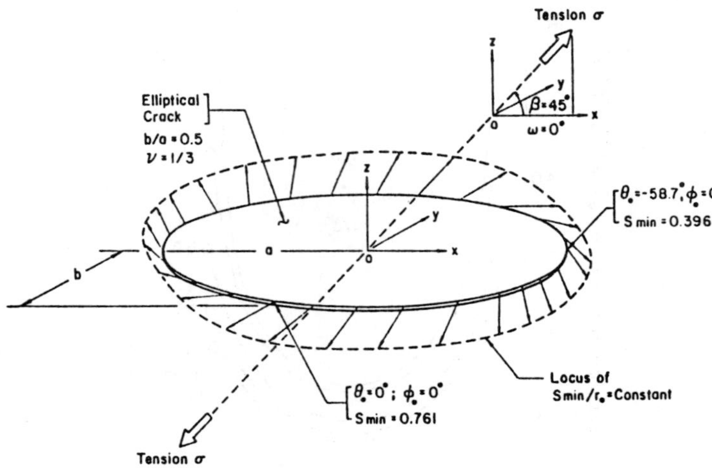

Figure 2.14. Initial fracture surface of an elliptical crack ($b/a = 0.5$) under tension, $\beta = 45°$ and $\omega = 0°$.

about the shape and direction of the new fracture surface. Hence, the theory can treat three-dimensional cracks with irregular shapes provided that the stress state of the problem is known. As in the two-dimensional case (Sih [3, 5]), the S-concept assumes the nearest neighbor element fails when the stored energy per unit volume reaches a critical value. The position of the critical element (Section 2.5) is determined by the angles (θ_0, ϕ_0) corresponding to S_{min} and a radial distance r_0 measured from the crack border. Sih [9] refers to r_0 as the radius of the core region within which the properties of the material and the nature of deformation differ significantly from those assumed in the continuum theory of elasticity. The size of r_0 around a three-dimensional crack is assumed to vary such that* $S_{min}/r_0 = (dW/dV)_{cr}$ remains constant, i.e. $r_0 \sim S_{min}$.

For example, the initial segment of the new crack surface around an elliptical crack of $b/a = 0.5$ and $\beta = 45°$ with the applied tension σ in the plane of the major axis as shown in Figure 2.14 is obtained by connecting the original crack front to the points (r_0, θ_0, ϕ_0) in accordance with $S_{min}/r_0 =$ constant.The predicted surface of fracture initiation is symmetric about the xz-plane and anti-symmetric about the yz-plane.

It is assumed that all the critical elements dissipate the same amount of energy per unit volume. The process of energy dissipation occurs very quickly in creating a small segment of the new crack surface such that the

* Such a criterion has been applied (Kipp and Sih [7]) successfully for predicting the failure of an inclined crack in a stretched plate. The theoretical curve of $dW/dV =$ constant coincides almost exactly with the entire trajectory of fracture. It is because the phenomenon of brittle fracture occurs so quickly that it leaves little time for stress redistribution once the crack begins to propagate.

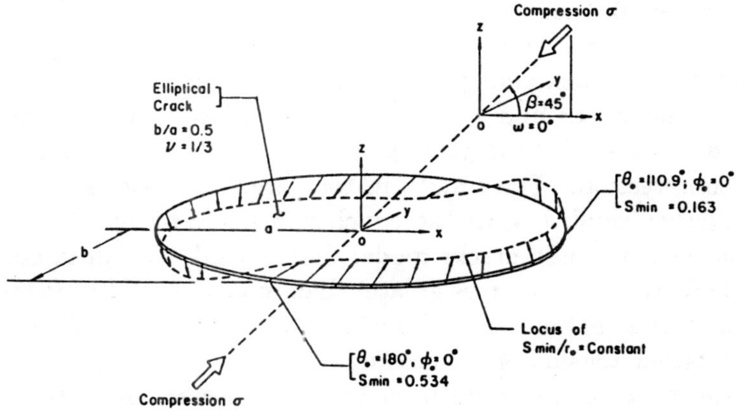

Figure 2.15. Initial fracture surface of an elliptical crack ($b/a = 0.5$) under compression, $\beta = 45°$ and $\omega = 0°$.

onset of rapid fracture follows immediately after initiation. This is known as brittle fracture. Since the calculation of dW/dV in the present work uses the asymptotic solution of the crack border stresses, the criterion of constant $S_{\min}/r_0$ can only yield the initial segment of the fracture surface. It is possible to predict the entire fracture surface which according to the strain energy density theory would be the locus of lines of constant dW/dV subject to the constraint that S be a local minimum. This requires a calculation of minimum dW/dV everywhere in the cracked solid.

By following the same procedure as that used in the case of tensile loading, the envelope of the new fracture surface for an elliptical crack, $b/a = 0.5$, compressed uniformly at an angle $\beta = 45°$ in the xz-plane is obtained and is shown in Figure 2.15. Unlike the tension case, the initial fracture surface is warped considerably and slants toward the direction of loading.

2.8. Concluding remarks

The strain energy density theory has been extended to study the growth characteristics of three-dimensional cracks. A considerable amount of knowledge has been gained in the understanding of fracture initiation from planes of discontinuities or cracks in solids. The S-criterion, when applied to an element in spherical coordinates near the crack border, suggests that fracture may not initiate in the normal plane. The directions of fracture are determined from the minimum values of S. For an elliptical crack, failure occurs at the ends of the minor axis for all the loadings considered. In general, the load at failure decreases as the angle β between σ_{cr} and the crack plane increases and the lowest σ_{cr} occurs at 90°. However, for the case of a narrow ellipse, the lowest failure stress lying in the plane of the major axis occurs at β approximately 60° for $\nu = \frac{1}{3}$. From the $S_{\min}$ distribution along the crack border, it is found that the elliptical crack grows noncoplanarly under loads not normal to the crack plane; a larger amount of extension occurs near the ends of the minor axis and a smaller amount of extension occurs near the ends of major axis. Similar findings are reported for an inclined elliptical crack in compression. The results in the three-dimensional internal crack problem further confirm with the conclusion made by Sih [3, 5] using the two-dimensional through cracks model that a solid can fail in compression at an applied stress many times greater than in tension. The model assumed that the material contains a crack inclined at an angle β with the direction of loading, either tension or compression.

In brittle fracture, the time interval between crack initiation and final separation of the material is so short that the material does not have sufficient time to redistribute the stresses during the fracture process. This is why the fracture surface around the crack border can be reasonably well determined by the original crack geometry. The present analysis can only predict crack growth in a region close to the crack border since it uses the asymptotic stress expressions. To find the entire fracture surface, the exact three-dimensional crack solution should be employed in accordance with the condition that $\mathrm{d}W/\mathrm{d}V$ is constant at the site of the prospective fracture surface. Such calculations for two-dimensional crack problems have been made by Kipp and Sih [7] who found for an inclined crack in a stretched plate that the theory matches with experiments very well for a distance of at least one-half the crack length from the initial crack tip. For the three-dimensional embedded crack, some experiments should be performed so that a similar comparison with the prediction of the S-theory can be made.

The effects of a finite crack border radius of curvature is of interest. In this case, the S-criterion should be applied to an ellipsoid in the same way as Sih [9] has formulated the notch sensitive S-theory for determining the core region radius r_0 in two dimensions. The crack is approximated by a narrow elliptical hole with a root radius of curvature $\rho = b^2/a$ where b and a are semi-minor and semi-major axes respectively. He estimated that for cracks with $b/a \leq 0.01$ and $r_0 \simeq 25\rho$, the crack and sharp notch solutions are in close agreement. Sih and Kipp [6] have given another interpretation to the core region radius r_0 on the basis of the experimental data (Williams and Ewing [15]) of polymethylmethacrylate plates. They proposed that the fracture angle θ_0 should be reported with reference to a critical distance r_0, which measures the radius of a core region surrounding the crack tip. The critical distance r_0 serves the purpose of defining a local region in which the continuum solution ceases to represent the physical properties of the material.

2.9. Appendix 2.1: Elliptical crack in tension

Variation of the fracture angle $(-\theta_0)$ with the angle β for $\omega = 0$, $\nu = \frac{1}{3}$ and $b/a = 0.1$ is shown in Figure 2.16. The curves that decrease monotonically with β correspond to values of α for which Mode I and II dominate while the bell-shaped curves correspond to Mode I and III loading. For a narrow ellipse, $b/a = 0.1$, Figure 2.16 shows that Mode I and III loading prevails on most of the crack border as the curves start to become bell-shaped at much smaller values of α in comparison with those shown in Figure 2.8 for $b/a = 0.5$. The angle between the fracture surface and original crack plane is greatest at the end of the major axis $\alpha = 0°$. A similar set of curves is obtained by varying the ratio b/a instead of α. For fixed position, $\alpha = 30°$, on the crack border, Figure 2.17 shows that when b/a is decreased the local stress state switches from a Mode I and II domination to that of Mode I and III. The effect of ω on fracture angles at the ends of the minor axis ($\alpha = 90°$) is illustrated in Figure 2.18 for a slender ellipse $b/a = 0.1$ and $\nu = \frac{1}{3}$. The angle ω determines the amount of Mode II shear transmitted to a region near the crack front. Note that for a given β the greatest fracture angle $(-\theta_0)$ is obtained at $\omega = 90°$ where the shear component of loading is directed along the minor axis of the ellipse.

The normalized strain energy density factor $16\mu S_{\min}/\sigma^2 b$ as a function of β for $\nu = \frac{1}{3}$, $\omega = 90°$ and $\alpha = 90°$ is given in Figure 2.19. It is seen that the curves attain a peak for small b/a at $\beta \simeq 60°$ and increase monotonically with β for large b/a. Figure 2.20 illustrates the influence of the load angle ω on the variations of $S_{\min}$ with β at $\alpha = 90°$. Note that the presence of Mode II shear, as measured by ω, tends to lower $S_{\min}$. The maximum reduction occurs at $\omega = 90°$.

The curves in Figure 2.21 represent a plot of the failure loads σ_{cr} required to fracture the solid increases as ω is increased. This suggests that it is less desirable to have the loading in the plane of the major axis, $\omega = 0°$, for in this case the solid fractures at the lowest value of applied stress.

2.10. Appendix 2.2: Combined Mode I and III loading

Mode I crack extension does not always yield the lowest allowable failure stress. An example of this is the case of Mode I and III loading of a plane strain through crack as shown in Figure 2.22. The stress intensity factors for this problem

$$k_1 = \sigma b^{1/2} \sin^2 \beta \tag{2.25a}$$

$$k_2 = \sigma b^{1/2} \sin \beta \cos \beta \sin \omega \tag{2.25b}$$

$$k_3 = \sigma b^{1/2} \sin \beta \cos \beta \cos \omega \tag{2.25c}$$

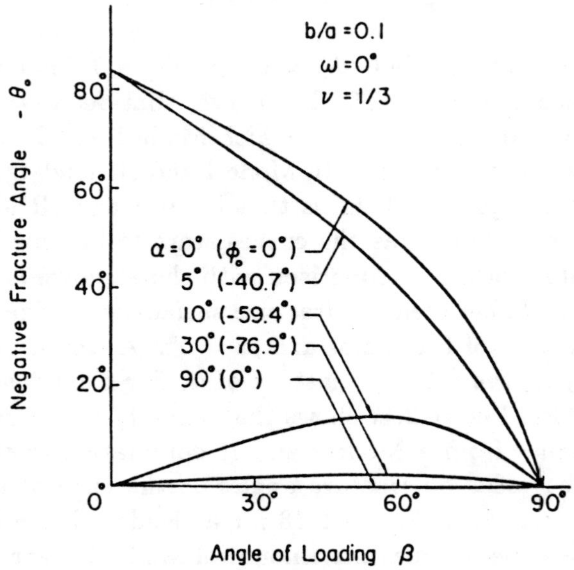

Figure 2.16. Negative fracture angles for $b/a = 0.1$ and $\omega = 0°$.

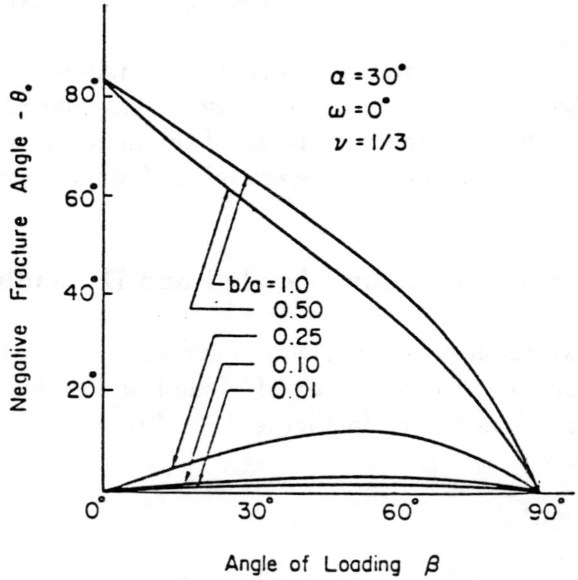

Figure 2.17. Effect of b/a on negative fracture angles for $\alpha = 30°$ and $\omega = 0°$.

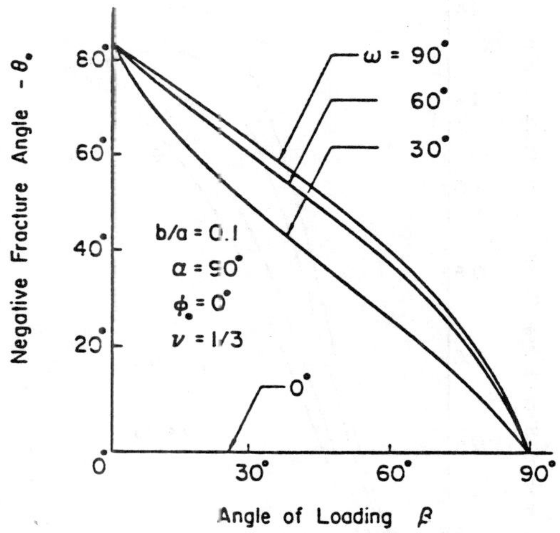

Figure 2.18. Negative fracture angles at end of minor axis for $b/a = 0.1$ and various ω.

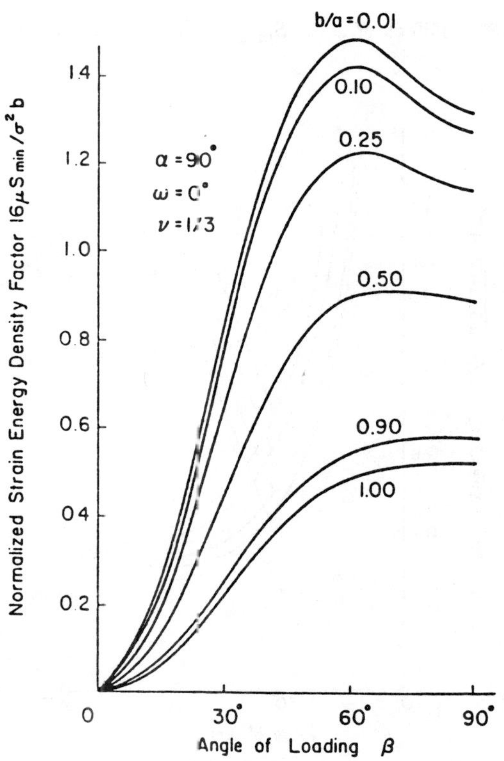

Figure 2.19. Variations of S_{min} with loading angle β for various b/a at $\alpha = 90°$ and $\omega = 0°$.

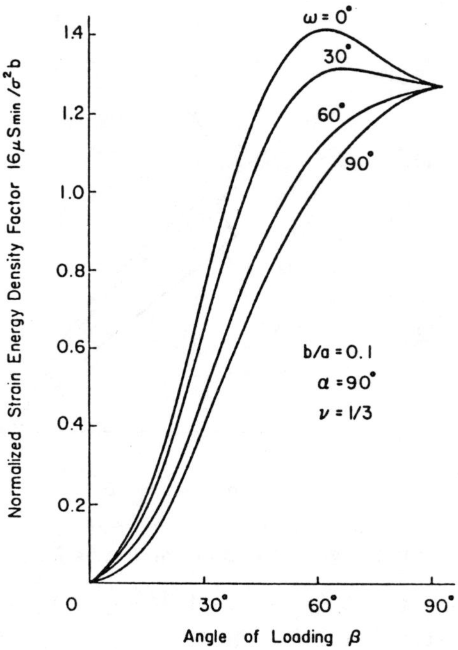

Figure 2.20. Effect of ω on $S_{\min}$ for $b/a = 0.1$ and $\alpha = 90°$.

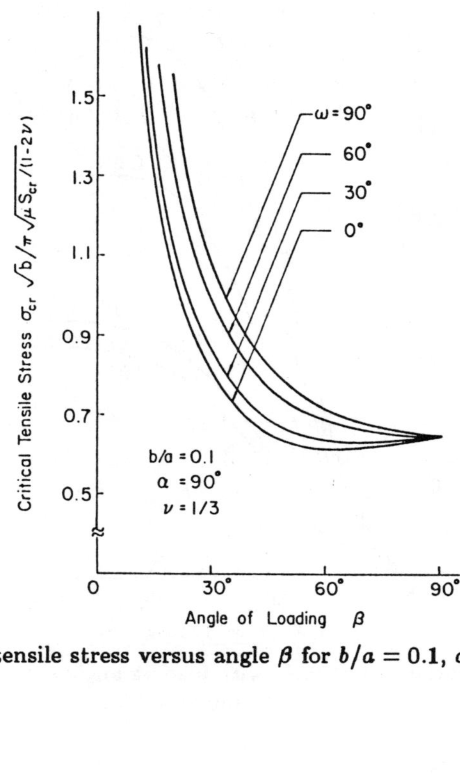

Figure 2.21. Critical tensile stress versus angle β for $b/a = 0.1$, $\alpha = 90°$ and various ω.

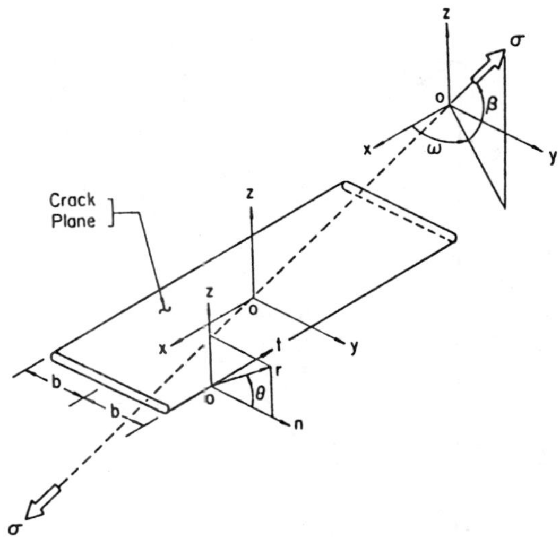

Figure 2.22. Through crack subjected to inclined uniform loading.

remain constant along the straight crack border. In equations (2.25), σ is the applied uniform stress, and b is the half crack length. The angles β and ω are defined in Figure 2.22. The stress field and the strain energy density field near the crack tip in the normal plane will be the same as those given by Sih [3]. When the shear component of the uniform load is parallel to the crack edge, i.e. $\omega = 0°$, the stress intensity factor k_2 in equation (2.25b) vanishes for the through crack. Equations (2.25) can also be obtained as a limiting case of the elliptical crack problem discussed earlier for $\omega = 0°$. Letting $\alpha = 90°$ and $\omega = 0°$ in equations (2.24), the stress intensity factors become

$$k_1 = \frac{1}{E(k)}\sigma b^{1/2} \sin^2 \beta \tag{2.26a}$$

$$k_2 = 0 \tag{2.26b}$$

$$k_3 = \frac{(1-\nu)[1-(b/a)^2]}{(k^2-\nu)E(k)+\nu k'^2 K(k)}\sigma b^{1/2} \sin\beta\cos\beta. \tag{2.26c}$$

In the limit as $b/a \to 0$, $k \to 1$, $k' \to 0$ and $E(k) \to 1$, equations (2.26) for $\alpha = 90°$ are the same as equations (2.25) for $\omega = 0°$.

The strain energy density theory may be used to find the applied stress σ_{cr} required to initiate fracture. The result is shown in Figure 2.10 for $b/a = 0.1$. Note that σ_{cr} at $\beta \simeq 60°$ is smaller than that at $\beta = 90°$.

2.11. Appendix 2.3: Elliptical crack in compression

The fracture angle θ_0 as a function of β for an elliptical crack under compres-

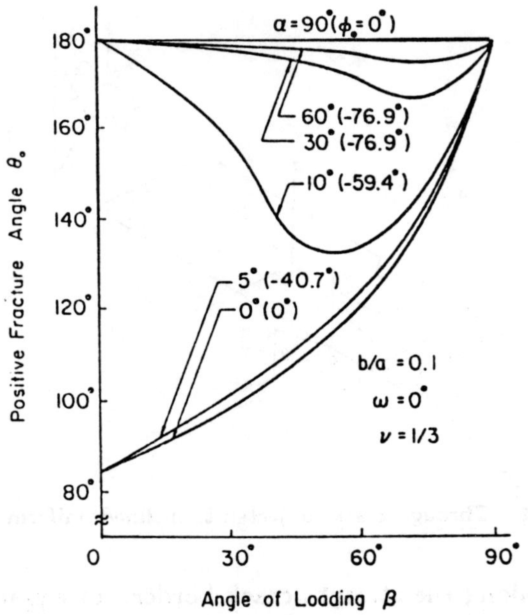

Figure 2.23. **Positive fracture angles for $b/a = 0.1$, $\omega = 0°$ and various α.**

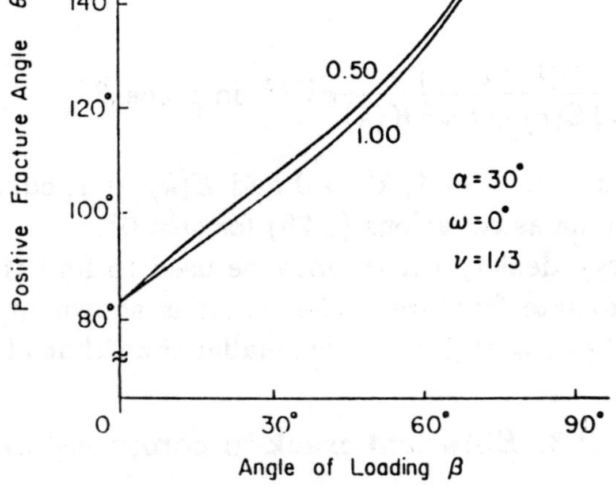

Figure 2.24. **Positive fracture angles for various b/a at $\alpha = 30°$ and $\omega = 0°$.**

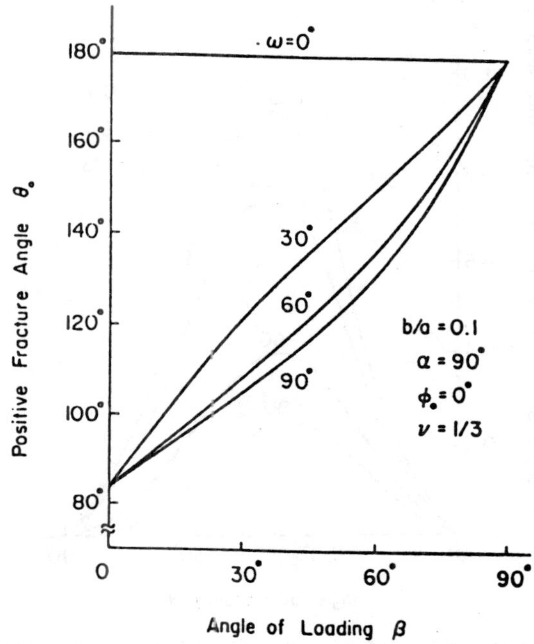

Figure 2.25. **Positive fracture angles at end of minor axis for $b/a = 0.1$ and various ω.**

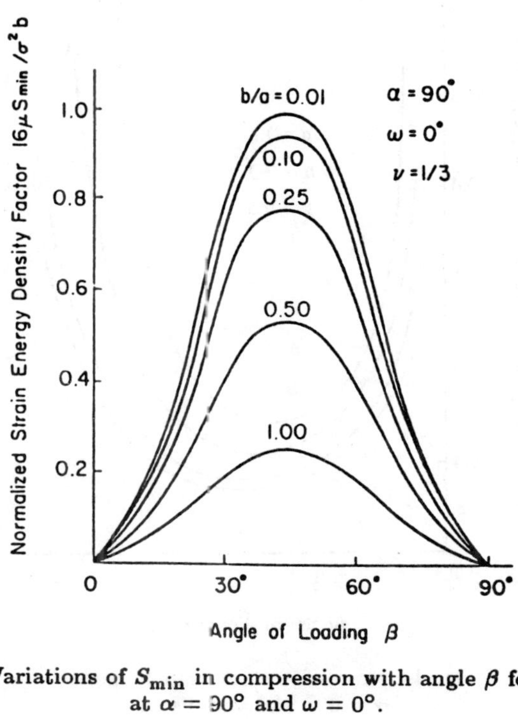

Figure 2.26. Variations of S_{min} in compression with angle β for various b/a at $\alpha = 90°$ and $\omega = 0°$.

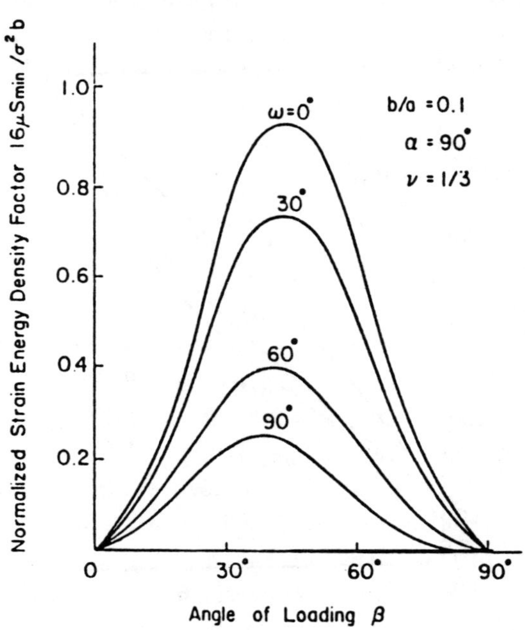

Figure 2.27. Effect of ω on $S_{\min}$ for $b/a = 0.1$ and $\alpha = 90°$.

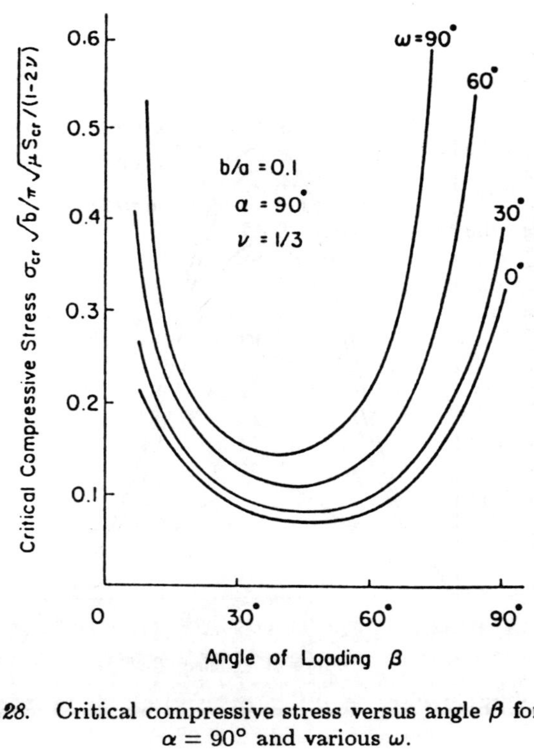

Figure 2.28. Critical compressive stress versus angle β for $b/a = 0.1$, $\alpha = 90°$ and various ω.

sion is shown in Figures 2.23 to 2.25. In Figure 2.23, the angle ω is taken as zero with $\nu = \frac{1}{3}$ and $b/a = 0.1$. Fracture resulting from mixed Modes I and II is predicted to initiate at $\theta_0 \simeq 85°$ while Mode I and III fracture starts at $\theta_0 = 180°$. Figure 2.24 illustrates the fracture direction at fixed $\alpha = 30°$ for various values of the ratio b/a. The angle θ_0 between the fracture surface and original crack plane is smallest for a penny-shaped crack $b/a = 1$. The variations of θ_0 with β for different values of ω and $b/a = 0.1$ are displayed in Figure 2.25. At the ends of the minor axis, $\alpha = 90°$, the fracture angle θ_0 is reduced with increasing ω.

A plot of $S_{\min}$ versus β for different values of b/a with $\alpha = 90°$ is given in Figure 2.26. With $\omega = 0$, the compressive load is in the plane of the major axis. The normalized strain energy density factor $16\mu S_{\min}/\sigma^2 b$ increases as b/a is decreased for a fixed value $\alpha = 90°$. By specifying the crack geometry $b/a = 0.1$ and $\nu = \frac{1}{3}$, the value of $S_{\min}$ is seen to vary with the angle ω as indicated in Figure 2.27. The curves decrease in amplitude as ω is decreased from 0° to 90°. The results presented in Figures 2.26 and 2.27 can best be interpreted in terms of the critical stress that triggers fracture.

For a constant S_{cr}, the applied compressive stress σ_{cr} required to fail the solid may be plotted against β as shown in Figure 2.28 for $\nu = \frac{1}{3}$, $\alpha = 90°$ and $b/a = 0.1$. Four curves corresponding to $\omega = 0°$, 30°, 60° and 90° are obtained. They first decrease in magnitude and then increase in magnitude passing through a minimum at $\beta \simeq 45°$.

2.12 References

[1] Griffith, A. A., The phenomena of rupture and flow in solids, *Philosophical Transactions, Royal Society of London, Series A221*, pp. 163–198 (1921).

[2] Sih, G. C., Some basic problems in fracture mechanics and new concepts, *Journal of Engineering Fracture Mechanics* 5, pp. 365–377 (1973).

[3] Sih, G. C., A special theory of crack propagation: methods of analysis and solutions of crack problems, *Mechanics of Fracture I*, edited by G. C. Sih, Noordhoff International Publishing, Leyden, pp. 21–45 (1973).

[4] Griffith, A. A., The theory of rupture, *Proceedings of First International Congress of Applied Mechanics*, Delft, pp. 55–93 (1924).

[5] Sih, G. C., Strain energy density factor applied to mixed mode crack problems, *International Journal of Fracture* 10, pp. 305–321 (1974).

[6] Sih, G. C. and Kipp, M. E., Discussion on fracture under complex stress: the angle crack problem by J. G. Williams and P. D. Ewing, *International Journal of Fracture* 10, pp. 261–265 (1974).

[7] Kipp, M. E. and Sih, G. C., The strain energy density failure criterion applied to notched elastic solids, *International Journal of Solids and Structures* Vol. II, No. 2, pp. 153–173 (1975).

[8] Sih, G. C. and Cha, B. C. K., A fracture criterion for three-dimensional crack problems, *Journal of Engineering Fracture Mechanics* Vol. 6, pp. 699–723, (1974).

[9] Sih, G. C., Applications of strain-energy-density theory to fundamental fracture problems, *Proceedings of the 10th Anniversary Meeting of the Society of Engineering Science* (1974).

[10] McClintock, F. A. and Walsh, J. B., Friction on Griffith cracks in rocks under pressure, *Proceedings of the 4th U.S. National Congress of Applied Mechanics*, pp. 1015–1021 (1962).
[11] Sih, G. C., A review of the three-dimensional stress problem for a cracked plate, *International Journal of Fracture Mechanics* **7.1**, pp. 39–61 (1971).
[12] Hartranft, R. J. and Sih, G. C., The use of eigenfunction expansions in the general solution of three-dimensional crack problems, *Journal of Mathematics and Mechanics* **19.2**, pp. 123–138 (1969).
[13] Kassir, M. K. and Sih, G. C., Three-dimensional stress distribution around an elliptical crack under arbitrary loadings, *Journal of Applied Mechanics* **33**; *Transactions of the ASME* **88**, Series E, pp. 601–611 (1966).
[14] Irwin, G. R., Crack-extension force for a part-through crack in a plate, *Journal of Applied Mechanics* **29**; *Transactions of the ASME* **84**, Series E, pp. 651–654 (1962).
[15] Williams, J. G. and Ewing P. D., Fracture under complex stress: the angle crack problem, *International Journal of Fracture Mechanics* **8**, pp. 441–446 (1972).

3

Strain energy density theory applied to plate-bending and shell problems

3.1. Introductory remarks

When a slender member is stretched gradually with consideration given only to the principal stress in the axial direction, then failure in the global sense is said to occur by yielding if this stress reaches the elastic limit or yield point and by fracturing if the ultimate strength of the material is reached. Material elements within this member, however, are in a multiaxial stress state. Hence, yielding and/or fracture in the local sense or at a point cannot be adequately described by considering just one of the six stress or strain components even if the remote loading is uniaxial. The distinction between localized and global yielding or fracture must be made before selecting a quantity whose critical value will be assigned to determine the load carrying capacity of the member. In principle, this quantity should involve some combination of the stresses or strains.

The ***energy*** quantity has been frequently associated with the development of failure theories. Two of the commonly used theories are the total energy or Beltrami–Haigh theory and the distortion energy or Huber–von Mises–Hencky theory. According to the former, failure by yielding in the material occurs when the total strain energy per unit volume

$$\frac{\mathrm{d}W}{\mathrm{d}V} = \frac{1+\nu}{2E}\left[\sigma_x^2 + \sigma_y^2 + \sigma_z^2 - \frac{\nu}{1+\nu}(\sigma_x + \sigma_y + \sigma_z)^2 + \right.$$
$$\left. + 2(\tau_{xy}^2 + \tau_{yz}^2 + \tau_{zx}^2)\right] \tag{3.1}$$

absorbed by the material is equal to the energy per unit volume stored in the material upon reaching the yield point under a uniaxial stress state, i.e., $\mathrm{d}W/\mathrm{d}V = \sigma_{\mathrm{yp}}^2/2E$. The latter theory can be stated in the same way by replacing the total strain energy density, $\mathrm{d}W/\mathrm{d}V$, with the distortional strain energy density:

$$\left(\frac{dW}{dV}\right)_d = \frac{1+\nu}{6E}[(\sigma_x - \sigma_y)^2 + (\sigma_y - \sigma_z)^2 + (\sigma_z - \sigma_x)^2 + \\ + 6(\tau_{xy}^2 + \tau_{yz}^2 + \tau_{zx}^2)] \tag{3.2}$$

where E is Young's modulus of elasticity and ν is Poisson's ratio. Equation (3.2) is obtained by decomposing dW/dV in equation (3.1) into two parts

$$\frac{dW}{dV} = \left(\frac{dW}{dV}\right)_v + \left(\frac{dW}{dV}\right)_d \tag{3.3}$$

in which

$$\left(\frac{dW}{dV}\right)_v = \frac{1-2\nu}{6E}(\sigma_x + \sigma_y + \sigma_z)^2 \tag{3.4}$$

represents that part of the strain energy density associated with volume change. It has been implicitly assumed in the two foregoing theories that the failure mode by yielding for an element in a multiaxial stress state is the same as that in a uniaxial stress state.

The Huber–von Mises–Hencky theory associates failure by yielding with $(dW/dV)_d$ only while $(dW/dV)_v$ is neglected.* Except for the extreme situation of a hydrostatic stress state, where $(dW/dV)_d = 0$, and a pure shear stress state, where $(dW/dV)_v = 0$, the failure of a unit volume of material ranging from incipient yielding to brittle fracture involves the release of both energy due to changing in volume and shape. For instance, equations (3.2) and (3.4) indicate that for a uniaxial stress state both $(dW/dV)_d$ and $(dW/dV)_v$ are present:

$$\left(\frac{dW}{dV}\right)_d = \frac{1+\nu}{3E}\sigma_x^2, \quad \left(\frac{dW}{dV}\right)_v = \frac{1-2\nu}{6E}\sigma_x^2. \tag{3.5}$$

Note that the distortional energy is only four times greater than the dilatational energy if $\nu = 0.2$. This difference, being less than one order of magnitude, cannot justify neglecting $(dW/dV)_v$ in comparison with $(dW/dV)_d$.**

A more general situation prevails in Figure (3.1) where the material element is in a triaxial stress state. Let r_0 be the radius of a core region centered around a convenient reference point, the origin. The element under consideration is located outside of this core region. In order to focus attention

---

* This assumption grew out of the test results of Bridgman [1] who showed that many materials did not exhibit yielding when subjected to very high hydrostatic pressure. In this case, the material elements experience volume change only with $(dW/dV)_d = 0$. It would be inaccurate to conclude in general that the energy absorbed in changing volume has no effect in causing failure by yielding for other stress states as well.

** An accurate description of the uniaxial extension of metal bars with ν between 0.2 and 0.3 should involve both changes in volume and shape.

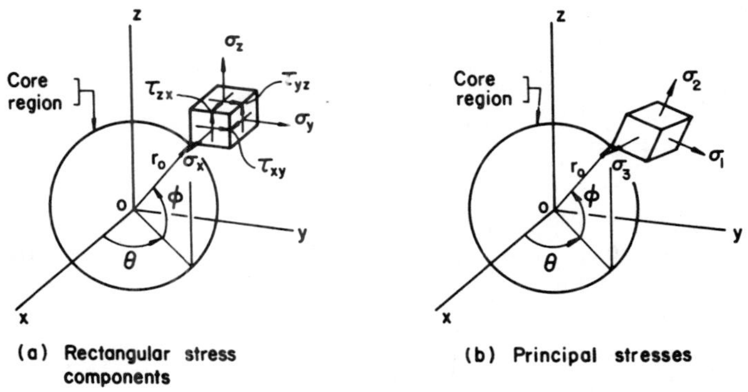

Figure 3.1. Stress element outside of core region.

on the location of failure, Sih [2] has defined a strain energy density factor $S = S_v + S_d$ such that

$$S_v = r_0\left(\frac{dW}{dV}\right)_v \tag{3.6}$$

is the factor associated with volume change and

$$S_d = r_0\left(\frac{dW}{dV}\right)_d \tag{3.7}$$

is the factor associated with shape change. With reference to the principal stresses σ_1, σ_2 and σ_3 in Figure 3.1(b), equations (3.4) and (3.6) may be written as

$$S_v = r_0\left(\frac{1-2\nu}{6E}\right)(\sigma_1+\sigma_2+\sigma_3)^2 \tag{3.8}$$

and equations (3.2) and (3.7) yield

$$S_d = r_0\left(\frac{1+\nu}{6E}\right)[(\sigma_1-\sigma_2)^2+(\sigma_2-\sigma_3)^2+(\sigma_3-\sigma_1)^2]. \tag{3.9}$$

For a state of plane strain where

$$\sigma_3 = \nu(\sigma_1+\sigma_2) \tag{3.10}$$

the ratio S_v/S_d obtained from equations (3.8) to (3.10) takes the form

$$\frac{S_v}{S_d} = \frac{(1-2\nu)(1+\nu)[(\sigma_1/\sigma_2)+1]^2}{[(\sigma_1/\sigma_2)-1]^2+[(1-\nu)-\nu(\sigma_1/\sigma_2)]^2+[(1-\nu)(\sigma_1/\sigma_2)-\nu]^2}. \tag{3.11}$$

Figure 3.2 displays a plot of S_v/S_d versus the ratio of the principal stresses σ_1/σ_2 in equation (3.11) for ν = 0.0, 0.1, 0.2, 0.3 and 0.4. The greatest volume change takes place when $\sigma_1 = \sigma_2$ corresponding to a two-dimensional

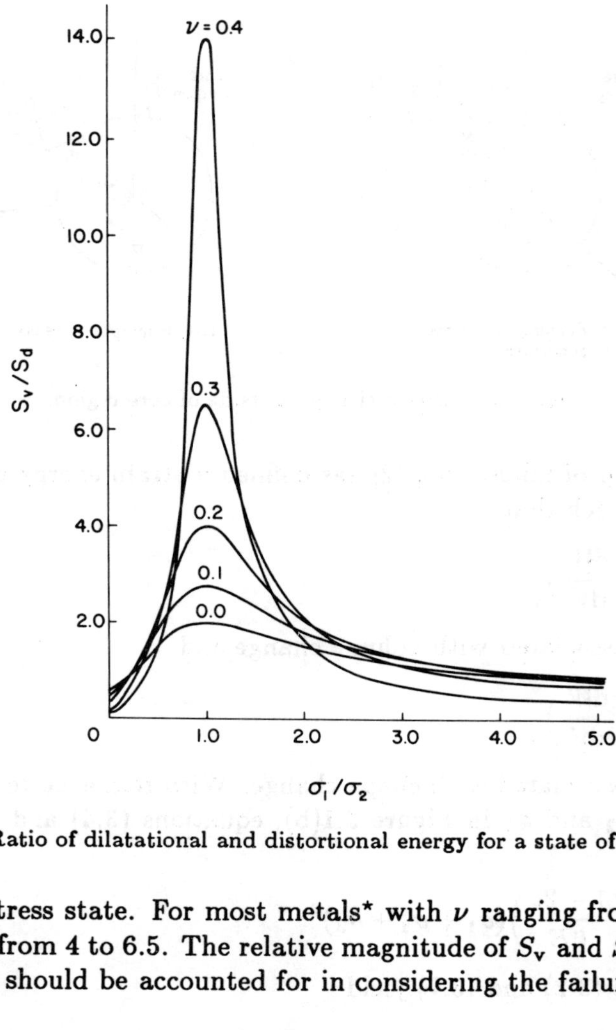

Figure 3.2. **Ratio of dilatational and distortional energy for a state of plane strain.**

hydrostatic stress state. For most metals* with ν ranging from 0.2 to 0.3, S_v/S_d varies from 4 to 6.5. The relative magnitude of S_v and S_d shows that both of them should be accounted for in considering the failure of material elements.

More recently, Sih [3, 4] has proposed a failure theory based on the stationary value of the strain energy density factor

$$S = S_v + S_d \tag{3.12}$$

reaching some critical value. Unlike that of Beltrami–Haigh, Sih's theory possesses the additional feature of being able to locate the element initiating yielding or brittle fracture. This is accomplished by obtaining the stationary values of S with respect to the angles θ and ϕ shown in Figure 3.1(a) or 3.1(b). The point (θ_0, ϕ_0) making S a minimum ($= S_{min}$) gives the position

* Keep in mind that the present discussion holds only if yielding or fracture occurs under small strain. It does not apply to rubber-like or polymer materials undergoing large deformation.

of the element that initiates fracture while the location of S_{max} determines the onset of yielding. The initiation of yielding and/or fracture usually occurs in the vicinity of a reentrant corner, a crack front or a notch border from which r_0, the radius of the core region in Figure 3.1, is measured. The radius r_0 serves as a limiting distance within which the microstructural effect of the material comes into play. In contrast to the classical theory of fracture mechanics [5] which utilizes the concept of a critical stress intensity factor k_{1c} or an energy release rate quantity G_{1c}, the strain energy density factor theory, or S-theory, is not restricted to loading symmetric with respect to the crack plane and does not require the material to possess an initial crack. In this chapter, the S-theory will be further employed in plate bending problems involving the interaction of crack size with plate thickness.

3.2. Strain energy density factor theory

A rational procedure for the design of structural members requires that the most likely of several possible failure modes of a member in service be determined and that a quantity be assigned to predict failure. Fracture mechanics deals with the conditions under which a member attains uncontrollable fracture by crack propagation. The critical value of the selected quantity which limits the allowable load must be sensitive to changes in the nature of loading, the type of geometry, etc., and should be a characteristic of the material. The k_{1c} value in the classical theory of fracture mechanics [5] is one such example, where $k_{1c} = \sigma\sqrt{a}$ for a given material should remain constant at fracture while σ, the applied stress, and a, the half crack length, can vary individually at crack instability. Suitable experiments may be designed to test a given theory by using two different loading conditions and comparing the experimental results with those predicted by the theory. Erdogan and Sih [6] have measured the critical tensile stress σ_c in uniaxial loading and critical shear stress τ_c in pure shear loading for plexiglass plate specimens containing a central crack. Their measured values of σ_c and τ_c yield an average of 0.915 for τ_c/σ_c and agree favorably with Sih's prediction [2] of $\tau_c/\sigma_c = 0.905$ for $\nu = 0.33$. The theoretical prediction based on the strain energy density theory was obtained by using the same critical value of S or S_c for the two different stress states. Detailed calculations of σ_c and τ_c and additional comments will follow subsequently.

Before proceeding further, it is necessary to state the failure mode to be treated in this discussion. In situations where the phenomenon of material separation occurs suddenly without noticeable warning such as slow crack growth, the fracture process is classified as *brittle* or simply as brittle

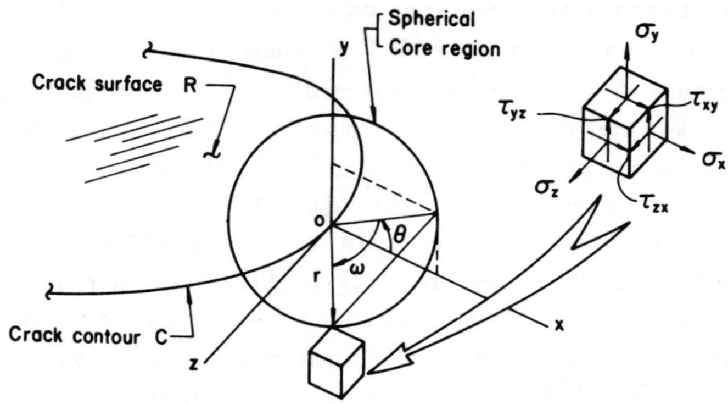

Figure 3.3. A special spherical coordinate system.

fracture.* The application of the S-theory to ductile fracture is discussed elsewhere and will not be dealt with here. Within the framework of brittle fracture where failure initiates from an element nearby the crack border, it suffices to compute the strain energy density S from the local stresses, σ_x, $\sigma_y, \ldots, \tau_{zx}$ as shown in Figure 3.3. Without loss of generality, the crack surface R bounded by an arbitrary contour C is assumed to coincide with the xz-plane. As the point 0 travels around the crack border, the x-axis is directed normal to C, the y-axis is perpendicular to R and the z-axis is tangent to C. The element under consideration is located at (r, θ, ω) which is related to (x, y, z) through the relationships

$$x = r\cos\theta\cos\omega \tag{3.13a}$$

$$y = r\sin\theta\cos\omega \tag{3.13b}$$

$$z = r\sin\omega \tag{3.13c}$$

where $-\pi/2 \leq \omega \leq \pi/2$ and $-\pi \leq \theta \leq \pi$. The choice of spherical coordinates (r, θ, ω) in Figure 3.3 yields a simple form** for the local stresses

* Brittle fracture includes those failures which occur without warning and give no evidence of ductility or stretching prior to rapid crack propagation but does not exclusively imply that the material must be inherently brittle in the metallurgical sense. Materials with the same chemical composition and microstructure can behave in a brittle fashion under one set of conditions (i.e., temperature, rate of loading, mechanical constraint, etc.) and in a ductile fashion under other conditions. Since the continuum mechanics model of the S-theory excludes metallurgical effects, it would not be accurate to say that the theory is restricted to brittle materials but rather it deals with an instability phenomenon that exhibits ***brittle behavior***.

** An alternative choice of spherical coordinates (r, θ, ϕ) such as those shown in Figures 3.1 lead to more complicated stress expressions [7] which are equivalent to equations (3.14). An error in the asymptotic expression for one of the ellipsoidal coordinates in [7] needs to be corrected.

$$\sigma_x = \frac{k_1}{(2r\cos\omega)^{1/2}}\cos\frac{\theta}{2}\left(1-\sin\frac{\theta}{2}\sin\frac{3\theta}{2}\right) -$$
$$-\frac{k_2}{(2r\cos\omega)^{1/2}}\sin\frac{\theta}{2}\left(2+\cos\frac{\theta}{2}\cos\frac{3\theta}{2}\right)+O(1) \tag{3.14a}$$
$$\sigma_y = \frac{k_1}{(2r\cos\omega)^{1/2}}\cos\frac{\theta}{2}\left(1+\sin\frac{\theta}{2}\sin\frac{3\theta}{2}\right) +$$
$$+\frac{k_2}{(2r\cos\omega)^{1/2}}\sin\frac{\theta}{2}\cos\frac{\theta}{2}\cos\frac{3\theta}{2}+O(1) \tag{3.14b}$$
$$\sigma_z = 2\nu\left(\frac{k_1}{(2r\cos\omega)^{1/2}}\cos\frac{\theta}{2}-\frac{k_2}{(2r\cos\omega)^{1/2}}\sin\frac{\theta}{2}\right)+O(1) \tag{3.14c}$$
$$\tau_{xy} = \frac{k_1}{(2r\cos\omega)^{1/2}}\sin\frac{\theta}{2}\cos\frac{\theta}{2}\cos\frac{3\theta}{2}+$$
$$+\frac{k_2}{(2r\cos\omega)^{1/2}}\cos\frac{\theta}{2}\left(1-\sin\frac{\theta}{2}\sin\frac{3\theta}{2}\right)+O(1) \tag{3.14d}$$
$$\tau_{xz} = -\frac{k_3}{(2r\cos\omega)^{1/2}}\sin\frac{\theta}{2}+O(1) \tag{3.14e}$$
$$\tau_{yz} = \frac{k_3}{(2r\cos\omega)^{1/2}}\cos\frac{\theta}{2}+O(1). \tag{3.14f}$$

Substituting equations (3.14) into (3.1) and using the relation $dW/dV = S/r$, it is found that

$$S = a_{11}k_1^2 + 2a_{12}k_1k_2 + a_{22}k_2^2 + a_{33}k_3^2. \tag{3.15}$$

The coefficients a_{ij} $(i,j = 1,2)$ depend on the material constants and the angles (θ,ω) as given by

$$16\mu\cos\omega\, a_{11} = (3-4\nu-\cos\theta)(1+\cos\theta) \tag{3.16a}$$
$$16\mu\cos\omega\, a_{12} = 2\sin\theta(\cos\theta-1+2\nu) \tag{3.16b}$$
$$16\mu\cos\omega\, a_{22} = 4(1-\nu)(1-\cos\theta)+(3\cos\theta-1)(1+\cos\theta) \tag{3.16c}$$
$$16\mu\cos\omega\, a_{33} = 4 \tag{3.16d}$$

in which μ stands for the shear modulus of elasticity.

On physical grounds, material elements are capable of storing energy by experiencing volume change (dilatation) and shape change (distortion). Excessive dilatation may result in brittle fracture while excessive distortion may tend to yield the material. The proportion of energy absorbed in dilatation and distortion can be determined for various material properties at locations of the stationary values of the strain energy density factor. In fracture mechanics, interest is centered on the conditions that trigger crack instability. The assumptions made in the strain energy density theory [3, 4] may be stated as

(1) ***Crack growth is directed along the line from the center of the spherical core (Figure 3.3) to the point on the spherical surface with the minimum strain energy density factor*** S_{min}.
(2) ***Growth along this direction begins when*** S_{min} ***reaches the maximum critical value*** S_c ***which the material will tolerate.***

The values of (θ_0, ω_0) corresponding to S_{min} are obtained by taking $\partial S/\partial\theta = 0$ and $\partial S/\partial\omega = 0$. At these locations, the elements tend to experience more volume change ΔV than those located at S_{max} where yielding is assumed to take place. The path along which S is a relative minimum has coincided accurately with the experimentally observed trajectory of brittle fracture [8, 9].

Experience indicates that failure can seldom be associated solely with volume change or shape change. It normally involves some of each depending on the stress state. The amount of dilatation and distortion for each material element can be determined by decomposing equation (3.15) into

$$S_v = b_{11}k_1^2 + 2b_{12}k_1k_2 + b_{22}k_2^2 + b_{33}k_3^2 \tag{3.17}$$

for dilatation and

$$S_d = c_{11}k_1^2 + 2c_{12}k_1k_2 + c_{22}k_2^2 + c_{33}k_3^2 \tag{3.18}$$

for distortion. The sum of the coefficients b_{ij} and c_{ij} $(i, j = 1, 2, 3)$ in equations (3.17) and (3.18) equal a_{ij} given by equations (3.16). Without going into details, it can be shown that

$$12\mu \cos\omega\, b_{11} = (1-2\nu)(1+\nu)(1+\cos\theta) \tag{3.19a}$$
$$12\mu \cos\omega\, b_{12} = (1-2\nu)(1+\nu)\sin\theta \tag{3.19b}$$
$$12\mu \cos\omega\, b_{22} = (1-2\nu)(1+\nu)(1-\cos\theta) \tag{3.19c}$$
$$12\mu \cos\omega\, b_{33} = 0 \tag{3.19d}$$

and the result of subtracting equations (3.19) from (3.16) yields

$$16\mu \cos\omega\, c_{11} = (1+\cos\theta)[\tfrac{2}{3}(1-2\nu)^2 + 1 - \cos\theta] \tag{3.20a}$$
$$16\mu \cos\omega\, c_{12} = 2\sin\theta[\cos\theta - \tfrac{1}{3}(1-2\nu)^2] \tag{3.20b}$$
$$16\mu \cos\omega\, c_{22} = \tfrac{2}{3}(1-2\nu)^2(1-\cos\theta) + 4 - 3\sin^2\theta \tag{3.20c}$$
$$16\mu \cos\omega\, c_{33} = 4. \tag{3.20d}$$

Once the location (θ_0, ω_0) for a given element is known, S_v and S_d can be computed from equations (3.17) and (3.18) for determining the effect of dilatation and distortion on the onset of fracture or yielding.

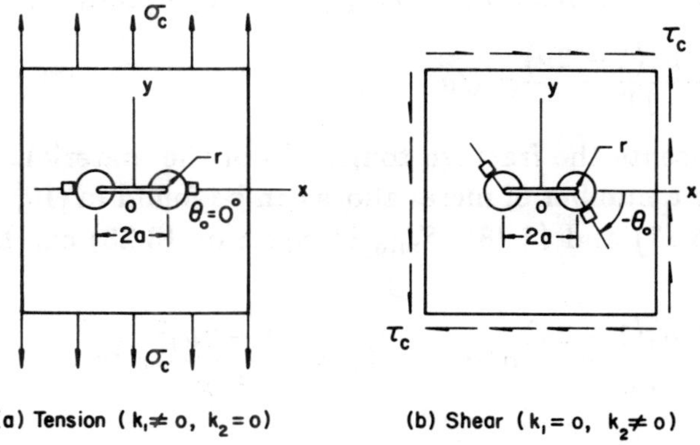

Figure 3.4. Plane strain specimens.

Now, if failure is confined to the plane normal to the crack front, then $\omega = 0$ and the core region in Figure 3.3 becomes a circle of radius r centered around the crack edge. Moreover, if the load is applied such that the same stress state prevails along the crack border, then the problem is said to be two-dimensional. The assumption of plane strain leads to additional simplification since in this case k_3 drops out of equations (3.14) and (3.15). Consider the plane strain specimens in Figures 3.4(a) and 3.4(b) for the uniaxial tension and in-plane shear of a solid containing a crack of length $2a$. It will be shown subsequently that, according to the strain energy density theory, the ratio of the critical stress τ_c/σ_c at incipient fracture is

$$\frac{\tau_c}{\sigma_c} = \left[\frac{3(1-2\nu)}{2(1-\nu)-\nu^2}\right]^{1/2}. \tag{3.21}$$

For $\nu = 0.33$, this ratio gives $\tau_c/\sigma_c = 0.905$ which agrees well with the experimentally measured value of 0.915 for cracked plexiglass plates [6] as discussed earlier.

Uniaxial tension test. When a tensile load is applied normal to the crack plane, only k_1 is nonzero and equation (3.15) reduces to

$$S = \frac{1}{16\mu}(3 - 4\nu - \cos\theta)(1 + \cos\theta)k_1^2 \tag{3.22}$$

in which $k_1 = \sigma\sqrt{a}$ for a large plate. Referring to Figure 3.4(a), σ_c denotes the critical value of σ at which crack propagation begins. The direction of crack initiation is found from the condition $\partial S/\partial\theta = 0$ that makes S a relative minimum. This corresponds to $\theta_0 = 0$ implying that the crack grows in a self-similar manner and

$$S_{\min} = \frac{1-2\nu}{4\mu}k_1^2. \tag{3.23}$$

At the point of crack instability, $S_{min} \to S_c$ and equation (3.23) becomes

$$S_c = \frac{(1+\nu)(1-2\nu)}{2E}\sigma_c\sqrt{a} \tag{3.24}$$

where S_c represents the fracture toughness of the material. The measured values of S_c for a number of metal alloys can be found in [10]. With the help of equations (3.17) and (3.18), S_{min} in equation (3.23) can be decomposed into

$$S_v = \frac{(1+\nu)(1-2\nu)}{6\mu}\sigma^2 a, \qquad S_d = \frac{(1-2\nu)^2}{12\mu}\sigma^2 a. \tag{3.25}$$

The ratio

$$\frac{S_v}{S_d} = \frac{2(1+\nu)}{1-2\nu} \tag{3.26}$$

shows that S_v is always greater than S_d for $\theta_0 = 0$. Hence, the element assumed to trigger fracture undergoes more volume change than shape change.

The relative maximum of S can be found in the same way. It occurs at $\theta_0 = \cos^{-1}(1-2\nu)$ which may be inserted into equation (3.22) to give

$$S_{max} = \frac{(1-\nu)^2}{4\mu}\sigma^2 a. \tag{3.27}$$

Separating equation (3.27) into S_d and S_v renders

$$S_d = \frac{(1-\nu)(1-\nu+4\nu^2)}{12\mu}\sigma^2 a, \qquad S_v = \frac{(1-\nu^2)(1-2\nu)}{6\mu}\sigma^2 a. \tag{3.28}$$

It is easily seen from

$$\frac{S_d}{S_v} = \frac{2(1-\nu+4\nu^2)}{(1+\nu)(1-2\nu)} \tag{3.29}$$

that $S_d > S_v$ corresponds to elements located at $\theta_0 = \cos^{-1}(1-2\nu)$ where yielding is assumed to take place.

Pure shear test. The specimen in Figure 3.4(b) is subjected to a state of in-plane shear where $k_1 = 0$ and $k_2 = \tau\sqrt{a}$. With this information, equation (3.15) can be written as

$$S = \frac{1}{16\mu}[4(1-\nu)(1-\cos\theta) + (1+\cos\theta)(3\cos\theta - 1)]\tau^2 a. \tag{3.30}$$

Equation (3.30) may be minimized with respect to θ giving

$$\theta_0 = \cos^{-1}\left(\frac{1-2\nu}{3}\right) \tag{3.31}$$

Table 3.1. Fracture angles.

ν	0.0	0.1	0.2	0.3	0.4	0.5
$-\theta_0$	70.5°	74.5°	78.5°	82.3°	86.2°	90°

as the direction of crack initiation,* Figure 3.4(b). Values of θ_0 for different Poisson's ratios are given in Table 3.1. Putting equation (3.31) into (3.30) renders

$$S_{\min} = \frac{2(1-\nu)-\nu^2}{12\mu}\tau^2 a. \tag{3.32}$$

Crack instability starts when $S_{\min} = S_c$ or

$$S_c = \frac{(1+\nu)[2(1-\nu)-\nu^2]}{6E}\tau_c^2 a. \tag{3.33}$$

If the specimens in Figures 3.4(a) and 3.4(b) are made of the same material, then S_c in equations (3.24) and (3.33) may be eliminated to give the ratio τ_c/σ_c as shown in equation (3.21). As in the case of tensile loading, equation (3.32) may also be separated into S_v and S_d. Under in-plane shear load, the critical element prior to fracture undergoes more distortion than the element situated along the line of symmetry in Figure 3.4(a).

3.3. Bending and twisting of cracked plates

The accurate translation of physical phenomena into mathematical terms often presents many difficulties. Too much importance cannot be attached to the initial stages of sifting the evidence to enable a useful analytical treatment. Some of these difficulties and complexities are brought out in chapter 2 of this book dealing with three-dimensional crack problems. The strict requirement of satisfying all the governing differential equations and boundary conditions in the theory of three-dimensional theory of elasticity cannot always be met. Assumptions and/or approximations are frequently made to relax either the governing differential equations or boundary conditions or both so as to make the problem manageable. Since the discipline of fracture mechanics requires an accurate description of the stress state near the crack tip region, it is of primary concern to preserve the exact nature of the crack surface boundary conditions in formulating approximate theories. More specifically, the basic character of the local stresses in equations (3.14)

* It should be cautioned that equation (3.31) applies to a perfectly sharp crack of zero radius of curvature, ρ. Furthermore, the distance r_0 at which θ_0 is measured has not been defined. In a more refined analysis, Sih and Kipp [11] have shown that small values of ρ and r_0 can greatly affect the calculated and measured values of the fracture angle θ_0. Comparison of theory and experiment on θ_0 without considering the effect of ρ and r_0 could lead to premature conclusions.

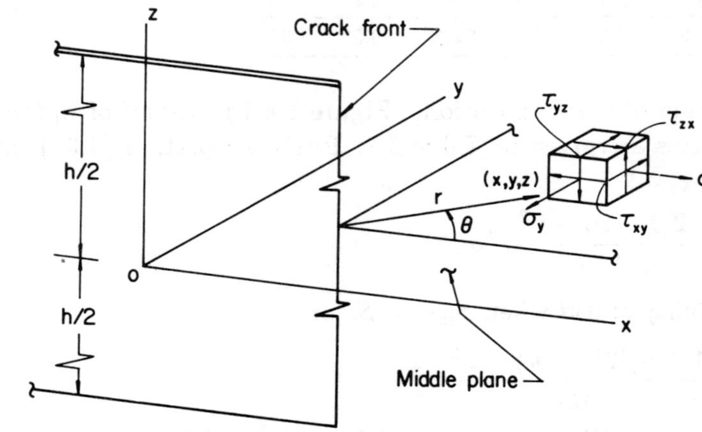

Figure 3.5. **Stresses on an element ahead of crack in a thin plate.**

as determined from the exact three-dimensional theory of elasticity should be retained. Such a preference implies that approximations will be made only in a quantitative sense through stress intensity factors k_1, k_2 and k_3.

Character of local moments and shear forces. Let the thickness h of an elastic body with a crack be small when compared with the other dimensions such as crack length, width of the body, etc. Then the body is said to be a plate. The conventional plate theories* assume that the stress variations in terms of the thickness coordinate are known as a priori. For thin plates, the transverse normal stress component σ_z is assumed to be zero throughout the plate. The middle plane bisecting the plate thickness as shown in Figure 3.5 is usually taken as reference. With respect to this plane $z = 0$, the in-plane stresses σ_x, σ_y and τ_{xy} on the element located at (x, y, z) are assumed to vary linearly in z, i.e.,

$$\sigma_x = \frac{12z}{h^3} M_x, \qquad \sigma_y = \frac{12z}{h^3} M_y, \qquad \tau_{xy} = \frac{12z}{h^3} M_{xy} \tag{3.34}$$

where M_x and M_y are the bending moments and M_{xy} the twisting moment per unit length acting on any section of the plate parallel to the xz and yz planes. They can be obtained by integration as

$$M_x(x, y) = \int_{-h/2}^{h/2} \sigma_x(x, y, z) z \, dz \tag{3.35a}$$

---

* The theory of Hartranft and Sih [12] does not preassume the stress distribution across the plate thickness. They determine this distribution from the plane strain condition ahead of the crack.

$$M_y(x,y) = \int_{-h/2}^{h/2} \sigma_y(x,y,z) z \, dz \tag{3.35b}$$

$$M_{xy}(x,y) = \int_{-h/2}^{h/2} \tau_{xy}(x,y,z) z \, dz. \tag{3.35c}$$

With σ_z already assumed to vanish everywhere, the free surface conditions at $z = \pm h/2$ can be satisfied by further requiring that

$$\tau_{xz}(x,y,\pm h/2) = \tau_{yz}(x,y,\pm h/2) = 0. \tag{3.36}$$

It follows that the equations of equilibrium require both τ_{xz} and τ_{yz} to vary parabolically in z as

$$\tau_{xz} = \frac{3}{2h}\left[1 - \left(\frac{2z}{h}\right)^2\right] V_x \tag{3.37a}$$

$$\tau_{yz} = \frac{3}{2h}\left[1 - \left(\frac{2z}{h}\right)^2\right] V_y. \tag{3.37b}$$

In equations (3.37), V_x and V_y are the shearing forces per unit length on sections parallel to the y and x axes and they are given by

$$V_x(x,y) = \int_{-h/2}^{h/2} \tau_{xz}(x,y,z) \, dz \tag{3.38a}$$

$$V_y(x,y) = \int_{-h/2}^{h/2} \tau_{yz}(x,y,z) \, dz. \tag{3.38b}$$

Since the z-dependence of the stress components is assumed to be known, the plate-bending problem has essentially been reduced to two dimensions and the boundary conditions will be specified in terms of the moments and shear forces.

For a crack whose surfaces are free from tractions, the bending moment, twisting moment and shear force must all vanish individually. If the crack is placed on the x-axis, then the conditions*

$$M_y(x,0) = M_{xy}(x,0) = V_y(x,0) = 0 \tag{3.39}$$

must prevail on the segment of the x-axis where the crack is located. To this end, the Reissner sixth-order plate-bending theory [13] is used to satisfy

* These conditions cannot be completely satisfied by the Poisson–Kirchhoff theory of plate bending which neglects the effect of transverse shear deformation. As a result, the two conditions on M_{xy} and V_y in equation (3.39) are replaced by a single one on $V_y + \partial M_{xy}/\partial x$, referred to as the equivalent or Kirchhoff shear. The consequences are that the angular variations of the local stresses become dependent on the Poisson's ratio and differ from those in equations (3.14) and the transverse shear stresses τ_{xz} and τ_{yz} attain a singularity of the order $r^{-3/2}$. This is in contrast to the exact three-dimensional solution of $r^{-1/2}$-singularity for τ_{xz} and τ_{yz}.

equations (3.39) and yields the desired character of the crack tip moments and shear forces:

$$M_x = \frac{K_1}{(2r)^{1/2}} \cos\frac{\theta}{2}\left(1 - \sin\frac{\theta}{2}\sin\frac{3\theta}{2}\right) - \\ - \frac{K_2}{(2r)^{1/2}} \sin\frac{\theta}{2}\left(2 + \cos\frac{\theta}{2}\cos\frac{3\theta}{2}\right) + O(1) \tag{3.40a}$$

$$M_y = \frac{K_1}{(2r)^{1/2}} \cos\frac{\theta}{2}\left(1 + \sin\frac{\theta}{2}\sin\frac{3\theta}{2}\right) + \\ + \frac{K_2}{(2r)^{1/2}} \sin\frac{\theta}{2}\cos\frac{\theta}{2}\cos\frac{3\theta}{2} + O(1) \tag{3.40b}$$

$$M_{xy} = \frac{K_1}{(2r)^{1/2}} \sin\frac{\theta}{2}\cos\frac{\theta}{2}\cos\frac{3\theta}{2} + \\ + \frac{K_2}{(2r)^{1/2}} \cos\frac{\theta}{2}\left(1 - \sin\frac{\theta}{2}\sin\frac{3\theta}{2}\right) + O(1) \tag{3.40c}$$

$$V_x = -\frac{K_3}{(2r)^{1/2}} \sin\frac{\theta}{2} + O(1) \tag{3.40d}$$

$$V_y = \frac{K_3}{(2r)^{1/2}} \cos\frac{\theta}{2} + O(1) \tag{3.40e}$$

in which K_1 and K_2 are the moment intensity factors and K_3 the shear force intensity factor. Note that the $1/\sqrt{r}$ singularity and the θ-dependence in equations (3.40) are the same as those shown in equations (3.14).

Moment and shear force intensity factors. Consider the problem of a through crack in a bent plate. The crack is tilted at an angle β with reference to the plane about which a bending moment of magnitude M is applied. This loading as shown in Figure 3.6(a) is equivalent to the application of bending moments M_1, M_2 and twisting moment M_{12} in Figure 3.6(b) through the transformation properties of the stresses:

$$M_1 = M\sin^2\beta \tag{3.41a}$$

$$M_2 = M\cos^2\beta \tag{3.41b}$$

$$M_{12} = M\sin\beta\cos\beta \tag{3.41c}$$

For this problem, Hartranft and Sih [12] and Wang [14] have provided the K_j ($j = 1,\ 2,\ 3$) expressions and they are

$$K_1 = \Phi(1)M_1\sqrt{a} \tag{3.42a}$$

$$K_2 = \Psi(1)M_{12}\sqrt{a} \tag{3.42b}$$

$$K_3 = -\frac{\sqrt{10}}{(1+\nu)h}\Omega(1)M_{12}\sqrt{a}. \tag{3.42c}$$

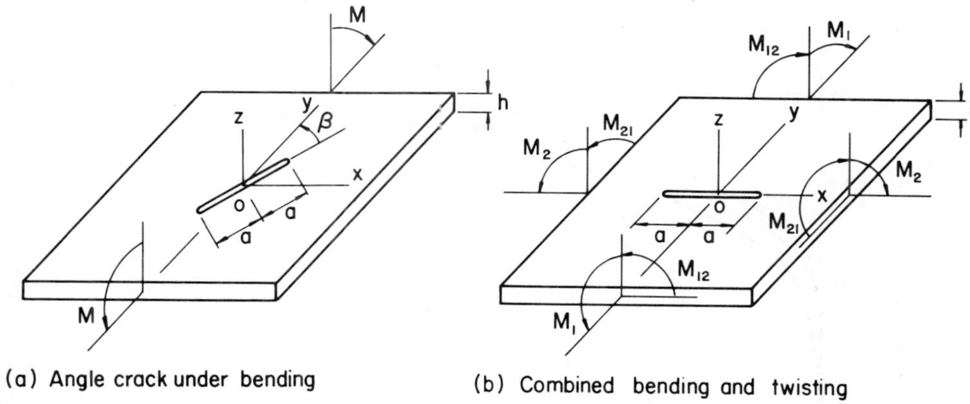

Figure 3.6. Loading configuration for bending and twisting of a cracked plate.

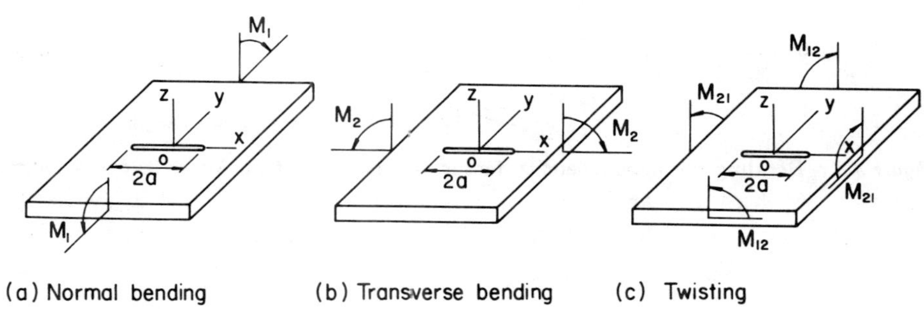

Figure 3.7. Three separate problems of plate bending.

Equations (3.42) show that only M_1 and M_{12} affect the moment intensity near the crack tip while M_2 has no contribution. The separate effects of M_1, M_2 and M_{12} can be best illustrated by Figures 3.7(a) to 3.7(c) where normal bending produces K_1, parallel bending gives rise to no moment singularity or intensity and pure twisting leads to both K_2 and K_3. It is apparent from equations (3.41) and (3.42) that

$$K_1 = \Phi(1) M\sqrt{a}\,\sin^2\beta \tag{3.43a}$$

$$K_2 = \Psi(1) M\sqrt{a}\,\sin\beta\cos\beta \tag{3.43b}$$

$$K_3 = -\frac{\sqrt{10}}{(1+\nu)h}\Omega(1) M\sqrt{a}\,\sin\beta\cos\beta \tag{3.43c}$$

where the functions $\Phi(1)$, $\Psi(1)$ and $\Omega(1)$ are computed numerically from integral equations [12, 14]. Their values as a function of $h/a\sqrt{10}$ for different Poisson's ratios can be found in Figures 3.8 to 3.10.

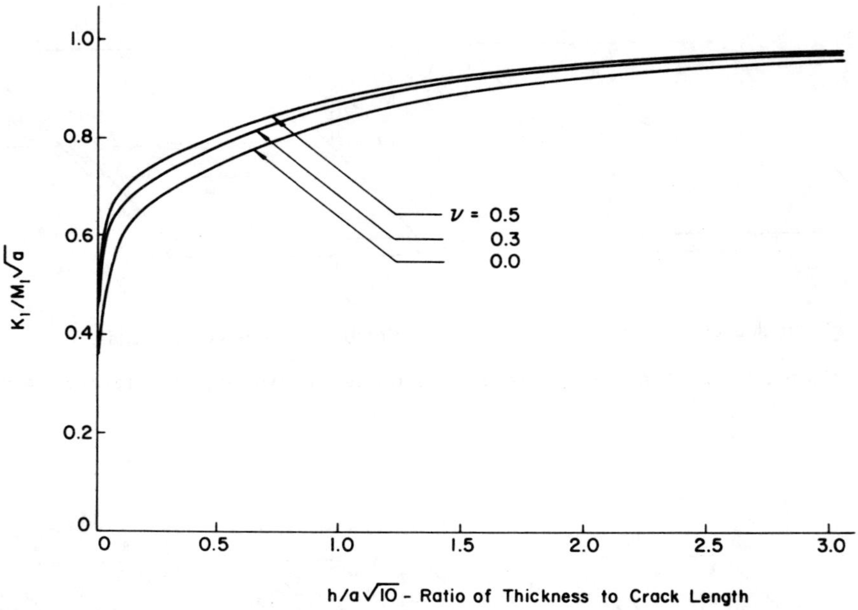

Figure 3.8. Bending moment intensity factor versus ratio of thickness to crack length.

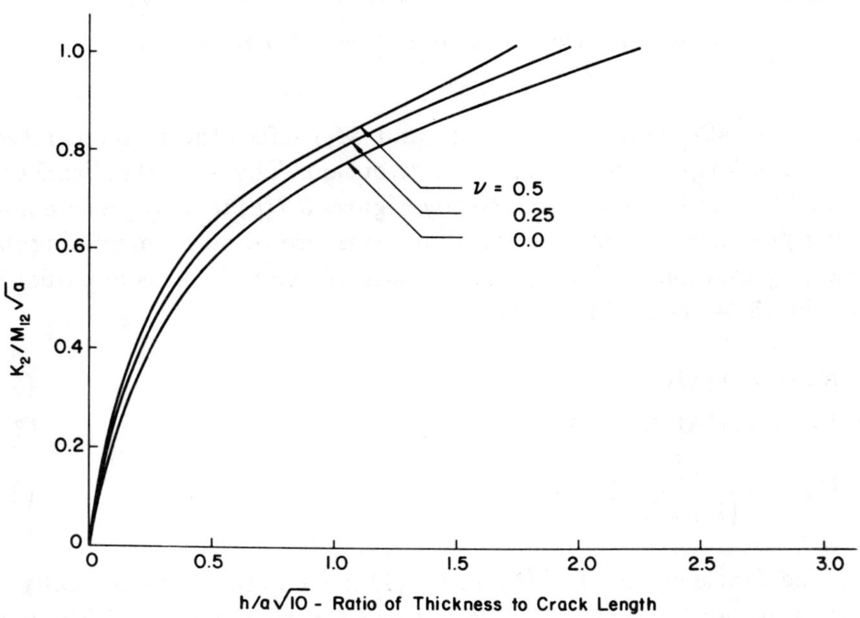

Figure 3.9. Twisting moment intensity factor versus ratio of thickness to crack length.

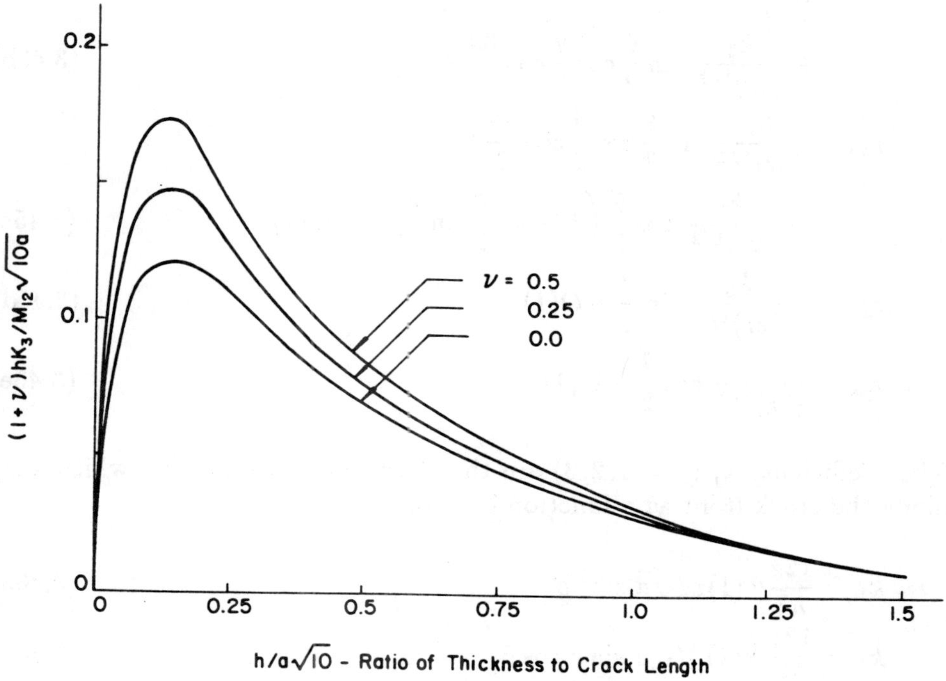

Figure 3.10. Shear intensity factor versus ratio of thickness to crack length.

3.4. Direction of crack growth

In the absence of symmetry, the path of crack propagation is no longer obvious since it may not coincide with its initial plane. The direction of crack initiation as predicted from the strain energy density theory depends on the intensity of the $1/r$ energy field near the crack tip. This intensity measured by S varies as a function of r and θ through the stresses on the element shown in Figure 3.5. With $\sigma_z = 0$, equation (3.1) simplifies to

$$\frac{dW}{dV} = \frac{1}{2E}[\sigma_x^2 + 2\tau_{xy}^2 + \sigma_y^2 + 2\nu(\tau_{xy}^2 - \sigma_x\sigma_y) + 2(1+\nu)(\tau_{xz}^2 + \tau_{yz}^2)]. \quad (3.44)$$

If the radius of the core region is kept sufficiently small, then consideration need only be given to the singular stresses. Substitution of equations (3.40) into (3.34) and (3.37) yields

$$\sigma_x = \frac{k_1}{(2r)^{1/2}}\cos\frac{\theta}{2}\left(1 - \sin\frac{\theta}{2}\sin\frac{3\theta}{2}\right) - $$
$$- \frac{k_2}{(2r)^{1/2}}\sin\frac{\theta}{2}\left(2 + \cos\frac{\theta}{2}\cos\frac{3\theta}{2}\right) + O(1) \quad (3.45a)$$

$$\sigma_y = \frac{k_1}{(2r)^{1/2}} \cos\frac{\theta}{2}\left(1+\sin\frac{\theta}{2}\sin\frac{3\theta}{2}\right)+$$
$$+\frac{k_2}{(2r)^{1/2}}\sin\frac{\theta}{2}\cos\frac{\theta}{2}\cos\frac{3\theta}{2}+O(1) \tag{3.45b}$$
$$\tau_{xy} = \frac{k_1}{(2r)^{1/2}}\sin\frac{\theta}{2}\cos\frac{\theta}{2}\cos\frac{3\theta}{2}+$$
$$+\frac{k_2}{(2r)^{1/2}}\cos\frac{\theta}{2}\left(1-\sin\frac{\theta}{2}\sin\frac{3\theta}{2}\right)+O(1) \tag{3.45c}$$
$$\tau_{xz} = -\frac{k_3}{(2r)^{1/2}}\sin\frac{\theta}{2}+O(1) \tag{3.45d}$$
$$\tau_{yz} = \frac{k_3}{(2r)^{1/2}}\cos\frac{\theta}{2}+O(1). \tag{3.45e}$$

The coefficients k_j $(j = 1, 2, 3)$ are the stress intensity factors* which vary along the crack front as a function of z, i.e.,

$$k_1 = \frac{12z}{h^3}\Phi(1)M\sqrt{a}\sin^2\beta \tag{3.46a}$$
$$k_2 = \frac{12z}{h^3}\Psi(1)M\sqrt{a}\sin\beta\cos\beta \tag{3.46b}$$
$$k_3 = -\frac{3\sqrt{10}}{2(1+\nu)h^2}\left[1-\left(\frac{2z}{h}\right)^2\right]\Omega(1)M\sqrt{a}\sin\beta\cos\beta. \tag{3.46c}$$

Inserting equations (3.45) into (3.44), the result may be cast into the form $\mathrm{d}W/\mathrm{d}V = S/r$ where the strain energy density factor S is given by

$$S = A_{11}k_1^2 + 2A_{12}k_1k_2 + A_{22}k_2^2 + A_{33}k_3^2. \tag{3.47}$$

The quantities A_{ij} $(i, j = 1, 2, 3)$ stand for

$$A_{11} = \frac{1}{8E}(1+\cos\theta)[3-\nu-(1+\nu)\cos\theta] \tag{3.48a}$$
$$A_{12} = \frac{1}{4E}\sin\theta[(1+\nu)\cos\theta-(1-\nu)] \tag{3.48b}$$
$$A_{22} = \frac{1}{8E}[4(1-\cos\theta)+(1+\nu)(3\cos\theta-1)(1+\cos\theta)] \tag{3.48c}$$
$$A_{33} = \frac{1+\nu}{2E}. \tag{3.48d}$$

---

* Note from equations (3.43) and (3.46) that k_j $(j = 1, 2, 3)$ are related to K_j $(j = 1, 2, 3)$ as $k_1 = (12z/h^3)K_1$, $k_2 = (12z/h^3)K_2$, $k_3 = (3/2h)[1-(2z/h)^2]K_3$.

For the plate-bending problem at hand, equation (3.47) becomes

$$S = \frac{2a}{E}\left[\frac{3\Phi(1)M\sin\beta}{h^2}\right]^2 \left\{\left(\frac{z}{h}\right)^2 F(\theta,\beta) + \left[1-\left(\frac{2z}{h}\right)^2\right]^2 G(\beta)\right\} \tag{3.49}$$

with the functions $F(\theta,\beta)$ and $G(\beta)$ being defined as

$$F(\theta,\beta) = B_{11}\sin^2\beta + 2\lambda B_{12}\sin\beta\cos\beta + \lambda^2 B_{22}\cos^2\beta \tag{3.50a}$$

$$G(\beta) = \frac{5}{8(1+\nu)}\gamma^2\cos^2\beta \tag{3.50b}$$

where $B_{ij} = 8EA_{ij}$ $(i,j = 1,2,3)$. The parameters λ and γ are

$$\lambda = \frac{\Psi(1)}{\Phi(1)}, \quad \gamma = \frac{\Omega(1)}{\Phi(1)}. \tag{3.51}$$

The numerical values of λ and γ can be obtained from the graphs in Figures 3.8 to 3.10.

The strain energy density theory [3, 4] assumes that the crack grows in a direction corresponding to $\theta = \theta_0$ which makes S in equation (3.49) a minimum:

$$\frac{\partial S}{\partial\theta} = 0; \quad \frac{\partial^2 S}{\partial\theta^2} > 0 \text{ at } \theta = \theta_0. \tag{3.52}$$

The vanishing of the first derivative of S with respect to θ gives

$$\begin{aligned}(1+\nu)\{2(1-3\lambda^2)\sin 2\theta_0 - (1-\lambda)(1-3\lambda)\sin[2(\theta_0+\beta)] - \\ - (1+\lambda)(1+3\lambda)\sin[2(\theta_0-\beta)]\} - 2(1-\nu)[2(1-\lambda^2)\sin\theta_0 - \\ - (1-\lambda)^2\sin(\theta_0+2\beta) - (1+\lambda)^2\sin(\theta_0-2\beta)] = 0.\end{aligned} \tag{3.53}$$

There exist two minimum values of S. One corresponds to the negative angles θ_0 on the tension side of the plate and the other corresponds to the positive angles on the compression side. If the crack is always assumed to initiate from the tension side, then only negative fracture angles will be considered. Figure 3.11 gives a plot of $-\theta_0$ in equation (3.53) as a function of the crack angle β for a fixed plate thickness to half crack length ratio of $h/a = 1.265$.

Under normal bending $\beta = 90°$, the predicted fracture angle is $\theta_0 = 0$ which implies that the crack grows in a self-similar manner. As the crack is tilted into the direction of bending by decreasing the angle β, the direction of crack initiation deviates more and more away from its initial plane as the fracture angle $-\theta_0$ increases. The values of $-\theta_0$ are also seen to be affected by the Poisson's ratio of the material. The variation appears to be more pronounced for crack angle β in the middle range. Unlike the plane theory of elasticity, the plate thickness to crack length ratio exerts an influence on the direction of crack initiation. For a fixed crack length and angle β,

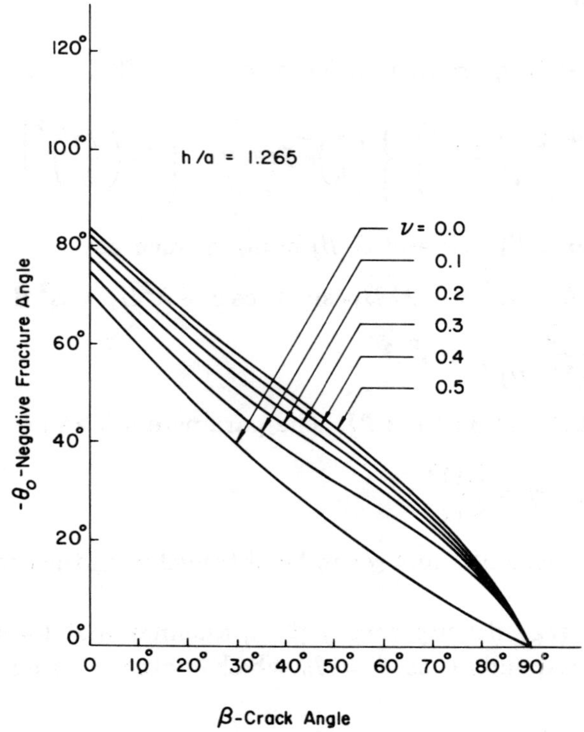

Figure 3.11. **Variations of fracture angle with crack angle for constant h/a.**

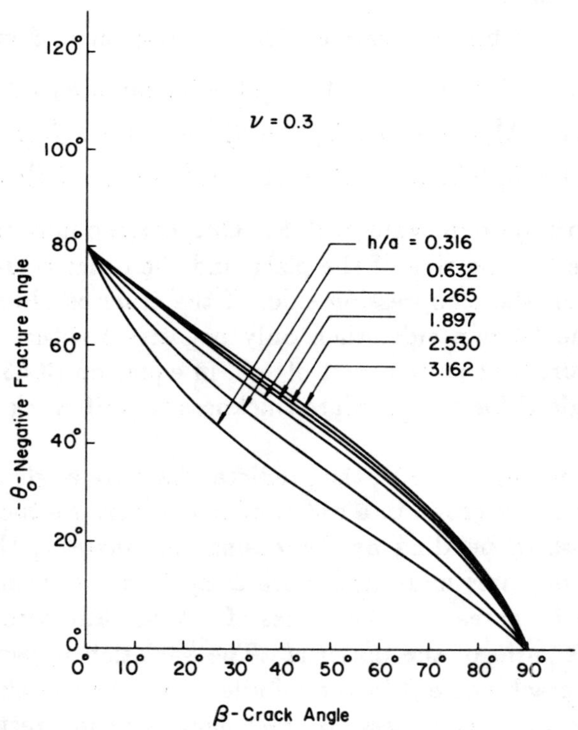

Figure 3.12. **Variations of fracture angle with crack angle for constant ν.**

Figure 3.13. Normalized strain energy density factor versus crack angle for constant h/a.

Figure 3.12 shows that the fracture angle $-\theta_0$ decreases with increasing plate thickness. Such an influence tends to diminish for β close to 0° and 90°. Hence, changes in the h/a ratio* should be accounted for when comparing experimental data with theory.

3.5. Minimum strain energy density factors and allowable bending moments

Once the fracture angles $-\theta_0$ are known, they may be inserted into equation (3.49) to obtain the minimum S values or S_{min}. Since $F(\theta, \beta)$ is the only function in equation (3.49) that depends on θ, its minimum values $F_{\text{min}}(\theta_0, \beta)$ are displayed graphically in Figure 3.13 for $h/a = 1.265$. The

* Variations of fracture angle $-\theta_0$ with h/a also occurs in the stretching of plates. The higher order theory of Hartranft and Sih [12] when applied to the mixed mode crack problem in plane extension can predict such an effect.

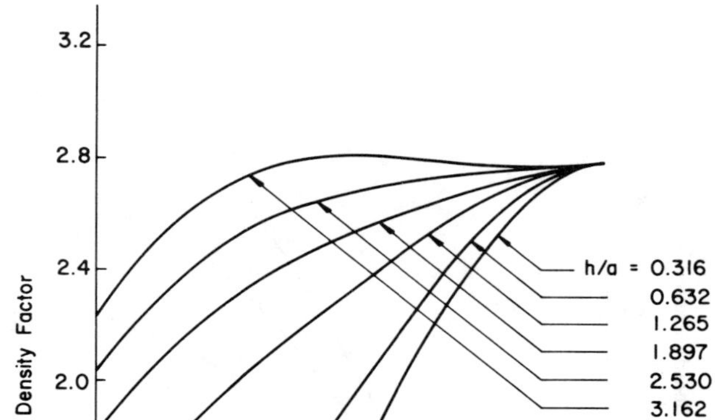

Figure 3.14. Normalized strain energy density factor versus crack angle for constant ν.

general trend of the curves for $F_{\min}$ or $S_{\min}$ tends to increase with the crack angle β and the rate of this increase becomes greater as Poisson's ratio is decreased. A similar plot for $F_{\min}(\theta_0, \beta)$ versus β with $\nu = 0.3$ is given in Figure 3.14. Here, the ratio h/a is varied from 0.316 to 3.162. For small values of β, significant change of $F_{\min}$ is observed as h/a is varied. This change diminishes as β approaches 90° where all the curves converge to the same value.

In order to estimate the allowable moment, M_c, that a cracked plate can sustain without causing fracture, a normalized moment quantity $M_c/(Eh^3 S_{\min})^{1/2}$ is defined by using equation (3.49):

$$\frac{M_c}{(Eh^3 S_{\min})^{1/2}} =$$

$$= \frac{\sqrt{h/a}}{3 \sin\beta\{2\Phi(1)[(z/h)^2]F_{\min}(\theta_0, \beta) + [1 - (2z/h)^2]G(\beta)\}^{1/2}} \tag{3.54}$$

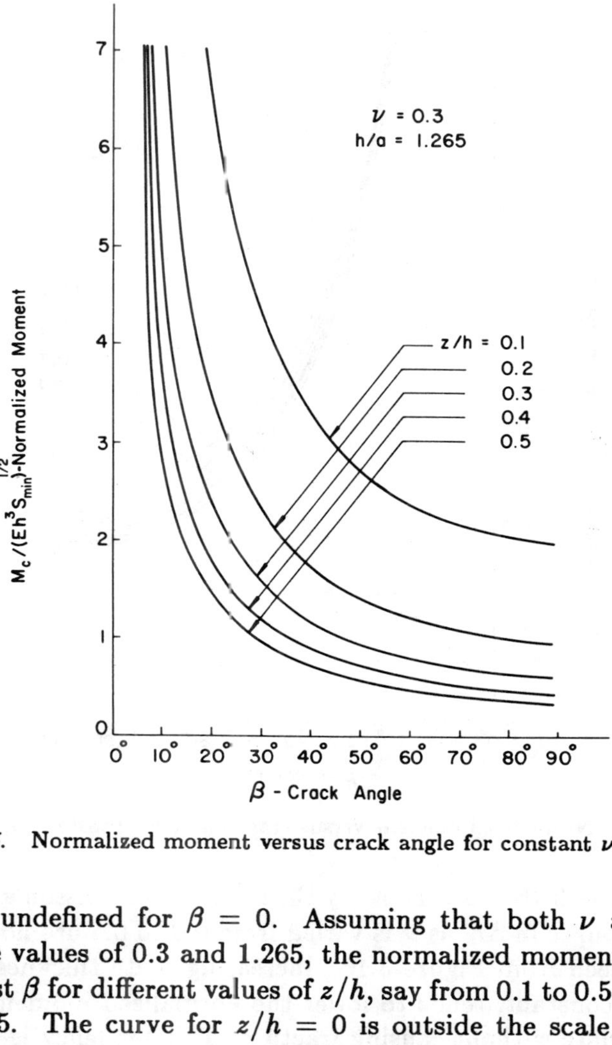

Figure 3.15. Normalized moment versus crack angle for constant ν and h/a.

where M_c is undefined for $\beta = 0$. Assuming that both ν and h/a take the respective values of 0.3 and 1.265, the normalized moment can then be plotted against β for different values of z/h, say from 0.1 to 0.5 as illustrated in Figure 3.15. The curve for $z/h = 0$ is outside the scale range of the graph. For a fixed S_{min}, the moment M is seen to decrease monotonically with β. The minimum value of M_c occurs at a position where the bending moment is applied normal to the crack plane, i.e., $\beta = 90°$. Each value of z/h corresponds to a layer of material in the plate with $z = 0$ being the middle layer and $z = 0.5h$ the outer or surface layer. According to the S-theory [3, 4], unstable crack propagation commences when $S_{min} \to S_c$. The critical value S_c is a constant representing the fracture toughness of the material. When $S_{min} = S_c$ in equation (3.54), M becomes M_c or the critical moment at incipient fracture. With this understanding, the results in Figure 3.15 indicate that fracture always starts in the surface layer of the plate ($z/h = 0.5$) which corresponds to the lowest critical moment M_c for all crack angles β.

With attention focused on the surface layer ($z/h = 0.5$) where fracture

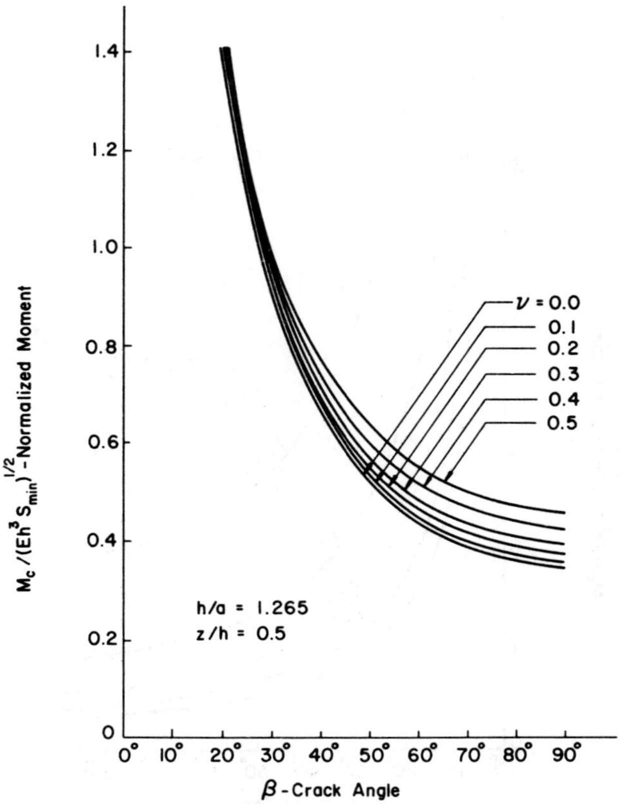

Figure 3.16. Normalized moment versus crack angle for constant h/a and z/h.

initiates, Figures 3.16 and 3.17 study the influence of Poisson's ratio and h/a ratio. The changes in M_c as ν is varied from 0.0 to 0.5 are not appreciable. This can be seen from Figure 3.16. Increasing plate thickness or h/a, say with a being constant, tends to lower the normalized moment that can be applied to a plate without causing fracture. This tendency is clearly shown in Figure 3.17.

3.6. Additional results on plates

When plate-like structural members are subjected to bending loads, the in-plane stresses are required to change sign in the thickness direction and the problem acquires a three-dimensional character. Since the through crack geometry* prohibits any exact three-dimensional solutions, the elasticity theory must be approximated by appealing to minimum principles in the

* The major drawback arises from an incomplete knowledge of the stress singularity on the plate surface where it intersects with the crack. Further discussions on this subject can be found in [15, 16].

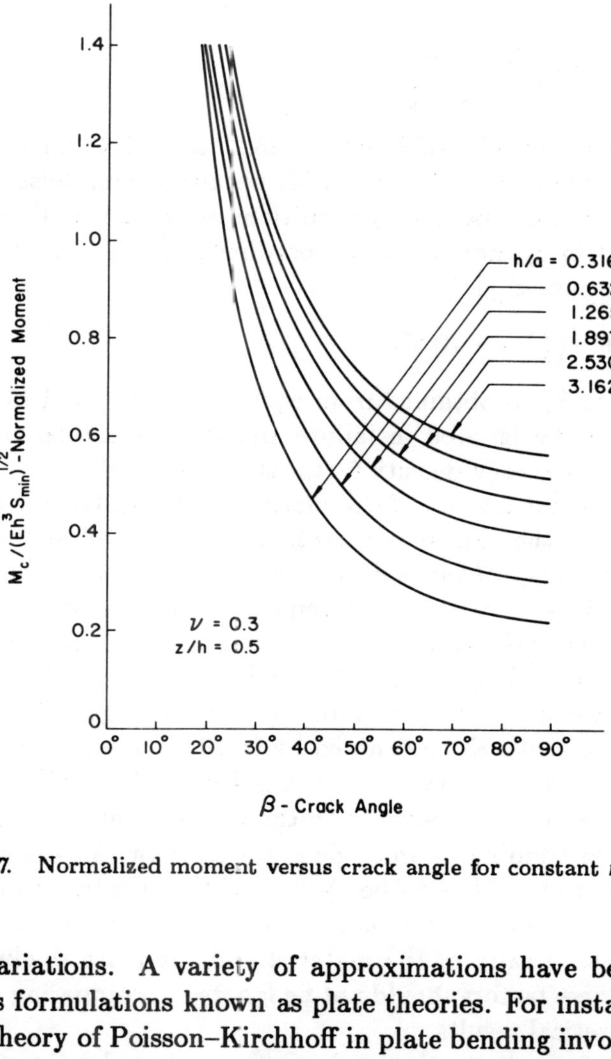

Figure 3.17. Normalized moment versus crack angle for constant ν and z/h.

calculus of variations. A variety of approximations have been made and led to various formulations known as plate theories. For instance, the most widely used theory of Poisson–Kirchhoff in plate bending involves alteration of the crack surface boundary conditions. Such a change has a drastic influence on the qualitative character of the stresses near the crack tip and will seriously affect the direction of crack initiation predicted from any assumed fracture criterion. As an example, the maximum principal stress criterion* states that crack growth takes place in a direction normal to the plane on which the tangential normal stress σ_θ is a maximum or the shear stress $\tau_{r\theta}$ vanishes. Consider the pure twisting case shown in Figure 3.7(c). The $\tau_{r\theta} = 0$ plane based on the Poisson–Kirchhoff theory can be found from Sih's work [15] as

* This criterion yields sufficiently accurate results only when one of the principal stresses dominates.

$$\theta_0 = -2\cos^{-1}\left[\frac{2(2+\nu)}{5+3\nu}\right]^{1/2} \tag{3.55}$$

which gives fracture angles of θ_0 from $-53.2°$ to $-57.4°$ for ν ranging from 0.0 to 0.5. This differs from the value of θ_0 obtained from Reissner's theory in which the crack surface boundary conditions are satisfied. Equation (3.45c) when expressed in cylindrical polar coordinates (r, θ) and $\tau_{r\theta}$ set equal to zero gives a fracture angle*

$$\theta_0 = -\cos^{-1}(\tfrac{1}{3}) = -70.5°. \tag{3.56}$$

The difference in θ_0 as obtained from equations (3.55) and (3.56) is significant and cannot be ignored in failure analysis. A similar discrepancy is found if the strain energy density factor theory is used.

Another important feature of the plate bending problem is that the material elements in the thickness direction may not fail simultaneously as assumed in the theory of plane elasticity. This is because the stress intensity factors k_1, k_2 and k_3 no longer remain constant along the crack front. The location of the critical element that triggers fracture must depend on some combination of k_j $(j = 1, 2, 3)$ reaching a critical value. Referring to the cylindrical polar coordinate system (r, θ, z) in Figure 3.5, the position of the critical element will be denoted by (r_0, θ_0, z_0), where r_0 is the radius of core region, θ_0 the fracture angle obtained from the condition of S reaching a minimum and z_0 corresponds to the greatest value of $S_{\min}$ along the crack front. The bending of an angle crack as shown in Figure 3.6(a) was analyzed. Predictions of the S-theory showed that the fracture angle θ_0 not only depended on the crack position relative to the direction of bending but also on the Poisson ratio of the material and the plate thickness to crack length ratio. These factors should not be ignored in comparing experimental data with theoretical results.

For the sake of simplicity, the in-plane stresses have been assumed to vary linearly through the plate thickness and the transverse shear stresses are parabolic functions of z. Higher order plate theories [12, 17] involving more generalized functions of z are available. Sih [17] has proposed to expand the z-dependence, $f'(2z/h)$, for the transverse shear stresses τ_{xz} and τ_{yz} in a trigonometric series

$$f'(2z/h) = b_0 - \sum_{n=1}^{m}(-1)^n b_n \cos[(2\pi n z/h)] \tag{3.57}$$

* The experimental result of Erdogan and Sih [6] on the twisting of cracked plates gives values of θ_0 closer to 70.5° than those predicted by equation (3.55).

which satisfies the conditions that $\tau_{xz} = \tau_{yz} = 0$ on the plate surfaces $z = \pm h/2$, i.e., $f'(\pm 1) = 0$. The constants b_0 and b_n are given by

$$b_0 = \sum_{n=1}^{m} (-1)^n b_n, \qquad b_n = \frac{1}{1 + (an/h)^2} \tag{3.58}$$

where a is the half crack length. The derivatives of $f'(2z/h)$ with respect to z or $f''(2z/h)$ gives the variations of the in-plane stresses σ_x, σ_y and τ_{xy} with z. A series of experiments on the bending of plexiglass plates [18] gave data checking remarkably well with the predictions made by Sih [17]. The stress intensity factor solution for this problem may also be used in conjunction with the strain energy density factor theory to predict the fracture behavior of plates under bending and/or twisting. The theoretical results will differ but not significantly from those based on the Reissner theory.

Another shortcoming of the conventional plate theory is that σ_z is neglected throughout the plate and hence the plane strain condition

$$\sigma_z = \nu(\sigma_x + \sigma_y) \tag{3.59}$$

ahead of the crack cannot be realized. Hartranft and Sih [12] have constructed a theory of plate stretching or bending in which the function $f(2z/h)$ in

$$\sigma_z = f(2z/h) Z_z(x, y) \tag{3.60}$$

will not be assigned arbitrarily but be determined from equation (3.59), a condition derived from the exact three-dimensional theory of elasticity [19]. Villarreal, Sih and Hartranft [20] have used a frozen stress photoelastic technique to measure the stress variation through the thickness of a thick plate containing a crack and the results were in excellent agreement with those postulated by the theory of Hartranft and Sih.

3.7. Shell theory with shear deformation

Although much work has been done on the stress analysis of cracks in initially flat plates [15], theoretical treatment of cracks in initially curved plates or shells has received attention only in recent years. The presence of curvature in a shell generates deviation from behavior of flat plates, in that stretching loads will induce both extensional and bending stresses, while bending loads will also lead to both types of stresses. In the neighborhood of geometric defects such as man-made flaws or cracks, the stresses are redistributed and can lead to a reduction of the load-carrying capacity of shell structures.

One of the simplest shell geometries is that of a spherical shell whose curvature radius R is everywhere constant. The first investigation on the stresses in a spherical shell containing a crack was made by Ang, Folias and Williams [21], who associated the shell problem with that of an initially flat plate resting on an elastic foundation. The equivalence of the two problems was made by identifying the foundation modulus with Eh/R^2 where E is Young's modulus and h the shell thickness. The general character of the crack tip stress field in a shell was studied by Sih and Setzer [22] who pointed out that the functional relationships of the local (extension-bending) stresses are identical with those obtained by superimposing the individual extensional and bending stresses of an initially flat plate. They further emphasized the coupling effects of extension and bending through the intensity of the local stress field. A separate treatment of the spherical shell problem was later given by Folias [23] who utilized singular integral equations with Cauchy-type kernels, as devised by Knowles and Wang [24], for solving crack problems involving flat plates. The formulation in [23] relies on certain approximations in the kernels for small values of the curvature parameter. Other methods for solving shell problems with cracks have been presented by Erdogan and Kibler [25], and Sih and Dobreff [26]. Their results covered a significantly wider range of the curvature parameter. Problems of cracks in cylindrical shells have also been treated by Folias [27, 28], Copley and Sanders [29], and Erdogan and Ratwani [30].

One of the shortcomings inherent in all of the foregoing works, which are based on the classical shallow shell theory of Reissner [31], is that the boundary conditions on the crack surfaces are only satisfied approximately in the Kirchhoff sense. The drawback gives rise to a difference in the angular distribution of the extensional and bending stress fields near the crack tip.* This fundamental difficulty was overcome by Sih and Hagendorf [32]. They developed a tenth-order system of shell equations which accounted for the effect of transverse shear. The new results satisfy five individual boundary conditions on a free edge of the shell and acquired a three-dimensional character in the crack front stresses resembling those found by Sih [15] for a through crack in a thick plate.

This chapter will be concerned with a tenth-order system of shell equations applied to solve shell problems with cracks. Fourier transform is employed reducing each problem to the solution of a system of coupled Fredholm integral equations of the second kind. Both symmetric and skew-symmetric loadings are considered and the resulting stress solutions can differ substantially from those based on the classical theory of Reissner [31].

The fundamental equations of the three-dimensional theory of elasticity involve overwhelming difficulties when applied to solve the shell problem

---

* Since the extensional and bending stress intensity factors can no longer be conveniently combined, special caution should be exercised when the results are used to analyze failure.

unless simplifying assumptions are invoked. A reasonable description of the behavior of elastic shells can be made by assuming that the shell thickness is everywhere at least one order of magnitude smaller than the radius of curvature of the reference surface. On this basis, approximate theories of thin elastic shells may be developed and they generally can differ from one another essentially in the formulation of appropriate stress–strain relations.

A theory of thin elastic isotropic shells which accounts for transverse shear deformation and normal stress was derived by Naghdi [33]. He obtained the stress–strain relations by applying Reissner's [34] variational theorem of elasticity. Kalnins [35] has also proposed a shear deformation theory of shells. In order to obtain a system of shell equations, which are amenable to effective solutions of cracks or cavities in shells, Sih and Hagendorf [32] developed a theory that satisfies five physical boundary conditions on an edge of the shell. This general theory of elastic shells is based on the following three assumptions:

(1) the displacement vector is linear in the normal coordinate curve to the middle surface;
(2) the distance between points on a normal to the undeformed middle surface do not change during deformation; and
(3) the stresses are replaced by a system of stress resultants and stress couples.

The stress–strain relations are deduced for an elastic orthotropic material satisfying a generalized Hooke's law that is consistent with the restriction imposed on the kinematics of deformation by assumption (2). The resulting system of equations is invariant with respect to rigid-body motion.

A shell is said to be shallow (i.e., slightly curved) if the middle surface is sufficiently smooth and all points on this surface are sufficiently close to some plane. In a theory of shallow shells, the material particles on the middle surface are located by the Cartesian coordinates of their projection on the XY-plane, i.e.,

$$X = x, \quad Y = y, \quad Z = f(x, y) \tag{3.61}$$

in which x and y are the curvilinear coordinates and $f(x, y)$ is the equation of the middle surface of the shell. The points in the shell are described by (x, y, z) with reference to the undeformed middle surface, where z is the distance measured along the normal to the middle surface and is such that $-h/2 \le z \le h/2$. Here, h stands for the thickness of the shell as shown in Figure 3.18.

The theory of shallow shells is based on three simplifying assumptions [36]:

(1) Neglect the squares of the derivatives $\partial f/\partial x$, $\partial f/\partial y$ and their products in comparison to unity.

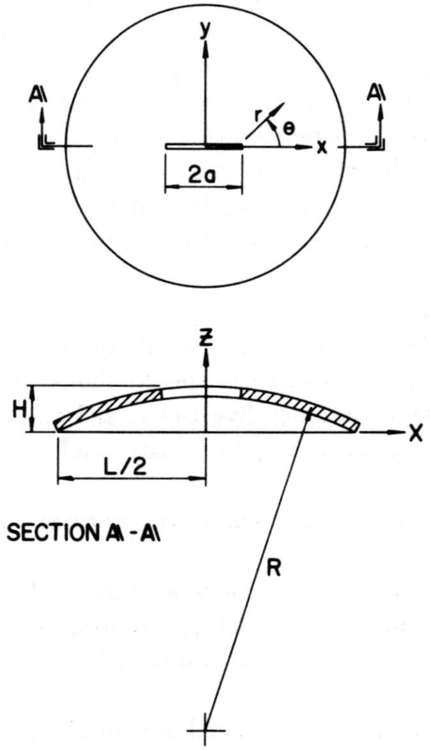

Figure 3.18. Spherical cap with a crack.

(2) Neglect the transverse shear resultants Q_x, Q_y in the first two equations of equilibrium.

(3) Neglect the tangential displacements u_x, u_y in the expressions for the transverse shearing strains γ_{xz}, γ_{yz}.

In what follows, only the first assumption will be incorporated:

$$\left(\frac{\partial f}{\partial x}\right)^2 \ll 1, \quad \left(\frac{\partial f}{\partial y}\right)^2 \ll 1, \quad \frac{\partial f}{\partial x}\frac{\partial f}{\partial y} \ll 1. \tag{3.62}$$

Let the reference surface be the middle surface of a shallow spherical shell of constant thickness, h. Then the principal radii of curvature are $R_x = R_y = R$. A system of five coupled partial differential equations can be obtained that govern the equilibrium of the middle surface of a shallow spherical shell in terms of the five displacement measures u_x, u_y, u_z, β_x and β_y. Referring to Figure 3.18, u_z is the displacement in the direction of the normal. The quantity β_x is the angle through which the normal $\mathbf{n}$ rotates in the direction of the x-coordinate line. A corresponding interpretation holds for β_y. The *force stress resultants* N_{xx}, N_{xy}, etc.; the *moment stress*

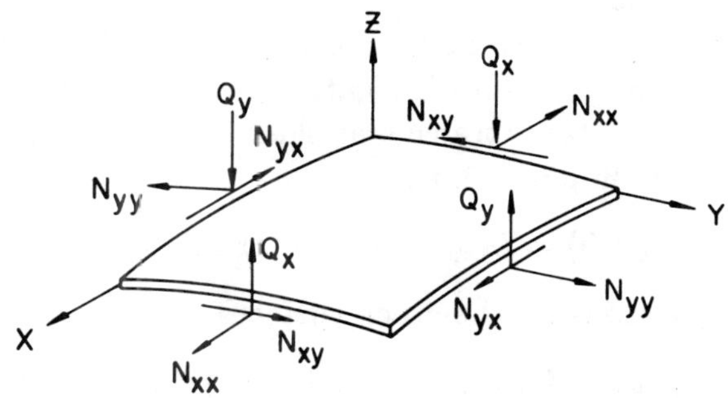

Figure 3.19(a). Notations for stress resultants and shear forces.

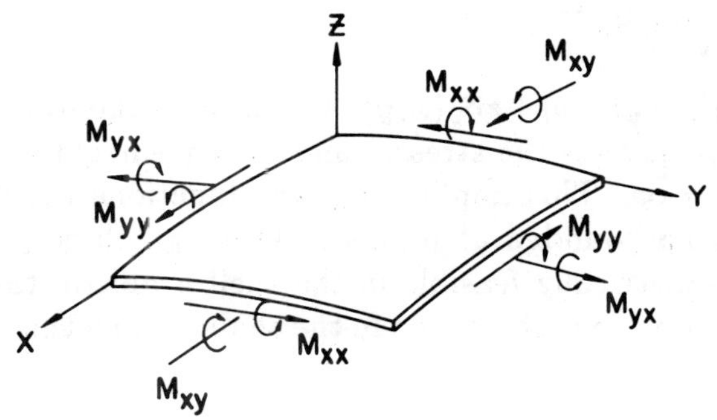

Figure 3.19(b). Notations for bending and twisting moments.

resultants M_{xx}, M_{xy}, etc.; and the ***transverse shears*** Q_x, Q_y at each point of the middle surface can all be determined once u_x, u_y, u_z, β_x and β_y are known. Refer to Figures 3.19(a) and 3.19(b) for the notations.

3.8. Symmetric loading on shell

Consider a portion of a spherical shell of constant thickness h containing a through crack at the apex of the shell. The shallowness condition requires that the crack of length $2a$ in Figure 3.18 is small in comparison with the radius of curvature R and hence the problem can be solved in the projected XZ-plane. The solution to this problem may be considered as the sum of two separate problems. The first concerns with loads applied along the edge of an uncracked shell. The membrane and moment stresses will be different from zero along the prospective crack site: $y = 0$, $|x| \leq a$. The second problem deals with a cracked shell with tractions specified on the crack surfaces. These tractions are equal and opposite to those found along $y = 0$, $|x| \leq a$ in the first problem. Since the solution for the uncracked shell is straightforward, attention will be focused only on the solution for a cracked shell subjected to surface tractions.

Let the shell in Figure 3.18 be loaded symmetrically with respect to the X-axis along which the line crack is located. From the symmetry conditions, the in-plane shear resultant, transverse shear and twisting moment vanish everywhere along the line $y = 0$, i.e.,

$$N_{xy}(x,0) = Q_y(x,0) = M_{xy}(x,0) = 0 \text{ for all } x. \tag{3.63}$$

The boundary conditions on the crack surfaces are

$$N_{yy}(x,0) = N(x); \quad M_{yy}(x,0) = M(x), \quad |x| < a \tag{3.64}$$

and the continuity requirements are given by

$$\lim_{|y|\to 0} \beta_y = \lim_{|y|\to 0} u_y = 0, \quad |x| > a \tag{3.65}$$

where $u_y(x,y,z) = u_y(x,y) + z\beta_y(x,y)$. Imposed on the solution are also the regularity conditions that the stresses and displacements must be bounded as $(x^2 + y^2)^{1/2} \to \infty$. This implies that the functions u_z, ψ, ϕ_0 and their derivatives must be bounded at infinity. Although these conditions at infinity are not geometrically feasible in the shell problem, they are satisfied within the approximation introduced in the shallow shell theory by equations (3.62).

Stress intensity factor. The stress intensity factor may be extracted from the near field solution as shown in the work of Sih and Hagendorf [32]. Let r and θ denote the polar coordinates measured from the line of expected crack extension, the positive X-axis in Figure 3.18. For symmetric loading case, the asymptotic expansions for the stress resultants in the neighborhood of the crack tip are found to be (Sih and Hagendorf [32])

$$N_{xx} = -\Phi_1(1)(a/2r)^{1/2}\cos(\theta/2)[1 - \sin(\theta/2)\sin(3\theta/2)] + \cdots \tag{3.66a}$$
$$N_{yy} = -\Phi_1(1)(a/2r)^{1/2}\cos(\theta/2)[1 + \sin(\theta/2)\sin(3\theta/2)] + \cdots \tag{3.66b}$$
$$N_{xy} = -\Phi_1(1)(a/2r)^{1/2}\cos(\theta/2)\sin(\theta/2)\cos(3\theta/2) + \cdots \tag{3.66c}$$

and the moments are

$$M_{xx} = -\frac{h}{6}\Phi_2(1)(a/2r)^{1/2}\cos(\theta/2)[1 - \sin(\theta/2)\sin(3\theta/2)] + \cdots \tag{3.67a}$$

$$M_{yy} = -\frac{h}{6}\Phi_2(1)(a/2r)^{1/2}\cos(\theta/2)[1 + \sin(\theta/2)\sin(3\theta/2)] + \cdots \tag{3.67b}$$

$$M_{xy} = -\frac{h}{6}\Phi_2(1)(a/2r)^{1/2}\cos(\theta/2)\sin(\theta/2)\cos(3\theta/2) + \cdots. \tag{3.67c}$$

Note that the angular distribution of the stress resultants and moments around the crack tip are identical. Equations (3.66) and (3.67) differ only in the coefficients or amplitude of the singular stress field determined by $\Phi_1(1)$

and $\Phi_2(1)$. This feature of the solution permits the membrane and moment stress fields to be combined in a natural fashion by the expressions:

$$\left(1+\frac{Z}{R}\right)\sigma_{xx}=\frac{1}{h}N_{xx}+\frac{12Z}{h^3}M_{xx} \tag{3.68a}$$

$$\left(1+\frac{Z}{R}\right)\sigma_{yy}=\frac{1}{h}N_{yy}+\frac{12Z}{h^3}M_{yy} \tag{3.68b}$$

$$\left(1+\frac{Z}{R}\right)\tau_{xy}=\frac{1}{h}N_{xy}+\frac{12Z}{h^3}M_{xy} \tag{3.68c}$$

$$\left(1+\frac{Z}{R}\right)\tau_{xz}=\frac{1}{h}Q_x \tag{3.68d}$$

$$\left(1+\frac{Z}{R}\right)\tau_{yz}=\frac{1}{h}Q_y. \tag{3.68e}$$

This is not the case in the classical shell theory [23, 25, 26] where the corresponding expressions for M_{xx}, M_{yy} and M_{xy} are incompatible with those for N_{xx}, N_{yy} and N_{xy} in that the θ-dependence for the stress resultants and moments are entirely different. It is obviously more meaningful to consider the combined stress field:

$$\sigma_{xx}=\frac{k_1(z)}{(2r)^{1/2}}\cos(\theta/2)[1-\sin(\theta/2)\sin(3\theta/2)]+O(1) \tag{3.69a}$$

$$\sigma_{yy}=\frac{k_1(z)}{(2r)^{1/2}}\cos(\theta/2)[1+\sin(\theta/2)\sin(3\theta/2)]+O(1) \tag{3.69b}$$

$$\tau_{xy}=\frac{k_1(z)}{(2r)^{1/2}}\cos(\theta/2)\sin(\theta/2)\cos(3\theta/2)+O(1) \tag{3.69c}$$

$$\tau_{xz}=\tau_{yz}=O(1). \tag{3.69d}$$

The stress intensity factor $k_1(z)$ in equations (3.69) depends on the thickness coordinate z and is defined as

$$k_1(z)=-\left[\frac{\Phi_1(1)+(2z/h)\Phi_2(1)}{1+(z/R)}\right]\frac{\sqrt{a}}{h}. \tag{3.70}$$

It is seen that the effects of shell curvature enter into the local stress field only through the stress intensity factor $k_1(z)$. Furthermore, it is clear that in the shell problem, both the extensional and bending effects are always coupled regardless of the nature of the loading. The numerical results for three specific cases will be discussed subsequently.

Extensional load: $N(x) = N_0$. Let the crack surfaces be opened out by uniform extensional loads of constant magnitude N_0. The loading situation is illustrated in Figure 3.20. Of interest are the stress intensity factors at the top and bottom surface layers $z = \pm h/2$ of the shell. Equation (3.70) will be rewritten for the extensional case as

$$k_1^{(e)}(\pm h/2) = \left[\frac{\psi_1^{(e)}(1) \pm \psi_2^{(e)}(1)}{1 \pm (h/2R)}\right] \frac{N_0\sqrt{a}}{h} \tag{3.71}$$

where $\psi_j^{(e)}(1)$ are related to $\Phi_j(1)$ $(j = 1, 2)$ by

$$\Phi_j(\xi) = -N_0\psi_j^{(e)}(\xi) \quad \text{for} \quad \xi = 1. \tag{3.72}$$

The numerical values of $k_1^{(e)}$ at the top layer of the shell are shown graphically in Figure 3.20 as a function of $h/2a$ for a Poisson's ratio of $\nu = 0.3$ and h/R ranging from 0.01 to 0.30 inclusive. These curves depict the effect of increasing radius of curvature, R. Note that for a given value of $h/2a$, the normalized stress intensity factor $hk_1^{(e)}(h/2)/N_0\sqrt{a}$ decreases with increasing radius of curvature. This effect is more pronounced as R is increased. The dotted curves in Figure 3.20 represent lines of constant $R/2a$, and hence show the thickness effect. A similar set of curves for $k_1^{(e)}$ at the bottom layer of the shell $z = -h/2$ are given in Figure 3.21. For large values of $R/2a$, say 3, $k_1^{(e)}(-h/2)$ does not change appreciably with the shell thickness to crack length ratio, $h/2a$, i.e., with increasing thickness. A comparison of the stress intensity factors $k_1^{(e)}$ for $z = \pm h/2$ is made in Figure 3.22 for $\nu = 1/3$ and a particular value of $h/R = 0.2$. For the given loading condition, the $k_1^{(e)}$ value at the top layer is slightly higher than that at the bottom layer. The point of intersection corresponds to the case of a short crack which falls outside the validity of the present theory of thin shells. Figure 3.23 illustrates the quantitative difference in the values of $k_1^{(e)}$ at the inner layer of the shell $z = -h/2$ between the proposed theory and the classical shell theory* [25, 26]. The proposed theory predicts larger values of $k_1^{(e)}$ and the difference becomes more pronounced for longer cracks, i.e., as $h/2a \to 0$.

Bending load: $M(x) = M_0$. When the loading on the crack surfaces are of the bending type, say equal and opposite uniform moments with magnitude M_0, the results are quite different. The functions $\Phi_j(1)$ in equation (3.70) are equal to

$$\Phi_j(\xi) = \frac{6M_0}{h}\psi_j^{(b)}(\xi) \quad \text{for} \quad \xi = 1 \tag{3.73}$$

---

* This comparison can only be made in the region ahead of the crack, i.e., for $\theta = 0$, as the extensional and bending effects in the classical theory cannot be combined for other values of θ.

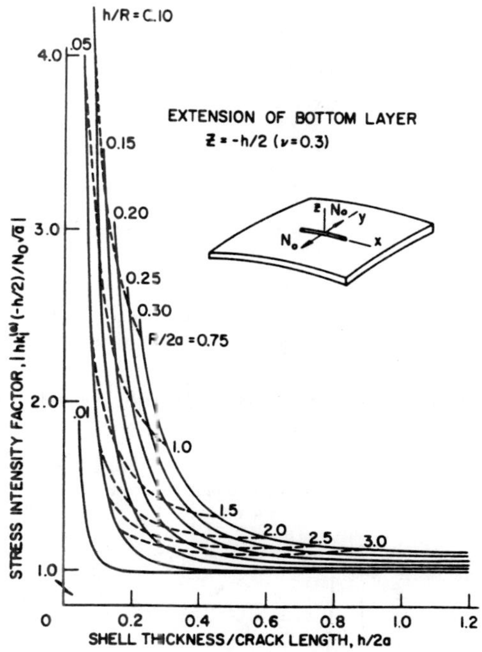

Figure 3.20. Variations of extensionally loaded stress intensity factor at outer layer with shell thickness to crack length ratio.

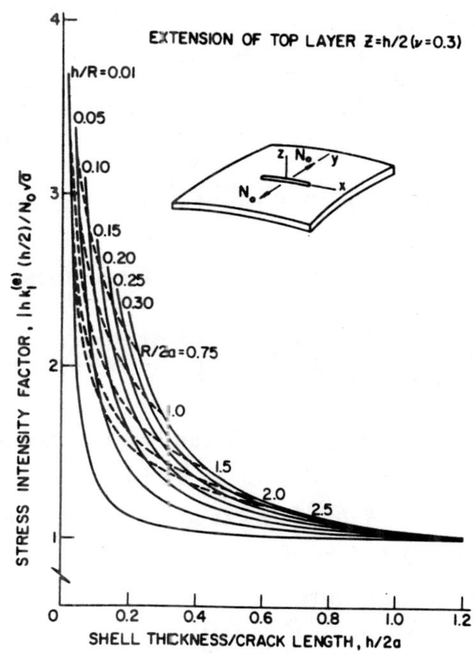

Figure 3.21. Variations of extensionally loaded stress intensity factor at inner layer with shell thickness to crack length ratio.

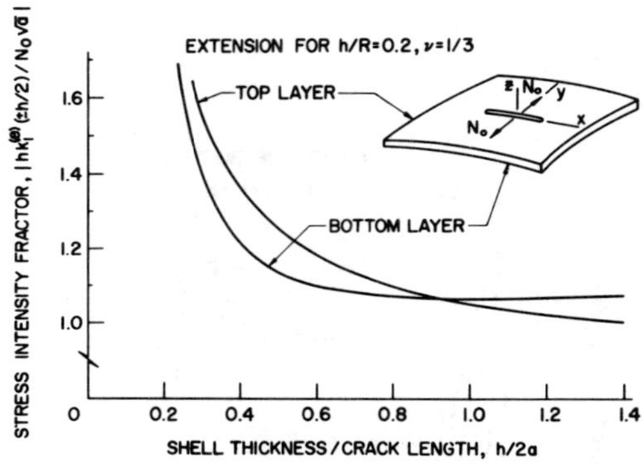

Figure 3.22. Comparison of extensionally loaded results at outer and inner layer of shell.

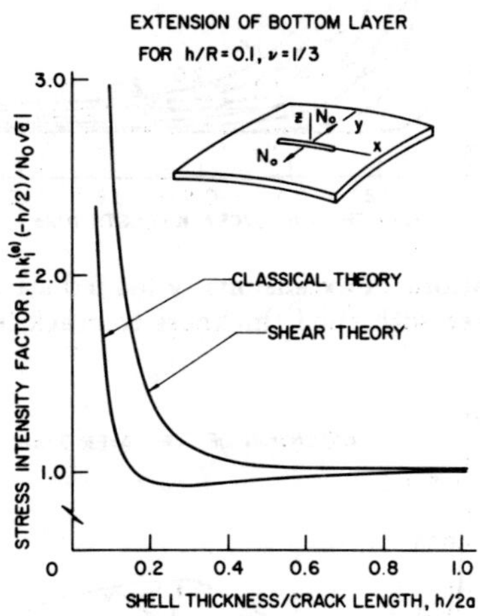

Figure 3.23. Classical theory versus shear theory for stretching load.

while the stress intensity factor for $z = \pm h/2$ becomes

$$k_1^{(b)}(\pm h/2) = \mp \left[\frac{\psi_2^{(b)}(1) \pm \psi_1^{(b)}(1)}{1 \pm (h/2R)} \right] \frac{6M_0}{h} \sqrt{a}. \tag{3.74}$$

Under bending, one side of the crack tends to close as it is subjected to compression. The maximum value of $k_1^{(b)}$ will always occur on the tension side. Hence, $k_1^{(b)}(h/2)$ and $k_1^{(b)}(-h/2)$ at the outer and inner surfaces of the shell will correspond to couples $-M_0$ and M_0, respectively. With the tension side at $z = h/2$, Figure 3.24 gives a plot of $h^2 k_1^{(b)}(h/2)/6M_0\sqrt{a}$ versus $h/2a$ for $\nu = 0.3$ and $h/R = 0.01$, 0.05, etc. Unlike the results for extensional loading, the magnitude of $k_1^{(b)}$ increases monotonically with $h/2a$ for a fixed value of h/R. This behavior was also observed in the flat plate

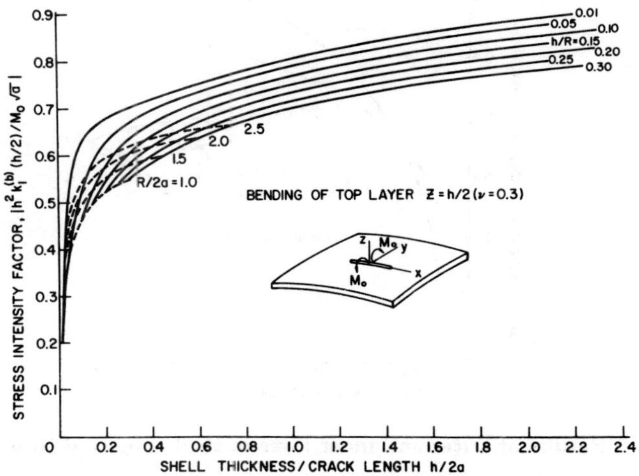

Figure 3.24. **Stress intensity factor at outer layer induced by bending load.**

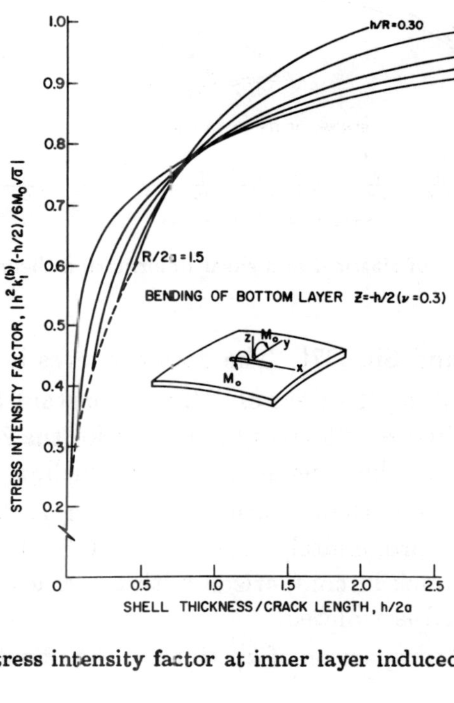

Figure 3.25. **Stress intensity factor at inner layer induced by bending load.**

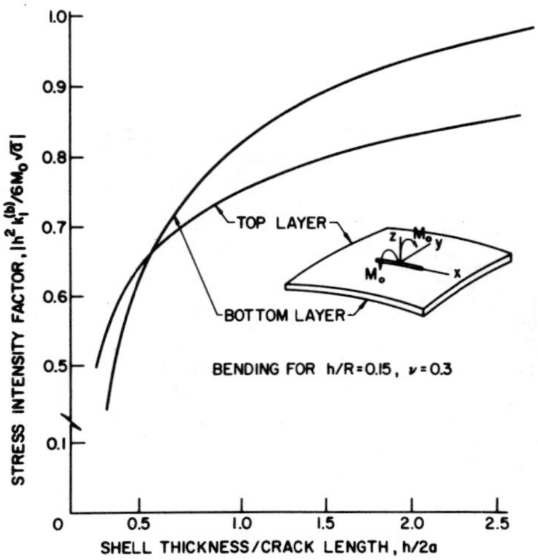

Figure 3.26. Results at outer and inner layer of shell subjected to bending load.

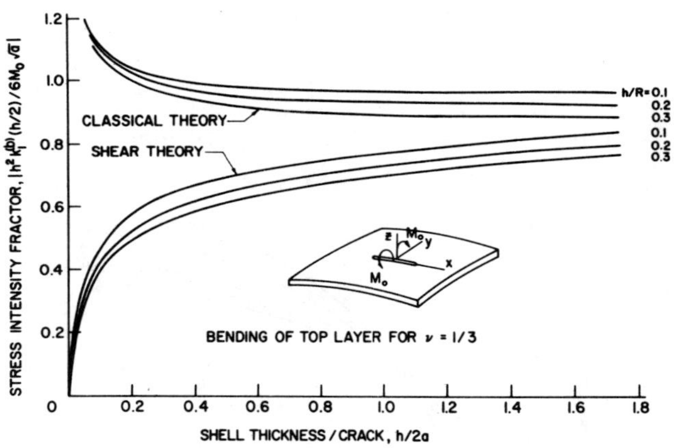

Figure 3.27. Comparison of classical and shear deformation theory for bending load.

theory by Hartranft and Sih [37]. The dotted curves in Figure 3.24 show the variations of $k_1^{(b)}$ along the lines of $R/2a =$ constant as the thickness of the shell is changed. Figure 3.25 gives the results for the situation where the tension side is reversed to the inner surface of the shell, $z = -h/2$. By fixing $h/2a < 0.8$, $k_1^{(b)}$ increases with decreasing values of h/R. The opposite effect is seen when $h/2a$ is approximately greater than 0.8. At this ratio of $h/2a$ the shell is relatively thick in comparison with the crack length $2a$ and the assumption of thin shell is violated.

Figure 3.26 makes a comparison of $k_1^{(b)}$ for $z = \pm h/2$ with $h/R = 0.15$ and

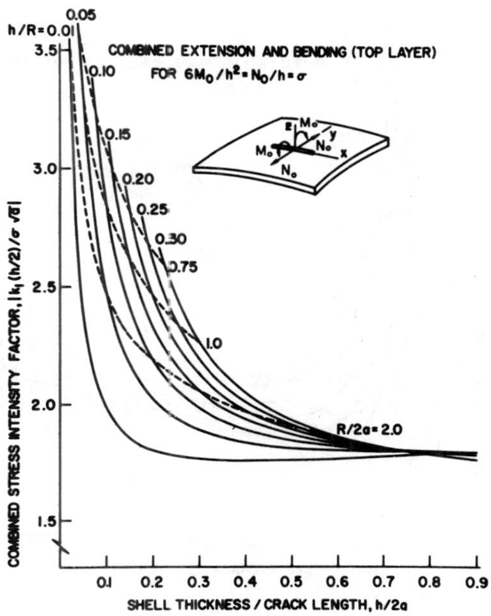

Figure 3.28. Results at outer layer for combined extensional and bending loads.

$\nu = 0.3$. The two curves intersect at $h/2a \simeq 0.48$ meaning that $k_1^{(b)}(h/2) > k_1^{(b)}(-h/2)$ for $h/2a < 0.48$ and $k_1^{(b)}(h/2) < k_1^{(b)}(-h/2)$ for $h/2a > 0.48$. It should be borne in mind that these results correspond to two different loading conditions, namely applied moments $\pm M_0$, i.e., opposite in sign.

A dramatic departure of the present results from those obtained by the classical theory is shown in Figure 3.27 where the trend of the variations of $k_1^{(b)}$ with $h/2a$ is, in fact, opposite. The difference becomes more and more significant as the ratio $h/2a$ decreases. The results based on the classical theory is known to be approximate since the boundary conditions on the crack surfaces are satisfied only in the Kirchhoff sense. A more detailed discussion on the approximate nature of the classical shell theory has been given by Sih and Dobreff [26].

Combined loading: $6M_0/h^2 = N_0/h = \sigma$. In general, both extensional and bending loads may be present and the total normal tractions applied to the crack surfaces consists of $N_{yy}(x,0)$ and that due to the couple $M_{yy}(x,0)$. Consider the special case where the crack surfaces are not permitted to come into contact. This condition can be satisfied by taking

$$|-N_0/h| \geq |\pm 6M_0/h^2|$$

where both N_0 and M_0 are constants. According to the above equation,

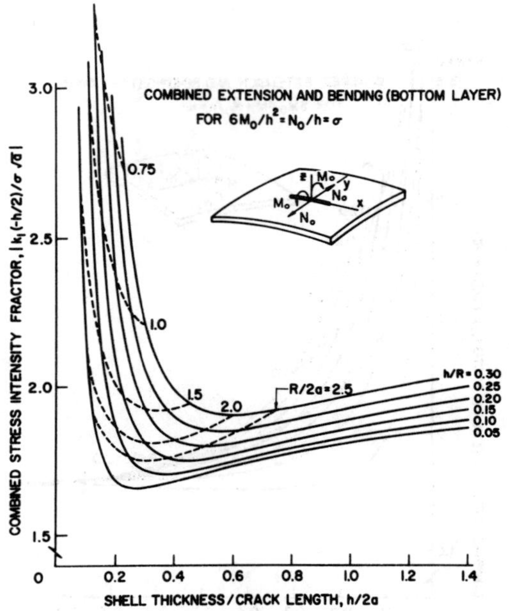

Figure 3.29. Results at inner layer for combined extensional and bending loads.

there arise two possible conditions of combined loading which are shown in Figures 3.28 and 3.29.

In Figure 3.28, the maximum tension occurs at the outer layer of the shell, $z = h/2$ with the combined stress intensity factor $k_1(h/2)$. For convenience, let the crack surfaces be subjected to the combined load such that $|-N_0/h| = |\pm 6M_0h^2|$. A plot of $k_1(h/2)/\sigma\sqrt{a}$ versus $h/2a$ is shown in Figure 3.28 and shows that $k_1(h/2)$ is reduced sharply with decreasing crack length. Curves for $h/R = 0.01$ to 0.30 inclusive are given. By following the dotted curve for $R/2a = 2$, it can be seen that an increase in the shell thickness from $h = 0.04a$ to $h = 0.2a$ corresponds to a reduction in the stress intensity factor of approximately 45%. If the direction of the applied couple is reversed, then the side of maximum tension switches to the inner layer of the shell at $z = -h/2$. The numerical results for $k_1(-h/2)$ are shown graphically in Figure 3.29. In the range $1.5 \leq R/2a \leq 2.5$, $k_1(-h/2)$ first decreases reaching a minimum at $h \simeq 0.6a$ and then increases as $h/2a$ is raised.

3.9. References

[1] Bridgman, P. W., *Studies in Large Plastic Flow and Fracture with Special Emphasis on the Effects of Hydrostatic Pressure*, First edition, McGraw-Hill Book Co., New York (1952).

[2] Sih, G. C., Strain energy density factor applied to mixed mode crack problems, *International Journal of Fracture* **10**, pp. 305–321 (1974).
[3] Sih, G. C., A special theory of crack propagation: methods of analysis and solutions of crack problems, *Mechanics of Fracture I*, edited by G. C. Sih, Noordhoff International Publishing, Leyden, pp. 21–45 (1973).
[4] Sih, G. C., A three-dimensional strain energy density factor theory of crack propagation: three-dimensional crack problems, *Mechanics of Fracture II*, edited by G. C. Sih, Noordhoff International Publishing, Leyden, pp. 15–53 (1975).
[5] *Linear Fracture Mechanics*, edited by G. C. Sih, R. P. Wei and F. Erdogan, Envo Publishing Co., Inc. Lehigh Valley, Pennsylvania (1976).
[6] Erdogan, F. and Sih, G. C., On the crack extension in plates under plane loading and transverse shear, *Journal of Basic Engineering* **85**, pp. 519–527 (1963).
[7] Sih, G. C. and Cha, B. C. K , A fracture criterion for three-dimensional crack problems, *Journal of Engineering Fracture Mechanics*, pp. 699–723 (1974).
[8] Kipp, M. E. and Sih, G. C., The strain energy density failure criterion applied to notched elastic solids, *International Journal of Solids and Structures* **2**, pp. 153–173 (1975).
[9] Sih, G. C., Discussion on some observations on Sih's strain energy density approach for fracture prediction by I. Finnie and H. O. Weiss, *International Journal of Fracture* **10**, pp. 279–283 (1974).
[10] Sih, G. C. and Macdonald, B., Fracture mechanics applied to engineering problems – strain energy density fracture criterion, *Journal of Engineering Fracture Mechanics* **6**, pp. 261 – 265 (1974).
[11] Sih, G. C. and Kipp, M. E., Discussion on fracture under complex stress – the angle crack problem by J. G. Williams and P. D. Ewing, *International Journal of Fracture* **10**, pp. 261–265 (1974).
[12] Hartranft, R. J. and Sih, G. C., An approximate three-dimensional theory of plates with application to crack problems, *International Journal of Engineering Science* **8**, pp. 711–729 (1970).
[13] Reissner, E., On bending of elastic plates, *Quarterly of Applied Mathematics* **5**, pp. 55–68 (1947).
[14] Wang, N. M., Twisting of an elastic plate containing a crack, *International Journal of Fracture Mechanics* **6**, pp. 367–373 (1970).
[15] Sih, G. C., A review of the three-dimensional stress problem for a cracked plate, *International Journal of Fracture Mechanics* **7**, pp. 39–61 (1971).
[16] Benthem, J. P., *Three-dimensional State of Stress at the Vertex of a Quarter-Infinite Crack in a Half-Space*, Delft University Report No. 563 (1975).
[17] Sih, G. C., Bending of a cracked plate with an arbitrary stress distribution across the thickness, *Journal of Engineering for Industry* **92**, pp. 350–356 (1970).
[18] Rubayi, N. A. and Ved, R., Photoelastic analysis of a thick square plate containing central crack and loaded by pure bending, *International Journal of Fracture* **12**, pp. 435–451 (1976).
[19] Hartranft, R. J. and Sih, G. C., The use of eigenfunction expansions in the general solution of three-dimensional crack problems, *Journal of Mathematics and Mechanics* **19**, pp. 123–138 (1969).
[20] Villarreal, G., Sih, G. C. and Hartranft, R. J., Photoelastic investigation of a thick plate with a transverse crack, *Journal of Applied Mechanics* **42**, pp. 9–14 (1975).
[21] Ang, D. D., Folias, E. S. and Williams, M. L., The bending stress in a cracked plate on an elastic foundation, *Journal of Applied Mechanics* **30**, pp. 245–251 (1964).
[22] Sih, G. C. and Setzer, D. E., Discussion of the bending stress in a cracked plate on an elastic foundation, *Journal of Applied Mechanics*, **31**, pp. 365–367 (1964).
[23] Folias, E. S., The stresses in a cracked spherical shell, *Journal of Mathematics and Physics* **44**, pp. 164–176 (1965).
[24] Knowles, J. K. and Wang, N. M., On the bending of an elastic plate containing a crack, *Journal of Mathematics and Physics* **39**, pp. 223–236 (1960).
[25] Erdogan, F. and Kibler, J. J., Cylindrical and spherical shells with cracks, *International Journal of Fracture Mechanics* **5**, pp. 229–237 (1969).

[26] Sih, G. C. and Dobreff, P. S., Crack-like imperfections in a spherical shell, *Glasgow Mathematical Journal* **12**, pp. 65–88 (1971).

[27] Folias, E. S., An axial crack in a pressurized cylindrical shell, *International Journal of Fracture Mechanics* **1**, pp. 104–113 (1965).

[28] Folias, E. S., A circumferential crack in a pressurized cylindrical shell, *International Journal of Fracture Mechanics* **3**, pp. 1–12 (1967).

[29] Copley, L. G. and Sanders, J. L., A longitudinal crack in a cylindrical shell under internal pressure, *International Journal of Fracture Mechanics* **5**, pp. 113–131 (1969).

[30] Erdogan, F. and Ratwani, M., A circumferential crack in a cylindrical shell under torsion, *International Journal of Fracture Mechanics* **8**, pp. 87–95 (1972).

[31] Reissner, E., On some problems in shell theory, *Structural Mechanics*, Proceedings First Symposium on Naval Structural Mechanics, pp. 74–113 (1958).

[32] Sih, G. C. and Hagendorf, H. C., A new theory of spherical shells with cracks, *Thin-Shell Structures: Theory, Experiment, and Design*, Edited by Fung, Y. C. and Sechler, E. E., Prentice-Hall, New Jersey, pp. 519–545 (1974).

[33] Naghdi, P. M., On the theory of thin elastic shells, *Quarterly Applied Mathematics* **54**, pp. 369–380 (1957).

[34] Reissner, E., On a variational theorem in elasticity, *Journal of Mathematics and Physics* **29**, pp. 90–95 (1950).

[35] Kalnins, A., On the derivation of a general theory of elastic shells, *Indian Journal of Mathematics* **9**, pp. 381–425 (1967).

[36] Novozhilov, V. V., *Thin Shell Theory*, Second edition, Noordhoff, Groningen, Netherlands (1964).

[37] Hartranft, R. J. and Sih, G. C., Effect of plate thickness on the bending stress distribution around through cracks, *Journal of Mathematics and Physics* **47**, pp. 276–291 (1968).

4

Dynamic crack problems – strain energy density fracture theory

4.1. Preliminary remarks

The dynamic behavior of cracked systems prior and subsequent to the critical condition of mechanical instability has been the subject of many past discussions. This instability corresponds to the point at which a sudden change occurs in the creation of free surfaces. Physical examples include the initiation of stationary cracks and the bifurcation of moving cracks traveling at high speed. In general, two different aspects of dynamic fracture have been considered.

In the first, attempts have been made [1–3] to deduce the kinetics of fracture from the equations of motion of a running crack by an extension of Griffith's energetic treatment [4]. The work required to create the new surfaces of the moving crack is assumed to be known from a quantity, 2γ, the specific surface energy,* originally used for analyzing stationary cracks that undergo a small segment of crack extension at instability. The velocity and acceleration characteristics of the crack were established without a consideration of the details of the crack tip stress field. This raises some question as to the accuracy of the results. In particular, the precise condition of crack initiation as influenced by the crack tip radius of curvature at and after instability is still not known and can have a strong influence on the kinetic aspects of fracture.

The second aspect deals with defining the condition which governs the onset of crack instability, either initiation or bifurcation. For stationary cracks that are inclined to the axis of loading, the crack may extend in a noncollinear fashion as shown in Figure 4.1(a). The direction of crack initiation is an unknown in the problem and cannot be assumed arbitrarily. Bifurcation of cracks moving with a high velocity also requires the knowledge

---

* Sih [5] has emphasized the need to distinguish γ referred to material damage at the microscopic level from that at the macroscopic level.

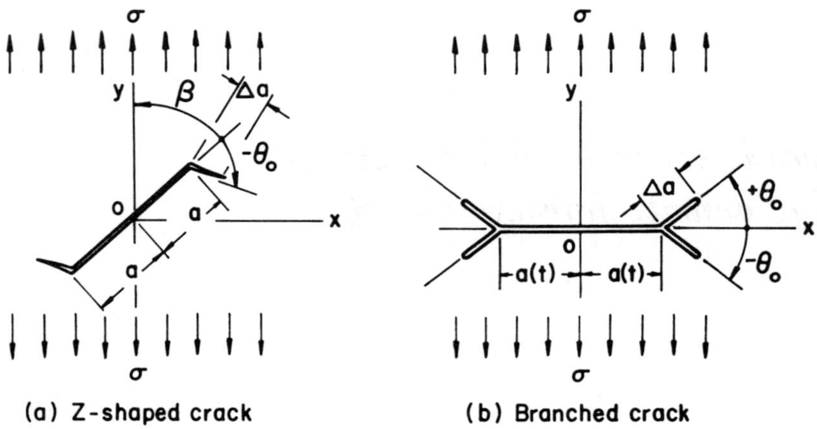

Figure 4.1. Cracks extended out of own planes.

of a criterion that can determine the branch angles $\pm\theta_0$, Figure 4.1(b), even when the load is applied symmetrically with respect to the crack plane. Strictly speaking, the phenomenon in Figure 4.1(b) will not occur in reality. That is to say a crack will not branch symmetrically about the path of crack propagation. The slightest amount of asymmetry in nonalignment of the load and material nonhomogeneity will turn the crack sideways at the point of dynamic instability. This tends to slow down the branched crack while another crack initiates from the bend giving the appearance of two branches. The mathematical assumption of symmetric branching* appears to be reasonable if the phenomenon is viewed on the macroscopic level although there is lack of careful experiments to justify this assumption.

In deducing the turning direction of the crack, there are two fundamentally different approaches that prevail in the literature. One is to take the post-instability configuration of the crack system and solve the problem for a small segment Δa of the extended crack, say in Figure 4.1(a) or (b). The direction of Δa will have to be further determined from a criterion such as minimization of the total energy of the system or maximization of the local energy release rate in the limit as Δa becomes very small. Such a procedure could be cumbersome and could involve the limit solution of two nearby singularities leading to a problem of uniqueness of solution. Alternatively, it is possible to use the solution of the pre-instability crack problem in conjunction with a criterion for determining the new direction of crack propagation. In such a case, attention is focused on an element ahead of the crack illustrated in Figures 4.2(a) and (b) as being the counterparts of Figures 4.1(a) and (b). Failure occurs when the element breaks. The ele-

---

* Branching often occurs unsymmetrically in that the two branch angles may be different and the branches may not have the same length.

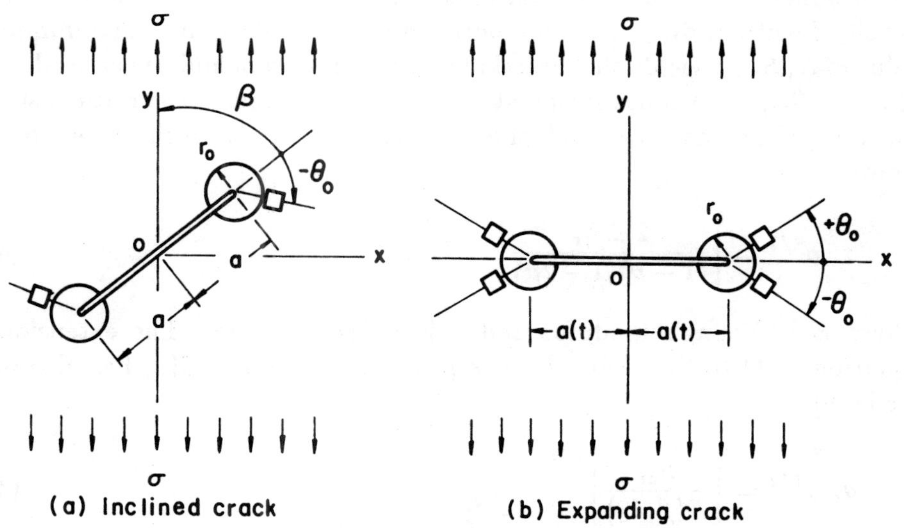

Figure 4.2. Cracks on the verge of turning.

ment to be analyzed is outside a small core region of radius r_0 around the crack tip.* For the static angle crack problem in Figure 4.2(a), Erdogan and Sih [6] assumed the crack to extend in a direction normal to the plane of maximum circumferential stress while Yoffé [7] found for the constant velocity crack that as the velocity increased the circumferential stress attains two separate maxima symmetrically disposed about the crack axis, Figure 4.2(b). This result seems to suggest the prospective sites of crack bifurcation. In principle, however, the maximum stress criterion, which assumed failure to occur when only one of the stress components reaches a maximum, cannot be expected to yield accurate results when another component is of approximately the same magnitude. It is apparent that the combination of the stress components plays a role and hence "energy" would be a more logical quantity to be associated with any criteria of failure in the presence of multiaxial stress state.

An easily usable, recently suggested fracture criterion has been applied quite successfully to several problems of crack growth initiation** under static loads [9, 10]. The basic theory uses the strain energy density factor S as the parameter for analyzing crack problems. This factor represents

* Conceptually, the postulated crack growth model in the classical theory implies a continuous separation of material along a plane ahead of the crack. There is experimental evidence [8] that, instead of continuous extension, voids are created and enlarged to join with the crack. The radius, r_0, discussed above may be regarded as the ligament distance between the first void and crack tip.

** The strain energy density theory can also be applied to predict crack trajectories [11], where the entire path of minimum strain energy density is determined.

the product of the radial distance, say r_0, multiplied by the strain energy density function, dW/dV, commonly employed in elasticity. The minimum value of S, $S_{\min}$, yields the direction of crack initiation and the critical value of $S_{\min}$, S_{cr}, determines incipient fracture. For instance, the critical stress, σ_{cr}, applied normal to a crack of length $2a$ in a large plate can be estimated from

$$\sigma_{cr} a^{1/2} = \left[\frac{2ES_{cr}}{(1+\nu)(1-2\nu)} \right]^{1/2} \tag{4.1}$$

where E is Young's modulus and ν is Poisson's ratio. The extension of equation (4.1) to the inclined crack problem in Figure 4.2(a) has also been made [9]:

$$\sigma_{cr} a^{1/2} = \left[\frac{S_{cr}}{F(\beta, \theta_0)} \right]^{1/2} \tag{4.2}$$

in which $F(\beta, \theta_0)$ is the function

$$F(\beta,\theta_0) = (a_{11} f_1^2 \sin^2\beta + 2a_{12} f_1 f_2 \sin\beta\cos\beta + a_{22} f_2^2 \cos^2\beta) \sin^2\beta \tag{4.3}$$

that depends on the crack position with respect to the load as measured by the angle β and the fracture angle, θ_0, corresponding to the location of $S_{\min}$ through the coefficients a_{ij} $(i, j = 1, 2)$:

$$16\mu a_{11} = (3 - 4\nu - \cos\theta_0)(1 + \cos\theta_0) \tag{4.4a}$$
$$8\mu a_{12} = \sin\theta_0[\cos\theta_0 - (1 - 2\nu)] \tag{4.4b}$$
$$16\mu a_{22} = 4(1-\nu)(1 - \cos\theta_0) + (1 + \cos\theta_0)(3\cos\theta_0 - 1) \tag{4.4c}$$

with $\mu = E/2(1+\nu)$ being the shear modulus of elasticity. Note that S_{cr}, which determines the onset of rapid crack propagation, in equation (4.1) for load applied normal to the crack is the same as that in equation (4.2) for load inclined to the crack. The functions f_1 and f_2 are unity for static loading.

The basic concept of the strain energy density theory applies equally well to stationary or moving cracks subjected to time dependent loading such as impact and vibration. As long as the crack remains stationary, the forms of equations (4.2) to (4.4) remain unchanged except that f_1 and f_2 in equation (4.3) must be modified to account for dynamic effects. If the load is applied suddenly to the crack, f_j $(j = 1, 2)$ are functions of time and σ_{cr} is evaluated at the instant of unstable fracture. For loads that vary periodically in time, f_j $(j = 1, 2)$ depend on the frequency of oscillation and the interaction of incident waves with the crack angle β. While the magnitude of σ_{cr} is expected to be different for static and dynamic loads, it is important to recognize that, in principle, the critical value S_{cr} or fracture toughness should remain unchanged* for a particular material.

One of the salient features of the strain energy density theory is that it can simultaneously predict the prospective locations of fracture ($S_v > S_d$) and distortion ($S_d > S_v$), where S_v is the energy density factor associated with volume change and S_d with shape change, where $S = S_v + S_d$. This is particularly useful for studying the stability behavior of cracks moving at high velocities with the assumption that cracks have a tendency to move into regions where dilatation exceeds distortion. Hence, it would be informative to inquire into the variations of the ratio S_v/S_d with the normalized crack speed c/c_2. In two dimensions, the ratio S_v/S_d can be expressed in terms of the principal stresses on an element directly ahead of the moving crack in Figure 4.3 as follows:

$$\frac{S_v}{S_d} = \frac{(1-2\nu)(1+\nu)[(\sigma_1/\sigma_2)+1]^2}{[(\sigma_1/\sigma_2)-1]^2+[(1-\nu)-\nu(\sigma_1/\sigma_2)]^2+[(1-\nu)(\sigma_1/\sigma_2)-\nu]^2} \tag{4.5}$$

in which σ_1/σ_2 takes the form

$$\frac{\sigma_1}{\sigma_2} = \frac{(1+s_2^2)(2s_1^2+1-s_2^2)-4s_1s_2}{4s_1s_2-(1+s_2^2)^2}. \tag{4.6}$$

The parameters

$$s_j = \left[1-(c/c_j)^2\right]^{1/2}, \quad j = 1,2 \tag{4.7}$$

are functions of the crack speed c, the compressional (dilatational) wave velocity c_1 and shear (equivoluminal) wave velocity c_2 in an infinitely extended region. The wave velocities c_j ($j = 1,2$) are

$$c_1 = \left[\frac{\lambda+2\mu}{\rho}\right]^{1/2}, \qquad c_2 = \left[\frac{\mu}{\rho}\right]^{1/2} \tag{4.8}$$

where λ and μ are the Lamé coefficients and ρ is the mass density. The numerical results of equation (4.5) as a function of c/c_2 are shown graphically in Figure 4.3 for Poisson's ratios ranging from 0.10 to 0.40. Along the path

* The reported differences between the statically and dynamically measured fracture toughness values can be attributed to ignoring the dynamic effect in the stress analysis and/or lacking of a precise definition of fracture toughness [12]. Strictly speaking, the fracture toughness parameter should represent a measure of the energy released by a process of material separation leading to instability of the macrocrack. As long as the macrocrack extends into a region within which no distinction is made between the static and dynamic material properties, then the fracture toughness value should be independent of the nature of loading. Any criterion incorporating a parameter that measures the total energy dissipation of the entire specimen, including both microscopic and macroscopic material damage, will be sensitive to changes in specimen size since it does not uniquely represent the available energy necessary to cause macrocrack instability.

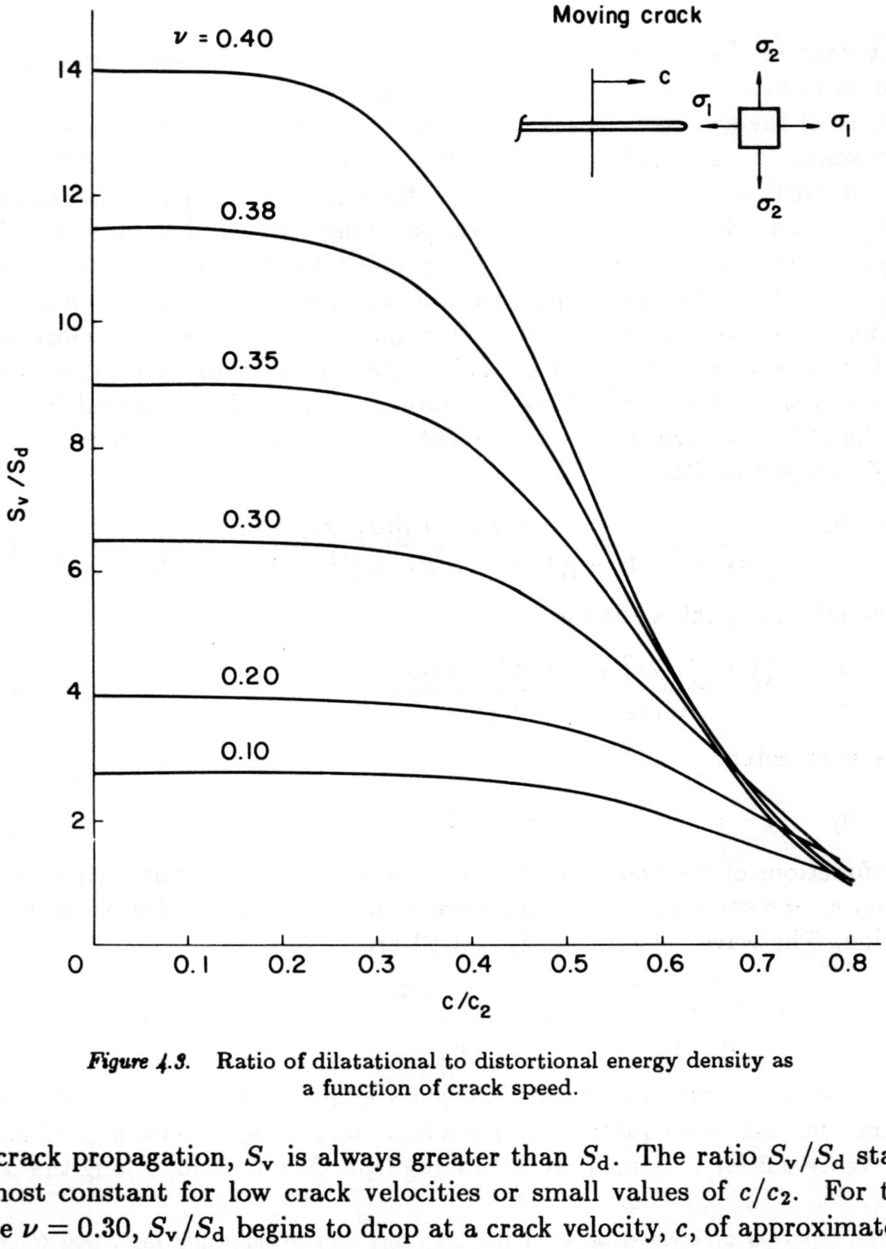

Figure 4.3. **Ratio of dilatational to distortional energy density as a function of crack speed.**

of crack propagation, S_v is always greater than S_d. The ratio S_v/S_d stays almost constant for low crack velocities or small values of c/c_2. For the case $\nu = 0.30$, S_v/S_d begins to drop at a crack velocity, c, of approximately $0.35c_2$. Whether this decrease of S_v/S_d is in any way associated with crack bifurcation requires further examination. The locations of the minimum value of S will shed some light on this in the work to follow.

4.2. Strain energy density criterion in elastodynamics

The strain energy density criterion [9, 10] grew out of the study of mixed mode crack extension under static loading and will be further extended to

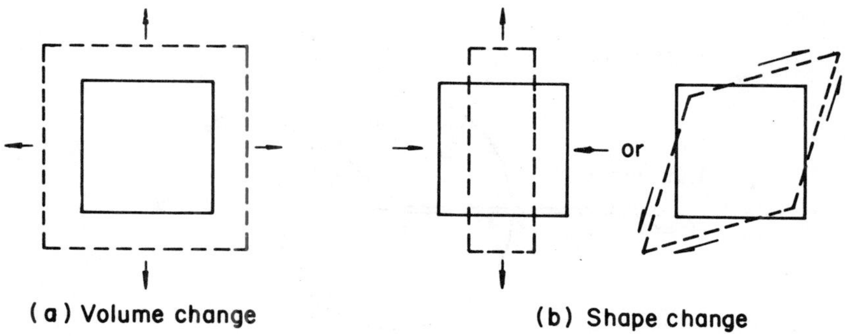

Figure 4.4. Dilatation and distortion of a material element.

situations where dynamic effects are important. Attention is focused on the strain energy density function, dW/dV, referred to a material element outside of the core region of radius r_0 around the crack tip. Such an element is shown in Figure 4.2(a). The function dW/dV is inversely proportional to r_0 such that it becomes increasingly large as r_0 is made smaller and smaller. The product $r_0\, dW/dV$ is a finite quantity and represents the intensity of the local energy field, i.e.,

$$S = r_0 \left.\frac{dW}{dV}\right|_{r=r_0} \tag{4.9}$$

in which dW/dV is known from the theory of elasticity. For two-dimensional problems in which $\sigma_{yz} = \sigma_{zx} = 0$, equation (4.9) may be expressed in terms of the stress components:

$$S = r_0 \left(\frac{1+\nu}{2E}\right)\left[\sigma_x^2 + \sigma_y^2 + \sigma_z^2 - \frac{\nu}{1+\nu}(\sigma_x + \sigma_y + \sigma_z)^2 + 2\sigma_{xy}^2\right]. \tag{4.10}$$

Sih has shown that the plane strain condition [13]

$$\sigma_z = \nu(\sigma_x + \sigma_y) \tag{4.11}$$

always prevails along the interior crack front and hence can be used without loss of generality.

Physical insight into the fracture problem can often be gained by dividing S into two parts, one part associated with volume change, S_v, and the other with shape change, S_d, such that $S = S_v + S_d$. A pictorial illustration of these changes on an element is given in Figures 4.4(a) and (b). The result of decomposing equation (4.10) gives

$$S_v = r_0 \left(\frac{1-2\nu}{6E}\right)(\sigma_x + \sigma_y + \sigma_z)^2 \tag{4.12}$$

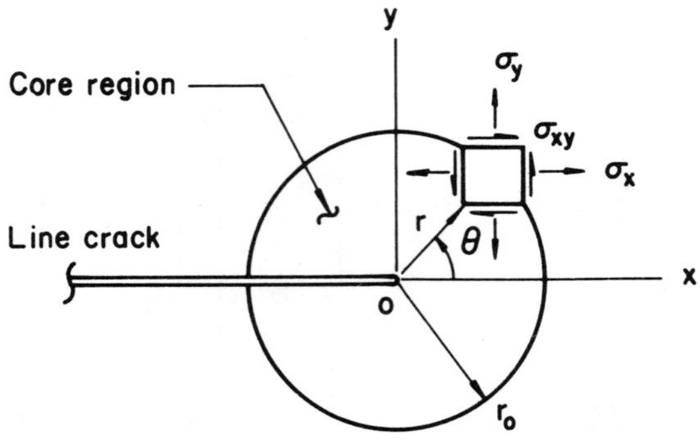

Figure 4.5. Local rectangular stress components.

and

$$S_{\rm d}=r_0\left(\frac{1+\nu}{6E}\right)\left[(\sigma_x-\sigma_y)^2+(\sigma_y-\sigma_z)^2+(\sigma_z-\sigma_x)^2+6\sigma_{xy}^2\right]. \quad (4.13)$$

With reference to the principal planes, where $\sigma_x=\sigma_1$, $\sigma_y=\sigma_2$ and $\sigma_{xy}=0$, equations (4.12) and (4.13) can be further simplified and the ratio of $S_{\rm v}/S_{\rm d}$ reduces to that of equation (4.5).

There remains to determine the details of the stress components σ_x, σ_y and σ_{xy} on an element near the crack tip. If the crack tip is stationary, the time variation of the local stresses is transmitted to the tip region through the stress intensity factors. The r and θ dependence of the local stresses in Figure 4.5 is the same as the static case, i.e.,

$$\sigma_x=\frac{k_1}{(2r)^{1/2}}\cos\frac{\theta}{2}\left(1-\sin\frac{\theta}{2}\sin\frac{3\theta}{2}\right)- \\ -\frac{k_2}{(2r)^{1/2}}\sin\frac{\theta}{2}\left(2+\cos\frac{\theta}{2}\cos\frac{3\theta}{2}\right) \quad (4.14a)$$

$$\sigma_y=\frac{k_1}{(2r)^{1/2}}\cos\frac{\theta}{2}\left(1+\sin\frac{\theta}{2}\sin\frac{3\theta}{2}\right)+ \\ +\frac{k_2}{(2r)^{1/2}}\sin\frac{\theta}{2}\cos\frac{\theta}{2}\cos\frac{3\theta}{2} \quad (4.14b)$$

$$\sigma_{xy}=\frac{k_1}{(2r)^{1/2}}\cos\frac{\theta}{2}\sin\frac{\theta}{2}\cos\frac{3\theta}{2}+ \\ +\frac{k_2}{(2r)^{1/2}}\cos\frac{\theta}{2}\left(1-\sin\frac{\theta}{2}\sin\frac{3\theta}{2}\right). \quad (4.14c)$$

Equations (4.14) may be substituted into equation (4.10) giving the coefficient of $1/r$ as

$$S=a_{11}k_1^2+2a_{12}k_1k_2+a_{22}k_2^2. \quad (4.15)$$

The form of S is the same as that in the static case [9] except that k_1 and k_2 are now dependent on time. In addition, the coefficients a_{ij} $(i,j = 1,2)$ given by

$$16\mu a_{11} = (3 - 4\nu - \cos\theta)(1 + \cos\theta) \tag{4.16a}$$

$$16\mu a_{12} = 2\sin\theta(\cos\theta - 1 + 2\nu) \tag{4.16b}$$

$$16\mu a_{13} = 4(1 - \nu)(1 - \cos\theta) + (1 + \cos\theta)(3\cos\theta - 1) \tag{4.16c}$$

are also the same.

The strain energy density criterion first requires an investigation of the variations of S in equation (4.15) as a function of θ. In general, there are two distinct local maxima and minima. Location of the two minima give the directions of crack extension and the two maxima give the directions of yielding provided that the loads are large enough to cause crack growth and/or yielding. One set of directions is associated with remote tensile loading and the other is for remote compressive loading. The second consideration in the S-criterion deals with prediction of the load required to produce fracture. Unstable fracture is assumed to occur when $S_{\min}$ reaches a certain critical value, S_{cr}, characteristic of the material under consideration.

The phenomenon of unstable fracture for a moving crack corresponds to a sudden change in the direction of propagation. Application of the S-criterion to such a situation is straightforward. The hypotheses originally stated in [9, 10] for stationary cracks can be modified slightly and applied to moving cracks:

Hypothesis (1): *The change of direction of crack propagation is determined from the location of the minimum value of the strain energy density factor,* $S_{\min}$:

$$\frac{\partial S}{\partial \theta} = 0 \quad at \quad \theta = \theta_0. \tag{4.17}$$

Hypothesis (2): *The moving crack changes its direction of propagation at the point when* $S_{\min}$ *reaches a critical value, i.e.,*

$$S_{\min} = S_{cr} \quad for \quad \theta = \theta_0. \tag{4.18}$$

Since the angular distribution of the stress field ahead of a moving crack is distorted and different from that in equations (4.14), the form of S will also be altered. With the element attached to the moving crack as in Figure 4.4, the local stresses in terms of r and θ can be written as [14]

$$\sigma_x = \frac{k_1}{(2r)^{1/2}} F_1(s_1, s_2)\left[\left(1+s_2^2\right)\left(2s_1^2+1-s_2^2\right)g(s_1) - 4s_1s_2g(s_2)\right] \tag{4.19a}$$

$$\sigma_y = \frac{k_1}{(2r)^{1/2}} F_1(s_1, s_2)\left[4s_1s_2g(s_2) - \left(1+s_2^2\right)^2 g(s_1)\right] \tag{4.19b}$$

$$\sigma_{xy} = \frac{k_1}{(2r)^{1/2}} 2s_1(1+s_2^2)F_1(s_1,s_2)[h(s_1)-h(s_2)] \tag{4.19c}$$

in which the functions $g(s_j)$ and $h(s_j)$ ($j = 1,2$) describe the θ-dependence of σ_x, σ_y and σ_{xy}*:

$$g^2(s_j) + h^2(s_j) = \sec\theta(1+s_j^2\tan^2\theta)^{-1/2} \tag{4.20a}$$

$$g^2(s_j) - h^2(s_j) = \sec\theta(1+s_j^2\tan^2\theta)^{-1/2} \tag{4.20b}$$

and s_j ($j = 1,2$) are given in equation (4.7). The stress intensity factor k_1 in equations (4.19) is chosen to be the same as in static loading while $F_1(s_1, s_2)$ reflects the dynamic effect. The results of equations (4.19) and (4.20) apply to a crack moving at a constant velocity along a straight line. It is not difficult to argue that angular distribution of the dynamic stresses in equations (4.19) remains valid for cracks propagating at nonuniform velocities provided that the region of investigation is limited to the immediate vicinity of the moving crack tip.

A straightforward substitution of equations (4.19) into (4.10) yields the dynamic strain energy density factor for a moving crack:

$$\begin{aligned} S = \frac{k_1^2}{8\mu}F_1^2(s_1,s_2)\{&2(1+s_2^2)^2[2(1-\nu)(1+s_1^2)^2 - \\ &-(1-2\nu)(2s_1^2+1-s_2^2)(1+s_2^2)]g^2(s_1) + 32s_1^2s_2^2g^2(s_2) - \\ &-16s_1s_2(1+s_1^2)(1+s_2^2)g(s_1)g(s_2) + 8s_1^2(1+s_2^2)[h(s_1)-h(s_2)]^2\}. \end{aligned} \tag{4.21}$$

It follows from equations (4.12) and (4.13) that S can be separated into

$$S_{\text{v}} = \frac{k_1^2}{6\mu}F_1^2(s_1,s_2)(1-2\nu)(1+\nu)(1+s_2^2)^2(s_1^2-s_2^2)^2g^2(s_1) \tag{4.22}$$

which is the dynamic energy density of dilatation and

$$\begin{aligned} S_{\text{d}} = \frac{k_1^2}{8\mu}F_1^2(s_1,s_2)\{&(1+s_2^2)^2[\tfrac{8}{3}(1-\nu+\nu^2)(s_1^2-s_2^2)^2 + \\ &+(1+s_2^2)(2s_1^2+1-s_2^2)]g^2(s_1) + 32s_1^2s_2^2g^2(s_2) - \\ &-16s_1s_2(1+s_2^2)(1+s_1^2)g(s_1)g(s_2) + 8s_1^2(1+s_2^2)^2[h(s_1)-h(s_2)]^2\}. \end{aligned} \tag{4.23}$$

the dynamic energy density of distortion. Equations (4.21) to (4.23) will be used subsequently to analyze crack bifurcation that occurs when the crack moves at high speed.

---

* Along the direction $\theta = 0$, σ_{xy} vanishes and σ_x and σ_y become the principal stresses σ_1 snd σ_2. Their ratio takes form of equation (4.6).

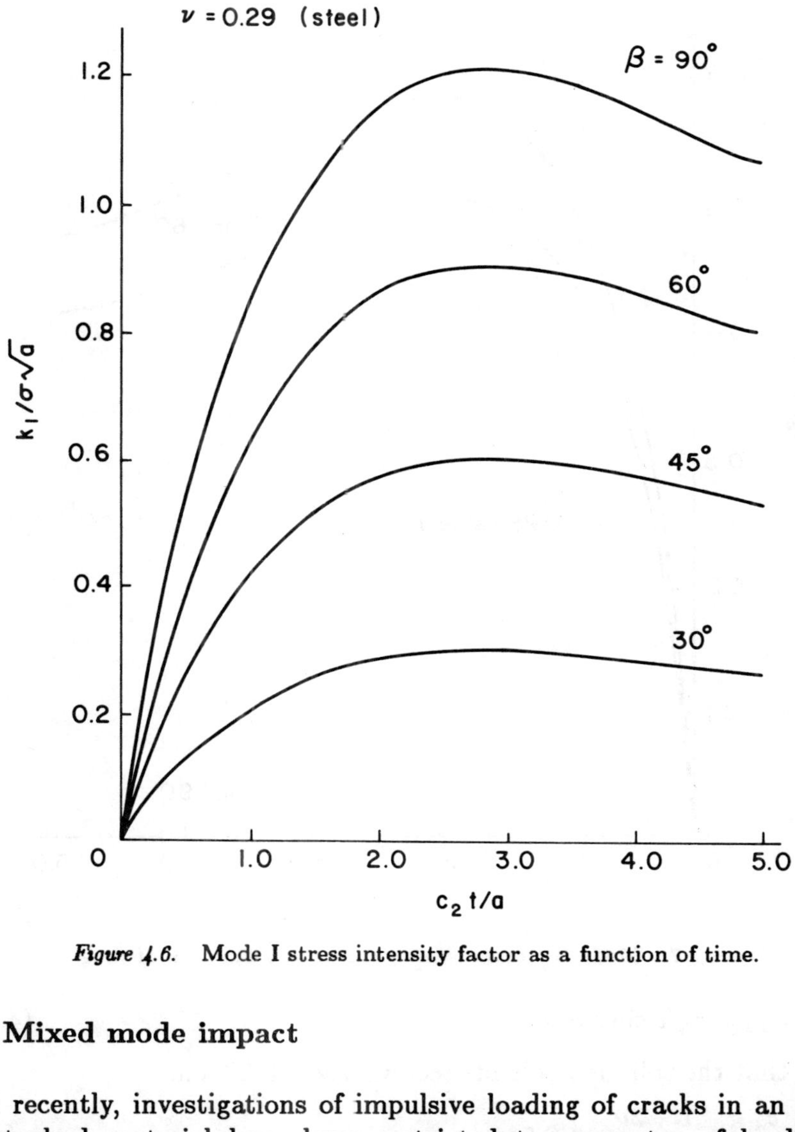

Figure 4.6. Mode I stress intensity factor as a function of time.

4.3. Mixed mode impact

Until recently, investigations of impulsive loading of cracks in an initially undisturbed material have been restricted to symmetry of loading and geometry such that the crack will propagate straight ahead. Situations in which lack of symmetry leads to a mixture of crack mode deformation are not so easily analyzed. Of primary interest are the direction of crack growth and the conditions under which it initiates.

Let the crack configuration in Figure 4.2(a) be subjected to a suddenly applied uniform load of magnitude σ which is then maintained at a constant level. The mixed mode stress intensity factors k_1 and k_2 for this problem are [15]:

$$k_1 = f_1(t)\sigma\sqrt{a}\sin^2\beta \tag{4.24a}$$

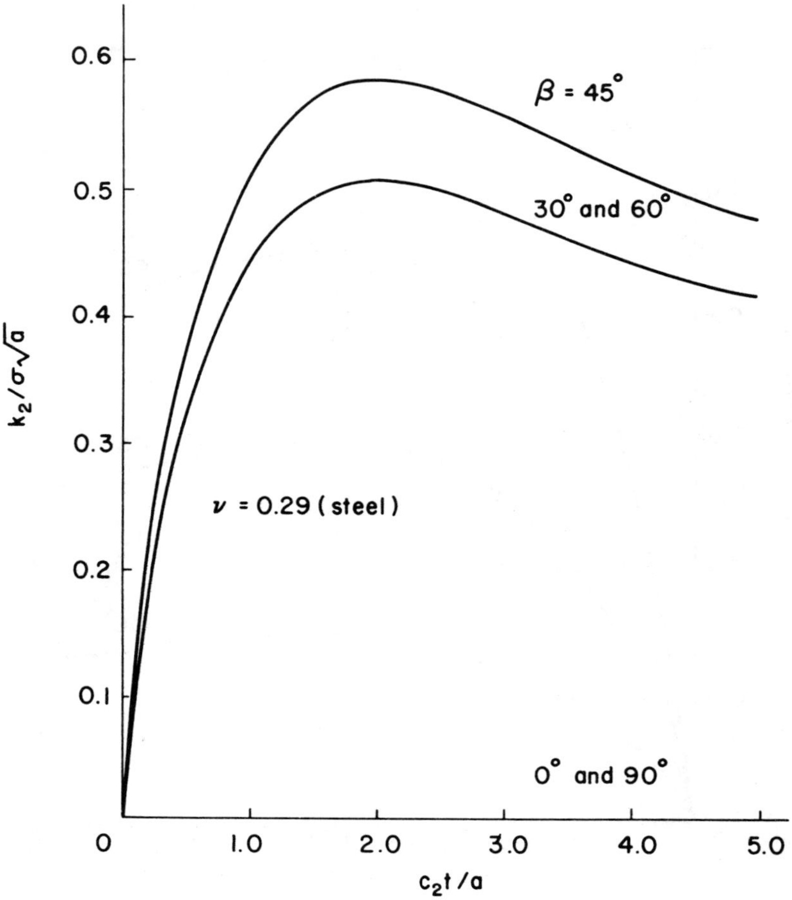

Figure 4.7. **Mode II stress intensity factor as a function of time.**

$$k_2 = f_2(t)\sigma\sqrt{a}\sin\beta\cos\beta. \tag{4.24b}$$

Suppose that the solid is made of steel with $\nu = 0.29$ and

$$\kappa = \frac{c_2}{c_1} = \frac{1-2\nu}{2(1-\nu)} = 0.54. \tag{4.25}$$

With these values, the numerically determined functions f_1 and f_2 in equations (4.24a) and (4.24b) are given in Figures 4.6 and 4.7 for cracks aligned at different angles to the axis of impact. The dynamic stress intensity factor k_1 is seen to increase at first reaching a maximum of $c_2t/a \simeq 3.0$ and then to decrease in amplitude. This effect becomes more pronounced as the crack is moved into a position normal to the applied load, i.e., $\beta = 90°$. Figure 4.7 shows that k_2 is not as large in magnitude as k_1 but reaches a maximum sooner with $c_2t/a \simeq 2.0$. The largest influence of k_2 appears at $\beta = 45°$ while the same results apply to $\beta = 30°$ and $60°$. The way in which these

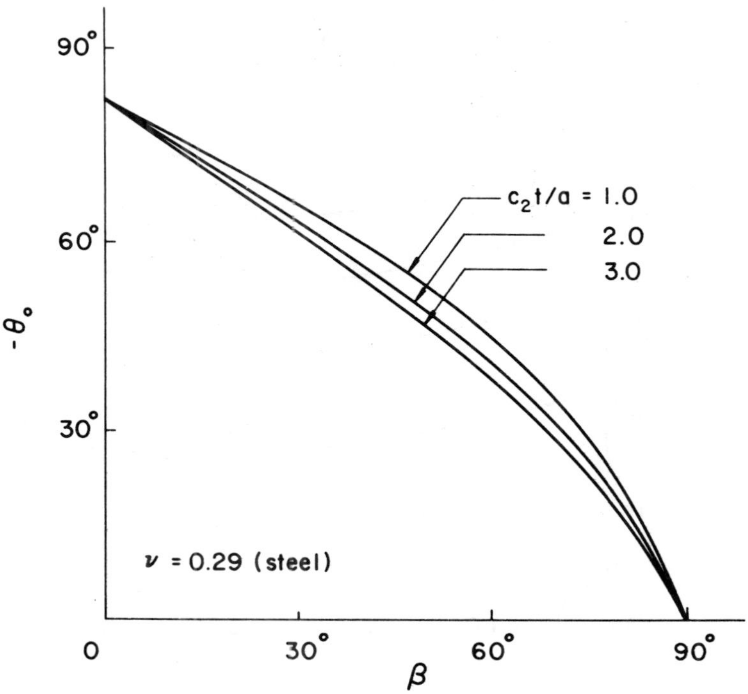

Figure 4.8. Fracture angle versus crack angle at different times after impact.

stress intensity factors affect the critical stress σ_{cr} can be determined by application of the strain energy density factor criterion.

Substituting equations (4.24) into (4.15), it is found that

$$S=\sigma^2 a \sin^2\beta\left[a_{11}f_1^2\sin^2\beta+2a_{12}f_1f_2\sin\beta\cos\beta+a_{22}f_2^2\cos^2\beta\right] \quad (4.26)$$

where the quantities a_{ij} $(i,j=1,2)$ are defined in equations (4.16) with the fracture angle, θ_0, determined from the condition $\partial S/\partial\theta = 0$ that makes S a minimum. The values of θ_0 plotted against β for three different values of $c_2t/a = 1.0$, 2.0 and 3.0 representing the time elapsed after impact are shown in Figure 4.8. An alternative plot of θ_0 versus time for $\beta = 15°$, $30°$, etc., is shown in Figure 4.9. These results imply that $S_{\min}$ is a function of time for impulsively loaded cracks. Figure 4.10 gives the variations of $16\mu S_{\min}/\sigma^2 a$ with c_2t/a where the curves first rise to a maximum and then taper off.

According to the strain energy density fracture criterion, the crack extends in the direction θ_0 when $S_{\min}$ reaches the critical value, S_{cr}, of a given material. Depending on the angle β and the amplitude of the impact load, $S_{\min}$ may be equal to S_{cr} on any part of the curves shown in Figure 4.10 or never reach S_{cr}. As the time increases, the strain energy density increases. For example, at $t \simeq a/c_2$, $16\mu S = 0.847\ \sigma^2 a$ is obtained from Figure 4.10

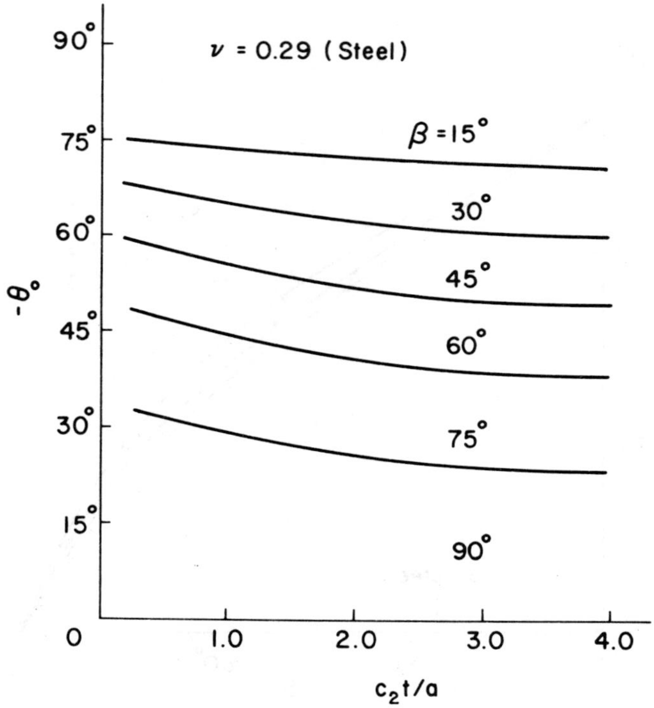

Figure 4.9. **Fracture angle versus time for different crack angle.**

for $\beta = 45°$. Whether or not this exceeds the critical value, S_{cr}, depends on the amplitude of the impact stress, σ. It is convenient to express the critical value of strain energy density in terms of the failure stress, σ_s (referred to as σ_{cr} in equation (4.1)): $16\mu S_{cr} = 4(1-2v)\sigma_s^2 a$. Then for this particular time, failure will take place at a load of $\sigma_m = 1.408\sigma_s$. In general, the failure load is given by

$$\left(\frac{\sigma_m}{\sigma_s}\right)^2 = \frac{4(1-2\nu)}{16\mu S_{\min}/\sigma^2 a} \tag{4.27}$$

where the denominator is the function of time and the other parameters shown in Figure 4.10. Figure 4.11 shows this result in the form of the failure load required for crack growth to occur at various times and for various crack orientations. The curves represent the stress levels just required for crack growth. A very high critical stress is needed for failure to take place for small time or immediately after impact. Each curve passes through a minimum that occurs between $c_2t/a \simeq 2.0$ and 3.0. For $\beta = 15°$, this minimum is at $c_2t/a \simeq 2.0$ and gives the smallest value of σ_m to cause fracture.

The angle of fracture also depends on the time at which it occurs. For $\beta = 45°$ and $t \simeq a/c_2$, Figure 4.9 shows that failure takes place in the direction, $\theta_0 \simeq -55.5°$.

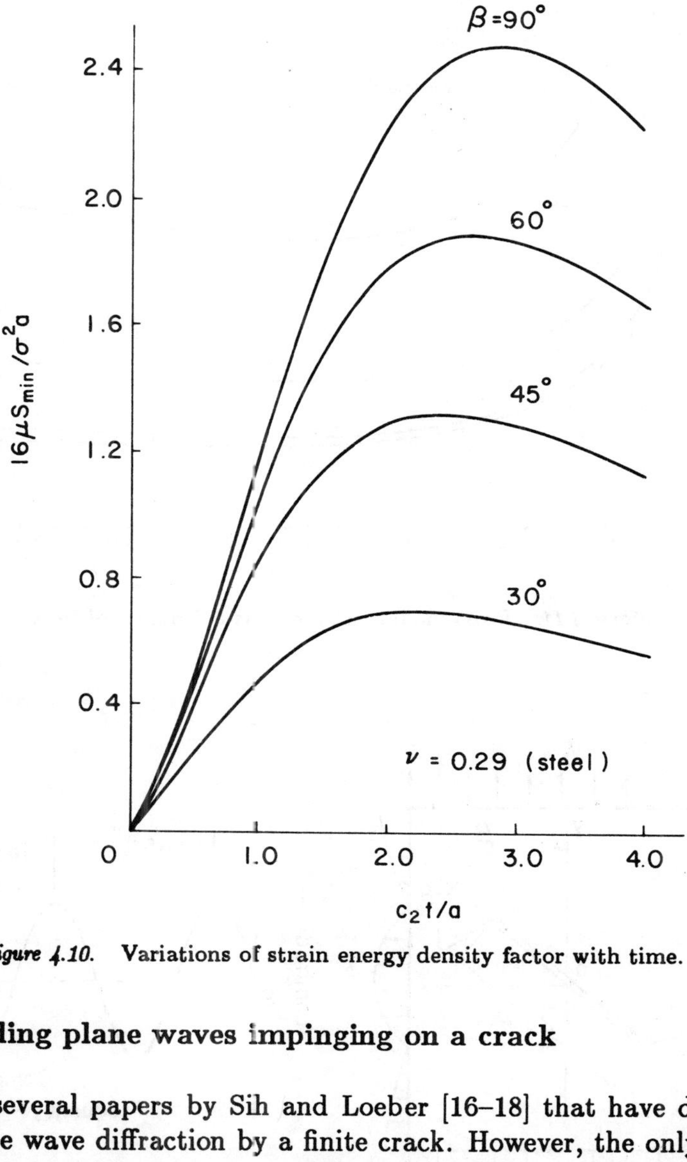

Figure 4.10. Variations of strain energy density factor with time.

4.4. Standing plane waves impinging on a crack

There are several papers by Sih and Loeber [16–18] that have dealt with steady plane wave diffraction by a finite crack. However, the only fracture analysis of such problems was made by Hartranft and Sih [19] on the basis of the strain energy density fracture criterion. They considered a plane wave impinging nonorthogonally on a flat crack in a plane strain analysis involving both stress intensity factors k_1 and k_2 and determined the direction of crack growth as well as the amplitude of the wave necessary to produce the crack growth.

For the sake of simplicity, a biaxial stress field

$$\sigma_y = \sigma \cos \alpha y \cos \omega t, \qquad \sigma_x = \frac{\nu}{1-\nu}\sigma_y \tag{4.28}$$

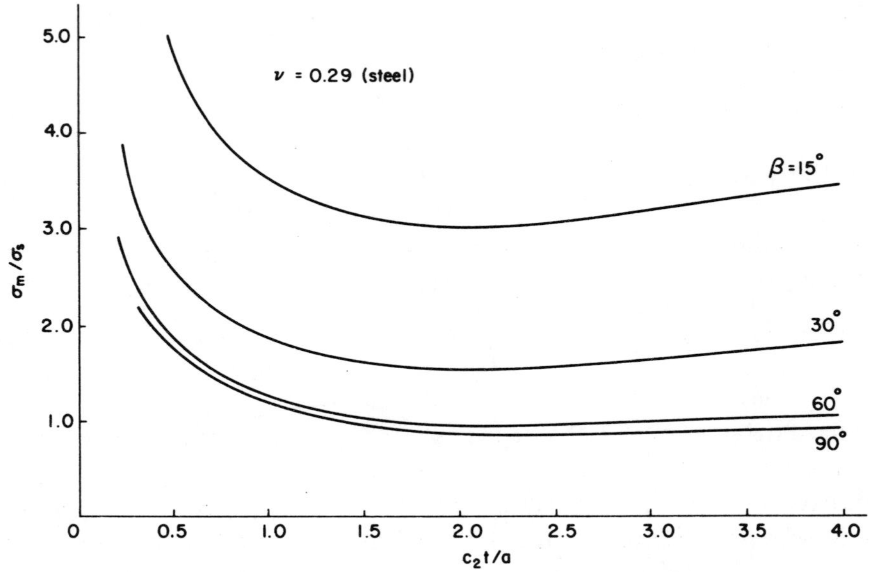

Figure 4.11. Dynamic impact stress as a function of time.

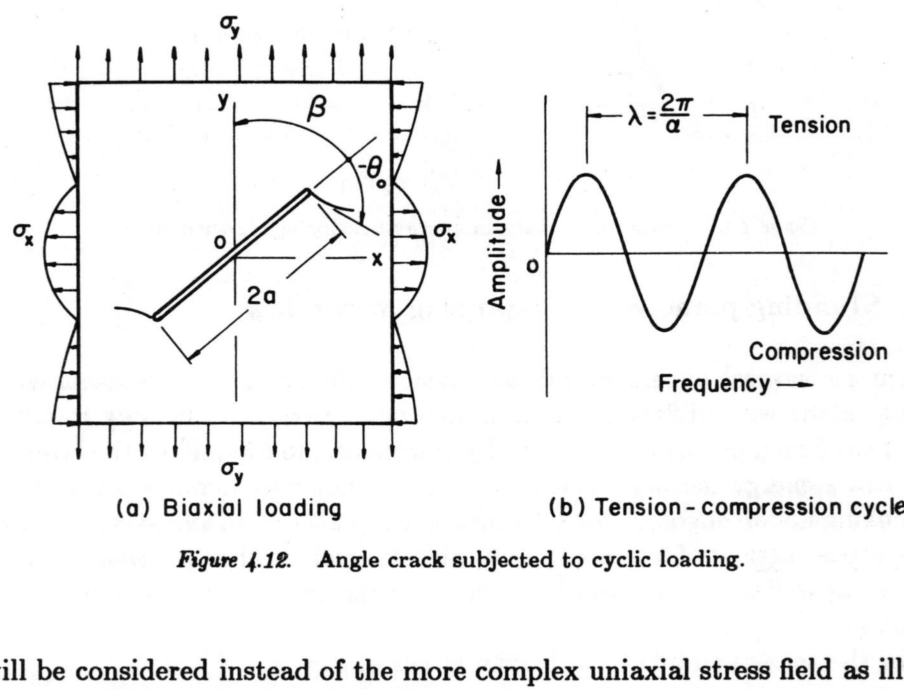

Figure 4.12. Angle crack subjected to cyclic loading.

will be considered instead of the more complex uniaxial stress field as illustrated in Figure 4.12. The stress wave amplitude σ is the same for both the tension and compression cycle. The nonuniform stresses σ_x acting on the vertical sides prevent the infinite body from displacing in the horizontal direction such that at locations remote from the crack, only vertical displacement prevails

$$u_y = \frac{\alpha}{\rho\omega^2}\sigma \sin \alpha y \cos \omega t \tag{4.29}$$

where the wave number, α, and the circular frequency, ω, are related by $\omega = \alpha c_1$.

The time-dependent stress intensity factors for this problem can be extracted from [16] and put into the forms

$$k_1 = |k_1| \cos \omega(t - \delta_1) \tag{4.30a}$$

$$k_2 = |k_2| \cos \omega(t - \delta_2). \tag{4.30b}$$

The phase delay δ_j $(j = 1, 2)$ are present as the peak applied stress and peak stress intensity factor do not occur at the same time. The dependence of the normalized k_1 and k_2 amplitudes on input frequency, ω, and crack angle, β, is depicted in Figures 4.13 and 4.14. These amplitudes initially increase as the frequency increases from zero and then decrease in magnitude after reaching their maximum peaks. The largest value of k_1 occurs when the crack is normal to the oscillating stress σ_y. Figure 4.14 illustrates that the peak values of k_2 do not coincide with those of k_1 and $\beta = 45°$ is the position for which k_2 attains its maximum. The phase difference between k_1 and k_2 becomes more and more significant as the frequency is increased.

Once the numerical values of k_1 and k_2 are known, equation (4.15) may be used to obtain the S-factor. The direction, θ_0, determined from the condition of $\partial S/\partial\theta = 0$, varies with time as the load oscillates from tension to compression. Figure 4.15 plots θ_0 against t/τ for $\alpha a = 0.10$ and 0.70, where the period $\tau = 2\pi/\omega$. At the lower frequency, for which the phase difference between k_1 and k_2 is small, θ_0 switches abruptly from negative (tensile loading) to positive (compressive loading) following closely to the load cycle. As the frequency is raised to $\alpha a = 0.70$, this change of angle with time becomes more gradual. Inserting the values of θ_0 into equation (4.15) yields $S_{\min}$ that also fluctuates in time as illustrated in Figure 4.16 for three different normalized wave numbers $\alpha a = 0.10$, 0.43 and 0.70. The value of $S_{\min}$ decreases as the load passes from tension to compression and remains small until the tension cycle begins again.

The condition to cause crack growth will depend on the frequency, stress amplitude, and crack orientation. Let this stress amplitude be σ_m which, when normalized with respect to the static failure stress, σ_s, as given in equation (4.1), can be computed in accordance with equation (4.27). Substituting the results of Figure 4.16 into the denominator of the right-hand side of equation (4.27) gives the plot of σ_m/σ_s versus $\omega a/c_1$ in Figure 4.17. As the frequency increases, the failure stress, σ_m, decreases at first, going

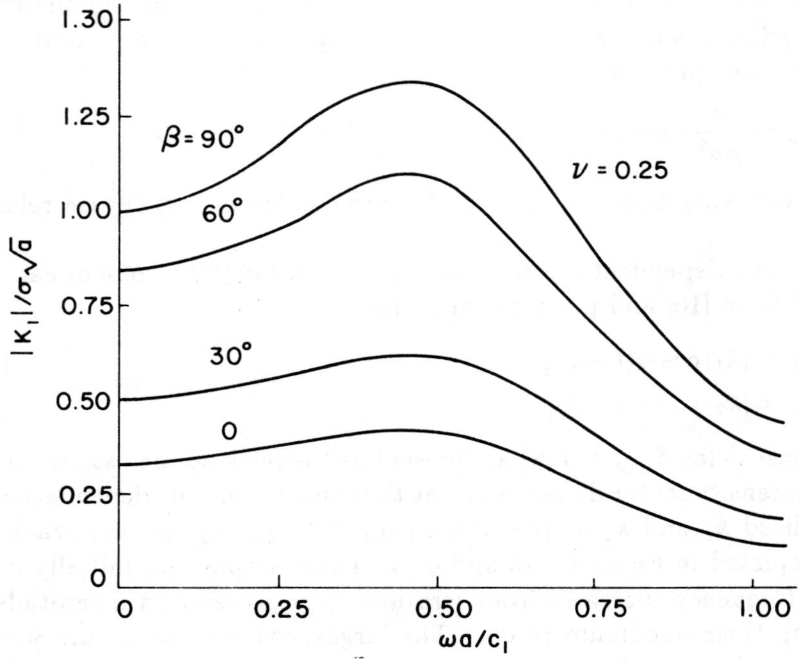

Figure 4.13. **Normalized amplitude of mode I stress intensity factor.**

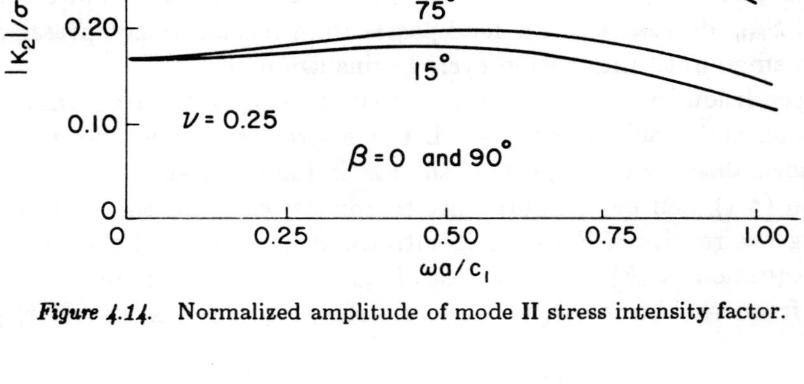

Figure 4.14. **Normalized amplitude of mode II stress intensity factor.**

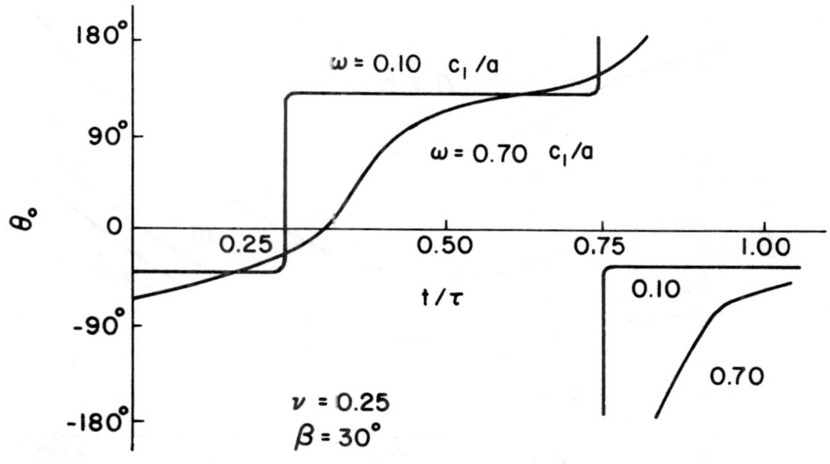

Figure 4.15. **Fracture angle as a function of frequency.**

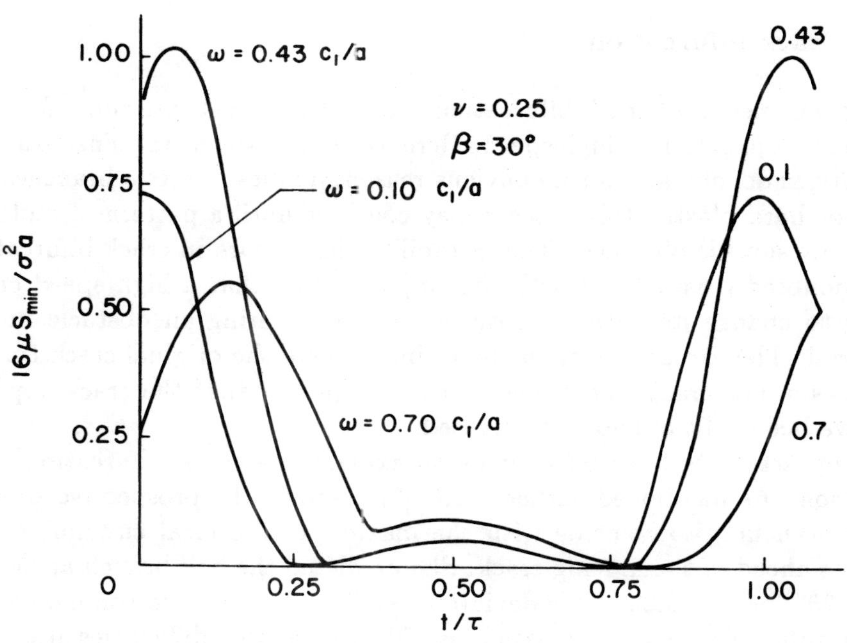

Figure 4.16. **Variations of dynamic strain energy density factor with time.**

through a minimum, and then increases. Higher values of σ_m are needed to initiate crack growth as the crack position is moved into the direction ($\beta = 0°$) of the applied displacement. When the plane waves impinge at normal incidence, the frequency at which the crack would grow at the smallest σ_m is $\omega \simeq 0.5c_1/a$.

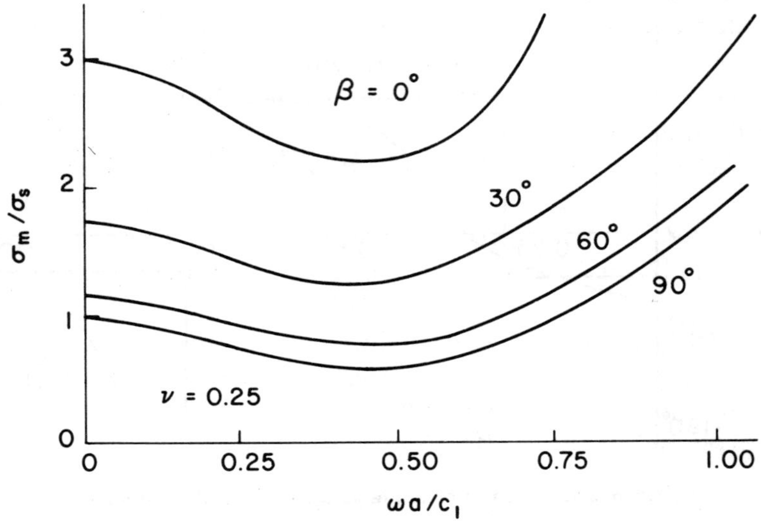

Figure 4.17. Normalized dynamic failure stress versus normalized wave number.

4.5. Crack bifurcation

The phenomenon of crack bifurcation is one of the most intriguing features of crack propagation at high speed. Here, the crack, when traveling at a high velocity, suddenly, and for no obvious reason, divides into two branches. In glass or hard plastic, this process may continue until a pattern of multiple crack divisions is obtained. The instability that occurs in crack bifurcation is undoubtedly associated with the empirical fact that a high speed crack tends to change its path abruptly when encountering an obstacle in the material. The excess energy in the vicinity where the original crack turned initiates a new crack. This event occurs so quickly that the crack appears to have been split in two or bifurcated.

Many attempts have been made to explain the crack bifurcation phenomenon. As mentioned earlier, Yoffé [7] assumed the prospective sites of crack branching to coincide with the maxima of the local circumferential stresses ahead of the moving crack. She predicted the half branch angle, θ_0, to be 26° for $\nu = 0.25$. This deviated significantly from the commonly observed value of $\theta_0 \simeq 15°$ in glass plates [20]. Conceptual difficulties also arise in applying the maximum stress criterion. Referring to equation (4.6), the ratio of the normal stresses, σ_1/σ_2, on an element ahead of the crack when calculated is always greater than unity. This means that prior to branching the crack propagated in a direction parallel to the maximum stress rather than normal to it, a condition that contradicts the original assumption. To alleviate this difficulty, Kerkhof [21] and Kalthoff [20] suggested including the nonalignment of load with crack through the angle β. For small values of

β, their static stress analysis and application of the maximum stress criterion gave a half branch angle of $\theta_0 \simeq -2(k_2/k_1)$, a result which depended on the selection of k_1 and k_2, and ignored dynamic effects that can be significant at high velocities of crack propagation. Being a static analysis, no prediction could be made on the terminal velocity of crack branching. Andersson [22] and Achenbach [23] analyzed the running crack in its branched state. The former assumes the crack to be stationary and the latter assumes the crack motion to be maintained by anti-plane shear. The half branch angle θ_0 reported in [22] is 39.2° which again deviated appreciably from the result in [20].

The model of a finite crack spreading at both ends at constant velocity will be used in conjunction with the strain energy density criterion to predict crack bifurcation. For this case, the function $F_1(s_1, s_2)$ in equation (4.21) for S is given by

$$F_1(s_1,s_2) = s_1\{[(1+s_2^2)^2 - 4s_1^2s_2^2]K_1(s_1) - 4s_1^2(1-s_2^2)K(s_2) - \\ - [4s_1^2 + (1+s_2^2)^2]E(s_1) + 8s_1^2E(s_2)\}^{-1} \tag{4.31}$$

in which K and E are complete elliptic integrals of the first and second kind, respectively. Application of Hypothesis (1) to equation (4.21) yields for each crack velocity c an angle θ_0 that makes S a minimum. The value S_{min} can then be computed from equation (4.21) as a function of c and the numerical results for $\nu = 0.33$ are shown in Figure 4.18 in which $(S_{\text{min}})_0$ represents the static value of $(1-2v)k_1/4\mu$ given in equation (4.1) with $k_1 = \sigma_{\text{cr}}\sqrt{a}$. The largest S_{min} occurs at $c = 0$ and then decreases smoothly as the crack velocity is increased. At low crack velocities, S_{min} coincided with the crack axis, $\theta_0 = 0$. As the velocity is increased, there appears a point at which two separate minima of S occur symmetrically about the crack axis. This velocity depends* sensitively on the Poisson's ratio as indicated in Table 4.1. Note that as ν is varied from 0.21 to 0.24, the half branch angle θ_0 changes from ±18.84° to ±15.52°. The case for $\nu = 0.25$ is summarized in Figure 4.19 and agrees very well with the experimental results of Kalthoff [20].

Further insight into the crack bifurcation phenomenon may be gained by computing S_{v} and S_{d} in equations (4.22) and (4.23). The results, when normalized with reference to their static counterparts,

$$(S_{\text{v}})_0 = \frac{(1-2\nu)(1+\nu)k_1^2}{6\mu} \tag{4.32a}$$

$$(S_{\text{d}})_0 = \frac{(1-2\nu)^2k_1^2}{6\mu} \tag{4.32b}$$

* It is also highly influenced by the crack tip radius of curvature which is assumed to be zero in the dynamic analysis. Hence, the theoretical result will tend to overestimate the most critical velocity.

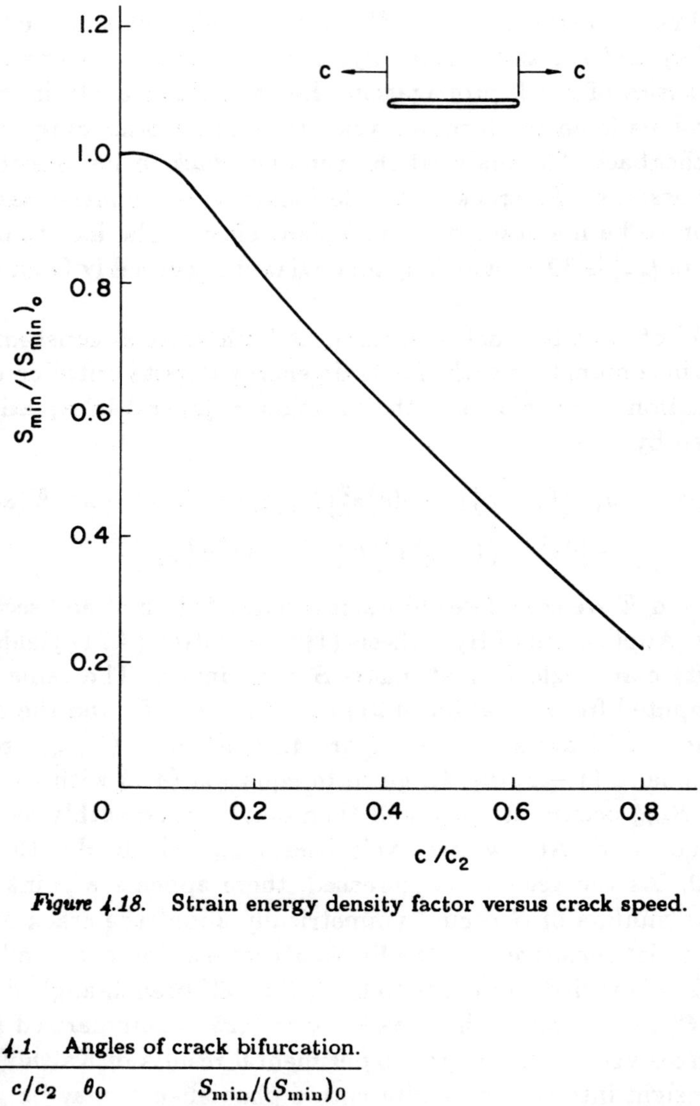

Figure 4.18. **Strain energy density factor versus crack speed.**

Table 4.1. **Angles of crack bifurcation.**

ν	c/c_2	θ_0	$S_{min}/(S_{min})_0$
0.21	0.45	0°	0.56209
	0.46	±18.84°	0.55199
0.22	0.46	0°	0.55196
	0.47	±17.35°	0.54204
0.23	0.47	0°	0.54185
	0.48	±16.27°	0.53210
0.24	0.48	0°	0.53177
	0.49	±15.52°	0.52217

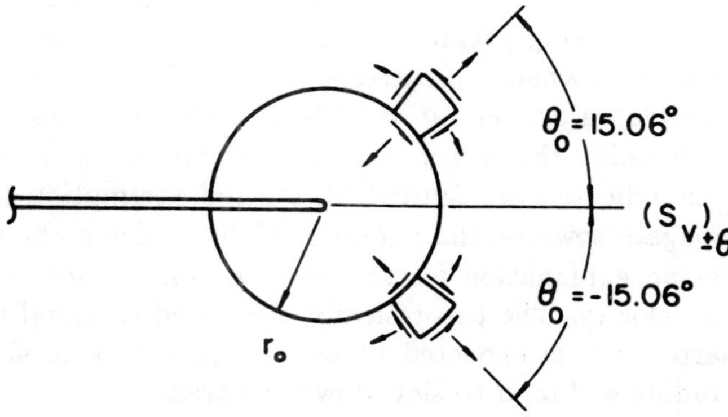

Figure 4.19. **Crack branching at $c/c_2 \simeq 0.5$ and $\nu = 0.25$.**

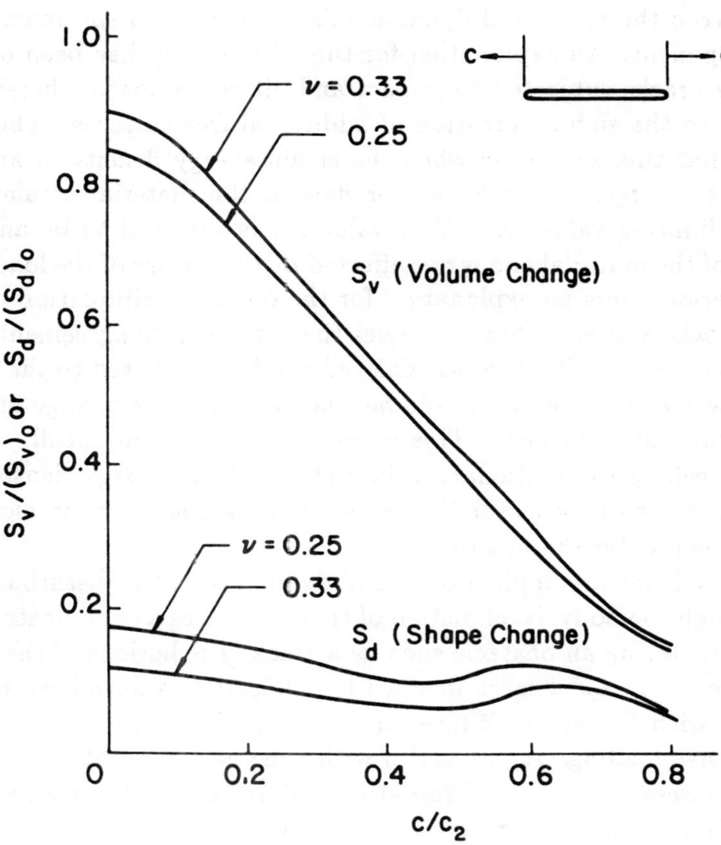

Figure 4.20. **Normalized dilatational and distortional energy density versus normalized crack velocity.**

are plotted against c/c_2 for $\nu = 0.25$ and 0.33 as shown in Figure 4.20. Both $S_v/(S_v)_0$ and $S_d/(S_d)_0$ tend to drop in magnitude as the crack picks up speed with S_v always being greater than S_d. The curves experience a bump in the neighborhood where θ_0, corresponding to S_{min}, switches from a single value to double value. For $\nu = 0.25$, this change takes place at $c/c_2 \simeq 0.5$ at which point the larger S_v values occur at $\theta_0 = \pm 15.06°$ instead of $\theta_0 = 0°$ as indicated in Figure 4.19. On the assumption that the crack tends to propagate towards the region of higher volume change, the S-criterion provides an explanation for the observed bifurcation of cracks moving with a high velocity. The theoretically calculated terminal velocity for a perfectly sharp crack is expected to be on the high side since the nonzero crack tip radius will tend to slow down the crack.

4.6. Summary and conclusions

One of the aims of dynamic fracture analysis is to account for the discrepancy between the static and dynamic failure loads for a specimen or structural component. An explanation for this discrepancy has been offered for stationary cracks subjected to impact and vibratory loads whereby failure is defined as the sudden creation of additional free surfaces. The analysis assumes that this will occur when the strain energy density in an element ahead of the largest imperfection or flaw in the material attains a characteristic limiting value, S_{cr}. This value S_{cr} is assumed to be an intrinsic property of the material and is not affected by the nature of the loading. The same criterion offers an explanation for the observed bifurcation of rapidly moving cracks and gives branch angles that are in close agreement with experiments on glass. The success achieved can be attributed to the ability of the S-criterion to account for volume change and shape change of the critical element simultaneously. This is accomplished automatically and most easily by seeking the stationary values of the strain energy density factor. In general, macrocrack instability tends to follow the direction along which dilatation is greater than distortion.

When the loads are applied to a solid dynamically, the disturbance, traveling through the body, is a function of time and takes a complicated pattern when encountering an obstacle such as a crack. Predictions of the dynamic failure stress, σ_m, for a crack inclined to the load axis have been made and compared with the static failure stress, σ_s, of the same physical system. For impulsive loading, σ_m/σ_s varies with time and orientation of crack. A very high stress level σ_m is required for failure to occur immediately after impact. Hence, there is usually a delay between the time when the load is applied and that when crack begins to propagate. The theoretical solution suggests a minimum stress amplitude, σ_m, for each crack orientation and a corresponding time delay. The actual point of failure depends on the magnitude of the applied load and the material properties. Similar findings are obtained for cracks encountering plane harmonic waves that arise in steady state vibration. In this case, σ_m/σ_s tends to vary with frequency and to approach unity at very low frequencies. Minimum values of σ_m are again found for certain combinations of frequency and crack orientation.

No attempts have been made to exhaust all the examples in dynamic fracture. There remains a variety of problems that can be analyzed by direct application of the strain energy density criterion as outlined for the few key problems mentioned earlier. Among the ones that can be analyzed in the future are:

1. Interaction of moving cracks with harmonic waves [25, 26].
2. Impact and vibration of solids containing penny-shaped cracks [27, 28].
3. Cracked plates subjected to dynamic extensional [29] and bending [30] loads.
4. Dynamically loaded composite materials [31].

A few remarks with regard to the limiting velocity at which crack bifurcates are in order. Experiments on glass [24] have shown that this velocity can be strongly affected by the material composition, a factor that has been left out of the stress analysis. Schardin [24] measured the limiting crack velocity of 29 different glasses and found them to vary between 0.347 to 0.614 times the shear wave speed, c_2, defined in equation (4.8). He also investigated the chemical composition of these glasses by controlling the content of the network-modifier which in turn alters the flow characteristics of the glass. Since continuum mechanics theory cannot easily incorporate the effect of chemical composition into a dynamic fracture analysis, it appears that, perhaps, this influence can be accounted for through the crack tip radius of curvature parameter in the elastodynamic crack problem. The way in which the network-modifier affects the crack tip radius could be investigated experimentally. There is no doubt that the sharpness or bluntness of a crack can either increase or decrease the limiting crack velocity.

While much of the current research and development effort is devoted to expanding fracture mechanics into dynamically loaded specimens and structures, there has been little progress made in understanding the various aspects of dynamic fracture. Whether there should be a separate parameter for characterizing the dynamic fracture toughness of a given material or not is a subject that requires some serious considerations. The basic premise of any fracture analysis that does not account for changes in material structure is to assume a constant fracture toughness regardless of loading. This is valid as long as the same physical phenomenon of instability is observed and the appropriate stress analysis of the system is made.

References

[1] Berry, J. P., Some kinetic considerations of the Griffith criterion for fracture (I& II), *Journal of the Mechanics and Physics of Solids* **8**, pp. 194–216 (1960).

[2] Sih, G. C. and Macdonald, B., Effect of material nonlinearity on crack propagation, *International Journal of Engineering Science* **12**, pp. 61–77 (1974).

[3] Kanninen, M. F., An augmented double-cantilever beam model for studying crack propagation and arrest, *International Journal of Fracture* **9**, pp. 83–92 (1973).

[4] Griffith, A. A., The phenomena of rupture and flow in solids, *Philosophical transactions, Royal Society of London, Series 221*, pp. 163–198 (1921).

[5] Sih, G. C., Some basic problems in fracture mechanics and new concepts, *Journal of Engineering Fracture Mechanics* **5**, pp. 365–377 (1973).

[6] Erdogan, F. and Sih, G. C., On the crack extension of plates under plane loading and transverse shear, *Journal of Basic Engineering* **85**, pp. 519–527 (1963).

[7] Yoffé, E. H., The moving Griffith crack, *Philosophical Magazine* **42**, pp. 739–750 (1951).

[8] Smith, H. L., Ferguson, W. J., McLean, E., George, T. W. and Kies, J. A., Subcritical crack growth in aluminum 2S and pure zinc foils, *Naval Research Laboratory memorandum Report No. 2113* (1973).

[9] Sih, G. C., A special theory of crack propagation: methods of analysis and solutions of crack problems, *Mechanics of Fracture I*, edited by G. C. Sih, Noordhoff International Publishing, Leyden, pp. 21–45 (1973).

[10] Sih, G. C., A three-dimensional strain energy density theory of crack propagation: three-dimensional crack problems, *Mechanics of Fracture II*, edited by G. C. Sih, Noordhoff International Publishing, Leyden, pp. 15–53 (1975).

[11] Kipp, M. E. and Sih, G. C., The strain energy density failure criterion applied to notched elastic solids, *International Journal of Solids and Structures* **2**, pp. 153–173 (1975).

[12] Sih, G. C., Fracture toughness concept: properties related to fracture toughness, *American Society for Testing and Materials*, ASTM STP 605, pp. 3–15 (1976).

[13] Sih, G. C., A review of the three-dimensional stress problem for a cracked plate, *International Journal of Fracture Mechanics* **7**, 39–61 (1971).

[14] Sih, G. C., Dynamic aspects of crack propagation: inelastic behavior of solids, edited by M. F. Kanninen, W. F. Adler, A. R. Rosenfield and R. I. Jaffee, McGraw-Hill, pp. 607–639 (1970).

[15] Chen, E. P. and Sih, G. C., Transient response of cracks to impact loads: elastodynamic crack problems, *Mechanics of Fracture IV*, edited by G. C. Sih, Noordhoff International Publishing, Leyden, pp. 1–58 (1977).

[16] Loeber, J. F. and Sih, G. C., Diffraction of antiplane shear waves by a finite crack, *Journal of the Acoustical Society of America* **44**, pp. 90–98 (1968).

[17] Sih, G. C. and Loeber, J. F., Wave propagation in an elastic solid with a line of discontinuity or finite crack, *Quarterly of Applied Mathematics* **27**, pp. 193–213 (1969).

[18] Sih, G. C. and Loeber, J. F., A class of diffraction problems involving geometrically induced singularities, *Journal of Mathematics and Mechanics* **19**, pp. 327–350 (1969).

[19] Hartranft, R. J. and Sih, G. C., Applications of the strain energy density fracture criterion to dynamic crack problems, *Proceedings of Prospects of Fracture Mechanics*, edited by G. C. Sih, H. C. van Elst and D. Broek, Noordhoff International Publishing, Leyden, pp. 281–298 (1974).

[20] Kalthoff, J. F., On the propagation direction of bifurcated cracks, *Dynamic Crack Propagation*, edited by G. C. Sih, Noordhoff International Publishing, Leyden, pp. 449–458 (1973).

[21] Kerkhof, F., Wave fractographic investigations of brittle fracture dynamics, *Dynamic Crack Propagation*, edited by G. C. Sih, Noordhoff International Publishing, Leyden, pp. 3–35 (1973).

[22] Andersson, H., Stress intensity factors at the tips of a star-shaped contour in an

infinite tensile sheet, *Journal of Mechanics and Physics of Solids* **18**, pp. 405–417 (1969).

[23] Achenbach, J. D., Elastodynamic stress intensity factors for a bifurcating crack, *Proceedings of Prospects of Fracture Mechanics*, edited by G. C. Sih, H. C. van Elst and D. Broek, Noordhoff International Publishing, Leyden, pp. 319–336 (1974).

[24] Schardin, H., Velocity effects in fracture, *Fracture*, edited by B. L. Averbach, D. K. Felbeck, G. T. Hahn, and D. A. Thomas, John Wiley, pp. 297–330 (1959).

[25] Sih, G. C. and Loeber, J. E., Interaction of horizontal shear waves with a running crack, *Journal of Applied Mechanics* **37**, pp. 324–330 (1970).

[26] Chen, E. P. and Sih, G. C., Scattering of plane waves by a propagating crack, *Journal of Applied Mechanics* **42**, pp. 705–711 (1975).

[27] Sih, G. C. and Loeber, J. F., Normal compression and radial shear waves scattering at a penny-shaped crack in an elastic solid, *Journal of the Acoustical Society of America* **46**, pp. 711–721 (1969).

[28] Embley, G. T. and Sih, G. C., Response of a penny-shaped crack to impact waves, *Proceedings of the 12th Midwestern Mechanics Conference* **6**, pp. 473–487 (1971).

[29] Sih, G. C. and Chen, E. P., Dynamic analysis of cracked plates in bending and extension: plates and shells, *Mechanics of Fracture III*, edited by G. C. Sih, Noordhoff International Publishing, Leyden, pp. 231–272 (1976).

[30] Embley, G. T. and Sih, G. C., Sudden appearance of a crack in a bent plate, *International Journal of Solids and Structures* **9**, pp. 1349–1359 (1973).

[31] Loeber, J. F. and Sih, G. C., Torsional wave scattering about a penny-shaped crack lying on a bi-material interface, *Dynamic Crack Propagation*, edited by G. C. Sih, Noordhoff International Publishing, Leyden, pp. 513–528 (1973).

5

Strain energy density and surface layer energy for blunt cracks or notches

5.1. Background information

Although much effort has been spent in the past to develop failure theories, there is still considerable uncertainty involved in determining the strength of a material. A central problem facing the designer today is optimizing the use of the multivarious materials now known to him and available for application under various loading and environmental conditions. This calls for an ***a priori*** knowledge of the kind of failure expected before design limits can be established. To acquire this knowledge, it is essential to know, at a minimum, the nature of loading, the prevailing environments, the material behavior, and whether there are any damaging cracks, notches or other mechanical imperfections present in the solid.

In recent years, a bewildering number of theories have been proposed on the grounds that they 'fit the data' rather than that they rely on self-consistency, and hence it has not been difficult to find other sets of data for which the theories failed to apply. This has caused a great deal of confusion in the field. On the other hand, very little has been done to coordinate these theories so that the failure of solids can be explained on a more unified basis. Today, separate theories are still used for the failure of notches and cracks. A case in point is that of the failure condition of solid with notches which is usually based on the local maximum normal stress criterion. Griffith [1] employed such a stress criterion to predict the failure of a solid with an internal elliptical notch instead of the now well-known concept of energy release rate as he originally proposed [2] for a perfectly sharp crack, Figure 5.1(a). This is because of the conceptual difficulties* involved in extending blunt cracks or notches in a self-similar manner. Figure

* The application of the energy release rate concept to the notch geometry in Figure 5.1(b) would require the general solution of a small crack Δa extending from the notch boundary. Such an approach is inappropriate, for it delves into a complicated mathematical analysis whose accuracy will always be doubtful in the limit as $\Delta a \rightarrow 0$.

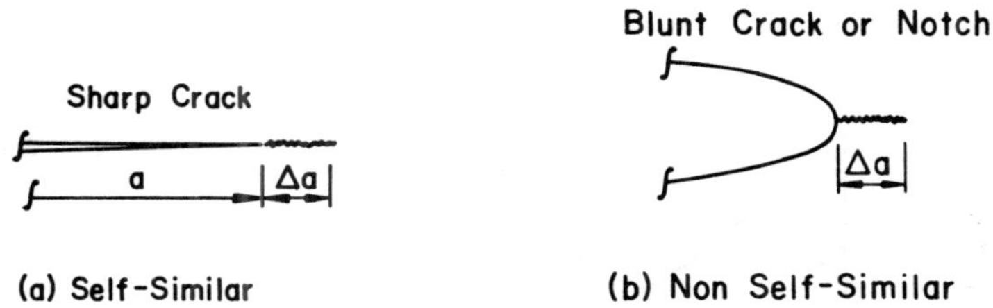

Figure 5.1. Crack and notch failure.

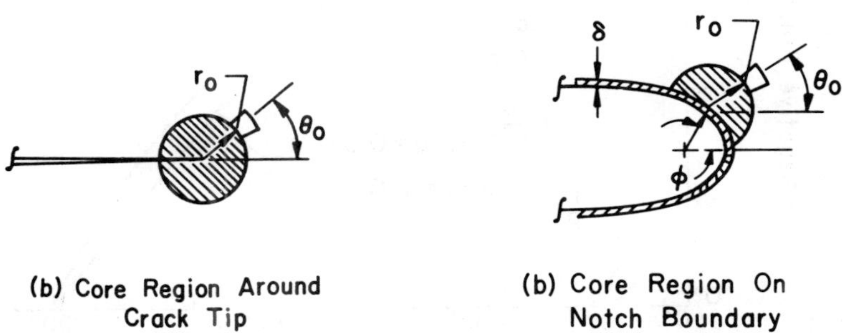

Figure 5.2. Core region near crack and notch front.

5.1(b) shows that the material ahead of the notch front fails by cracking and hence the notch geometry before and after material separation is not preserved. Therefore, Δa can be considered as crack extension only for the case of a sharp crack (Figure 5.1(a)) where the crack tip radius of curvature has a negligible influence.

A coherent theory of fracture that can consistently account for the fracture behavior of both sharp cracks and blunt notches has been advanced only recently [3, 4]. The basis of the theory is the hypothesis that fracture occurs when a small element of material near the notch or crack tip has absorbed a critical amount of energy, and then releases it to cause material separation. This material element is always kept at a finite distance, say r_0, from the crack tip, Figure 5.2(a), or notch front, Figure 5.2(b). The concept of a local core region in Figure 5.2 stems from a concern that in the immediate vicinity of the crack or notch tip, inhomogeneity of the material due to grain boundaries, microcracks, and dislocations preclude an accurate analytical solution, whereas an analysis presuming a valid continuum mechanics solution external to the core region can provide a sufficiently accurate measure of the failure behavior of the material. The continuum description requires the size of the core region to be limited to the smallest macroscopic element. For the notch geometry, a surface layer failure criterion [4] is dealt with as a preliminary requirement to locate the position (ϕ) where fracture may be

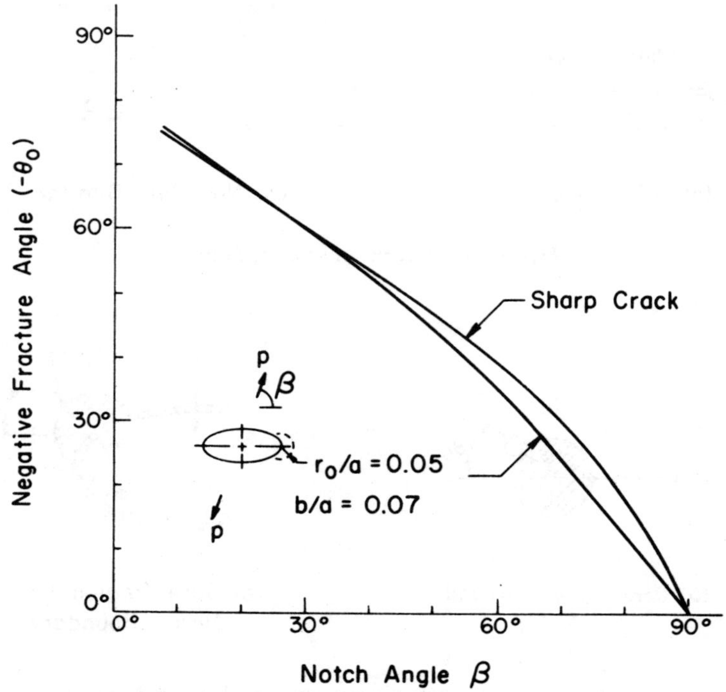

Figure 5.3. **Fracture angle versus notch angle for elliptical notch in tension.**

expected to initiate from the notch surface. Irregularities at the microscopic scale are bounded within this layer of thickness δ as indicated in Figure 5.2(b). Once the angle ϕ in Figure 5.2(b) is established, the knowledge of the energy stored in an element outside the core region is used as a means for establishing the location of failure at some point in the bulk of the solid. The angle θ_0 determines the initial direction of the surface of material separation. The calculation involves the determination of the stationary values of the strain energy density function, dW/dV.

The application of the strain energy density theory, involving an element outside the core region, makes possible the joining of the fracture analysis of cracks and notches. From the analysis point of view, the fracture of a sharp crack can be regarded simply as the limiting case of the elliptical notch, and needs, for the most part, no special consideration. Figures 5.3 and 5.4 provide some information as to how the initial direction of crack propagation would be affected by the crack tip radius of curvature, ρ. For an ellipse $\rho = a(b/a)^2$ where a and b are respectively the semi-major and semi-minor axes of the ellipse. Suppose that an elliptical notch in an infinite medium is stretched by a uniform stress p in a direction that makes an angle β with reference to the major axis of the ellipse. Figure 5.3 gives the fracture angle θ_0 (θ_0 is negative because it is in the clockwise direction) for tensile loading

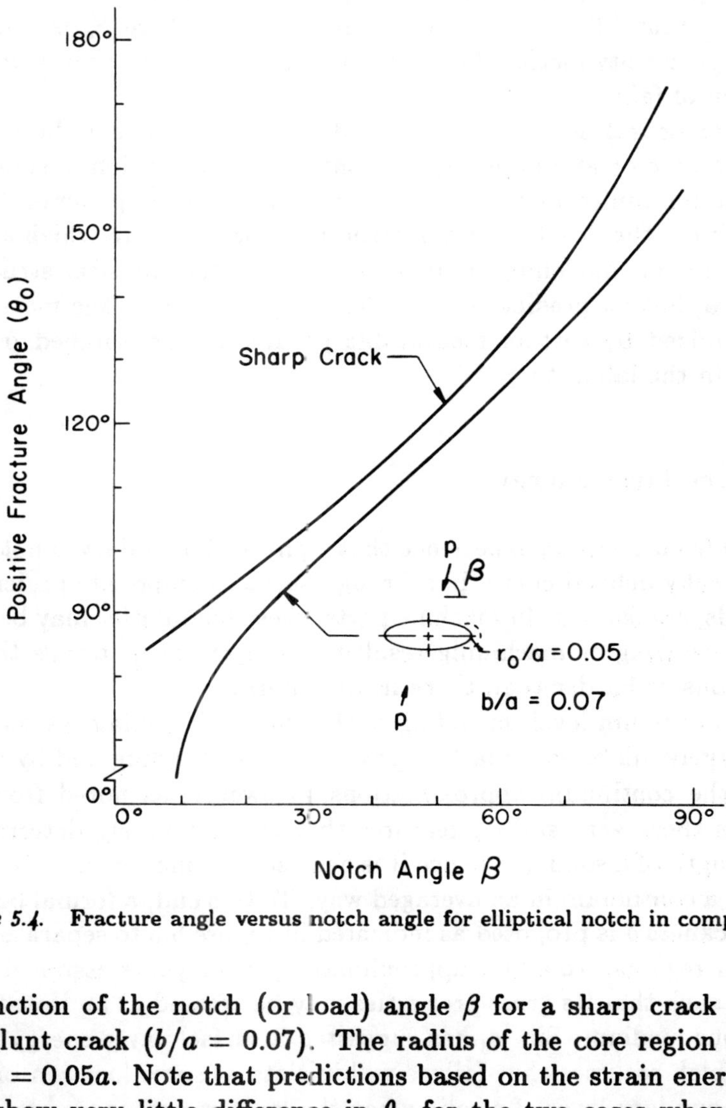

Figure 5.4. **Fracture angle versus notch angle for elliptical notch in compression.**

as a function of the notch (or load) angle β for a sharp crack ($b/a = 0$) and a blunt crack ($b/a = 0.07$). The radius of the core region is selected to be $r_0 = 0.05a$. Note that predictions based on the strain energy density theory show very little difference in θ_0 for the two cases when β is kept below 40°. Deviations are seen for β between 50° and 80°. A different set of results is obtained when the applied stress is compressive as shown in Figure 5.4. First of all, the direction of crack initiation tends toward the applied compressive stress and the difference between θ_0 for the sharp crack and notch is approximately the same for $20° \leq \beta \leq 80°$. Available experimental data [5, 6] on plexiglass plates with cracks show striking agreement with the results in Figure 5.3. This will be discussed in more detail subsequently.

The use of the critical value of strain energy density function, $(dW/dV)_c$ as a material constant has also been demonstrated by others [7, 8]. In particular, Gillemot [8] and his co-workers have performed many experiments

to measure $(\mathrm{d}W/\mathrm{d}V)_c$ for numerous engineering materials. Sih has pointed out in [4, 9] that $(\mathrm{d}W/\mathrm{d}V)_c$ is equivalent to S_c/r where S_c is the critical strain energy density factor while r can be used as a radius vector to identify the location of failure.

The fundamental assumption at hand is that although failure initiates at a microscopic or atomic level, the continuum description is sufficient to establish design limits for the instability of structural components. This has been the aim of the development of structural mechanics in which attention is focused not on modeling or understanding of the physical structure of the material, but on prediction. The desired properties of the material will be characterized by certain measurable parameters determined from test specimens in the laboratory.

5.2. Surface layer energy

It is known from common experience that a physical boundary is not a sharp mathematically defined curve, but a rough surface composed of microscopic ridges, voids, cracks, etc. In machine parts, these boundaries may be formed by rolling, drawing or machining resulting in layers of materials that may be less porous or harder than those in the interior.

On the continuum level, modeling of the surface irregularities mentioned above is largely inhibited by lack of physical understanding and by the very nature of the continuum approximations. However, as noted from experience, it is these very surface features that can ultimately determine the failure strength of a solid and hence it is necessary to incorporate the surface effects into a continuum in an averaged way. To this end, a formal boundary layer of thickness δ is proposed as indicated in Figure 5.5 to separate surface and interior regions. To a first approximation, this layer is assumed to be a continuum such that its gross properties may be treated as an isotropic and homogeneous medium. Sih [4] has suggested a surface layer energy* failure criterion which assumes that failure will initiate at some location or region along the boundary layer of a solid when the local energy caused by loading will exceed some experimentally determined material constant.

Mechanics of the surface layer. The energy associated with the surface layer can be derived from the mechanics of the thin layer shown in Figure 5.6. A small element, $\rho\Delta s$, of the surface layer for a free boundary is isolated and shown in Figure 5.7. If δ is of the order of magnitude of a continuum element, then σ_r and $\tau_{r\theta}$ are zero at both $r = \rho$ and $r = \rho + \delta$. Thus the

* This quantity is conceptually different from the surface tension for a liquid, nor is it related to the specific surface energy used in the Griffith criterion [2].

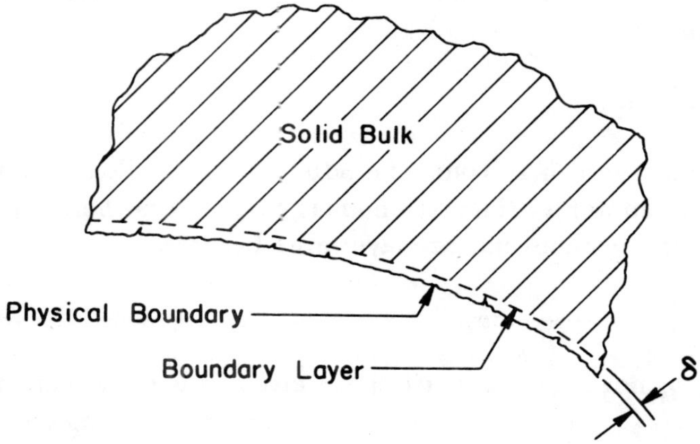

Figure 5.5. **Free surface boundary layer.**

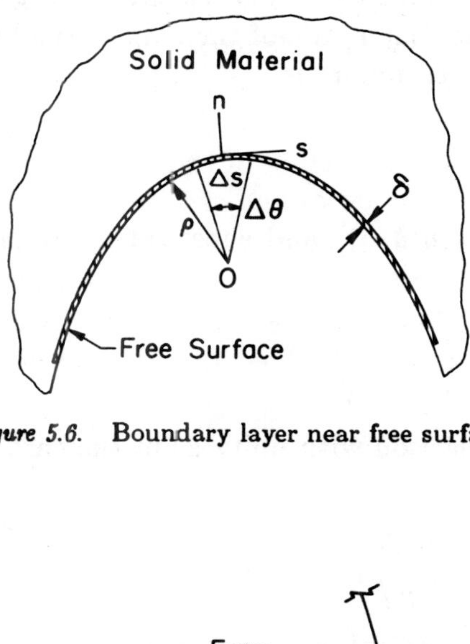

Figure 5.6. **Boundary layer near free surface.**

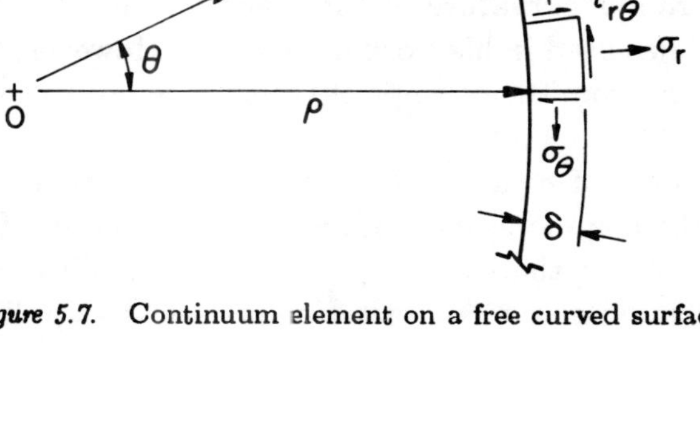

Figure 5.7. Continuum element on a free curved surface.

element is under uniaxial tension, $\sigma_\theta = \sigma_t$, in the tangential direction. The corresponding tangential strain is

$$\epsilon_t = \frac{1-\nu^2}{E}\sigma_t \tag{5.1}$$

for plane strain where E is Young's modulus and ν is Poisson's ratio for an elastic isotropic material. The strain energy density is therefore $\sigma_t\epsilon_t/2$ and the strain energy per unit surface layer area becomes

$$\gamma_e = \tfrac{1}{2}\sigma_t\epsilon_t\delta = \frac{1-\nu^2}{2E}\delta\sigma_t^2. \tag{5.2}$$

Since neither γ_e nor δ has explicitly been given, it is convenient to form the quantity $2E\gamma_e/\delta(1-\nu^2)$ and treat it as a material parameter. A notched specimen may be loaded to failure and from equation (5.1) $2E\gamma_e/\delta(1-\nu^2)$ computed; then for any notch geometry of this material, the failure load may be obtained. The location on the notch boundary at which failure occurs is assumed to coincide with the maximum value of γ_e.

For the simple case of an elliptical notch subjected to applied stress p in a direction normal to the major axis of the ellipse, the local tangential stress along the line of expected failure is

$$\sigma_t = p\left(1+\frac{2a}{b}\right)^2. \tag{5.3}$$

Using the relation $\rho = a(b/a)^2$ and substituting σ_t in equation (5.3) into (5.2) gives

$$\gamma_e = \frac{\delta(1-\nu^2)p^2}{2E}\left(1+2\sqrt{\frac{a}{\rho}}\right)^2. \tag{5.4}$$

If a/ρ is large in comparison with unity as in the case of a crack, equation (5.4) further reduces to

$$\gamma_e = \frac{2(1-\nu^2)p^2a}{E}\left(\frac{\delta}{\rho}\right), \quad \rho \ll a. \tag{5.5}$$

Note that γ_e is not the Griffith surface energy term and is derived from an entirely different physical basis. It emphasizes a characteristic length δ with reference to the radius curvature of the crack and is somewhat similar to what Neuber [10] has used in his work on notches. However, his argument is completely different from the concept of the surface layer energy introduced here.

Although the form of equation (5.2) may suggest that the γ_e-criterion is equivalent to the commonly recognized failure criterion of the maximum surface tangential stress, there is a major conceptual difference that must be emphasized. The maximum stress criterion is basically applied in a layer of zero thickness, at surface points, whereas the surface layer energy has been applied over an element of nonzero thickness, and resulted in an additional parameter δ, in equation (5.2). As indicated earlier, the presence of δ provides a means of absorbing the local surface inhomogeneities into the energy theory; that is, different values of δ may be assigned to quantify different surface conditions, although the bulk material properties remain unaffected. The use of the surface tangential stress alone is unable to include such local phenomena. Brittle materials are especially sensitive to surface characteristics, and unless the continuum model can in some way include parameters reflecting such characteristics, the prediction of failure is likely to be quite inaccurate.

An experimental knowledge of the surface characteristic can indeed improve the use of the maximum surface layer energy by assigning material parameters to the thin boundary layer different from those of the bulk material. This step removes some of the arbitrariness associated with assigning $2E\gamma_e/\delta(1-\nu^2)$ as a new parameter, when it appears at first glance merely to bear the image of the maximum tangential stress criterion.

Elliptical notch under tension. A numerical example illustrating the behavior of the surface layer energy will now be given for the geometry of an elliptical notch in a large plate loaded at an angle as shown in Figure 5.8(a). Around the notch surface, which is assumed to be rough, is depicted a boundary layer of uniform thickness, δ. At each point along the layer, represented by the equation of the ellipse in parametric form

$$x = a\cos\eta, \qquad y = b\sin\eta \tag{5.6}$$

where η is the eccentric angle for the ellipse in Figure 5.8(b), the magnitude of the surface layer energy, γ_e, may be calculated from equation (5.2) once the local tangential stress is known. Without going into details, σ_t for the problem in Figure 5.8(a) can be obtained from Muskhelishvili [11] and hence

$$\gamma_e = \frac{\delta(1-\nu^2)p^2}{2E} \times \left[\frac{(a+b)^2(\sin\eta\cos\beta - \cos\eta\sin\beta)^2 - a^2\sin^2\beta - b^2\cos^2\beta}{a^2\sin^2\eta + b^2\cos^2\eta}\right]^2 . \tag{5.7}$$

For $\beta = 90°$, the largest value of γ_e is found at the ends of the major axis $\eta = 0$ and equation (5.7) reduces to the form in equation (5.4).

In its present form, equation (5.7) obscures the difference between tensile and compressive loading. Along the boundary of an elliptical cavity, σ_t may be tensile or compressive, as indicated in Figure 5.9. However, γ_e remains positive at all points along the boundary, and hence, requires that the nature of the loading be dictated in order to select the region in which the surface

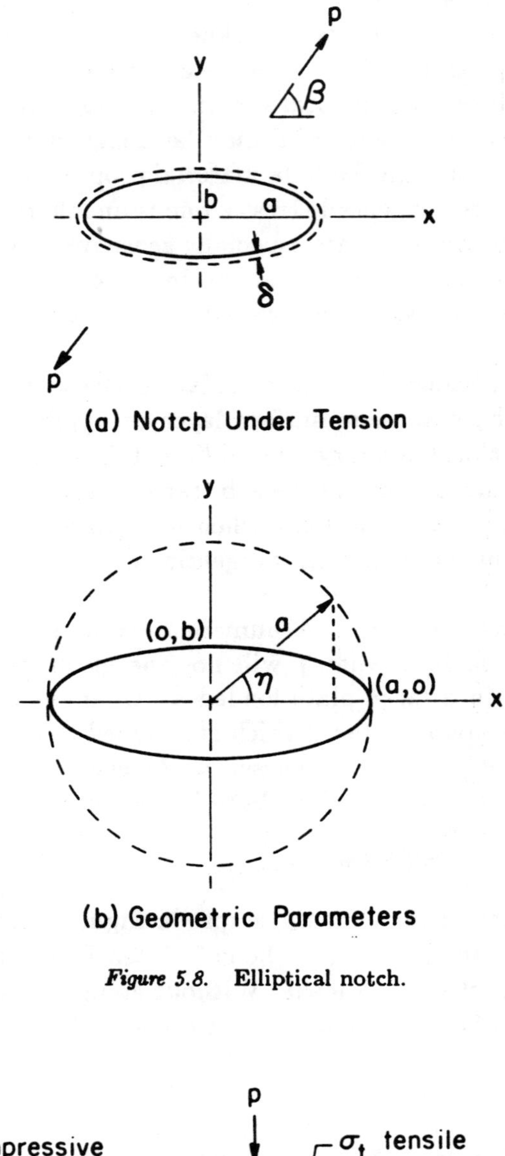

Figure 5.8. Elliptical notch.

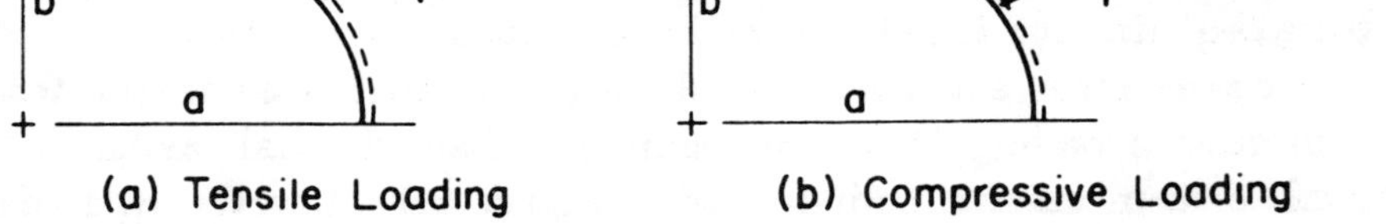

Figure 5.9. Tensile and compressive normal loading on an elliptical notch.

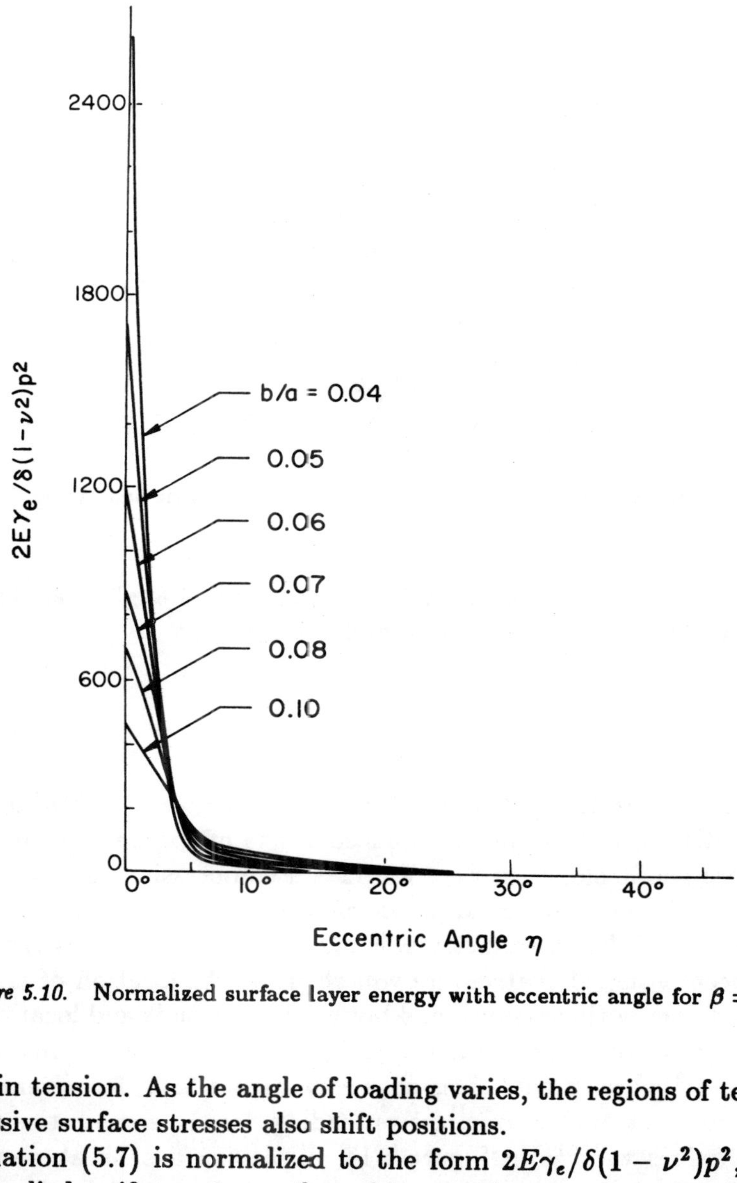

Figure 5.10. Normalized surface layer energy with eccentric angle for $\beta = 90°$.

layer is in tension. As the angle of loading varies, the regions of tensile and compressive surface stresses also shift positions.

If equation (5.7) is normalized to the form $2E\gamma_e/\delta(1-\nu^2)p^2$, where p is the applied uniform stress, then this quantity may be represented as a function of the loading angle β, semi-minor to semi-major axes ratio b/a, and the eccentric angle η.

Fixing $\beta = 90°$, and applying a uniform tensile load results in the surface layer energy varying as a function of the eccentric angle η as seen in Figure 5.10, for several values of b/a. It is observed that for small values of b/a, the surface layer energy displays a sharp peak near $\eta = 0$ (an angle of symmetry), and this peak rapidly disappears as b/a increases. This is in part due to the nature of the eccentric angle. The same values of the surface

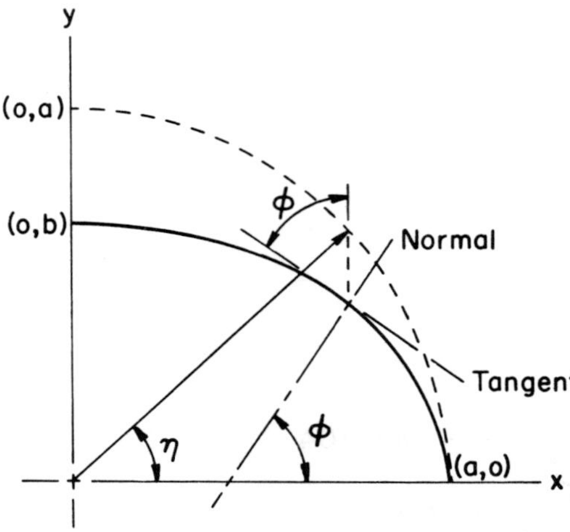

Figure 5.11. Relationship of normal and eccentric angles.

layer energy may be plotted against the local normal angle ϕ as defined in Figure 5.11. The relation between η and ϕ is given by

$$\tan\phi = \frac{a}{b}\tan\eta. \tag{5.8}$$

Referring to the results in Figure 5.12 in which $2E\gamma_e/\delta(1-\nu^2)p^2$ is plotted against ϕ, it is seen that the pronounced peaks of Figure 5.10 have given way to more gentle slopes in Figure 5.12. In compression, at $\beta = 90°$, the stresses around an ellipse are always such that the stress at $\eta = 90°$, equal to the negative of the applied stress, is unchanged for all ratios b/a.

It has been assumed that failure will occur at the location of maximum surface layer energy (in tension), and both the magnitude and location along the surface vary with angle of loading. In tension, the maximum values of $2E\gamma_e/\delta(1-\nu^2)p^2$ vary with β as shown in Figure 5.13, for several values of b/a. In contrast, under compressive loading, for the same ratios b/a, the maximum values vary as in Figure 5.14, displaying a peak at some angle β near 45°. The importance of these results may be more clearly observed if the curves in Figure 5.13 and 5.14 are replotted as $p\sqrt{\delta(1-\nu^2)/2E\gamma_e}$, so that the permissible loads appear on the ordinate. Here the quantity $\delta(1-\nu^2)/2E\gamma_e$ is assumed to be characteristic of the material. In Figure 5.15, under tension the minimum load to failure is at $\beta = 90°$, while under compression (Figure 5.16), the minimum load to failure occurs in the region $\beta \simeq 45°$. Since the notch size enters into the problem, there is no need to assume surface contact under compression as in the case of the crack configuration. This influence, however, has been exaggerated [12] in the past

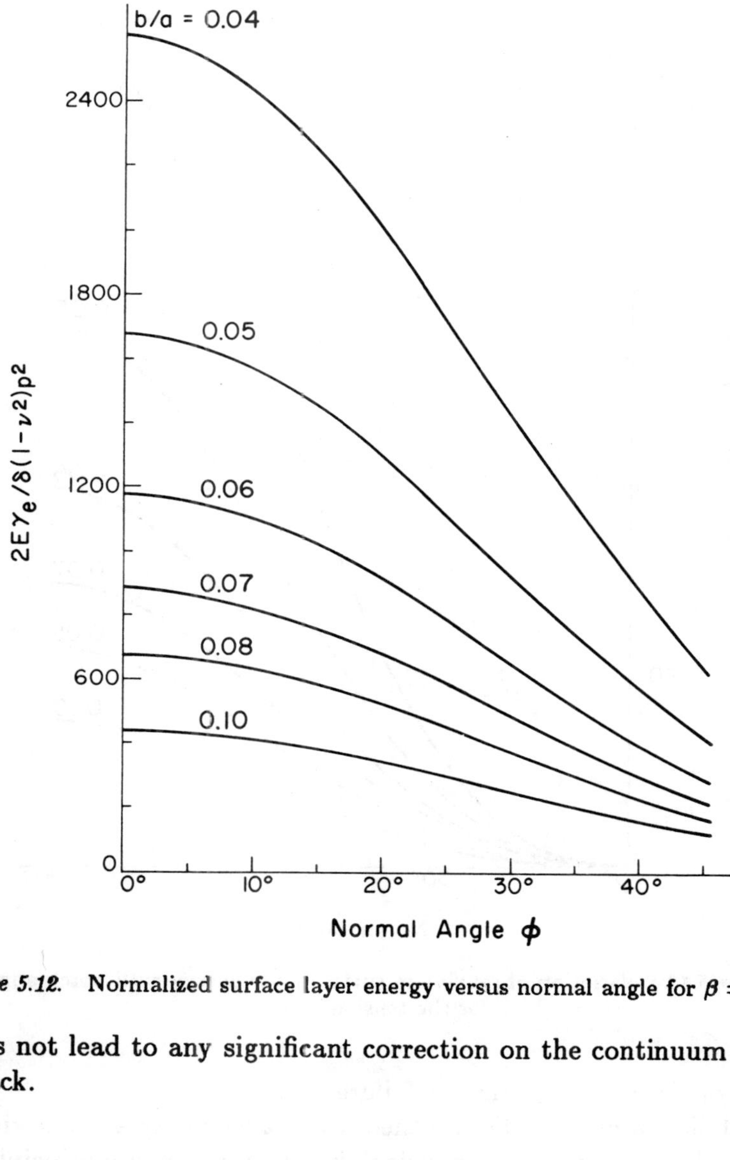

Figure 5.12. Normalized surface layer energy versus normal angle for $\beta = 90°$.

and does not lead to any significant correction on the continuum solution* for a crack.

5.3. Strain energy density theory

Having discussed the surface layer energy criterion that applies only to the surface of a notch, this section moves into the strain energy density criterion local to points interior to the surface. A knowledge of the location of failure at points on the notch surface and in the bulk of the solid sheds light on the

* Experimental data on cracks under compression will show considerable influence due to surface contact as the crack surfaces are not mathematically smooth and parallel but notched with a zig-zag pattern.

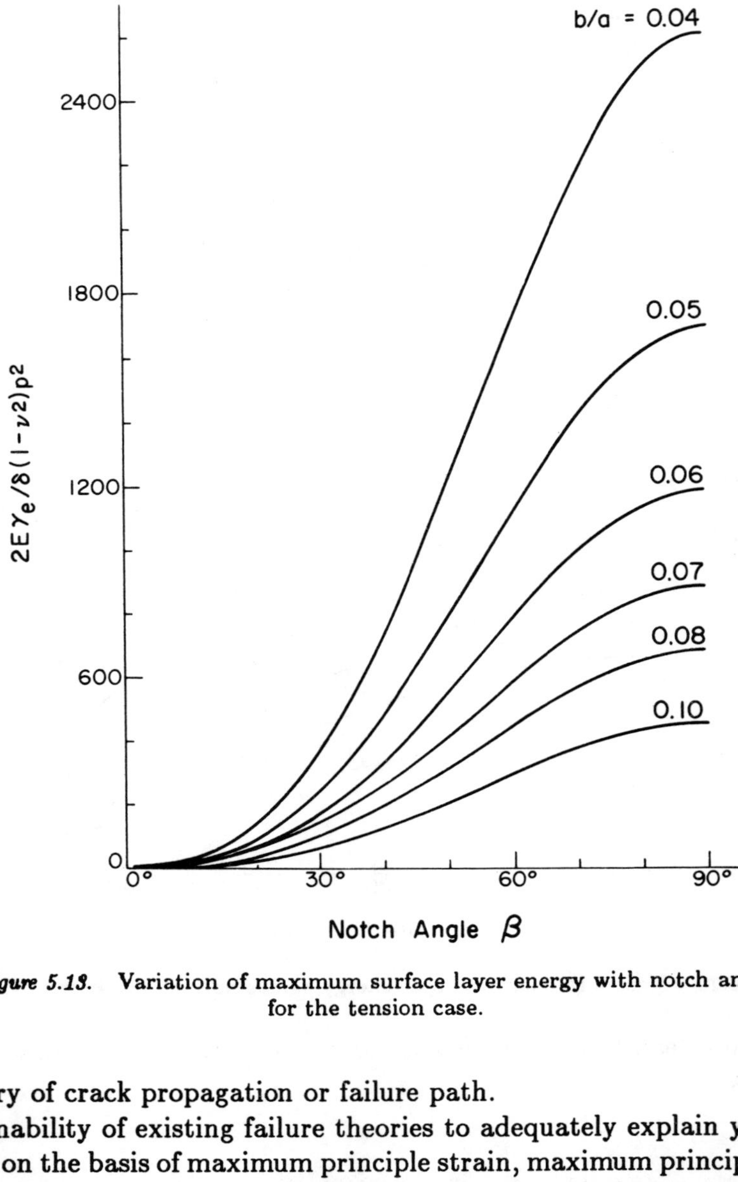

Figure 5.13. Variation of maximum surface layer energy with notch angle for the tension case.

trajectory of crack propagation or failure path.

The inability of existing failure theories to adequately explain yield phenomena on the basis of maximum principle strain, maximum principle stress, or maximum shear stress hypotheses prompted Haigh [13] to expound a theory based upon the elastic energy absorbing capacity of a material. This strain energy function was defended rationally on the basis of thermodynamic considerations, and for materials with rather well defined yield points, was substantiated by already existing data. The principal objective was to determine the "limiting strain energy" that could be absorbed per unit volume of material uniformly strained to its elastic limit, and to use this quantity as a material constant. It was also noted that the principles set forth probably applied only to materials exhibiting ductile behavior, relegating

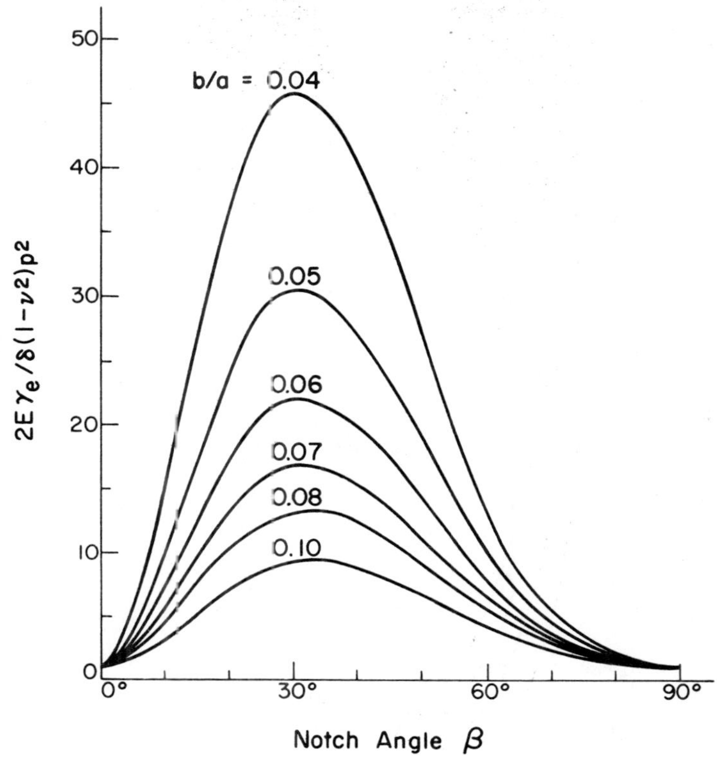

Figure 5.14. Variation of maximum surface layer energy with notch angle for the compression case.

brittle fracture phenomenon to be governed by other principles.

Recently, Sih [3, 4] has proposed a theory of fracture based on the field strength of the local strain energy density in an element of material ahead of the crack or notch. This theory can simultaneously account for yielding and fracture. The fundamental parameter in the new theory, the "strain energy density" factor S, is direction sensitive in that it predicts the direction of crack initiation and/or propagation and yielding. This is accomplished by calculating the stationary value of S or dW/dV, where $dW/dV = S/r$ with r being the radial distance measured from the crack front and dW/dV is the strain energy density function. The direction of S_{max} determines maximum distortion while S_{min} relates to dilatation. Figure 5.17 shows that distortion of material elements is associated with yielding which is to the sides of a crack under symmetric loading. Dilatation or volume change of material elements tends to be associated with the creation of free surface or fracture and occurs along the line of expected crack extension. As a fracture criterion, the critical value of S_{min}, S_c, is used as a material constant to evaluate the failure load of a material, whether at a crack tip, notch front, re-entrant

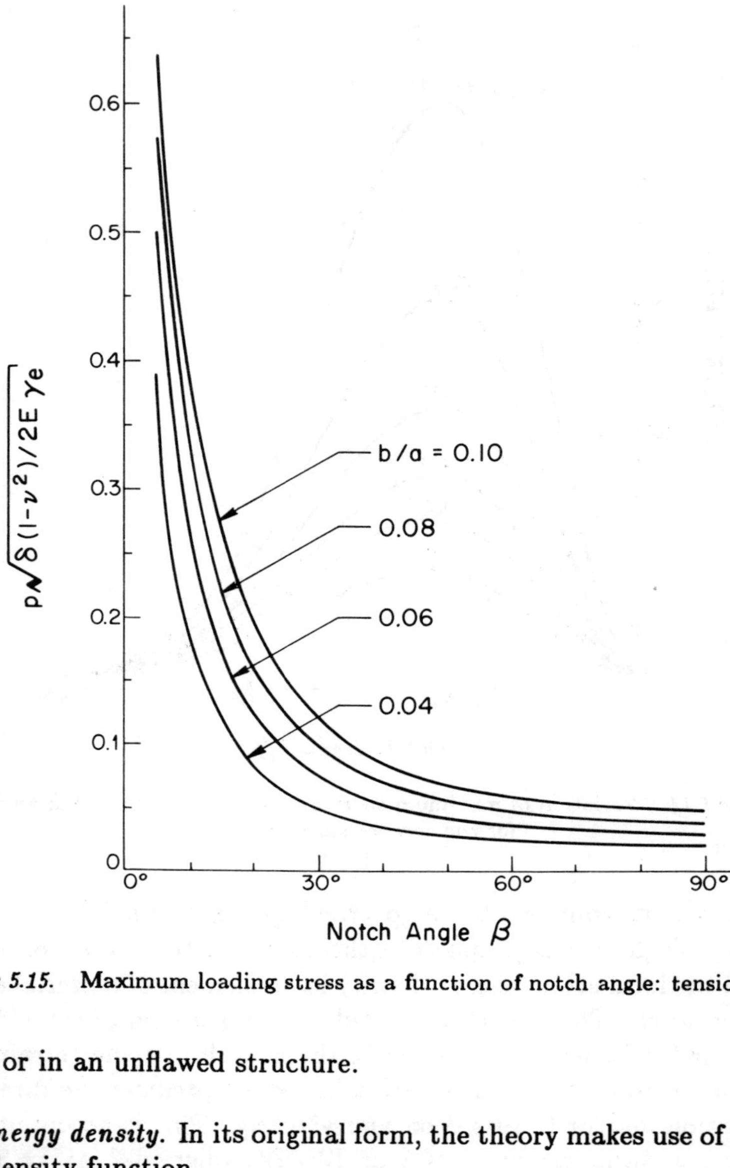

Figure 5.15. Maximum loading stress as a function of notch angle: tension case.

corners, or in an unflawed structure.

Strain energy density. In its original form, the theory makes use of the strain energy density function

$$\frac{dW}{dV} = \frac{dW}{dV}(\epsilon_{ij}) = \tfrac{1}{2}(c_0 + c_{ij}\epsilon_{ij} + c_{ijkl}\epsilon_{ij}\epsilon_{kl} + \cdots) \tag{5.9}$$

where ϵ_{ij} are the rectangular components of the strain and the coefficient c_0 can usually be disregarded if the body is not prestressed. For a linear elastic material, the stress components can be obtained from

$$\sigma_{ij} = \frac{\partial}{\partial \epsilon_{ij}}\left(\frac{dW}{dV}\right) = \tfrac{1}{2}\sigma_{ij}\epsilon_{ij}, \quad i, j = 1, 2, 3. \tag{5.10}$$

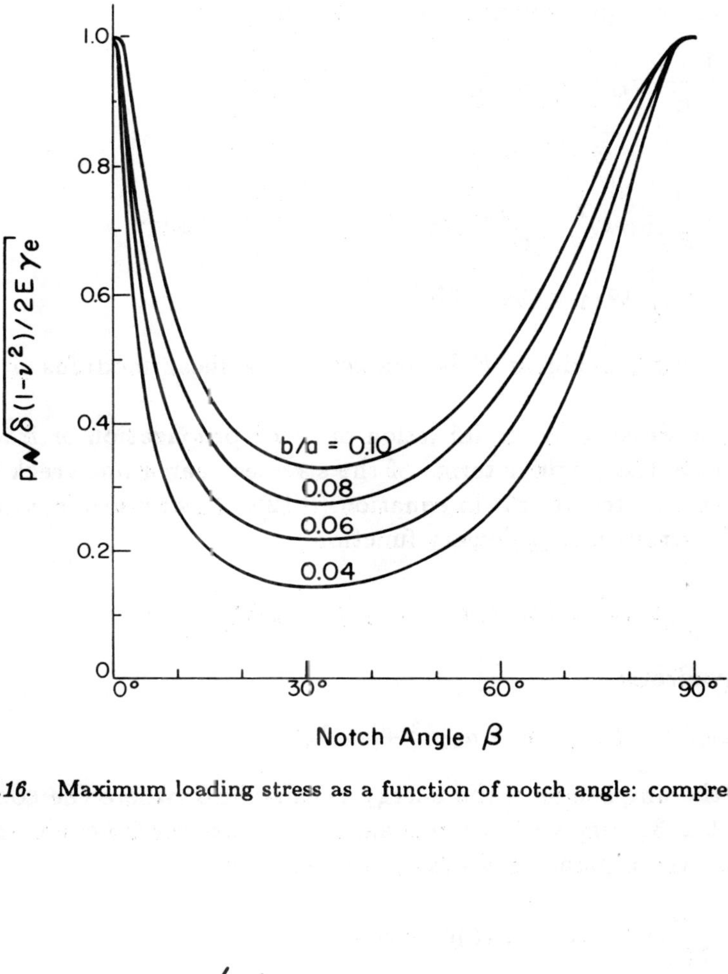

Figure 5.16. Maximum loading stress as a function of notch angle: compression case.

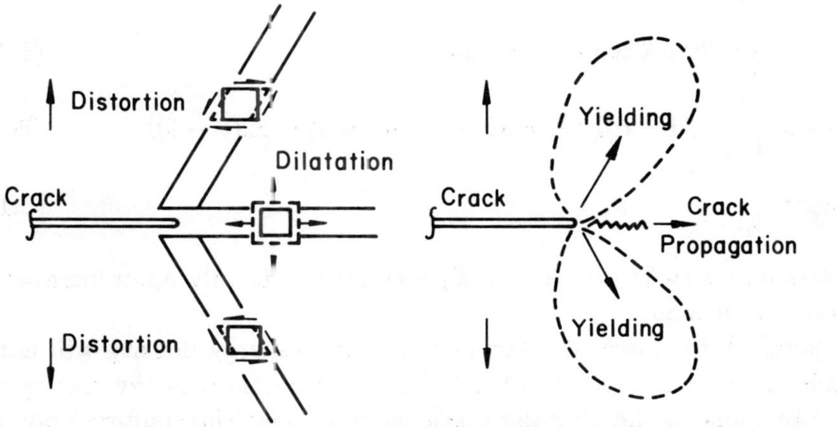

Figure 5.17. Dilatation and distortion of material elements ahead of a crack.

The components ϵ_{ij} in equation (5.10) may be eliminated by using the relation for an isotropic, homogeneous elastic medium:

$$\epsilon_{ij} = \frac{1+\nu}{E}\sigma_{ij} - \frac{\nu}{E}\sigma_{kk}\delta_{ij}. \tag{5.11}$$

This gives

$$\frac{\mathrm{d}W}{\mathrm{d}V} = \frac{1}{2E}(\sigma_{kk})^2 - \frac{\nu}{E}(\sigma_{11}\sigma_{22} + \sigma_{22}\sigma_{33} + \sigma_{11}\sigma_{33}) + \\ + \frac{1}{2\mu}(\sigma_{12}^2 + \sigma_{23}^2 + \sigma_{13}^2) \tag{5.12}$$

in which Young's modulus E is related to the shear modulus μ as $E = 2(1+\nu)\mu$.

In order to determine the initiation of crack propagation or fracture, it suffices to take the singular terms of the stresses near a line crack tip, and substitute them into $\mathrm{d}W/\mathrm{d}V$ in equation (5.12). This results in a quadratic form for the strain energy density function,

$$\frac{\mathrm{d}W}{\mathrm{d}V} = \frac{1}{r}(a_{11}k_1^2 + 2a_{12}k_1k_2 + a_{22}k_2^2 + a_{33}k_3^2) + \cdots \tag{5.13}$$

Here the quadratic

$$S = a_{11}k_1^2 + 2a_{12}k_1k_2 + a_{22}k_2^2 + a_{33}k_3^2 \tag{5.14}$$

represents the amplitude of the energy density field, where the coefficients a_{ij} $(i, j = 1, 2, 3)$ vary with a polar angle, θ, measured from the crack tip, and for the case of plane strain they are given by:

$$a_{11} = \frac{1}{16\mu}[(3 - 4\nu - \cos\theta)(1 + \cos\theta)] \tag{5.15a}$$

$$a_{12} = \frac{1}{16\mu}2\sin\theta[\cos\theta - (1 - 2\nu)] \tag{5.15b}$$

$$a_{22} = \frac{1}{16\mu}[4(1-\nu)(1-\cos\theta) + (1+\cos\theta)(3\cos\theta - 1)] \tag{5.15c}$$

$$a_{33} = \frac{1}{4\mu}. \tag{5.15d}$$

The stress intensity factors k_1, k_2, k_3 are dependent only upon loading and geometric conditions.

In general, for a notch, the form of the strain energy density will not be as concise as that of equation (5.13), but the evaluation of the energy field will be the same as that for the crack, and the resulting failure loads and angles also determined on the basis of the strain energy density assuming a critical constant value in a given material. For a relatively sharp elliptical notch, the stresses near the notch tip may be expanded in a series in terms of a set of local polar coordinates r and θ as follows:

$$\sigma_{11} = \frac{1}{\sqrt{2r}}\left[k_1 \cos\frac{\theta}{2}\left(1 - \sin\frac{\theta}{2}\sin\frac{3\theta}{2}\right) - k_1\left(\frac{\rho}{2r}\right)\cos\frac{3\theta}{2} - \right.$$
$$\left. - k_2 \sin\frac{\theta}{2}\left(2 + \cos\frac{\theta}{2}\cos\frac{3\theta}{2}\right) + k_2\left(\frac{\rho}{2r}\right)\sin\frac{3\theta}{2}\right] \tag{5.16a}$$

$$\sigma_{22} = \frac{1}{\sqrt{2r}}\left[k_1 \cos\frac{\theta}{2}\left(1 + \sin\frac{\theta}{2}\sin\frac{3\theta}{2}\right) + k_1\left(\frac{\rho}{2r}\right)\cos\frac{3\theta}{2} + \right.$$
$$\left. + k_2 \sin\frac{\theta}{2}\cos\frac{\theta}{2}\cos\frac{3\theta}{2} - k_2\left(\frac{\rho}{2r}\right)\sin\frac{3\theta}{2}\right] \tag{5.16b}$$

$$\sigma_{12} = \frac{1}{\sqrt{2r}}\left[k_1 \sin\frac{\theta}{2}\cos\frac{\theta}{2}\cos\frac{3\theta}{2} - k_1\left(\frac{\rho}{2r}\right)\sin\frac{3\theta}{2} + \right.$$
$$\left. + k_2 \cos\frac{\theta}{2}\left(1 - \sin\frac{\theta}{2}\sin\frac{3\theta}{2}\right) - k_2\left(\frac{\rho}{2r}\right)\cos\frac{3\theta}{2}\right] \tag{5.16c}$$

$$\sigma_{33} = \nu(\sigma_{11} + \sigma_{22}), \quad \sigma_{13} = \sigma_{23} = 0 \tag{5.16d}$$

which correspond to a state of plane strain. The radial distance r in equations (5.16) is measured from the focal point of the notch and always remains nonzero and θ is measured from a line extended from the major axis of the notch. The ratio ρ/r is such that it is not negligible in comparison with unity. Although the stress field in equations (5.16) differs from that for a sharp crack where $\rho = 0$, the stress intensity factors k_1, k_2 and k_3 are unchanged. For plane strain problems, equation (5.12) simplifies to the form

$$\frac{dW}{dV} = \frac{1}{2\mu}\left[\frac{1-\nu}{2}\left(\sigma_{11}^2 + \sigma_{22}^2\right) - \nu\sigma_{11}\sigma_{22} + \sigma_{12}^2\right] \tag{5.17}$$

where $dV = dA \cdot 1$ for two-dimensional problems. Substituting the stresses in equations (5.16) into (5.17) yields a strain energy density factor

$$S = r\frac{dW}{dV} = S_1 + \frac{S_2}{r} + \frac{S_3}{r^2} \tag{5.18}$$

in which

$$S_1 = a_{11}k_1^2 + 2a_{12}k_1k_2 + a_{22}k_2^2 \tag{5.19a}$$
$$S_2 = b_{11}k_1^2 + 2b_{12}k_1k_2 + b_{22}k_2^2 \tag{5.19b}$$
$$S_3 = c_{11}k_1^2 + 2c_{12}k_1k_2 + c_{22}k_2^2. \tag{5.19c}$$

The coefficients a_{ij} are the same as those in equations (5.15) for the crack problem while b_{ij} and c_{ij} are given by

$$b_{11} = 0, \quad b_{12} = -\frac{\rho}{8\mu}\sin\theta, \quad b_{22} = -\frac{\rho}{4\mu}\cos\theta \tag{5.20a}$$

$$c_{11} = \frac{\rho^2}{16\mu}, \quad c_{12} = 0, \quad c_{22} = \frac{\rho^2}{16\mu}. \tag{5.20b}$$

For more details, refer to the work of Sih [14] for the example of an elliptical notch subjected to in-plane shear loading.

Core region. A fundamental difficulty in fracture mechanics arises in the interpretation of high stresses at sharp corners in the analytical models which are physically unattainable. In part, this difficulty arises because analytically, one cannot expect the solution to be accurate much closer to the tip of a crack or notch than the minimum continuum element size. It must also be recognized that in a very local region at the crack tip, the physical behavior is unknown, and cannot be incorporated into a mathematical model. In this region, the material being highly strained, may become inhomogeneous, and in general, is not conducive to modeling. In polycrystalline materials, for example, the orientation of a grain may be significant, but unknown from a continuum viewpoint. These same limitations hold for sharp notches as well, and likely for blunted notches. But it must be emphasized that both the analytical and physical aberrations mentioned are confined to a very small region near the notch tip, and for this reason are not expected to significantly perturb the analytic solution external to this region.

The initial approach taken here is that of postulating the previously stated aberrant material behavior to be limited to a region local to the notch tip, Figure 5.2(b), characterized by a length dimension, r. This core region will remain temporarily unspecified in size. The main purpose is to avoid the requirements of an explicit continuum solution everywhere around the notch and hence evaluation of the fracture toughness will be done on the basis of material behavior external to the core region.

Unstable crack extension is assumed to occur when some small element just outside the core region (Figure 5.2(b)) has absorbed as much elastic energy as possible and releases it to allow material separation.* The failure of the core region remains unexplained in view of the lack of a mathematical description of its material properties. It is also necessary to state that the 'boundary' separating the elastic material from the core region is a fictitious one. Continuity must prevail at the interface. Rather, the core dimension is simply a measure stating that on a continuum level, closer approach to the notch tip renders the analytical model inaccurate. In reality, the size of this core region will reflect a basic property of the material at the microscopic scale level.

As discussed earlier, the location on the free surface of a notch of the maximum surface layer energy may be postulated to be the region of initial

---

* Brittle behavior is assumed in that the material and notch geometry do not deform appreciably prior to instability triggered by crack propagation.

failure of the notch. Subsequent behavior, such as the immediate post-failure crack direction, may be in part described by the local features of the strain energy density function. A radius vector may be attached to the point of maximum surface layer energy. The vector is of length r, and its position relative to a fixed axis is given by the angle θ, measured positive counterclockwise as shown in Figure 5.2(b). The origin of the vector is a function of both loading and notch geometry. Further, for each notch and loading geometry, there is usually a separate origin for the tension and compression cases. In the degenerate case, when the notch becomes a line crack, the concept of surface layer energy is inapplicable, and the origin moves to the crack tip, as in the S-theory [3].

Prediction of failure path. Prior to becoming involved in the details of an example, a few additional observations and conjectures on failure at notches and cracks will be advanced for consideration. Having already noted that while in principle the strain energy density theory may be extended with suitable constitutive and kinematic relationships to account accurately for behavior near the notch tip, the primary limitation placed on the present discussion of the theory, including that of the core region, is that the material or structural member under consideration behaves in a brittle fashion. Within such a framework, it does not seem unreasonable to assert that the initial notch and loading geometries (and material properties) determine not only the load required to precipitate fracture and the initial fracture angle, but also the subsequent path the fracture will follow. That is, upon being loaded, the notch must seek a path of release, through knowledge of the local properties of the strained material, seen from its ***present configuration***. And in view of the work of Kipp and Sih [15], it appears possible that in tension the fracture path follows the trajectory of points of minimum strain energy density remarkably closely. That is, once fracture is initiated, it is assumed that its path can be determined from the condition of the uncracked state. These, however, exclude displacement controlled situations for which crack growth is globally stable. This is because the input energy remains constant and the crack will initially grow, then stop, and remain stationary until a further increase in load or input energy occurs. Under such conditions, long-range path predictions would not be possible. In the situations where local and global instability occur almost simultaneously, such as notches loaded uniformly at distances far from the notch, crack growth is quite sudden and unstable. Then the material has no time to readjust to a new configuration, and as suggested above, quite possibly this path can be predetermined with sufficient accuracy. A primary argument in favor of such behavior comes from an understanding of the global energy field. For brittle behavior of blunt notches, or line cracks at low angles of loading, the loads required for initial fracture are relatively high (when compared to that of the line crack in symmetric loading) so that there already exists a high density energy field in the regions at both near and far distance from the point of incipient failure. Such a field will do much to contribute to rapid fracture and failure, when few available physical resources exist to inhibit the fracture from propagating.

In order to determine a path for the fracture, the family of curves generated for a wide range of radius vectors provides the locations of the sequence of points that indicate stationary values of the strain energy density function. It is suggested that this trajectory should be in close agreement with the actual fracture path* taken from a notch that behaves in a brittle fashion. The subsequent discussion for the specific example considered will bear out at least locally the accuracy of the path prediction capabilities.

5.4. The embedded elliptical notch

For the purpose of illustration, a specific notch geometry has been chosen to demonstrate the methods of analysis and the particular uses for which the resulting information about the strain energy density near the notch tip can be employed. The choice of a uniformly loaded solid containing an embedded elliptical notch is based primarily on the availability of an analytic solution for this geometry, and also on the availability of recent experimental work involving this geometry under both tensile and compressive loading conditions. Additionally, the ease with which the solution is reduced to that of the finite crack geometry permits comparison with this limiting case of blunted notches.

The details of the stress field around the elliptical cavity under uniform loading are provided by Muskhelishvili [11] for an isotropic and homogeneous material under plane strain. The elliptical notch is defined by the ratio of minor to major axis half-lengths, b/a, with all other length parameters normalized with respect to the major axis half-length, a. The uniform loading (at infinity) forms an angle of β with the major axis of the cavity, and the radial distance r is referenced from the point on the notch surface that forms the normal angle ϕ with the major axis and is measured positive counterclockwise. The position of r with respect to the horizontal line is determined by the angle θ (positive counterclockwise). Figure 5.18 indicates the geometry described. Only the right hand notch tip will be considered, as antisymmetry of the geometry implicitly allows description of the other notch tip as well.

---

* Hartranft and Sih [16] have predicted the fracture surface emanating from an embedded three-dimensional flat elliptical crack in a combined stress field. This surface is assumed to coincide with that of minimum strain energy density.

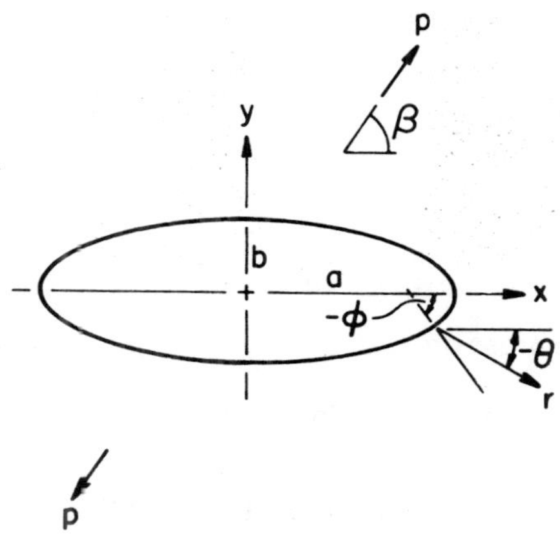

Figure 5.18. Elliptical notch under angle loading.

A brief exposition on the strain energy density function in linear elasticity theory has already been given in Section 5.3. As mentioned earlier, the stresses for the elliptical notch problem in Figure 5.18 can be obtained from Muskhelishvili [11]:

$$\sigma_{11} + \sigma_{22} = 2\left[\phi'(z) + \overline{\phi'(z)}\right] \tag{5.21a}$$

$$\sigma_{22} - \sigma_{11} + 2i\sigma_{12} = 2[\bar{z}\phi''(z) + \psi'(z)] \tag{5.21b}$$

in which σ_{11}, σ_{22} and σ_{12} are the rectangular stress components and $\phi(z)$ and $\psi(z)$ are complex functions of the variable $z = x + iy$. The subscripts 1 and 2 correspond to the directions x and y, respectively. These functions are given by

$$4\phi'(z) = \frac{p}{c^2}\left\{(a+b)^2\cos 2\beta + \left[c^2 - (a^2+b^2)e^{2i\beta}\right]\frac{z}{\sqrt{z^2-c^2}}\right\} \tag{5.22a}$$

$$4\psi'(z) = p(a^2+b^2-c^2\cos 2\beta)\frac{z}{(z^2-c^2)^{3/2}} - \frac{(a+b)^2}{c^4}\Big\{2(a^2+b^2)\cos 2\beta + \\ + 4iab\sin 2\beta + [2ab\cos 2\beta + i(a^2+b^2)\sin 2\beta]\frac{z(2z^2-3c^2)}{(z^2-c^2)^{3/2}}\Big\} \tag{5.22b}$$

where $c^2 = a^2 - b^2$. Putting equations (5.22) into (5.21) and the subsequent results into equation (5.17) yields the strain energy field surrounding an elliptical notch. All stationary values of the strain energy density are obtained numerically, as the contours followed do not lend themselves to analytic evaluation.

Since the behavior of the strain energy field is load dependent, the cases of compression and tension are treated separately. There are some funda-

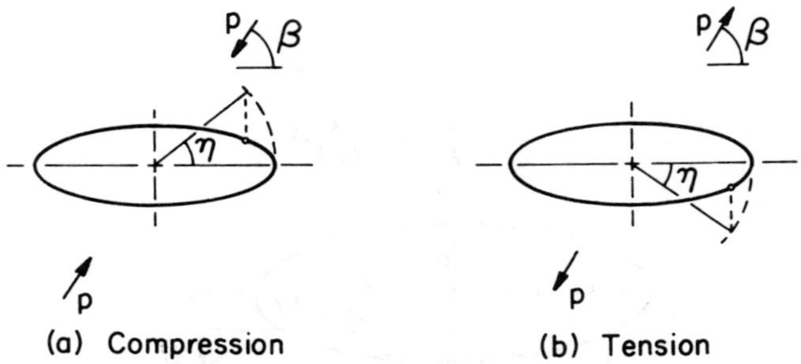

Figure 5.19. Location of failure points on notch in tension and compression.

mental differences in the two cases, and it is intended that separation will clarify these departures without redundancy of explanation. The strain energy density function is also dependent upon the value of Poisson's ratio, and it will be assigned a value of 0.250, unless otherwise stated. It is clear from the work of Sih [3, 4] that there are significant deviations attributable to a change in Poisson's ratio. These deviations are restricted to magnitudes only, however, and trends remain unchanged. Hence, subsequent discussion will center not on this effect, but on basic observable behavior, restricting changes in Poisson's ratio to matching experimental specimen material where necessary.

The choice of specific values for the ratio b/a has been governed to a large extent by the experimental data available. In other cases, the choice is simply for illustrative purposes, and any further comparison with new data requires new calculations to be performed. It is intended that the choices reflect fairly completely the range of possible behavior for this geometry of notch and loading.

Origin of failure on notch. The application of the local strain energy density field requires a knowledge of the position of maximum surface layer energy, γ_e, which determines the point on the notch boundary where failure may occur. For the elliptical notch, this position may be obtained analytically by setting the first derivative of γ_e in equation (5.7) with respect to eccentric angle η equal to zero, i.e., $\partial\gamma_e/\partial\eta = 0$. The resulting equation, quadratic in $\tan\eta$, may be solved to give

$$\tan\eta = \frac{b[a\sin^2\beta - b\cos^2\beta \pm \sqrt{a^2\sin^2\beta + b^2\cos^2\beta}]}{a(a+b)\sin\beta\cos\beta}. \tag{5.23}$$

The positive root provides the location for failure initiation under compressive loading, and the negative root for failure initiation under tensile loading (Figure 5.19).

It will be convenient to have the normal angle to the surface at both these points; the transformation is provided in equation (5.8). These equations are meaningful only for $b/a \neq 0$. The special case of $b/a = 0$ corresponds to that of a line crack, and in such a case, the radius vector is attached to the crack tip. This has been treated to considerable extent by Sih [3], where the asymptotic expansion is considered. Rather than repeat the results of that work, it will be instructive here to discuss possible behavior further from the crack tip.

Elliptical notch in tension. If an elliptical cavity in an isotropic, homogeneous, linearly elastic solid is uniformly loaded as shown in Figure 5.18, at an angle β to the major axis, the maximum surface layer energy occurs in the surface region in the fourth quadrant, where $-90° \leq \phi \leq 0°$ (considering right-hand tip only). Loading behavior as a function of crack angle, notch size, and position in the medium is the subject of the initial discussion. Then considerations will focus on the initial and subsequent crack trajectories. This order is necessary in light of the earlier criteria established for the expected nature of failure and its relationships to the core region size.

For purposes of initial analysis, the applied uniform stress, p, will be assumed to be small, with no core region present, to establish the behavior of the strain energy field. Although normalized slightly differently, the strain energy behavior at the surface coincides with that shown in Figure 5.15, since at the surface, the strain energy is a function only of the one nonzero stress component, the tangential stress, the same one involved in the surface layer energy. As the radius (normalized to the major axis half-length) is made to increase, the stationary values of the strain energy begin to decrease, indicating that analytically (exclusive of surface layer energy considerations) the strain energy density assumes its absolute maximum on the surface of the notch (Figures 5.20–5.22). In each of the figures, a family of curves appears, each curve representing the possible loading behavior of a particular notch at a different distance from the notch surface. It may be observed that in all cases, the minimum load required for fracture occurs in the normal loading case ($\beta = 90°$). Intuitively, such a result appears to be reasonable, although some other fracture criteria have been proposed in which the minimum load appears in the vicinity of $\beta = 70°$. Palaniswamy [17] attempted an energy release rate procedure for complex loading, in which a small extension to the main crack was varied in direction until the maximum strain energy release rate for small crack extension was determined; in this case, a minimum was located near 72°. It must be mentioned that in order to solve the stated problem, approximations in the numerical procedures altered the strain energy calculation by unknown amounts, assumed small, casting some doubt on the validity of the results, and the extent to which any implications can be made concerning fracture load behavior. A second case in which the minimum load is predicted to appear near $\beta = 70°$ occurs in the criterion of

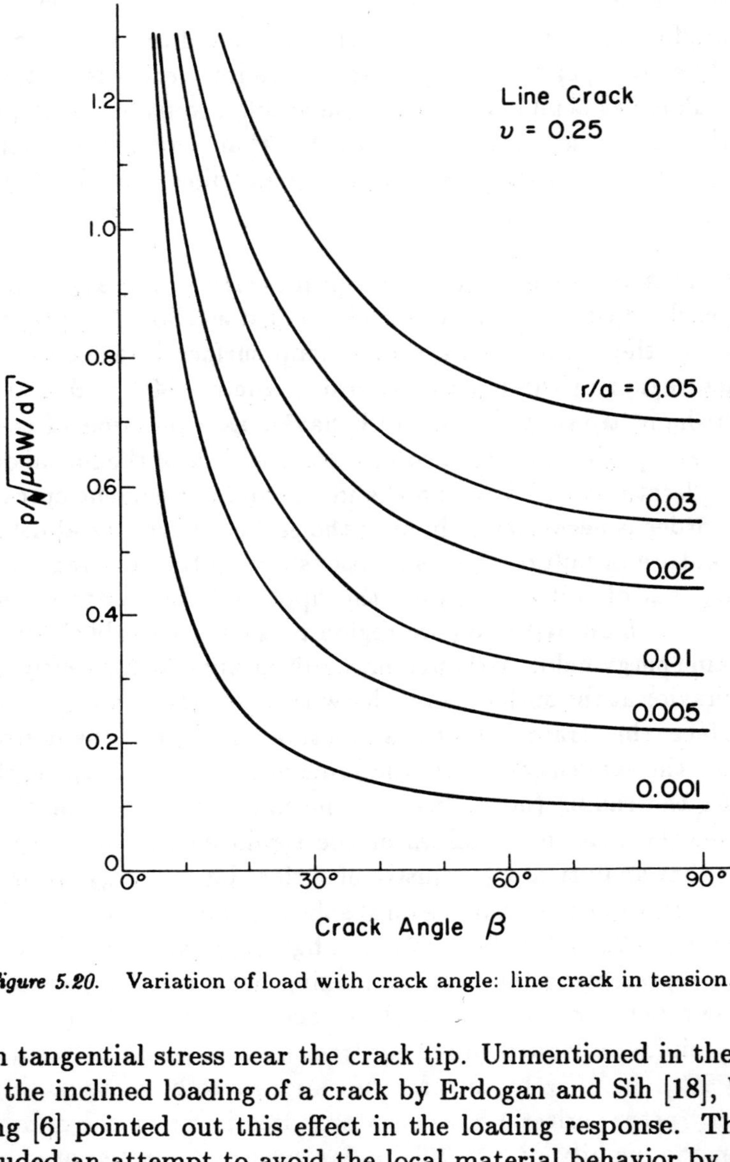

Figure 5.20. Variation of load with crack angle: line crack in tension.

maximum tangential stress near the crack tip. Unmentioned in the original paper on the inclined loading of a crack by Erdogan and Sih [18], Williams and Ewing [6] pointed out this effect in the loading response. This latter work included an attempt to avoid the local material behavior by applying the criterion a short distance from the crack tip using an additional term in the asymptotic solution. Unfortunately, truncation errors are the cause of the discrepancy, as is manifested when compared to the exact solution (Sih and Kipp [5] and Kassir [19]). The data presented in [6] appear to display a trend towards minimum load near $\beta = 70°$; however, much of the data had to be questioned because of inability to distinguish the initial direction of crack growth. In addition to the interaction of Mode I and II stress intensity factors, Mode III may also be present if the crack propagates at an angle in the thickness direction, and so the data [6] is not truly representative of the

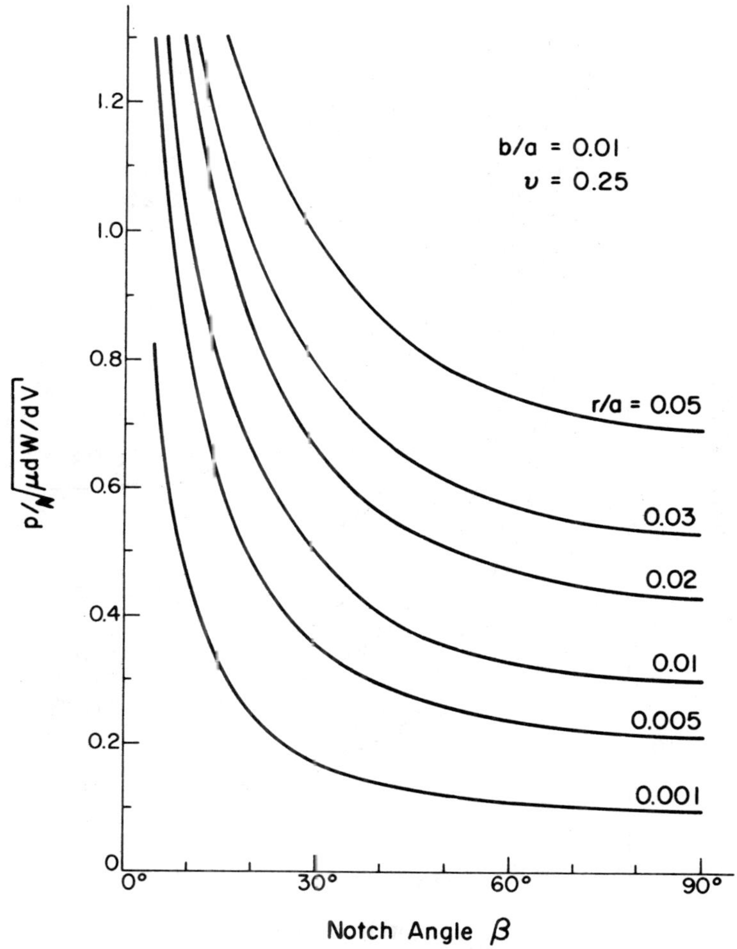

Figure 5.21. Variation of load with notch angle: $b/a = 0.01$ and tensile loading.

failure load as a function of the angle of loading. As Sih [9] has shown, the minimum failure would indeed be near $\beta = 70°$ when Mode III is present in addition to in-plane extension.

Figures 5.20–5.22 also reflect the trend of the increased loading permitted as the notch tip becomes more blunted. If an elliptical notch is experimentally loaded to failure at several crack angles, then from the corresponding family of curves for the same size notch, the curve that matches the data may be extracted, and that choice will provide the characteristic dimension of the core region, assuming that the elastic strain energy density constant is known. A knowledge of these two parameters has been earlier postulated to be sufficient to describe the failure of this material in the presence of any geometric configuration, whether line crack or notch. In the case at hand, given the core dimension, a family of curves may now be established

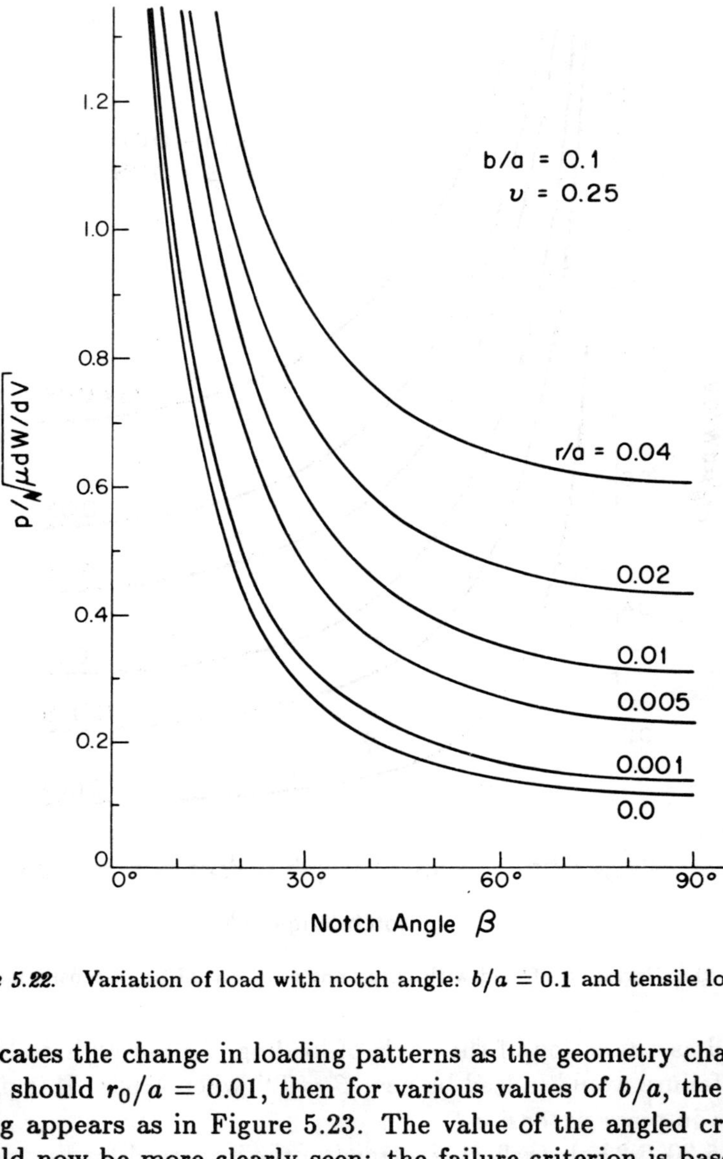

Figure 5.22. Variation of load with notch angle: $b/a = 0.1$ and tensile loading.

that indicates the change in loading patterns as the geometry changes. For example, should $r_0/a = 0.01$, then for various values of b/a, the variation in loading appears as in Figure 5.23. The value of the angled crack problem should now be more clearly seen: the failure criterion is based not on a single load to failure, but a function, and while more difficult to obtain, in principle, the prediction reliability should be considerably greater than presently is the case.

Although interest from a safety standpoint centers on loading capacity, and more specifically, on the worst case if possible, the actual post-failure behavior of the fracture provides another insight into the validity of the theory under study. As briefly mentioned earlier, Griffith had suggested that a crack would extend in a direction normal to the maximum tangential stress, and Erdogan and Sih [18], in applying this criterion obtained striking

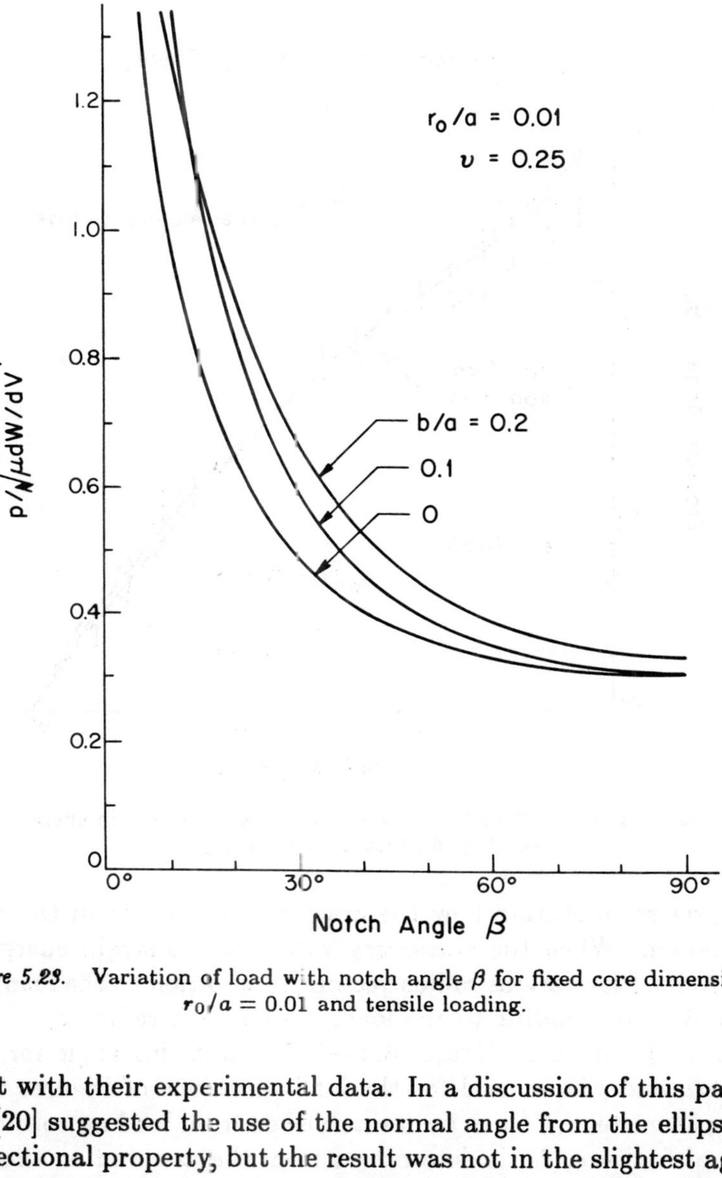

Figure 5.23. Variation of load with notch angle β for fixed core dimension: $r_0/a = 0.01$ and tensile loading.

agreement with their experimental data. In a discussion of this paper, McClintock [20] suggested the use of the normal angle from the ellipse surface as the directional property, but the result was not in the slightest agreement with the observed behavior. The published data of Williams and Ewing [6] for initial crack angle (with respect to the plane of the crack) corraborates that of [18]. Figure 5.24 illustrates the predictions of the strain energy density function for the core region dimensions suggested by the experimental data. At $\beta = 90°$, the crack is expected to propagate in its own plane, but as β becomes small, the direction becomes less well defined. At $\beta = 0$, the material reacts as if (in theory) no crack at all were present, and while the material would be expected to break at a normal to the load, the crack solution cannot predict this. Before proceeding further, it is necessary to

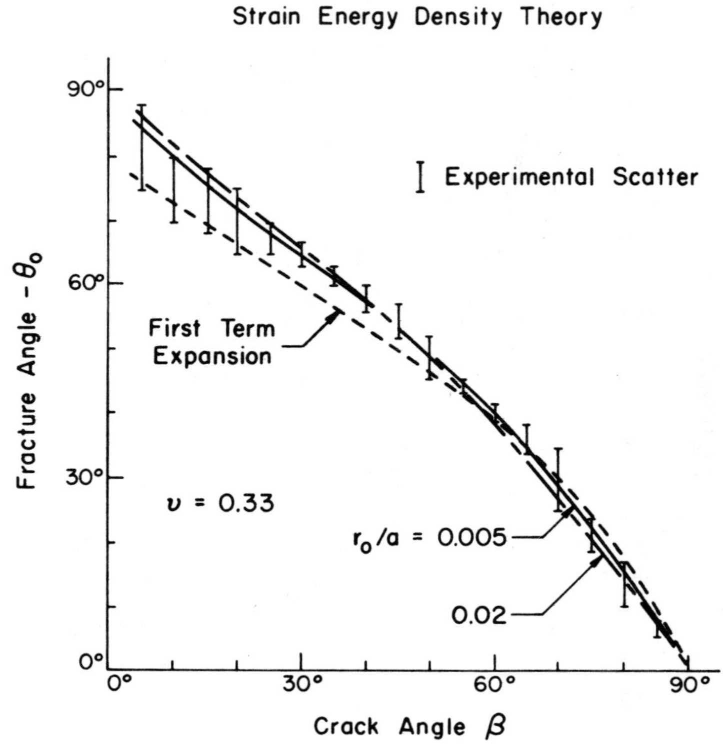

Figure 5.24. **Variation of fracture angle with crack angle: strain energy density solution for line crack in tension.**

be more precise in stating how the crack will extend from the tip of the crack or notch. When the stationary values of the strain energy density were determined for various radius vectors, in addition to the load, an angle θ was found corresponding to the load. So for each radius r/a, a curve is generated in the angle θ; Figures 5.25–5.27 reflect this angle for the same ratios of b/a and r/a as used for the load variations of Figures 5.20–5.22. Once the parameter r_0/a has been established from loading considerations, the angle of fracture, θ_0, is determined from the assumption that there is separation of the material from the surface to the edge of the core region in the direction θ_0, with subsequent fracture proceeding from there. If r_0/a is taken to be 0.01, corresponding to the load example, then for varying ratios of b/a, the resulting alterations in initial fracture angle would appear as in Figure 5.28.

One must remain cognizant of the nonstationary position of fracture initiation along the notch surface. As the loading angle changes, so does the position of initial failure. One result is that now, when $\beta = 0$, there is a well-defined point of failure (because the geometry is now nontrivial), and

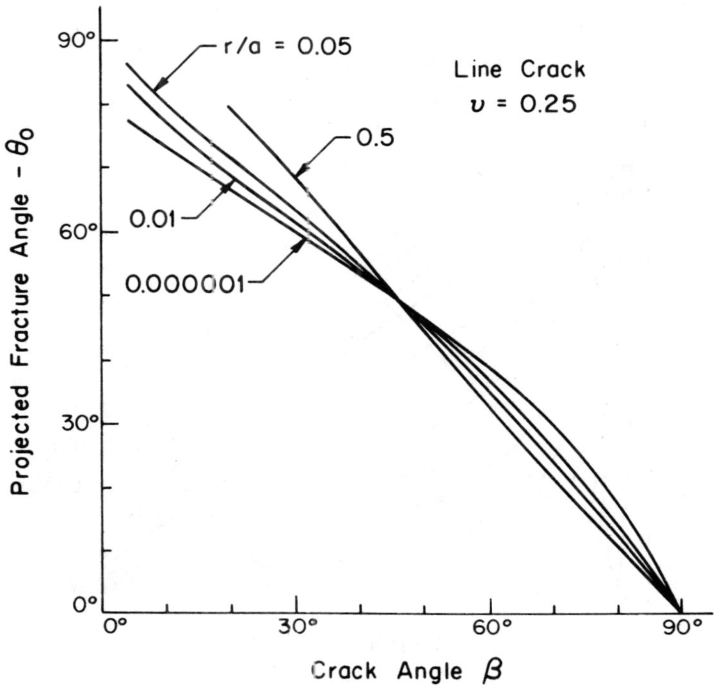

Figure 5.25. **Variation of fracture angle with crack angle: line crack in tension.**

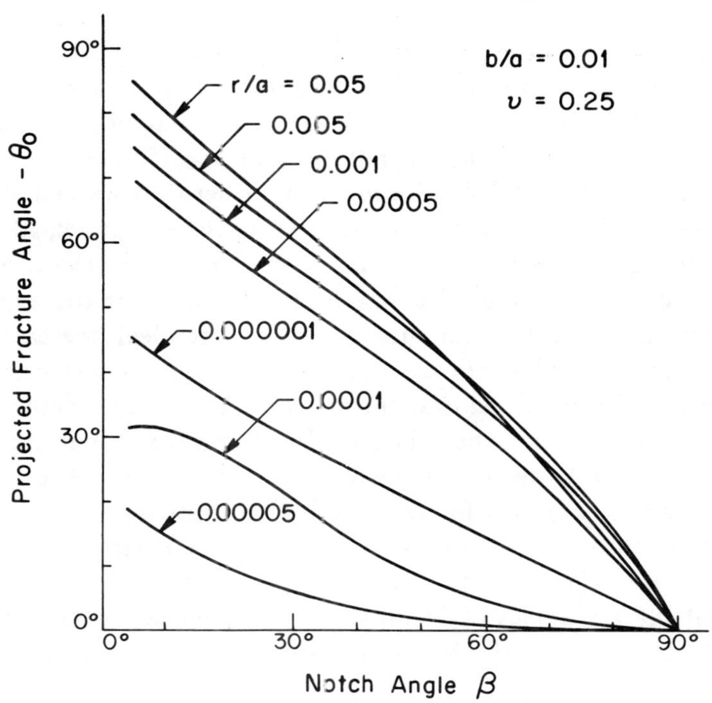

Figure 5.26. Variation of fracture angle with notch angle: $b/a = 0.01$ and tensile loading.

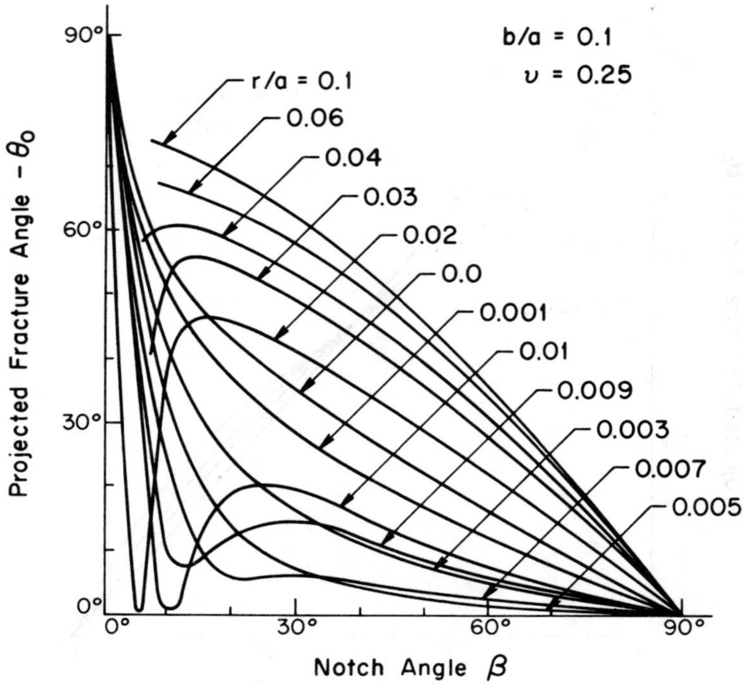

Figure 5.27. Variation of fracture angle with notch angle: $b/a = 0.1$ and tensile loading.

it is to be expected that the fracture will extend from the minor edge of the cavity along the minor axis direction. The path is, of course, perpendicular to the load, and the behavior, apart from load magnitude, corresponds exactly to that from the major axis notch tip at $\beta = 90°$.

Cotterell [21] has observed that for notches, rather than immediately turning into a path of fracture coinciding with that of the line crack, the fracture trajectory extends forward from the notch tip, roughly in the same plane, for a distance of approximately one notch tip radius, before turning into the path direction matching that for the line crack. The ideal line crack has, of course, no tip radius, and would immediately assume its characteristic direction of failure. If it is assumed that the crack path can be predetermined for unstable crack propagation, then the path of minimum strain energy density should trace out the trajectory along which fracture is expected to occur. In effect, the trajectory will follow a path that restores symmetry to the geometry of fracture. For a line crack, Figure 5.29 illustrates the projected paths for several angles of loading. Photographs of actual crack trajectories [15] are available for angles of loading $\beta = 30°$ and $60°$ which agree very well with the predicted path shown in Figure 5.30, out to a distance of at least one-half crack length.

If the behavior of a rather blunt notch, say $b/a = 0.1$ is examined, an

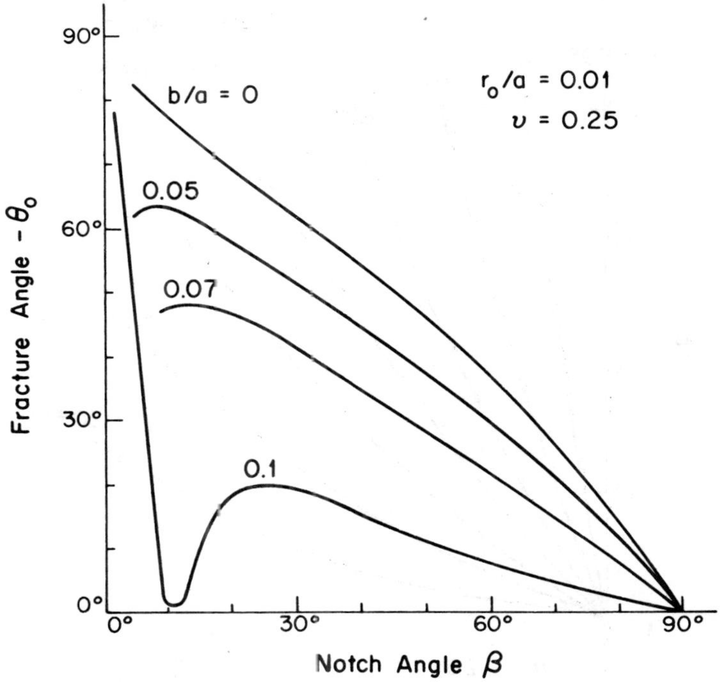

Figure 5.28. **Variation of fracture angle with notch angle for fixed core dimension: $r_0/a = 0.01$ and tensile loading.**

overall view (Figure 5.30) suggests that there have been at least some local changes in behavior from that of the line crack. Upon close observation, the trajectories near the surface reveal a behavior that connects some previously advanced, but analytically unsupported, ideas. When the surface of the notch separates, the fracture begins its course tangentially to the normal angle at the surface; i.e., initial fracture is perpendicular to the surface, coinciding with McClintock's [20] hypothesis. But this is only a local tangent, for the path then bends sharply towards the horizontal (with respect to the major axis of the notch), and extends in this way for a short distance. Again, a sharp turn occurs, now into a direction whose tangent matches that for the initial fracture angle of the line crack. The short extension distance is approximately equal to the notch tip radius, in accordance with the observations of Cotterell [21]. Subsequent to turning onto the path of fracture coinciding with that of the line crack, the trajectory continues in much the same manner as that of the line crack. Experimentally, surface conditions likely would obscure much of this local behavior, yet clearly, for small core regions, the separation would occur linking two regions somewhere before the trajectory turns into that of the line crack, and the local horizontal extension is still expected to occur.

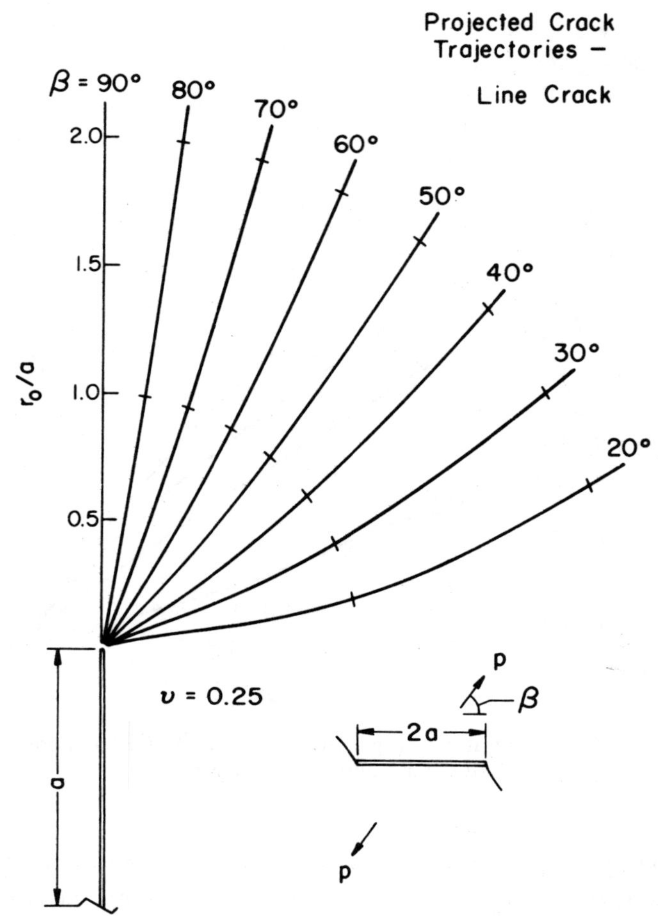

Figure 5.29. **Fracture trajectories predicted from the strain energy density theory for a line crack under tension.**

These latter observations have been made assuming prevailing linear elastic response both locally and globally. Should there be extensive localized inhomogeneity, or unaccountable grain size effects, it is expected that few of the above phenomena will be observable, and a more refined model must be created to determine the local behavior. Also, if the fracture is slow, then the material has time to react to the new geometry, and the details listed above are no longer valid.

Elliptical notch in compression. Although much of the behavior to be described in this section corresponds to that of the tensile case, there are a few additional difficulties that must be dealt with in compression. Rather than reiterating the arguments already presented in detail, emphasis will be placed upon the features that tend to complicate the compressive analysis.

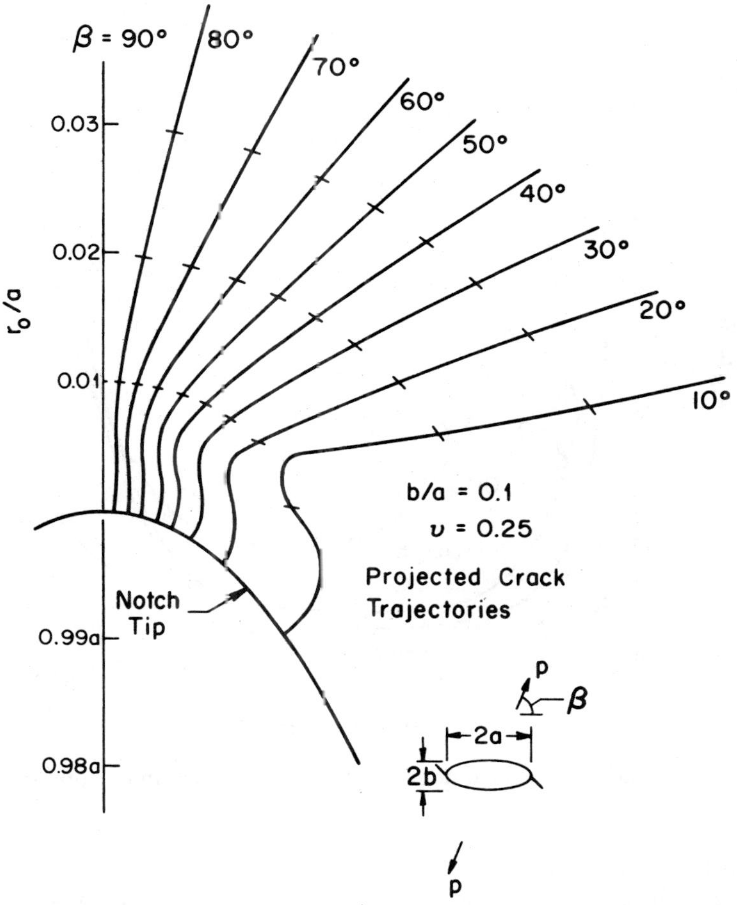

Figure 5.30. Fracture trajectories for a notch with $b/a = 0.1$ under tension.

Caution must be exercised in requiring that no interpenetration of notch faces occurs. The line crack, for instance, must be reformulated to ensure that sufficient normal stresses occur on the faces to prevent penetration, and this has been attempted by McClintock and Walsh [12]. In addition, it is clear that functional stresses will be developed on the faces if any slip is to occur to allow fracture to progress. In the development here, the notch geometry will be restricted to such dimensions that contact between opposing faces cannot occur.

As before, assuming a small applied uniform stress, the surface position of maximum surface layer energy may be located, and an origin established to evaluate the load behavior as a function of radial distance from the surface. Figures 5.31 and 5.32 describe the load behavior as a function of notch angle, for various constant radii, r/a. The curves for $r/a = 0$ correspond to those of the maximum surface layer energy in Figure 5.16, apart from a different

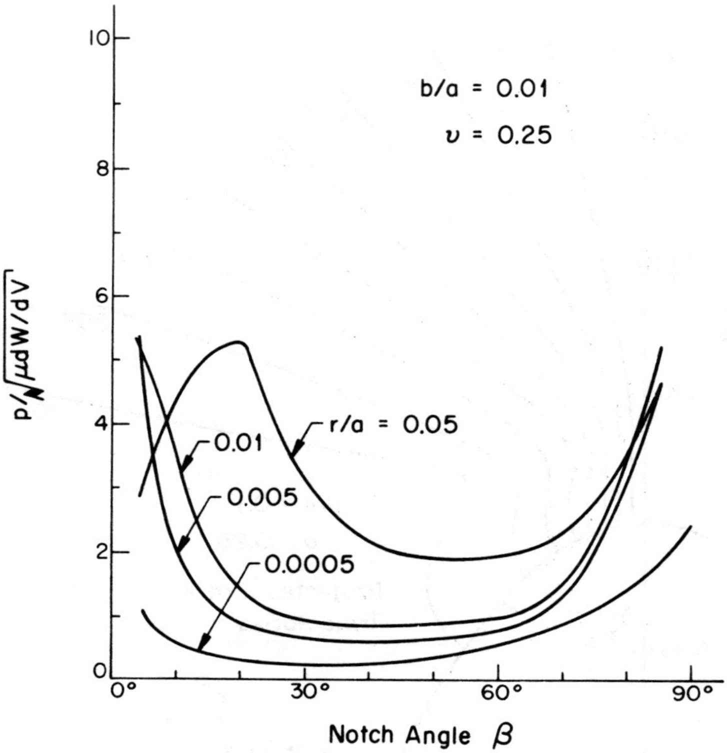

Figure 5.31. Variation of load with notch angle for $b/a = 0.01$ and compressive loading.

normalization of the load.

Once a core dimension has been established, say $r_0/a = 0.01$, the variation in load with geometry may be described, as in Figure 5.33. For a given material, and the present analysis, it is expected that a new core dimension, r_0/a, must be determined, since in general, the behavior of a material in compression differs considerably from that in tension. That is to say, another sequence of tests would be required to obtain the core dimension.

A characteristic of compressive loading is that the minimum load to fracture occurs at some angle between $\beta = 0°$ and $90°$ depending on the size of the core region. Cotterell [22] has published the experimental data that appear in Figure 5.34 for an elliptical cavity, $b/a = 0.1$, in glass. The curves in this figure are those of Figure 5.32, normalized to the load at $\beta = 90°$. The curve $r/a = 0$ is that of the maximum surface layer energy criterion. As the position of possible failure is allowed to move deeper into the material, there appears quite satisfactory agreement between $r/a = 0.005$ and 0.008. Hence, the energy criterion may be expected to provide quite reasonable failure predictions. On the basis of the data, one could choose a critical core dimension of, say, $r_0/a = 0.005$, and then use this dimension as set forth in

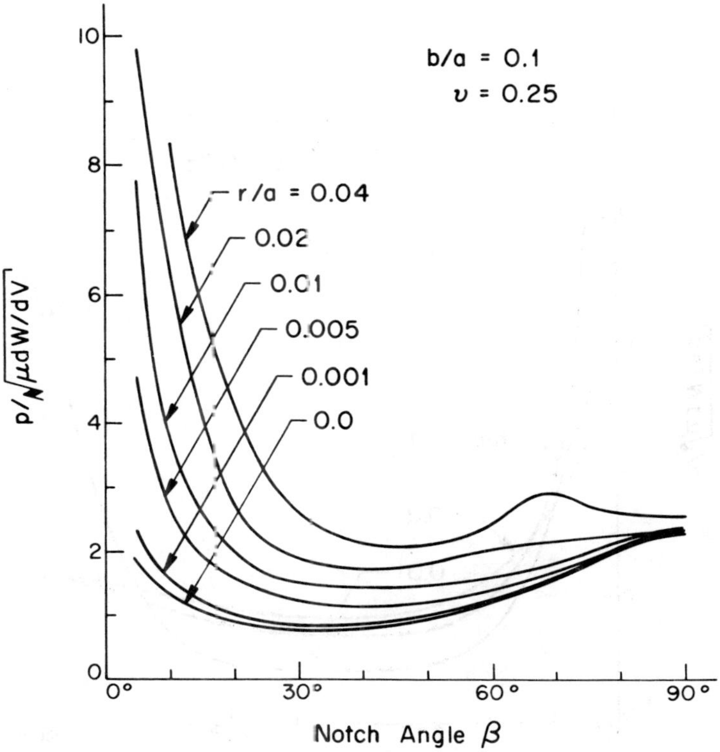

Figure 5.32 Variation of load with notch angle for $b/a = 0.1$ and compressive loading.

previous sections for load predictions for other geometries in this material.

Under uniform loads, it has been observed that propagating cracks tend to restore symmetry by running normal to tensile loads, and parallel to compressive loads. Then, it is expected that for the elliptical cavity under compression, for loads parallel to the major axis ($\beta = 0$), the fracture will initiate at the notch tip on the major axis, and extend along that same axis toward the applied load, while for loads parallel to the minor axis, fracture will initiate at the face intersection with the minor axis, and again extend towards the applied load; i.e., for $\beta = 0°$, $\theta_0 = 0°$ and for $\beta = 90°$, $\theta_0 = 90°$. Between these two limiting cases, the behavior is determined as in the tensile loading case; the stationary values of θ are plotted as functions of crack angle β, for various magnitudes of r/a (Figures 5.35 and 5.36). It must be emphasized that not until a specific core dimension has been obtained can any of these curves represent the initial fracture angle. If, for example, $r_0/a = 0.01$, then the initial fracture angle, as it varies with geometry, is given by Figure 5.37. The data shown in Figure 5.38 is from Cotterell [22] for elliptical notches in glass plates loaded in compression. The curves are those from Figure 5.36, since the data is for cavities of ratio $b/a = 0.1$.

The lower curves presented in Figure 5.38 are for small values of r/a, since

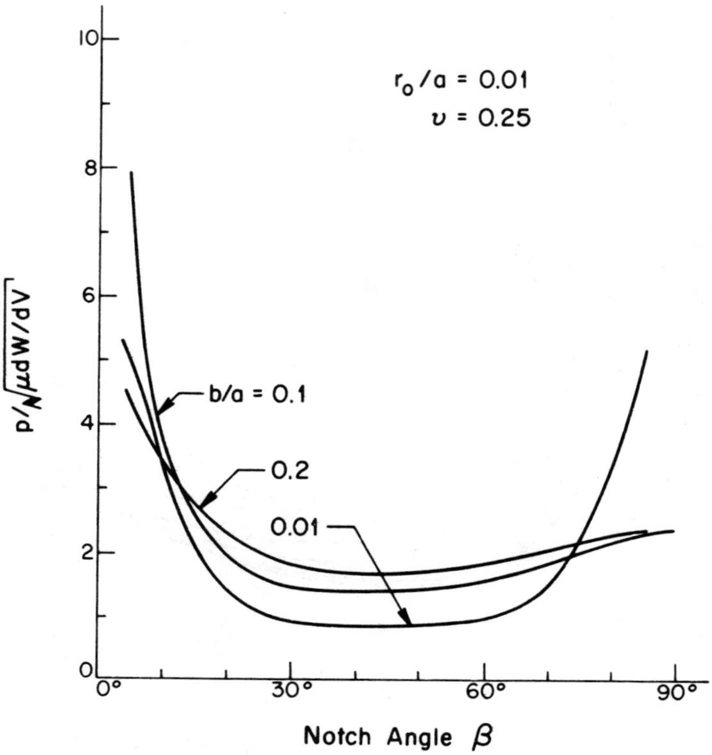

Figure 5.33. Variation of load with notch angle β for fixed core dimension: $r_0/a = 0.01$ and compressive loading.

the core region of necessity must be small compared to the notch length. The upper curve represents the tangent angle taken from the plots in Figure 5.39 and 5.40, of the crack trajectory after it makes the sharp turn into its final path.

In general, the loads required to cause failure in compression are substantially higher than those in tension, and there results throughout the member severely strained material, so that there is prevailing a high energy density; once a crack is initiated, it is expected to propagate catastrophically in this particular geometry. Then the locus of stationary values of the strain energy density may be plotted to attempt to predict the subsequent path taken by the fracture (Figures 5.39 and 5.40). As in the tension case, the path is initially tangent to the local normal before bending away; then there is a quick return into the general trend leading the path towards the uniform load. This describes the behavior for most of the angles of loading; apparently, as β increases, so does the length of the initial trajectory before its final turn into the load. This is also evident by following changes in r/a for constant β in Figures 5.35 and 5.36.

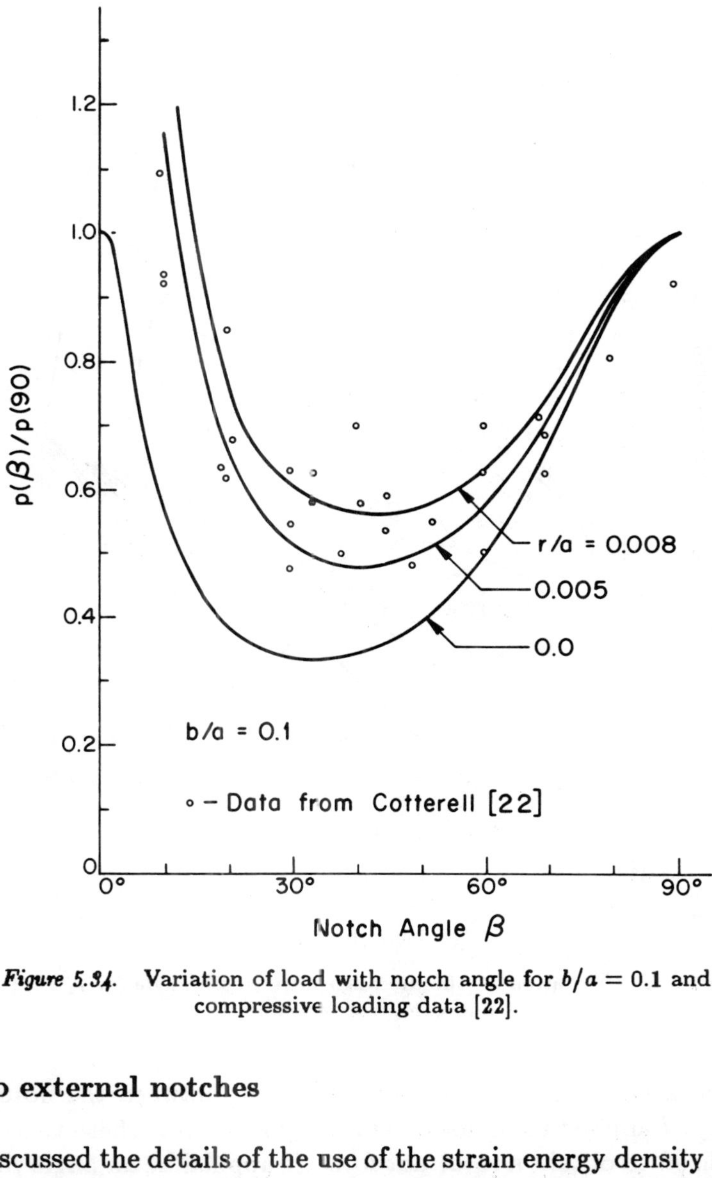

Figure 5.34. Variation of load with notch angle for $b/a = 0.1$ and compressive loading data [22].

5.5. Two external notches

Having discussed the details of the use of the strain energy density theory for the internal notch, the problem of two interacting external notches will be treated in this section to demonstrate certain features that are not intuitively obvious.

The elastic solution for plane extension of two external hyperbolic notches has been obtained by Griffith [23] and Neuber [10]. Two cases are considered, simple normal extension and pure shear, shown respectively in Figures 5.41(a) and 5.41(b). The thickness is taken to be h, and assumed large enough to have plane strain prevail. P is the net tensile force applied a large distance from the waist between the notches. Q is the net shear force

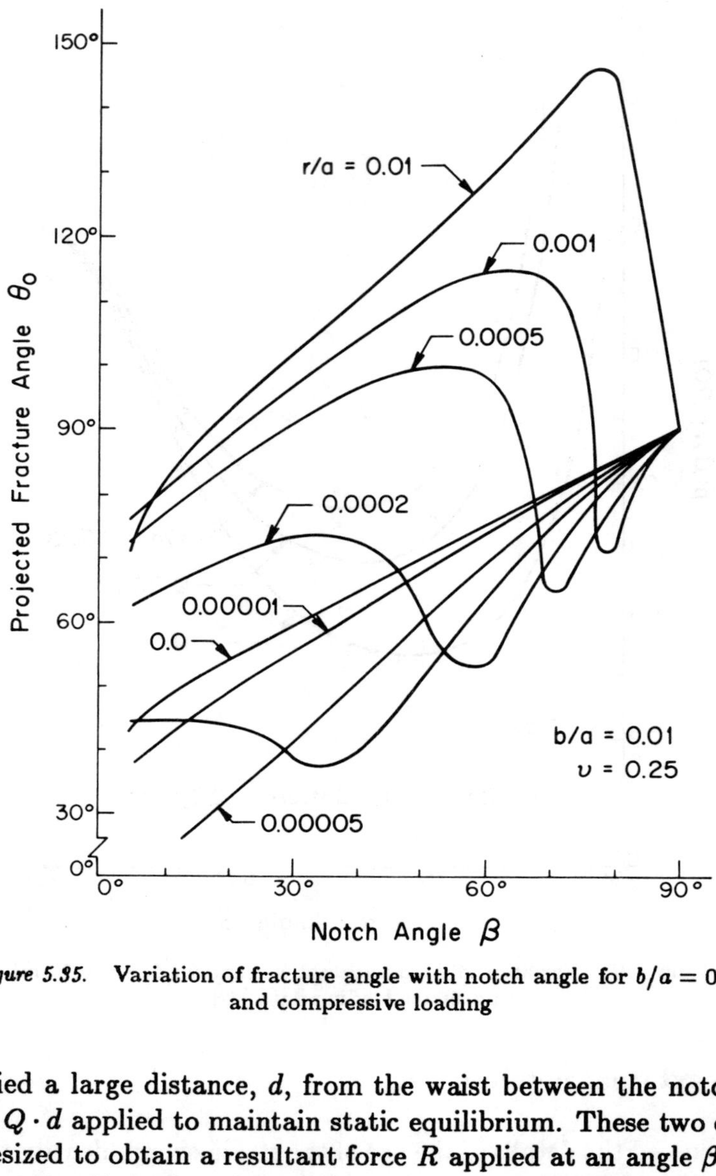

Figure 5.35. Variation of fracture angle with notch angle for $b/a = 0.01$ and compressive loading

also applied a large distance, d, from the waist between the notches, with moments $Q \cdot d$ applied to maintain static equilibrium. These two cases may be synthesized to obtain a resultant force R applied at an angle β to the x-axis, where $P = R \sin \beta$, and $Q = R \cos \beta$. Figure 5.42 illustrates the results of this combination of forces and couples to the single resultant force R. In addition, the geometry of the external notch, assuming the form of a plane hyperbola, is shown, with waist length $2a$, normal angle ϕ, and asymptotic slope(s) $\pm b/a$.

The solutions to the two basic cases are available in elliptic coordinates (Griffith [23] and Neuber [10]), but a more convenient form is that of rectangular cartesian coordinates, and the transformed solutions for the stresses follow in terms of the complex variable $z = x + iy$. In tension, the stress

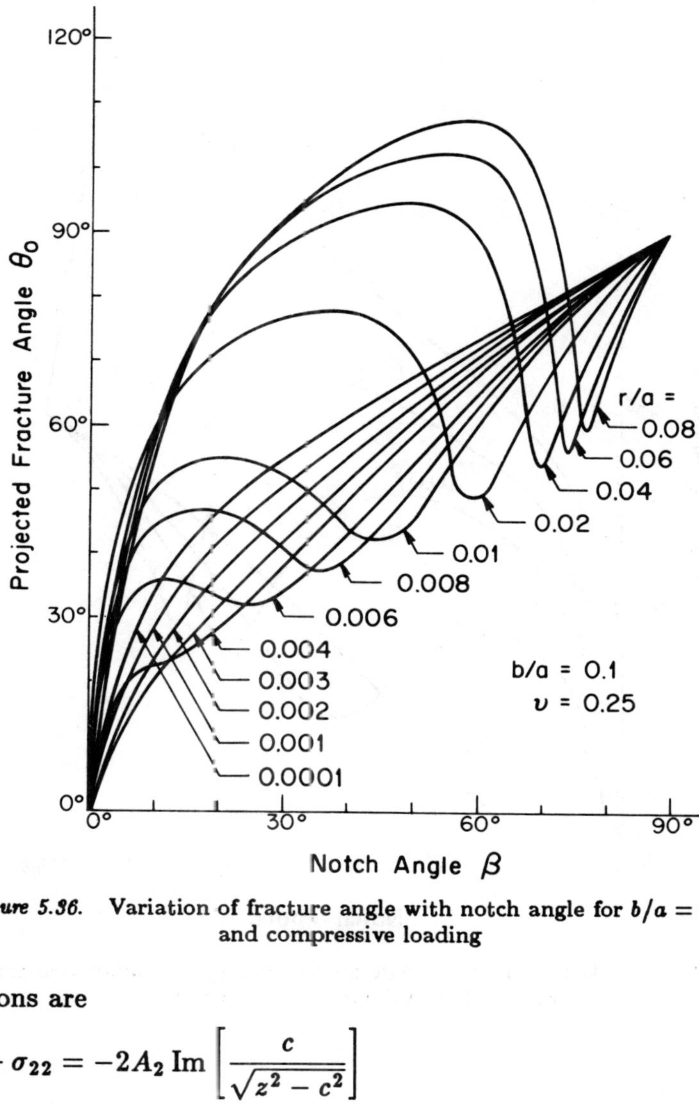

Figure 5.36. **Variation of fracture angle with notch angle for $b/a = 0.1$ and compressive loading**

combinations are

$$\sigma_{11} + \sigma_{22} = -2A_2 \operatorname{Im}\left[\frac{c}{\sqrt{z^2 - c^2}}\right] \tag{5.24a}$$

$$\sigma_{22} - \sigma_{11} + 2i\sigma_{12} = iA_2 \frac{c}{\sqrt{z^2 - c^2}}\left[1 + \frac{a^2 - b^2 - z\bar{z}}{z^2 - c^2}\right] \tag{5.24b}$$

whereas, in shear, the stresses can be found from

$$\sigma_{11} + \sigma_{22} = 2A_1 \operatorname{Re}\left[\frac{c}{\sqrt{z^2 - c^2}}\right] \tag{5.25a}$$

$$\sigma_{22} - \sigma_{11} + 2i\sigma_{12} = -A_1 \frac{c}{\sqrt{z^2 - c^2}}\left[1 + \frac{z\bar{z} - a^2 + b^2}{z^2 - c^2}\right]. \tag{5.25b}$$

The symbols Re and Im stand for the real and imaginary parts of a complex function. The constants A_1 and A_2 in equations (5.24) and (5.25) are given

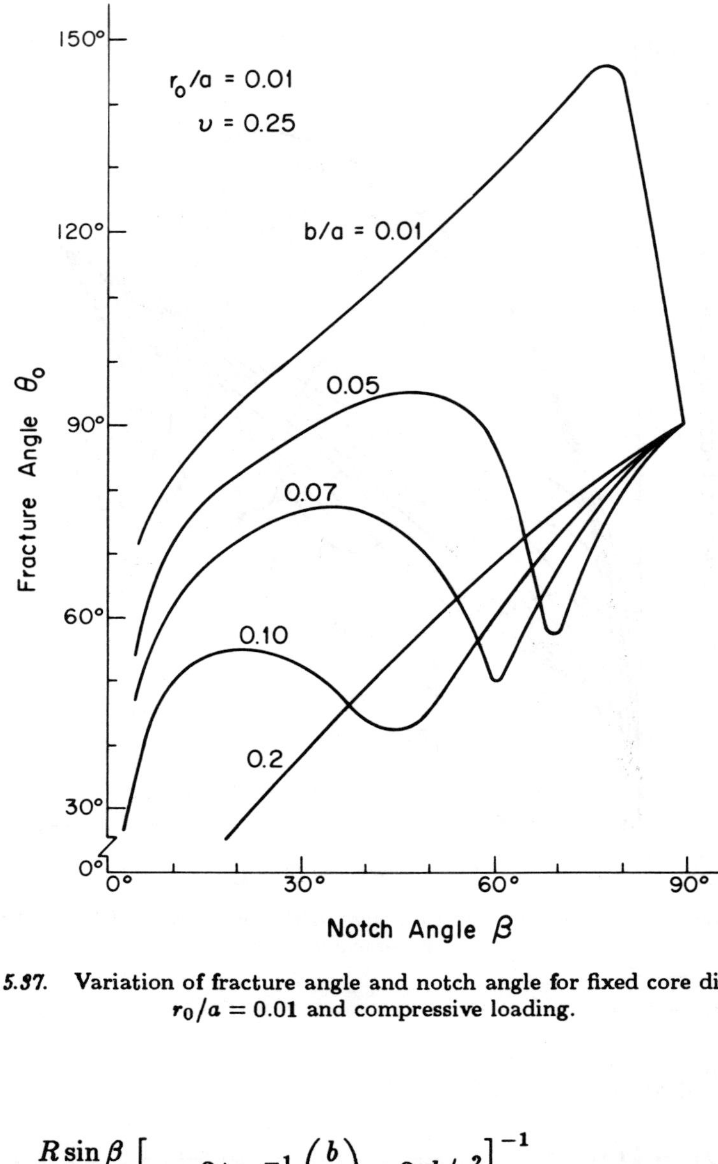

Figure 5.37. **Variation of fracture angle and notch angle for fixed core dimension: $r_0/a = 0.01$ and compressive loading.**

by

$$A_1 = \frac{R\sin\beta}{ch}\left[\pi - 2\tan^{-1}\left(\frac{b}{a}\right) - 2ab/c^2\right]^{-1} \tag{5.26a}$$

$$A_2 = \frac{R\cos\beta}{ch}\left[\pi - 2\tan^{-1}\left(\frac{b}{a}\right) + 2ab/c^2\right]^{-1}. \tag{5.26b}$$

In equations (5.26), R is the applied force, h the thickness of the solid, B the angle of inclination of R to the horizontal axis and $c^2 = a^2 + b^2$. Superposition of the tension and shear cases provides the stresses as a function of z. The branch cuts of the solution are such that the solution is correct only in the first and third quadrants. (The negative of the solutions is required

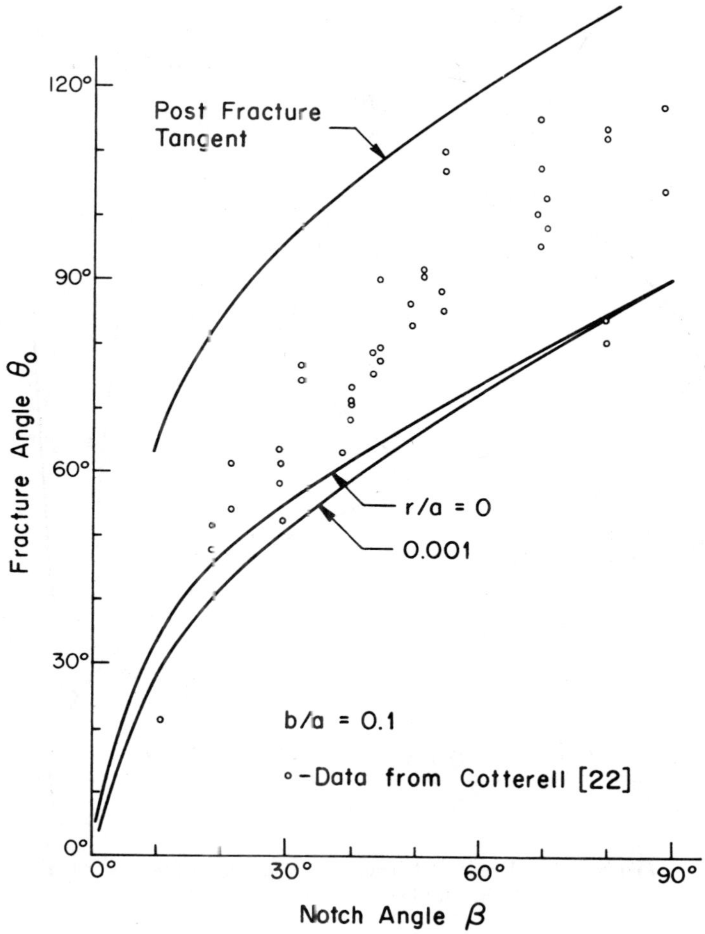

Figure 5.38. Variation of fracture angle with notch angle for $b/a = 0.01$ and compressive loading data [22].

for the remaining regions.)

Surface layer energy. For nonvanishing ratios of b/a (i.e., notches other than line crack), it is necessary to know the location of failure next to the notch boundary. As in the case of the internal notch, a formal boundary layer that separates the notch surface and interior regions is introduced. This boundary layer has a layer thickness, δ, of approximately continuum dimensions in thickness. The failure of this layer is to be analyzed using a surface layer energy quantity γ_e as given by equation (5.2) with σ_t given by

$$\sigma_t = \frac{2(-A_1 c \tan\phi + A_2)\sqrt{a^2 - b^2\tan^2\phi}}{ab(1 + \tan^2\phi)} \tag{5.27}$$

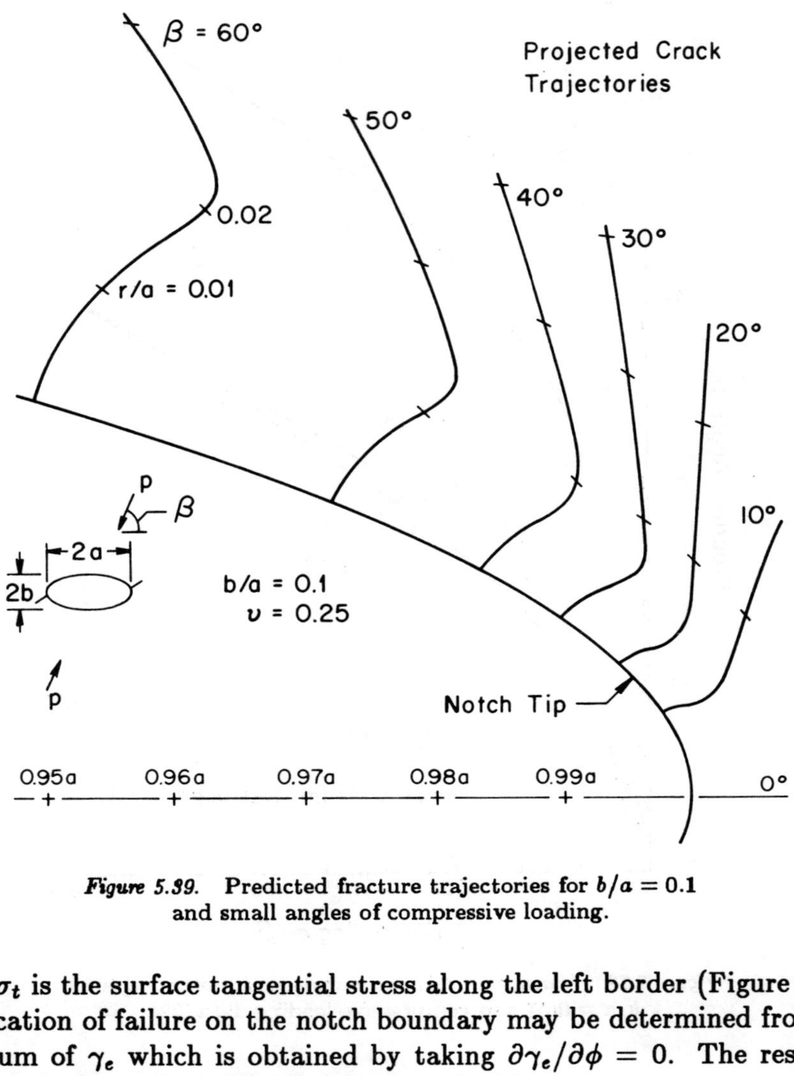

Figure 5.39. **Predicted fracture trajectories for $b/a = 0.1$ and small angles of compressive loading.**

where σ_t is the surface tangential stress along the left border (Figure 5.42). The location of failure on the notch boundary may be determined from the maximum of γ_e which is obtained by taking $\partial\gamma_e/\partial\phi = 0$. The resulting expression to be solved for the normal angle ϕ is

$$\tan^3\phi + \frac{A_1}{A_2}\left(2 + \frac{a^2}{b^2}\right)\tan^2\phi - \left(\frac{2a^2}{b^2} + 1\right)\tan\phi - \frac{A_1}{A_2}\frac{a^2}{b^2} = 0. \quad (5.28)$$

The coordinates (x_0, y_0) locating the point of failure can be found from

$$x_0 = -a^2\Big/\sqrt{a^2 - b^2\tan^2\phi} \quad (5.29a)$$

$$y_0 = b^2\tan\phi\Big/\sqrt{a^2 - b^2\tan^2\phi} \quad (5.29b)$$

For the case of the line crack ($b/a = 0$), the coordinates reduce to $(x_0, y_0) = (-a, 0)$, and the angle ϕ is not defined.

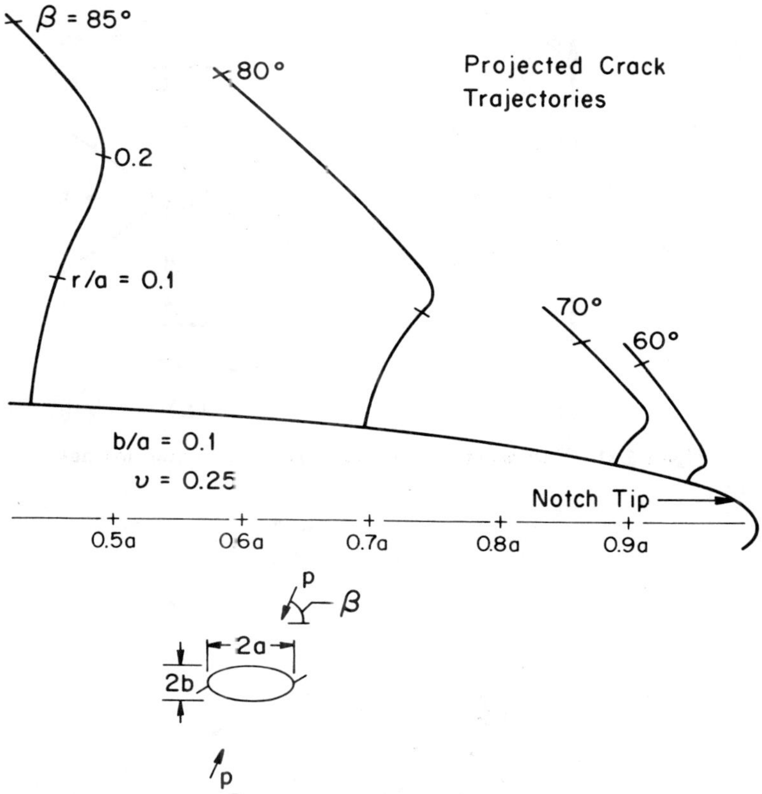

Figure 5.40. Predicted fracture trajectories for $b/a = 0.1$ and large angles of compressive loading.

Strain energy density criterion. As discussed earlier, failure is expected to initiate in the region where γ_e is a maximum, although the failure loads are expected to be governed not by γ_e, but by material response interior to the free surface. Specifically, application of the strain energy density failure criterion requires the use of a local polar coordinate system; the complex variable z is assigned the form

$$z = x_0 + iy_0 + r\,e^{i\theta} \tag{5.30}$$

i.e., the origin of the polar coordinate system is located where γ_e is maximum. The above expression for z is substituted into the stresses, and the stresses into the strain energy density function in equation (5.17).

Taking r/a to be constant, θ is varied until the strain energy density assumes its minimum value; failure is postulated to occur at this location, at a load corresponding to setting dW/dV equal to some critical constant, $(dW/dV)_c$. As r is changed, a family of curves, as functions of β, is generated, of which one in particular is expected to describe a particular material, i.e., $(dW/dV)_c$ and one r/a are to be constants, resulting in a two-parameter

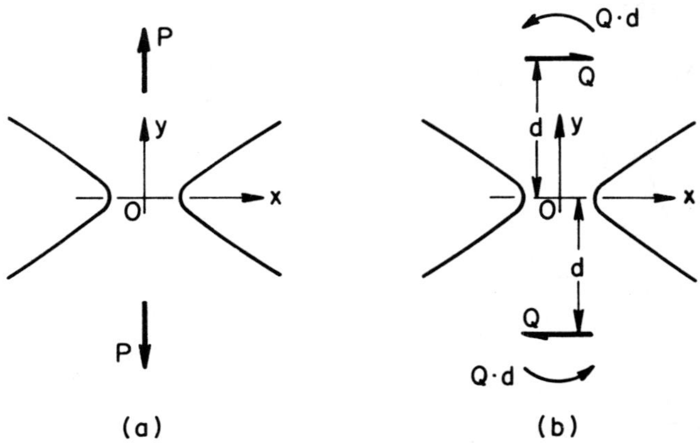

Figure 5.41 Extension and shearing of two external notches.

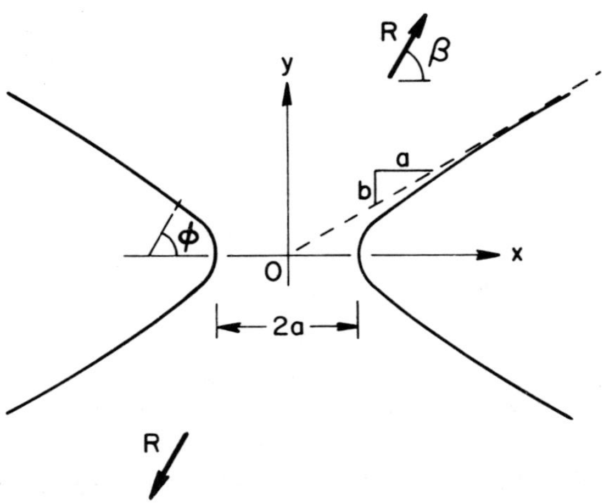

Figure 5.42 Geometry and loading of two hyperbolic notches.

failure criterion. For any value of r/a, with $\mathrm{d}W/\mathrm{d}V = (\mathrm{d}W/\mathrm{d}V)_c$, the right-hand side of equation (5.17) is of the form $(R/ah)^2$ times a nondimensional function of the variables r/a, β, θ and b/a. Then a required load to failure may be expressed as a nondimensional force, $R/ah\sqrt{\mu\,\mathrm{d}W/\mathrm{d}V}$. It is emphasized that on the subsequent discussion, no particular material is considered, so that the results are presented for a range of values of r/a and b/a.

Poisson's ratio has been set at 0.3 for convenience, and its variation will modify the numerical results slightly, but the tenor of the discussion is unaffected.

A characteristic of the internal notch is that the weakest position is in the normal loading configuration as shown earlier; in general, this is not the case for the external notch. Figure 5.43 indicates that for a wide range of r/a, (in fact, down to 10^{-6}), the minimum load occurs for $20° < \beta < 30°$,

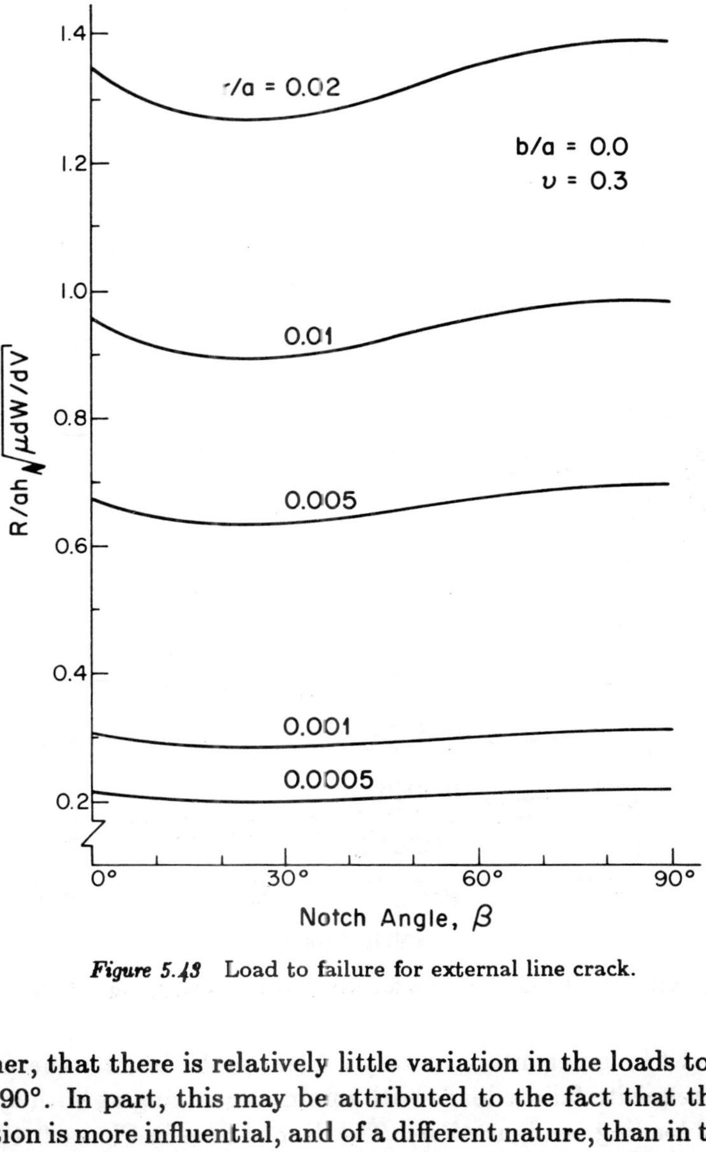

Figure 5.43 Load to failure for external line crack.

and further, that there is relatively little variation in the loads to failure for $0 \leq \beta \leq 90°$. In part, this may be attributed to the fact that the shearing contribution is more influential, and of a different nature, than in the internal notch case, i.e., net force loading, rather than uniform stress loading. As b/a is increased from zero to 0.01, Figure 5.44, the minimum is seen not to occur until $r/a \simeq 0.0005$. The minimum at distances $r/a \leq 0.0001$ does not reappear until $b/a \simeq 0.3$, Figure 5.47. Figures 5.45 and 5.46 show the variations of expected failure loads for $b/a = 0.1$, and 0.2. As may be observed, say in Figure 5.45, if r/a is chosen to be 0.005, the behavior differs substantially from that of $r/a = 0.01$. It is therefore imperative that for a given material, a value of r/a be found; this will allow, then, the description of several geometries, as described in Figure 5.48. In this figure, r/a is fixed

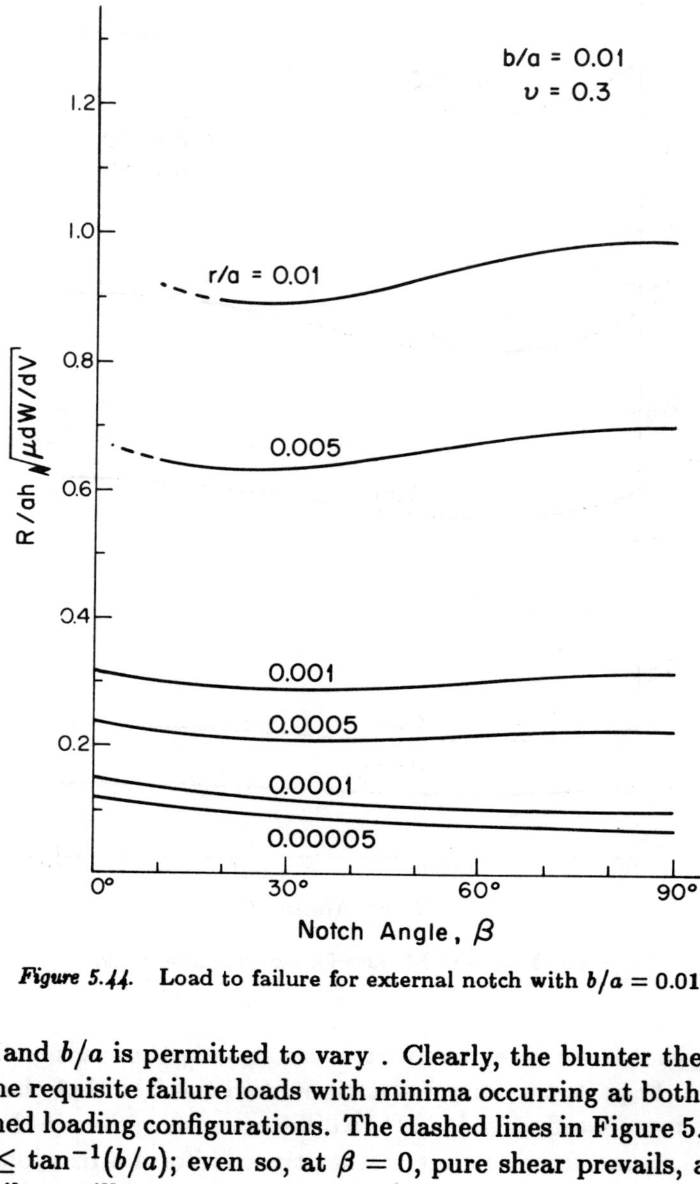

Figure 5.44. Load to failure for external notch with $b/a = 0.01$.

at 0.001, and b/a is permitted to vary . Clearly, the blunter the notch, the greater the requisite failure loads with minima occurring at both symmetric and inclined loading configurations. The dashed lines in Figure 5.48 indicate where $\beta \leq \tan^{-1}(b/a)$; even so, at $\beta = 0$, pure shear prevails, and a finite load to failure still occurs.

Predicted fracture angles. Preliminary observations on double-edge cracked specimens under inclined loading suggest that the longer range predicted fracture trajectories, that were quite satisfactory in the internal crack case, are no longer reliable. Equivalently, the external notch fracture trajectory is very dependent upon the growing-crack geometry. This is especially true for initially sharp cracks; for more blunted cracks or notches, initial failure

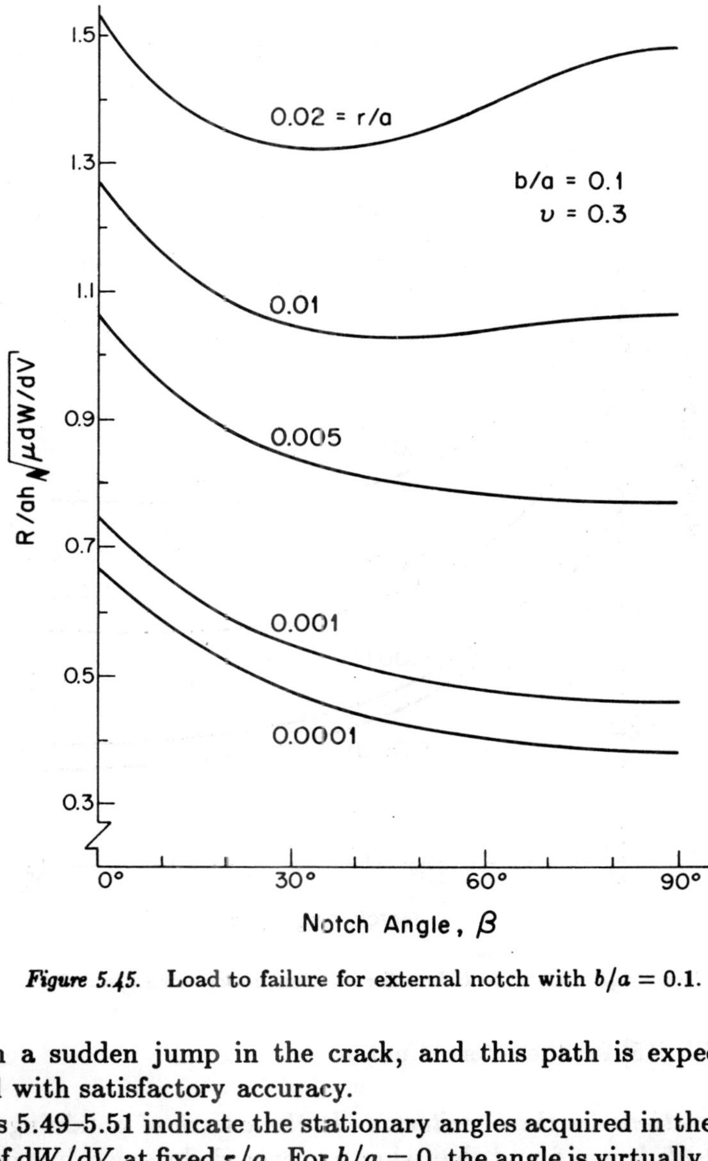

Figure 5.45. Load to failure for external notch with $b/a = 0.1$.

results in a sudden jump in the crack, and this path is expected to be predicted with satisfactory accuracy.

Figures 5.49–5.51 indicate the stationary angles acquired in the search for minima of dW/dV at fixed r/a. For $b/a = 0$, the angle is virtually unchanged for a wide range of r/a. However, for blunted cracks, the angles at fixed β are seen to first decrease, then increase with increasing r/a; $r/a = 0$ corresponds to the surface normal angle. If the tangent angle is taken after θ begins to increase, it is observed to correspond to the line crack angle.

The summary of plots such as Figure 5.51 may be seen in Figure 5.52 where for $b/a = 0.3$, at a fixed angle β, the sequence of locations of minima for increasing r/a are plotted. One observes a bending from the surface normal towards the x-axis, then away into a tangent that parallels those of the line crack. This was observed by Cotterell [21], and predicted later by

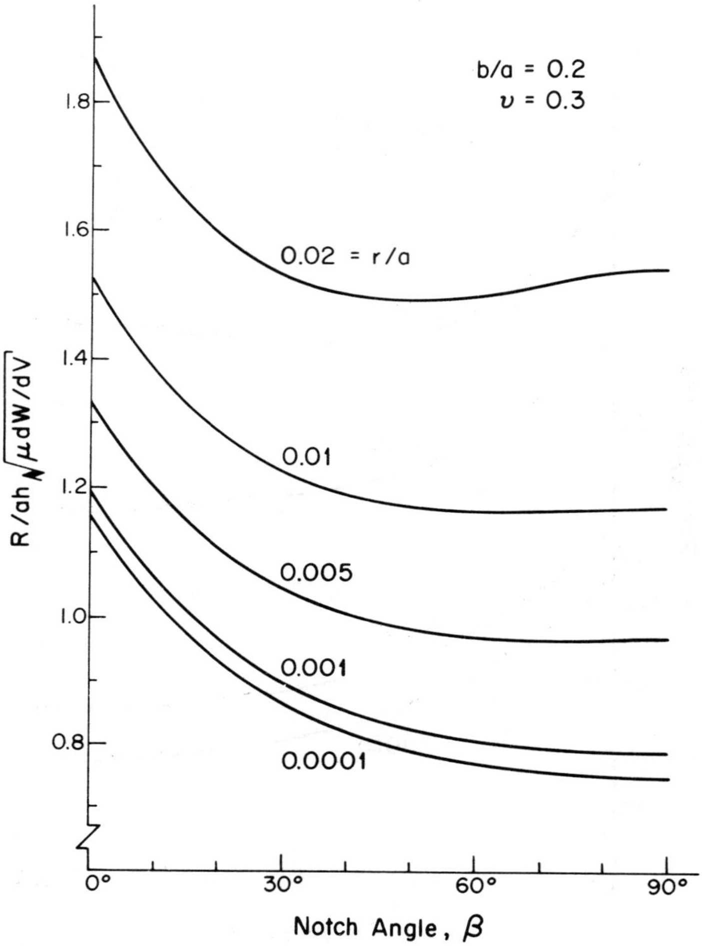

Figure 5.46. Load to failure for external notch with $b/a = 0.2$.

Kipp and Sih [15] for the internal notch.

Some observations on the S-criterion. In the course of obtaining the minima of dW/dV, it was found that when r/a exceeded some value, dependent on b/a, β, etc., a splitting of the locations of the minima occurred; that is, instead of acquiring a single minimum, a maximum value for dW/dV appeared, flanked by two minima. Initially, it was thought that bifurcation was being predicted, but it now appears that the function is merely stating its own limitations. Such a phenomenon was not observed for the internal notch, so it must indicate the limiting region where useful trajectory information can be found. If instead of fixing the origin at the surface of the notch, the origin is moved to the new minimum, at each stage, a floating origin is established that follows the path of predicted fracture. This scheme

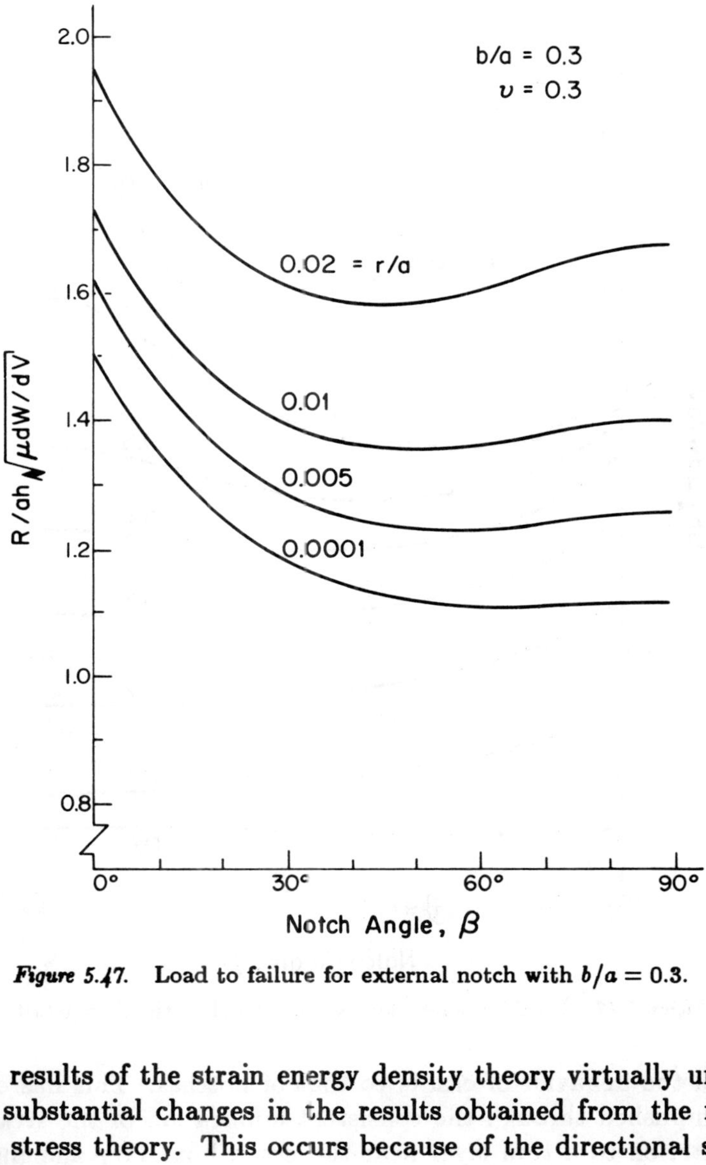

Figure 5.47. **Load to failure for external notch with $b/a = 0.3$.**

leaves the results of the strain energy density theory virtually unchanged; there are substantial changes in the results obtained from the maximum tangential stress theory. This occurs because of the directional sensitivity of the latter.

Most significant of the results previously discussed is the appearance of nonsymmetrical weak orientations for notches. On the premise that failure is governed by interior behavior of the solid, it is clear that the predicted failure loads are quite sensitive to the distance from the (free) surface, not only in magnitude, but also in character.

5.6. Concluding remarks

One of the most important purposes of the application of the theory at hand

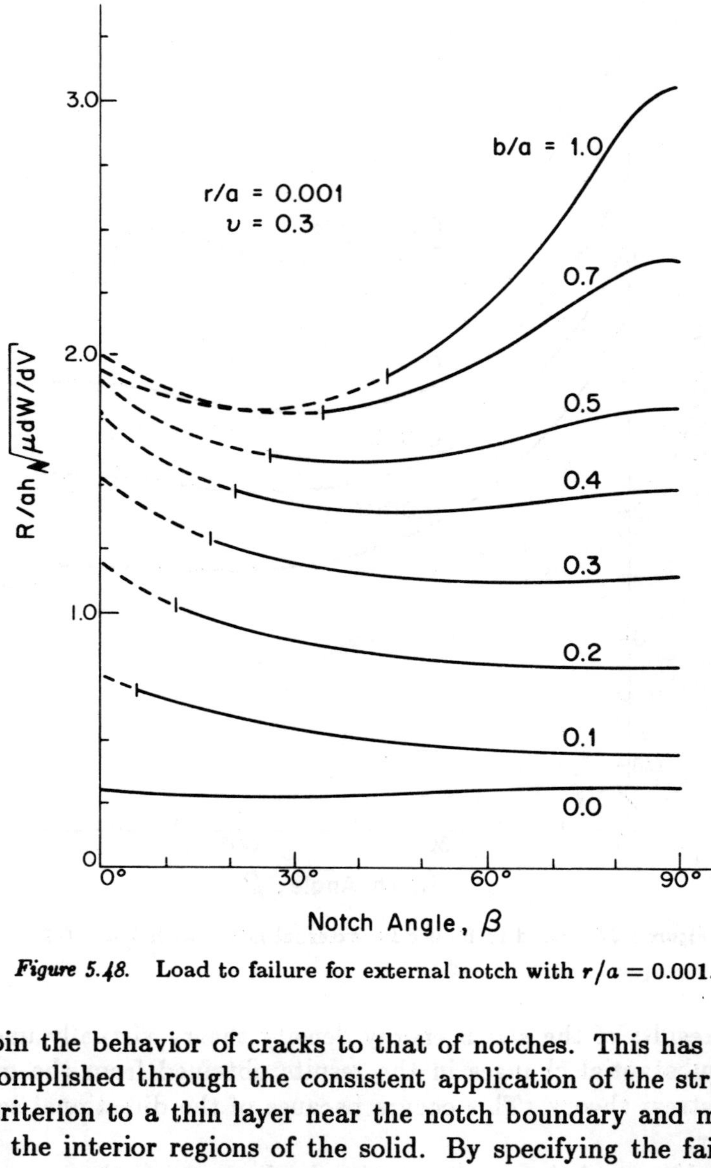

Figure 5.48. Load to failure for external notch with $r/a = 0.001$.

was to join the behavior of cracks to that of notches. This has essentially been accomplished through the consistent application of the strain energy density criterion to a thin layer near the notch boundary and material element in the interior regions of the solid. By specifying the failure to be dependent upon internal conditions being met before it can occur, the barrier that has separated notch and crack failure theories has been disposed. Another engaging result was the description of the phenomenon of post-fracture behavior. The local behavior has been shown to agree well with observable details of crack paths emanating from a notch tip.

The theory upon which all the results were based is that of total strain energy density. Bridgman [24] has observed that when hydrostatic pressure is applied to a specimen, the strain energy density may be raised to very high values, far exceeding those found under normal conditions, without ensuing

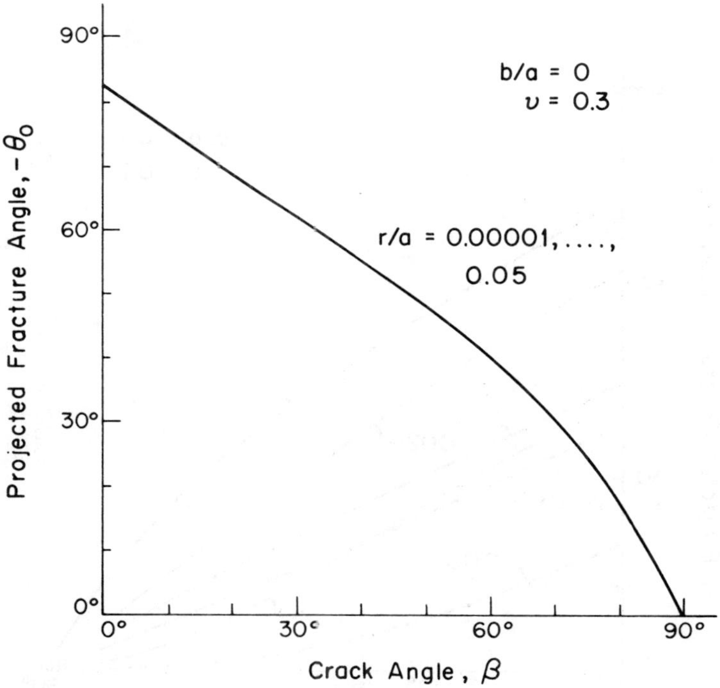

Figure 5.49. **Fracture angles for external line cracks.**

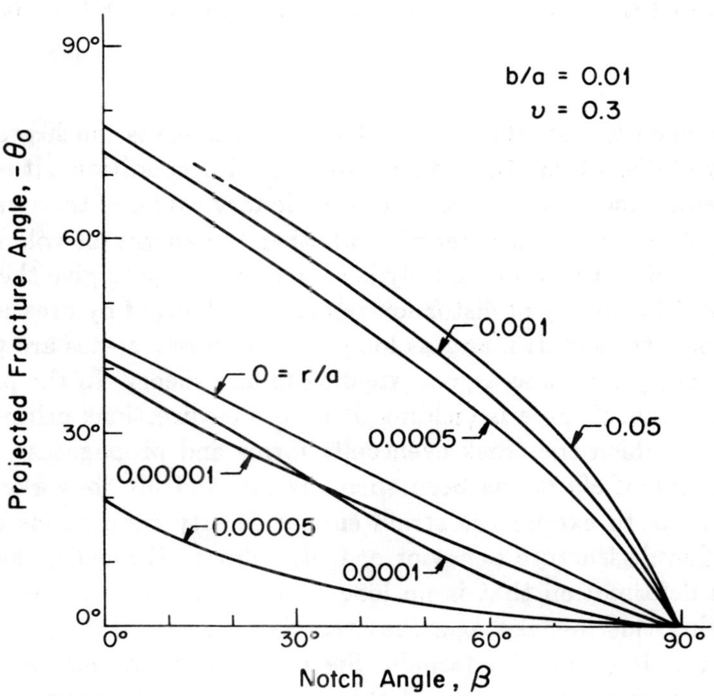

Figure 5.50. **Fracture angles for two external notches with $b/a = 0.01$.**

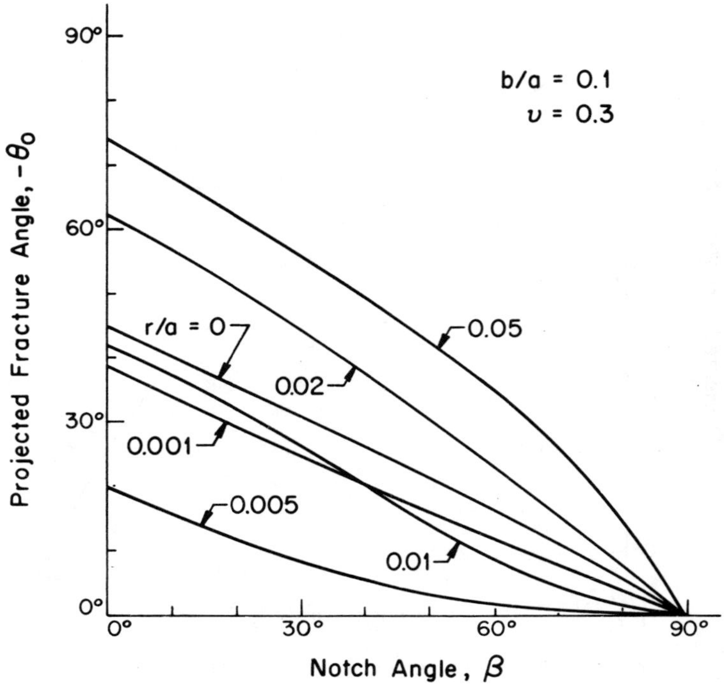

Figure 5.51. Fracture angles for two external notches with $b/a = 0.1$.

failure. In such a case, the energy of volume change is much greater than the energy of distortion. But under tensile loading conditions, it is still held that the requirements are to locate the stationary values of the strain energy density as described, and determine whether the energy of volume change exceeds that of distortion or simply is sufficiently large to give the direction of fracture. The energy of distortion remains unchanged by presence of uniform triaxial stress states, and as long as other stress states are present as well, apparently stands as a good yielding failure theory. In the presence of a crack or notch, if there is yielding, it occurs in directions primarily other than that in which the crack eventually forms and propagates. A considerable amount of effort has been spent by Sih and his co-workers [25–27] in recent years to extend the strain energy density concept for characterizing the ductile fracture behavior of metal alloys. Basically, the material undergoes deformation that is no longer negligible prior to unstable fracture. During this initial stage, the crack is constrained by plasticity and may grow slowly in a stable fashion. Such a phenomenon has been referred to as necking in a tensile bar and the thinning of a plate specimen. Analytical modeling of ductile fracture is inherently three-dimensional in nature

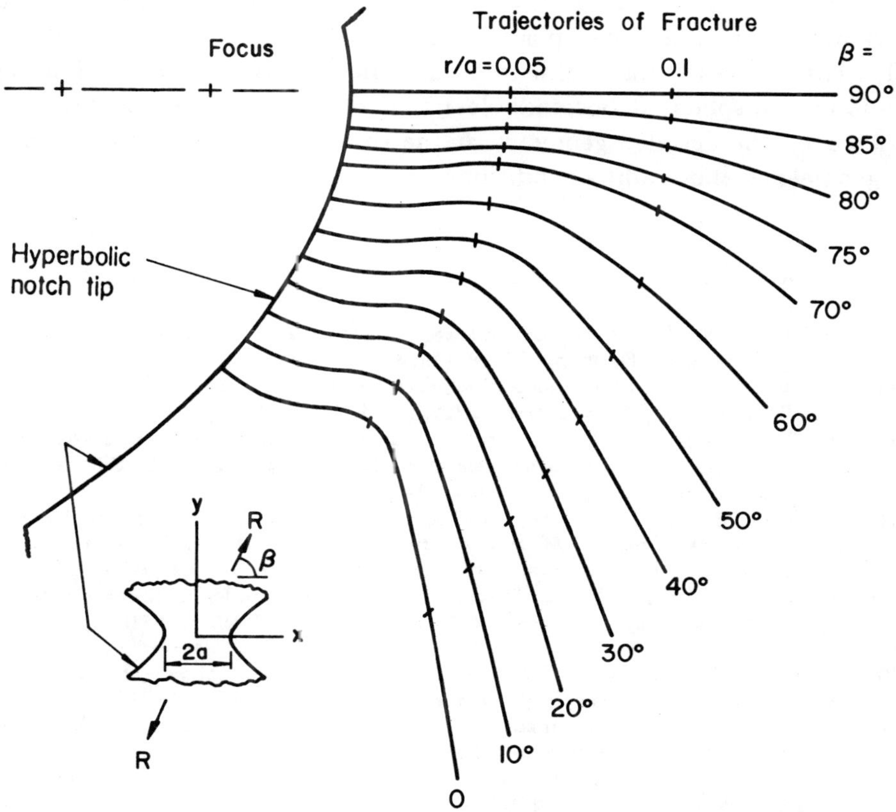

Figure 5.52. **Local fracture trajectories for external notch with $b/a = 0.3$.**

and is antisymmetrical such as the cup-and-cone failure. In most cases, even though the initial load is applied normal to the crack plane, the last ligament of failure may be at an angle with the specimen boundary which necessitates the use of a fracture criterion that can treat mixed mode crack extension. One of the basic views taken in the strain energy density approach is that the critical value of dW/dV is assumed to be a material constant at the very beginning and hence the objective of the analysis is to predict the failure load and the crack shape and size at instability. Preliminary results [27] obtained thus far have been very encouraging.

Finally, it is also appropriate to observe that failure theories must be developed that will allow continuity from nonflawed to flawed members. At present, once a flaw is postulated to be the mechanism whereby failure is expected to occur, then no matter how small the flaw becomes, it still dominates. In the case of prevailing fracture theories based on the crack tip singularity, although the strength of the singularity depends on the crack length, if the crack length becomes small, there is no other criterion that can

be coordinated to apply in conjunction with the postulated crack. Parameters are required that can place the size of the flaw in context with some other characteristic material length, and allow a more unified evaluation of the failure of solids. It becomes clear that the final criterion cannot be restricted by the detailed geometry at the tip of the flaw but must apply to all geometries and loading conditions.

5.7. References

[1] Griffith, A. A., The theory of rupture, *Proceedings of First International Congress of Applied Mathematics*, Delft, pp. 55–93 (1924).

[2] Griffith, A. A., The phenomena of rupture and flow in solids, *Philosophical Transactions, Royal Society of London, Series A221*, pp. 163–198 (1921).

[3] Sih, G. C., A special theory of crack propagation: methods of analysis and solutions of crack problems, *Mechanics of Fracture I*, edited by G. C. Sih, Noordhoff International Publishing, Leyden, pp. 21–45 (1973).

[4] Sih, G. C., Surface layer energy and strain energy density for a blunted crack or notch, *Prospects of Fracture Mechanics*, edited by G. C. Sih, H. C. van Elst and D. Broek, Noordhoff International Publishing, Leyden, pp. 85–102 (1974).

[5] Sih, G. C. and Kipp, M. E., Discussion on fracture under complex stress – the angled crack problem, *International Journal of Fracture Mechanics* **10**, pp. 261–265 (1974).

[6] Williams, J. G. and Ewing, P. D., Fracture under complex stress – the angled crack problem, *International Journal of Fracture Mechanics* **8**, pp. 441–446 (1972).

[7] Matthaes, K., Betrachtungen zur theorie der Werkstoff-Festigkeit, *Zeitschrift für Metallkunde* **43**, pp. 11–19 and 90–95 (1952).

[8] Gillemot, L. F., Criterion of crack initiation and spreading, *Journal of Engineering Fracture Mechanics* **8** pp. 239–253 (1976).

[9] Sih, G. C., A three-dimensional strain energy density factor theory of crack propagation: three-dimensional crack problems, *Mechanics of Fracture II*, edited by G. C. Sih, Noordhoff International Publishing, Leyden, pp. 15–53 (1975).

[10] Neuber, H., *Theory of Notches*, Edwards, Michigan (1946).

[11] Muskhelishvili, N. I., *Some Basic Problems of the Mathematical Theory of Elasticity*, Noordhoff, Groningen (1953).

[12] McClintock, F. A. and Walsh, J. B., Friction on Griffith cracks in rocks under pressure, *Proceedings of Fourth U.S. National Congress of Applied Mechanics*, pp. 1015–1021 (1962).

[13] Haigh, B. P., The strain energy function and the elastic limit, *British Association of Advancement of Sciences*, pp. 486–495 (1919).

[14] Sih, G. C., Application of strain energy density theory to fundamental fracture problems, *Proceedings of the 10th Annual Meeting of the Society of Engineering Science*, North Carolina State University, pp. 221–234 (1975).

[15] Kipp, M. E. and Sih, G. C., The strain energy density failure criterion applied to notched elastic solids, *International Journal of Solids and Structures* **2**, pp. 153–173 (1975).

[16] Hartranft, R. J. and Sih, G. C., Stress singularity for a crack with an arbitrarily curved front, *Journal of Engineering Fracture Mechanics* **9**, pp. 705–718 (1977).

[17] Palaniswamy, K., *Crack propagation under general in-plane loading*, Ph.D. Dissertation, California Institute of Technology (1972).

[18] Erdogan, F. and Sih, G. C., On the crack extension in plates under plane loading and transverse shear, *Journal of Basic Engineering*, pp. 519–525 (1963).

[19] Kassir, M. K., Three-dimensional notch problems: stress analysis of notches, *Mechanics of Fracture V*, edited by G. C. Sih, Noordhoff International Publishing, Alphen

a/d Rijn, pp. 173–194 (1978).

[20] McClintock, F. A., Discussion on crack extension in plates under plane loading and transverse shear, *Journal of Basic Engineering*, pp. 525–527 (1963).

[21] Cotterell, B., The paradox between the theories for tensile and compressive fracture, *International Journal of Fracture Mechanics* **5**, pp. 251–252 (1969).

[22] Cotterell, B., Brittle fracture in compression, *International Journal of Fracture Mechanics* **8**, pp. 195–208 (1972).

[23] Griffith, A. A., Stresses in a plate bounded by a hyperbolic cylinder, *Great Britain Aeronautical Research Council Technical Report* **2**, pp. 668–677 (1928).

[24] Bridgman, P. W., *Studies in Large Plastic Flow and Fracture*, McGraw-Hill, New York (1952).

[25] Sih, G. C., Elastic-plastic fracture mechanics, *Prospects of Fracture Mechanics*, edited by G. C. Sih, H. C. van Elst and D. Broek, Noordhoff International Publishing, Leyden, pp. 613–621 (1974).

[26] Sih, G. C., Fracture toughness concept, *American Society for Testing of Materials*, ASTM STP 605, pp. 3–15 (1976).

[27] Sih, G. C., Mechanics of ductile fracture, *Proceedings of Fracture Mechanics and Technology*, edited by G. C. Sih and C. L. Chow, Noordhoff International Publishing, Alphen a/d Rijn, pp. 767–784 (1977).

6

Thermoelastic and hygrothermoelastic behavior of cracks

6.1. Introduction

The failure behavior of solids is affected not only by mechanical loads but also by the change of temperature and/or moisture. These external disturbances can interact and their influence on the integrity of structural members can be reflected in the energy stored in a unit volume of material. Determination of the volume energy density will involve mechanical deformation in addition to changes in temperature and/or moisture.

In solids, the predominant effect of heat transfer is due to heat conduction while convection is negligibly small and radiation need not be considered unless the environment calls for it. Moreover, if the temperature gradients are not sufficiently great* to cause material transformation, then for all practical purposes the thermoelastic parameters may be assumed to be temperature independent. Coupling between the stress and temperature field may also be disregarded. The conventional theory of linear thermoelasticity can thus be applied such that the temperature field is determined independent of the stress or strain fields.

The magnitude of the temperature gradient near a crack tip was found by Sih [1] to increase indefinitely as a function of $1/\sqrt{r}$ as $r \to 0$, where r is the radial distance measured from the crack tip. This singular character is preserved in the thermoelastic stresses in the immediate vicinity of the crack tip as derived by Sih [2]. The corresponding strain energy density function dW/dV possesses the character of $1/r$. Unlike the stresses whose singular behavior depends on the constitutive relation, the $1/r$ functional form for dW/dV remains unchanged not only for all constitutive relations, but also for all shapes of planes of discontinuities. The same applies to moisture gradient regardless of whether it is coupled or not coupled to temperature.

---

* In regions where geometry undergoes sudden changes such as a plane of discontinuity or crack, Sih [1] has shown that the temperature gradient can have extremely high elevation and special attention must then be given to the local behavior.

Coupling could only affect the results quantitatively but not the singular character of dW/dV. This chapter is concerned with failure initiating from a crack in a solid that undergoes nonuniform changes in temperature and/or temperature gradient. The additional influences of mechanical loading and moisture will also be presented by application of the hygrothermal–elasticity theory of Sih, Michopoulos and Chou [3].

6.2. Linear thermoelasticity

When temperature gradients are present, the strain energy density function, dW/dV, is dependent not only on the state of stress and strain but also the level of thermal energy. If $T(x, y)$ represents the temperature at point (x, y) in a solid having a reference temperature of T_0, then dW/dV in the theory of linear thermal elasticity can be written as [4]

$$\frac{dW}{dV} = \tfrac{1}{2}\sigma_{ij}\epsilon_{ij} - \tfrac{1}{2}\alpha\,\Delta T\sigma_{kk} \tag{6.1}$$

in which $\Delta T = T - T_0$ and α is the linear coefficient of thermal expansion. The stress and strain components are denoted by σ_{ij} and ϵ_{ij} respectively. Making use of the linear isotropic forms of the stress–strain relation

$$\sigma_{ij} = 2\mu\left[\epsilon_{ij} + \frac{\nu}{1-2\nu}\epsilon_{kk}\delta_{ij} - \frac{1+\nu}{1-2\nu}\alpha\,\Delta T\delta_{ij}\right] \tag{6.2}$$

into equation (6.2), it is found that

$$\frac{dW}{dV} = \frac{1}{2E}\left[(1+\nu)\sigma_{ij}\sigma_{ij} - \nu\sigma_{kk}^2\right], \tag{6.3}$$

where ν is Poisson's ratio. In equation (6.2), μ is the shear modulus and δ_{ij} is the Kronecker delta. Note that the influence of ΔT does not appear explicitly in equation (6.3) when dW/dV is expressed in terms of the stress components. However, when the σ_{ij} in equation (6.3) are eliminated by means of the strain–stress relation

$$\epsilon_{ij} = \frac{1+\nu}{E}\sigma_{ij} - \frac{\nu}{E}\sigma_{kk}\delta_{ij} + \alpha\,\Delta T\delta_{ij}, \tag{6.4}$$

there results the expression

$$\frac{dW}{dV} = \mu\left[\epsilon_{ij}\epsilon_{ij} + \frac{\nu}{1-2\nu}\epsilon_{kk}^2 - \frac{2(1+\nu)}{1-2\nu}\alpha\,\Delta T\epsilon_{kk}\right] + \frac{3\mu(1+\nu)}{1-2\nu}(\alpha\,\Delta T)^2. \tag{6.5}$$

Either equation (6.3) or (6.5) may thus be used to compute the strain energy density function.

6.3. Fracture caused by nonuniform temperature changes

It was shown by Kipp and Sih [5] and Sih [6] that the initial notch geometry and loading (and material properties) may not only determine the failure load and initial fracture angle, but also the subsequent fracture path. For a line crack in tension, the fracture path appears to follow a trajectory of points of relative minimum strain energy density. When local and global instability occur simultaneously, the energy state with reference to the initial configuration may determine the subsequent fracture path in some neighborhood. In such cases, crack growth occurs rapidly and the material has little time to follow the change in crack geometry for small time intervals. Hence the prediction yields sufficiently accurate results for distances comparable to the half-length of the crack or notch.

Failure initiation and $\mathrm{d}W/\mathrm{d}V$ *path.* When the crack acquires a finite radius of curvature, the point of failure initiation is no longer obvious. The location on the blunted crack front or notch must be determined. To this end, the strain energy density in a surface layer has to be calculated from the expression derived by Sih [6]. If the thickness of this layer is δ, the strain energy per unit surface layer area for plane strain is given by*

$$\frac{\mathrm{d}W}{\mathrm{d}A} = \frac{1-\nu^2}{2E}\delta\sigma_t^2, \tag{6.6}$$

where σ_t is the local tangential stress. The location on the notch boundary at which failure occurs is assumed to coincide with the maximum value of $\mathrm{d}W/\mathrm{d}A$. The surface thickness δ provides a means of considering material properties in a surface layer that may differ from those in the bulk of the solid. Inspection of equation (6.6) shows that $\mathrm{d}W/\mathrm{d}A$ does not distinguish the difference between regions of tensile and compressive tangential stress along the notch boundary. The sign of the tangential stress at the point of maximum strain energy density along the notch boundary, however, may serve as a guide for determining the point of failure initiation. Positive tangential stress or tension may be assumed to coincide with the site of failure. Alternatively, experience shows that the fracture path tends to deviate in a direction normal to tension and to spread into the direction of compression. With this knowledge, the notch side at which failure initiates becomes obvious.

Once the point of failure initiation p on the notch is known, the relative minima of the strain energy density, or $(\mathrm{d}W/\mathrm{d}V)_{\min}$, may be calculated for determining the fracture trajectories. To be more specific, a set of local polar coordinates (r,θ) can be defined at the point of failure initiation, as shown

* The notation γ_e was used by Sih in [6] instead of $\mathrm{d}W/\mathrm{d}A$.

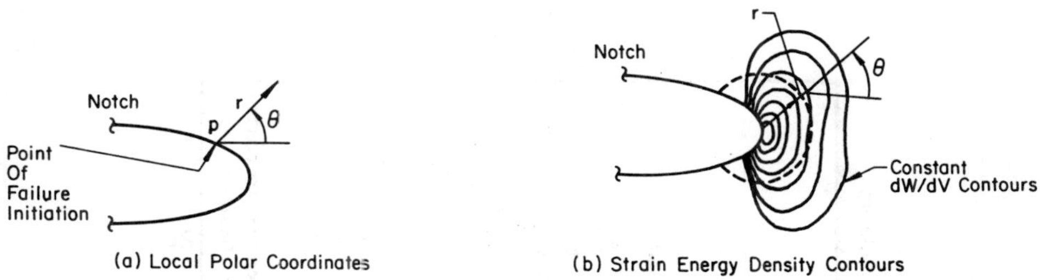

Figure 6.1. Schematic of state of affairs near notch front.

in Figure 6.1(a). The values of dW/dV at all points near the notch front can then be located. Figure 6.1(b) shows a typical set of constant dW/dV contours. The stationary values of dW/dV are thus obtained. In general, there might be several $(dW/dV)_{\min}$ values. The maximum relative minima, $(dW/dV)^{\max}_{\min}$, are selected as the locus of the predicted fracture path.

Heat transfer and stress analysis. A finite element procedure is used to solve the heat conduction portion of the problem. The steady-state temperature distribution is then input into a separate stress analysis computer program [8] that utilizes the 12-node isoparametric quadrilateral elements having cubic shape functions. These elements assume a bicubic variation of the displacement field. Numerical integration is performed using a 3×3 Gaussian quadrature. The program has a thermal stress analysis capability requiring the input of nodal temperatures. Correction on the out-of-plane stress σ_z for the case of plane strain problems with thermal loading is made:

$$\sigma_z = \nu(\sigma_x + \sigma_y) - E\alpha\,\Delta T. \tag{6.7}$$

Strains, stresses and strain energy densities are calculated at sampling points in the interior of the element and interpolated to other interior points or points on the element boundary. This results in improved accuracy and interelement continuity. The Gauss–Legendre quadrature points are used in the numerical integrations. Thus, values of dW/dV can be determined for points along the locus of a radial distance r of a given magnitude. Consequently, the $(dW/dV)_{\min}$ paths can be determined from the numerical solution.

Uniform heat flow across narrow elliptical notch. The problem of an insulated elliptical notch engulfed by a uniform heat flow is solved numerically. The ratio of the semi-minor to semi-major axis of the ellipse is taken to be $b/a = 0.05$ and a condition of plane strain is assumed. Refer to Figure 6.2 for the geometric configurations and thermal boundary conditions. More

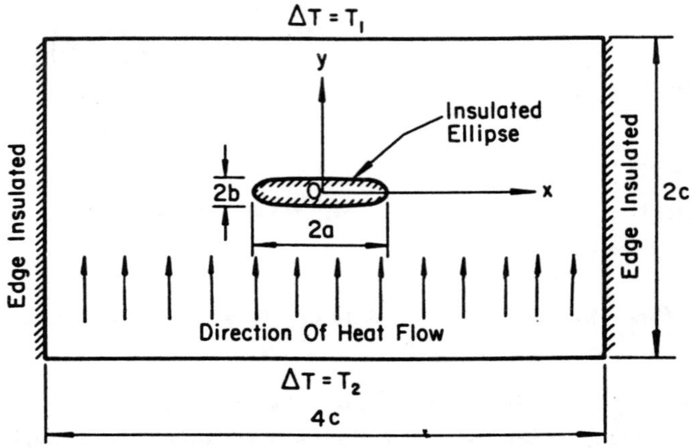

Figure 6.2. Uniform heat flow past an elliptical notch.

specifically, the thermal boundary conditions are

$$\left.\begin{aligned} T(x,c) - T_0 = T_1 \\ T(x,-c) - T_0 = T_2 \end{aligned}\right\} \quad \text{at} \quad y = \pm c \tag{6.8a}$$

$$\frac{\partial T}{\partial x} = 0 \quad \text{at} \quad x = \pm 2c \tag{6.8b}$$

and the elliptical notch is insulated from heat flow. This results in a temperature distribution approximating that of uniform heat flow across the notch. The resulting temperature field is symmetric with reference to both the x- and y-axis, as defined in Figure 6.2. By means of equation (6.3) or (6.5) the strain energy density functions dW/dV are calculated for determining their relative minima, i.e., values of (r, θ) corresponding to $(dW/dV)_{min}$ are obtained. The results are displayed in Figure 6.3 for the notch with $T_1 = -156.7°C$ and $T_2 = 121.1°C$ and Figure 6.4 for the crack with $T_1 = -T_2 = -121.1°C$. Two possible solutions of the $(dW/dV)_{min}$ path prevail because dW/dV is a positive definite function that does not depend on the direction of heat flow. The choice of the correct path can be made on physical grounds that the material on the side with the lower temperature is in tension while compression is experienced by the material on the hotter side. Hence, the solid lines correspond to the correct crack trajectories. The dotted curves are also possible solutions when the direction of heat flow is reversed. Alternatively, it is also possible to determine the location of dW/dA maximum in equation (6.6). In this case, point p in Figure 6.3 is thus selected to be the origin of failure and the upper path $(y > 0)$ is the probable fracture trajectory. Since the crack is the limiting case of the notch, the upper path is also selected for the case of a crack in Figure 6.4.

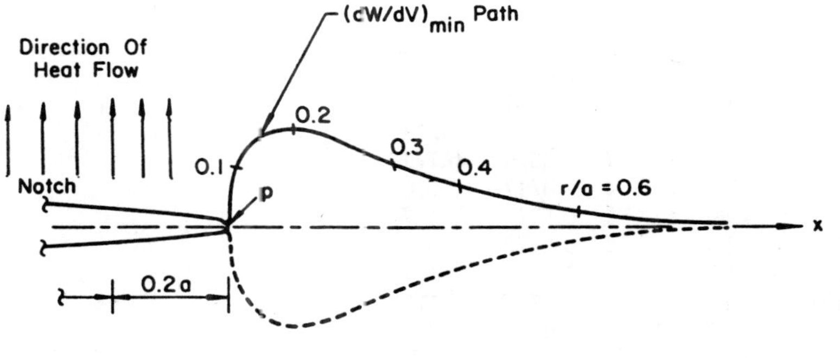

Figure 6.3. **Fracture trajectory emanating from a notch with $b/a = 0.05$, $T_1 = -156.7°C$ and $T_2 = 121.1°C$.**

$|T_2 - T_1| = 277.8\,°C$

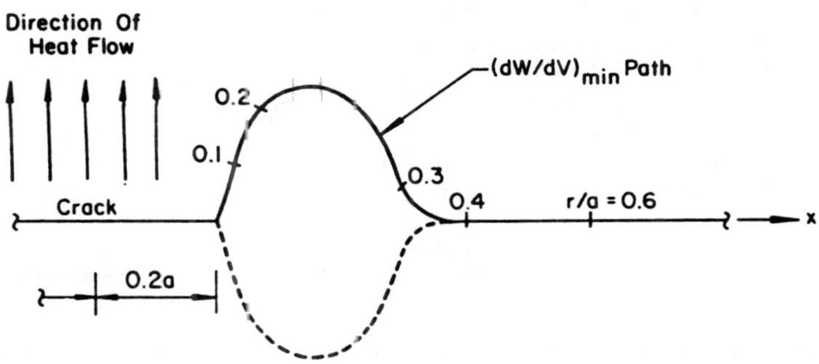

Figure 6.4. **Fracture trajectory emanating from a crack with $b/a = 0.0$, $T_1 = -T_2 = -121.1°C$.**

The paths of $(dW/dV)_{min}$ for the crack and notch in Figures 6.3 and 6.4 are seen to be similar. Particular values of dW/dV along the paths can be found in Tables 6.1 and 6.2. They are normalized with respect to

$$\left(\frac{dW}{dV}\right)_0 = \tfrac{1}{2}\mu\,[a\alpha(T_1 - T_0)/L]^2\,. \tag{6.9}$$

The values of $(dW/dV)_{min}$ for the crack are higher because of the influence of the $1/\sqrt{r}$ singularity for the crack. The curvature of the fracture trajectory in Figure 6.4 is more pronounced than that in Figure 6.3. It is encouraging that the temperature distribution for the crack agreed closely with the analytical solution in [7].

Table 6.1. Minimum strain energy densities for notch with $b/a = 0.05$.

r/a	θ	x/a	y/a	$(dW/dV)^{max}_{min}/(dW/dV)_0$
Initiation point		0.998	0.0032	12.9010
0.01	60	1.0029	0.0119	2.6430
0.02	60	1.0077	0.0207	1.8370
0.03	80	1.0032	0.0323	1.2140
0.05	80	1.0067	0.0524	0.7712
0.07	80	0.0104	0.0717	0.5907
0.1	80	1.0153	0.102	0.4570
0.2	55	1.1127	0.1673	0.3121
0.3	20	1.280	0.1064	0.2442
0.4	10	1.392	0.0745	0.1776
0.8	0	1.8	0.0	0.0572

Table 6.2. Minimum strain energy densities for crack with $b/a = 0.00$.

r/a	θ	x/a	y/a	$(dW/dV)^{max}_{min}/(dW/dV)_0$
0.01	−82	0.0014	−0.0099	18.8926
0.02	−82	0.0028	−0.0198	9.5015
0.03	−82	0.0042	−0.0297	6.3862
0.05	−82	0.0070	−0.0495	3.9220
0.07	−81	0.0110	−0.0691	2.8923
0.1	−80	0.0174	−0.0985	2.1553
0.2	−71	0.0651	−0.1891	1.4365
0.3	−35	0.2457	−0.1721	1.2371
0.4	0	0.4000	0.0	0.8982
0.8	0	0.8000	0.0	0.3849

6.4. Simultaneous application of thermal and mechanical loading

The interaction of mechanical loading with thermal effects is considered. Figure 6.5 shows that, in addition to a temperature gradient in the positive y-direction, the cracked body is subjected to a tensile stress of magnitude σ_0. The elliptical notch is again assumed to be insulated. A temperature gradient is applied across the notch with $b/a = 0.1$ such that $T_1 = -17.8°C$ and $T_2 = 260°C$ prevail respectively at $y = \mp c$. This corresponds to $\Delta T = 277.8°C$. The properties $\alpha = 6 \times 10^{-5}$ in./in.°C, $\nu = 0.3$ and 30×10^3 ksi are used to yield $\sigma_0/\alpha E\,\Delta T = 0.00$, 0.103 and 0.308 that correspond respectively to $\sigma_0 = 0.0$, 10×10^3 and 30×10^3 ksi. Refer to Tables 6.3 to 6.5 for values of normalized minimum strain energy density function.

Shown in Figure 6.6 are three $(dW/dV)_{min}$ constant lines. As $\sigma_0/\alpha E\,\Delta T$ increases, the predicted fracture path tends to shift toward the x-axis. This is to be expected since the fracture path is known by experience to coincide with the plane of symmetry for mechanical loading. Within a small region r_0, however, the failure path is seen to be highly distorted, which indicates the $(dW/dV)_{min}/(dW/dV)_0$ localized effect of thermal loading.

Table 6.3. Normalized minimum strain energy density function for $b/a = 0.1$ and $\sigma_0/\alpha E\,\Delta T = 0.0$.

r/a	θ	x/a	y/a	$(dW/dV)^{\max}_{\min}/(dW/dV)_0$
Initiation point		0.9926	0.0122	21.9173
0.01	50	0.9991	0.02	17.2204
0.02	55	1.0041	0.0286	16.5887
0.03	85	0.9952	0.0421	16.3672
0.05	90	0.9925	0.0622	15.3786
0.07	100	0.9804	0.0811	14.9110
0.1	110	0.9580	0.1062	14.3777
0.2	120	0.8919	0.1854	13.1963
0.4	120	0.7910	0.3583	11.7893

Table 6.4. Normalized minimum strain energy density function for $b/a = 0.1$ and $\sigma_0/\alpha E\,\Delta T = 0.103$.

r/a	θ	x/a	y/a	$(dW/dV)^{\max}_{\min}/(dW/dV)_0$
Initiation point		0.9999	0.0014	261.5063
0.01	25	1.0092	0.0056	47.5839
0.02	5	1.02	0.0030	36.1063
0.03	20	1.028	0.0115	29.0426
0.05	20	1.0469	0.0189	23.7755
0.07	30	1.0605	0.0363	21.2610
0.1	40	1.0767	0.0656	19.6509
0.2	40	1.1542	0.1285	17.1835
0.4	50	1.257	0.3079	14.9725

Table 6.5. Normalized minimum strain energy density function for $b/a = \sigma_0/\alpha E\,\Delta T = 0.308$.

r/a	θ	x/a	y/a	$(dW/dV)^{\max}_{\min}/(dW/dV)_0$
Initiation point		1.0	0.0	2158.0934
0.01	−20	1.0095	−0.0034	296.4969
0.02	0	1.02	0.0	175.3220
0.03	0	1.03	0.0	126.6716
0.05	0	1.05	0.0	81.3438
0.07	0	1.07	0.0	64.0742
0.1	0	1.1	0.0	47.9121
0.2	0	1.2	0.0	33.1446
0.4	10	1.3917	0.0653	26.6880

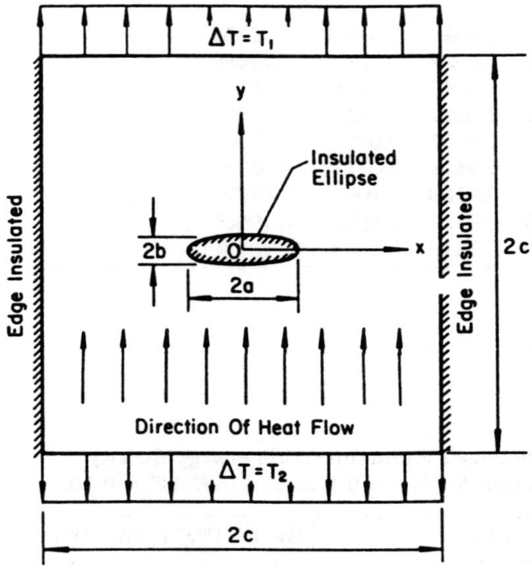

Figure 6.5. Notch under combined mechanical and thermal loading: geometry and boundary conditions.

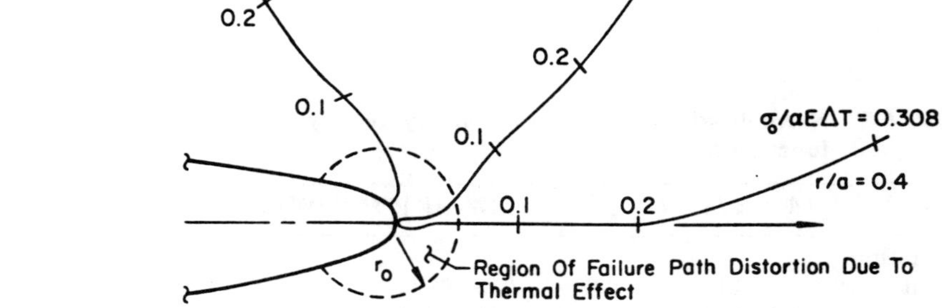

Figure 6.6. Effect of the proportion of mechanical to thermal loading $(\sigma_0/\alpha E\,\Delta T)$ on fracture trajectories for notch with $b/a = 0.1$.

Notch geometry is changed by varying the ratio of the minor axis of the notch to the major axis, b/a. An examination of the results in Tables 6.6 to 6.8 shows that the primary effect of a change from $b/a = 0.1$ to $b/a = 0.05$

Table 6.6. Normalized minimum strain energy density function for $b/a = 0.05$ and $\sigma_0/\alpha E\,\Delta T = 0.0$.

r/a	θ	x/a	y/a	$(dW/dV)^{max}_{min}/(dW/dV)_0$
Initiation point		0.998	0.0032	29.1164
0.01	60	1.0029	0.0119	18.4018
0.02	65	1.0065	0.0213	17.4338
0.03	80	1.00295	0.0328	16.4575
0.05	90	0.9982	0.0532	15.5796
0.07	100	0.9859	0.0722	15.1079
0.1	110	0.9638	0.0973	14.5418
0.2	120	0.8979	0.1750	13.3645
0.4	120	0.7980	0.3496	12.1421

Table 6.7. Normalized minimum strain energy density function for $b/a = 0.05$ and $\sigma_0/\alpha E\,\Delta T = 0.103$.

r/a	θ	x/a	y/a	$(dW/dV)^{max}_{min}/(dW/dV)_0$
Initiation point		1.0000	0.0002	464.2547
0.01	30	1.0086	0.0049	30.2404
0.02	15	1.0193	0.0053	36.3155
0.03	20	1.0284	0.0103	28.2755
0.05	25	1.0453	0.0211	23.4884
0.07	30	1.0606	0.0355	21.0231
0.1	40	1.0766	0.0643	19.5463
0.2	40	1.1542	0.1285	17.1712
0.4	55	1.2294	0.3279	14.9725

Table 6.8. Normalized minimum strain energy density function for $b/a = 0.05$ and $\sigma_0/\alpha E\Delta T = 0.308$.

r/a	θ	x/a	y/a	$(dW/dV)^{max}_{min}/(dW/dV)_0$
Initiation point		1.0		3819.0171
0.01	−25	1.0089	−0.0041	241.0789
0.02	−40	1.0152	−0.0127	214.3326
0.03	−30	1.0259	0.0150	153.7042
0.05	0	1.0500	0.0	78.8416
0.07	10	1.0689	0.0126	60.6695
0.1	0	1.1	0.0	46.9276
0.2	0	1.2	0.0	32.7180
0.4	5	1.3990	0.0349	26.7044

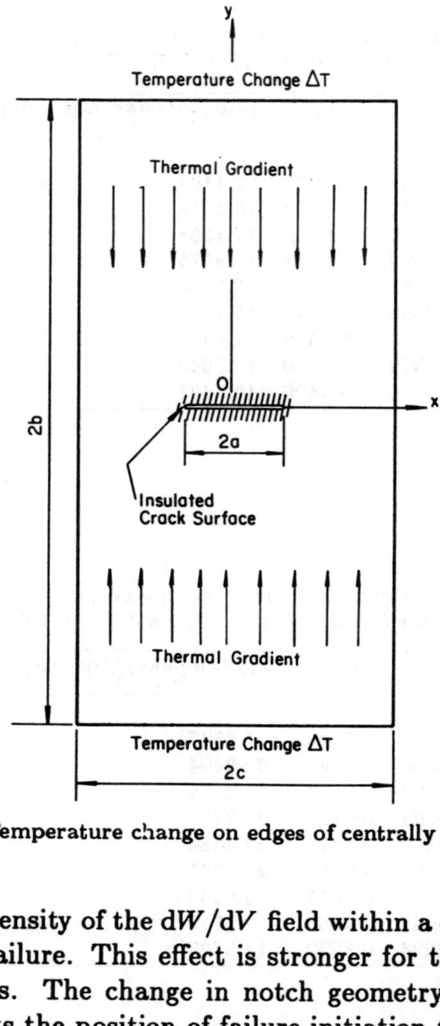

Figure 6.7. Temperature change on edges of centrally cracked plate.

is to increase the intensity of the dW/dV field within a distance of $r = 0.02a$ from the origin of failure. This effect is stronger for the cases with higher applied tensile loads. The change in notch geometry from $b/a = 0.1$ to $b/a = 0.05$ also shifts the position of failure initiation toward the notch tip for the combinations of loading and geometry considered here.

6.5. Thermally induced slow crack growth

Let a plate with dimensions $b = 5$ in. and $c = 2.5$ in. be subjected to temperature changes such that a thermal gradient is established across an insulated crack of length $2a$, as shown in Figure 6.7.

Boundary conditions. With reference to the Cartesian coordinates oxy in Figure 6.7, a temperature difference ΔT is applied in steps at $y = \pm b$ while the boundaries $x = \pm c$ experience no temperature gradient, i.e., $\partial T/\partial x = 0$. The resistance to heat on the crack surface $y = \pm 0$; $-a \leq x \leq a$ may vary through the normalized quantity:

Table 6.9. Thermal and mechanical properties of steel.

Thermal expansion coefficient α(in./in.°C)	Young's modulus E(ksi)	Poisson's ratio ν	Yield strength σ_{ys}(ksi)
6×10^{-3}	30×10^3	0.33	75

Table 6.10. Fracture properties of steel.

Stress intensity K_{1c}(ksi$\sqrt{\text{in.}}$)	Volume energy density $(dW/dV)_c$(psi)	Energy density factor S_c(lb/in.)
75	93.75	18.7

$$\left(\frac{\partial T}{\partial y}\right)^* = \frac{(\partial T/\partial y)_0}{(\partial T/\partial y)} \tag{6.10}$$

with $(\partial T/\partial y)_0$ being the undisturbed temperature gradient which is classically related to the ratio of heat flux and the coefficient of thermal conductivity. An initial crack of length $2a_0 = 1$ in. is taken. It grows in steps of a_1, a_2, etc., in accordance with the strain energy density criterion of Sih [9] as ΔT on $y = \pm b$ is increased in increments of ΔT_1, ΔT_2, etc. The foregoing conditions can thus be summarized as

$$T - T_0 = \Delta T_j \quad (j = 1, 2, \ldots), \quad y = \pm b, \tag{6.11a}$$

$$\frac{\partial T}{\partial x} = 0, \qquad x = \pm c, \tag{6.11b}$$

$$\frac{\partial T}{\partial y} = \left(\frac{\partial T}{\partial y_0}\right) \Big/ \left(\frac{\partial T}{\partial y}\right)^*, \qquad y = \pm 0 \quad \text{and} \quad -a \leq x \leq a \tag{6.11c}$$

Material and fracture properties. The thermal and mechanical properties of the plate are given in Table 6.9 while the corresponding fracture properties are listed in Table 6.10.

The yield strength σ_{ys} is assumed to be temperature independent in this analysis. As a result, the critical strain energy density function $(dW/dV)_c$ that may be regarded as the fracture toughness value of material will also be taken as constant for the temperature range considered. In linear thermoelasticity, K_{1c} and S_c for plane strain are related by the expression

$$S_c = \frac{(1+\nu)(1-2\nu)K_{1c}^2}{2\pi E} \tag{6.12}$$

while S_c and $(dW/dV)_c$ follow from

$$\frac{dW}{dV} = \frac{S}{r}. \tag{6.13}$$

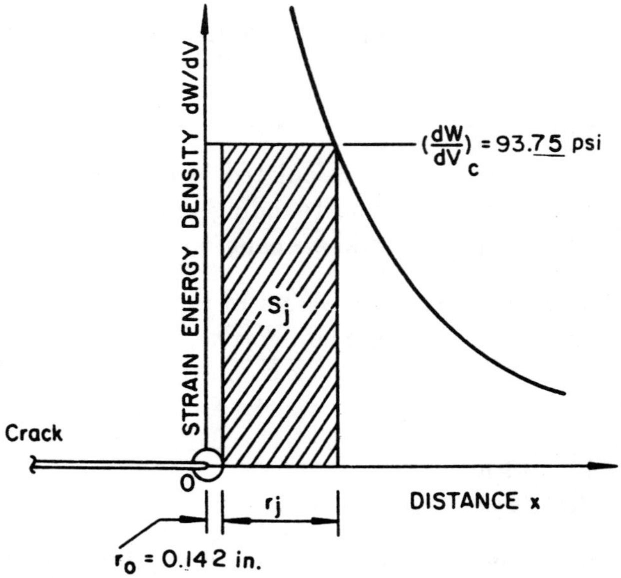

Figure 6.8. Schematic of strain energy density decay characteristic ahead of growing crack.

In equation (6.13), S is the strain energy density factor as in the case of isothermal elasticity. It can vary with crack length and provides an indication of the resistance to crack growth.

Slow crack growth. Depending on the rate at which the temperature is increased, the crack could extend slowly or quickly. The amount of growth at a given increment of temperature, say ΔT_j, as specified in equation (6.11a), can be determined from the near tip volume energy density distribution. Displayed in Figure 6.8 is the decay of dW/dV as a function of the distance x ahead of the crack, where S_j corresponds to the jth increment of crack growth r_j. According to the volume energy density criterion of Sih [9], crack growth is assumed to initiate when the maximum of the local minimum of dW/dV or $(dW/dV)^{\max}_{\min}$ reaches the critical value $(dW/dV)_c$. This establishes the area S_j for the jth step of crack growth and the growth increment r_j is thus obtained. The total segments of growth $r_1, r_2, \ldots, r_j, \ldots, r_c$ with their corresponding values of $S_1, S_2, \ldots, S_j, \ldots, S_c$ are that*

$$S_1 < S_2 < \cdots < S_j < \cdots < S_c, \tag{6.14a}$$

* The thermal loading in the present problem increases in each consecutive step such that crack growth steps increase monotonically leading to eventual rapid crack growth. The opposite can be achieved by decreasing the temperature steps. In such an event, crack arrest can result.

$$r_1 < r_2 < \cdots < r_j < \cdots < r_c. \tag{6.14b}$$

Global instability or rapid crack propagation occurs when the condition

$$\left(\frac{dW}{dV}\right)_c = \frac{S_c}{r_c} \tag{6.15}$$

is reached. All numerical data are calculated in regions outside of a small core region of radius $r_0 = 0.142$ in surrounding the crack tip. This distance represents the resolution of the finite element calculation.

Crack growth resistance curves. Four sets of numerical results have been obtained by Sih and Chen [10]. They correspond to $(\partial T/\partial y)^* = 1.00$, 1.43, 2.00 and 3.33. The case of a crack completely free from insulation is $(\partial T/\partial y)^* = 1.00$. Resistance to heat transfer from the crack increases with $(\partial T/\partial y)^*$. These data are summarized in Tables 6.11 to 6.14 inclusive.

Change in surface temperature and insulation. With a 15°C increment increase in temperature, it required only approximately four steps for the crack to become unstable when its surface is not insulated, i.e., for $(\partial T/\partial y)^* = 1.00$. More steps in ΔT are needed for S to reach S_c as $(\partial T/\partial y)^*$ is increased. For a crack whose surface is heavily insulated with $(\partial T/\partial y)^* = 3.33$, no crack growth could be detected for ΔT up to approximately 45°C. Even when ΔT is increased up to 150°C, the crack is still in the subcritical stage as $S \ll S_{cr}$.

The above results are displayed graphically in Figures 6.9 and 6.10. In general, the crack length increases with increasing ΔT applied to the plate edge. The curves rise very rapidly for $(\partial T/\partial y)^* = 1.00$ while they change much more gradual as $(\partial T/\partial y)^*$ is increased, Figure 6.9. This effect is also reflected by the plot of a versus $(\partial T/\partial y)^*$ for different ΔT. Note that small changes in ΔT can initiate crack motion if the crack surface insulation is not high. Otherwise ΔT has to be sufficiently high in order for crack growth to take place.

One of the salient features of the strain energy density criterion is that it linearizes the crack growth data in Figures 6.9 and 6.10. That is, all results in the strain energy factor versus crack growth plots are straight lines. Figure 6.11 gives the results for the strain energy density factor as a function of crack growth. The advantage of establishing the $dS/da = \text{const.}$ relationship is that results to other boundary conditions follow immediately. For instance, curves for $(\partial T/\partial y)^*$ other than 1.00, 1.42, 2.00 and 3.33 may also be found by interpolation. The dS/da curve in Figure 6.11 is seen to rotate in the counterclockwise direction as $(\partial T/\partial y)^*$ is decreased. This means that the crack can achieve longer length before it becomes critical if its surface insulation is held constant for the thermal loading conditions considered in this work.

Table 6.11. Crack growth data for $(\partial T/\partial y)^* = 1.00$.

ΔT(°C)	$\partial T/\partial y$(°C/in.)	a(in.)	S(lb/in.)
30°	15.0	1.022	2.030
45°	22.5	1.077	5.167
60°	30.0	1.177	9.411
75°	37.5	1.360	17.161

Table 6.12. Crack growth data for $(\partial T/\partial y)^* = 1.43$.

ΔT(°C)	$\partial T/\partial y$(°C/in.)	a(in.)	S(lb/in.)
30°	10.50	1.005	0.480
45°	15.75	1.030	2.284
60°	21.00	1.073	4.060
75°	26.25	1.142	6.459
90°	31.50	1.277	12.659
105°	36.75	1.472	18.305

Table 6.13. Crack growth data for $(\partial T/\partial y)^* = 2.00$.

ΔT(°C)	$\partial T/\partial y$(°C/in.)	a(in.)	S(lb/in.)
30°	7.50	1.002	0.185
45°	11.25	1.020	1.698
60°	15.00	1.044	2.214
75°	18.75	1.081	3.506
90°	22.50	1.138	5.351
105°	26.25	1.221	7.750
120°	30.00	1.347	11.810
135°	33.75	1.533	17.456

Table 6.14. Crack growth data for $(\partial T/\partial y)^* = 3.33$.

ΔT(°C)	$\partial T/\partial y$(°C/in.)	a(in.)	S(lb/in.)
30°	4.50	–	–
45°	6.75	–	–
60°	9.00	1.006	0.517
75°	11.25	1.015	0.923
90°	13.50	1.033	1.624
105°	15.75	1.060	2.583
120°	18.00	1.094	3.137
135°	20.25	1.141	4.429
150°	22.50	1.202	5.720

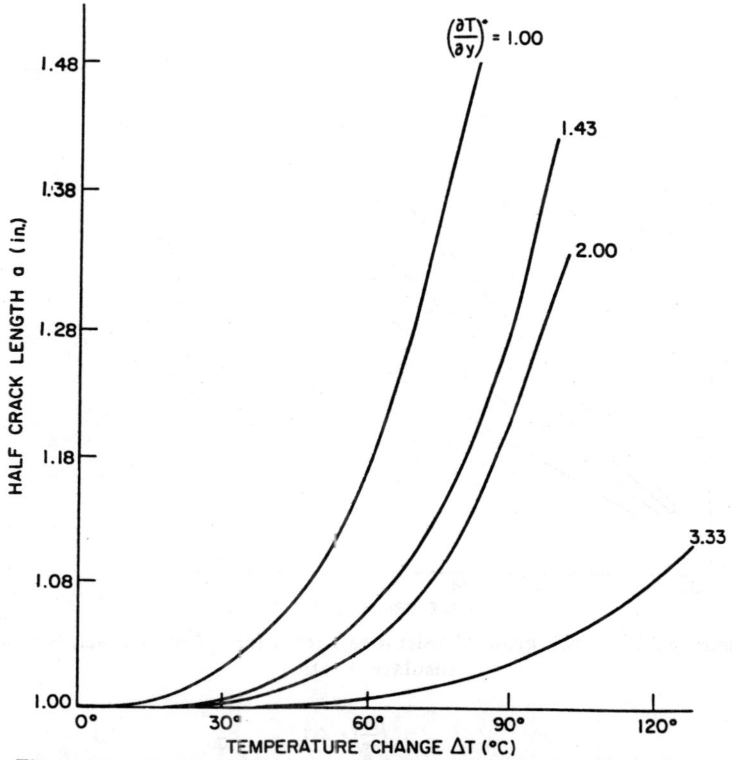

Figure 6.9. Crack growth versus temperature change for different insulation ratio.

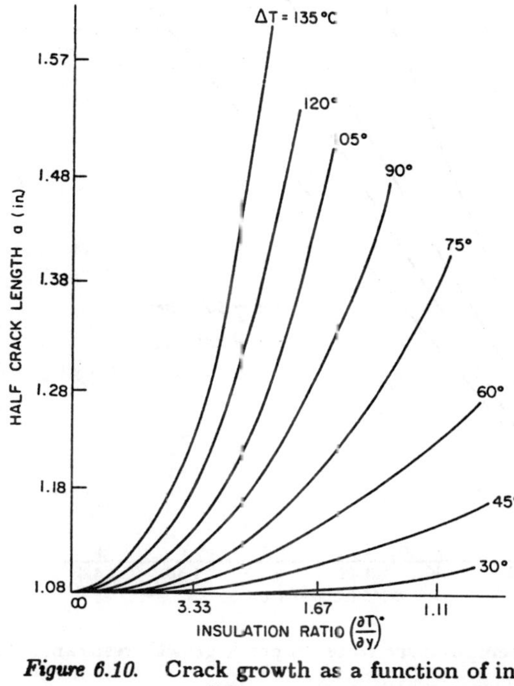

Figure 6.10. Crack growth as a function of insulation ratio for different temperature change.

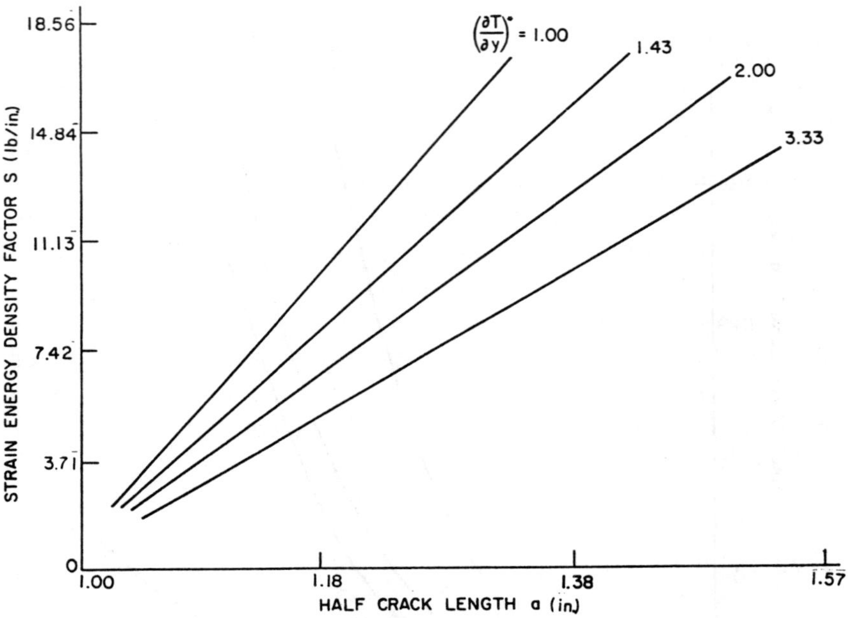

Figure 6.11. Crack growth resistance curves for different crack surface insulation ratio.

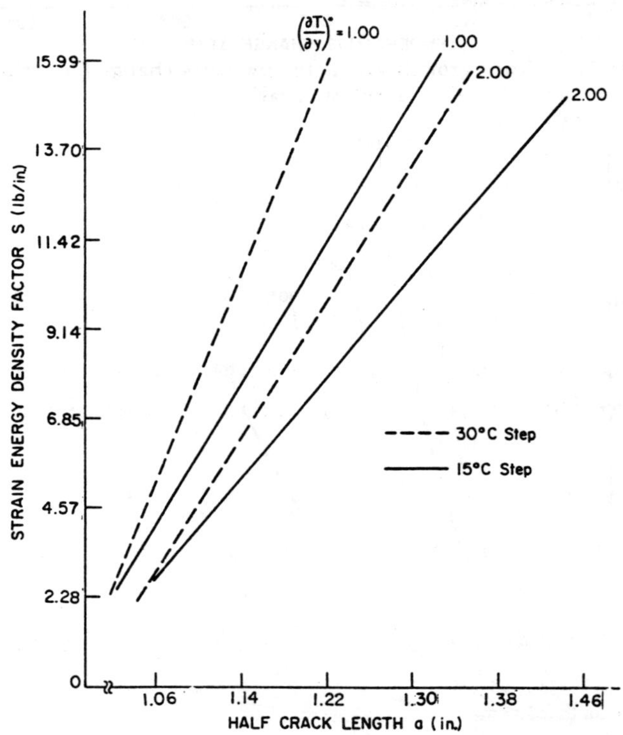

Figure 6.12. Influence of temperature steps on crack growth resistance curves.

Influence on temperature steps. When a crack extends, energy is released or dissipated in an irreversible manner for the creation of new surfaces even though the neighboring material does not undergo plastic or permanent deformation. This process is inherently load-time history dependent. In other words, the step increase in temperature has a direct effect on the subcritical crack growth characteristics. Suppose that the incrementally applied temperature change ΔT on the plate at $y = \pm b$ is increased to 30°C instead of 15°C. The crack growth resistance curves in Figure 6.11 will tend to rotate counterclockwise or attain steeper slopes. For a given material or S_c value, the crack would become critical at a shorter length. This effect is exhibited in Figure 6.12 for $(\partial T/\partial y)^* = 1.00$ and 2.00. The dashed lines correspond to ΔT increased in steps of 30°C and solid lines in steps of 15°C. Similar trends prevail for other values of $(\partial T/\partial y)^*$.

6.6. Hygrothermoelasticity

If the moisture in a solid also varies from point to point in addition to the temperature, then equations (6.1) through (6.5) must be modified. With reference to a thermodynamic state of uniform temperature T_0 and moisture C_0, Sih, Michopoulos and Chou [3] have derived the volume energy density:

$$\frac{\mathrm{d}W}{\mathrm{d}V} = \tfrac{1}{2}\sigma_{ij}\epsilon_{ij} - \tfrac{1}{2}[\alpha(T-T_0) + \beta(C-C_0)]\sigma_{kk} \tag{6.16}$$

where β is the linear moisture expansion coefficient while C is the moisture concentration. The hygrothermoelastic stress and strain relation is given by

$$\sigma_{ij}=2\mu\left\{\epsilon_{ij}-\frac{\nu}{1-2\nu}\epsilon_{kk}\delta_{ij}-\frac{1+\nu}{1-2\nu}[\alpha(T-T_0)+\beta(C-C_0)]\delta_{ij}\right\} \tag{6.17}$$

where μ is the shear modulus of elasticity. Equation (6.17) may be inserted into equation (6.16) to express $\mathrm{d}W/\mathrm{d}V$ in terms ϵ_{ij} alone, i.e.,

$$\frac{\mathrm{d}W}{\mathrm{d}V} = \mu\left\{\epsilon_{ij}\epsilon_{ij} + \frac{\nu}{1-2\nu}\epsilon_{kk}^2 - \frac{2(1+\nu)}{1-2\nu}[\alpha(T-T_0) + \beta(C-C_0)]\epsilon_{kk}\right\} + \\ + \frac{3\mu(1+\nu)}{1-2\nu}[\alpha(T-T_0) + \beta(C-C_0)]^2 \tag{6.18}$$

which reduces to equation (6.5) if $C = C_0$. The strain and stress relation is

$$\epsilon_{ij} = \frac{\nu}{E}\sigma_{kk}\delta_{ij} + \frac{1+\nu}{E}\sigma_{ij} + [\alpha(T-T_0) + \beta(C-C_0)]\delta_{ij}. \tag{6.19}$$

The appearance of $\mathrm{d}W/\mathrm{d}V$ in terms of σ_{ij} is the same as that of equation (6.3) if ϵ_{ij} in equation (6.19) is substituted into equation (6.16).

When ΔT, ΔC and σ_{ij} or ϵ_{ij} are assumed to be coupled, they can no longer be determined independently as in the classical theory of linear thermoelasticity for the temperature and stresses or strains. The governing equations derived by Sih, Michopoulos and Chou [3] are

$$\frac{D^*}{D_t}\nabla^2 C = \frac{\partial}{\partial \tau^*}(C - \lambda^* R) \tag{6.20a}$$

$$\frac{\mathcal{D}^*}{D_t}\nabla^2 R = \frac{\partial}{\partial \tau^*}(R - \nu^* C) \tag{6.20b}$$

in which R stands for

$$R = T + N_c^* \psi. \tag{6.21}$$

The function ψ is given by

$$\psi = \sigma_{kk} + \frac{2E}{1-\nu}[\alpha(T - T_0) + \beta(C - C_0)]. \tag{6.22}$$

The Laplacian operator ∇^2 in equations (6.20) is expressed in terms of the dimensionless space variables x/h and y/h, where h is the thickness dimension that can be taken as unity for plane strain. Time t is normalized in terms of the variable $\tau^* = tD_t/h^2$ with D_t being a parameter that scales the time depending upon whether moisture or temperature is specified. There are five coupling coefficients, namely D^*, $\mathcal{D}^*$, λ^*, ν^* and N_c^*. Their relationships with the physical parameters such as mass density, specific heat, etc., are too involved to be mentioned here. Reference can be made to the work of Sih, Michopoulos and Chou [3]. It is more pertinent to discuss the influence of heat/moisture/load on the crack behavior.

6.7. Effect of heat/moisture/load on crack behavior

To be considered are the sudden application of moisture ΔC and temperature ΔT to the surfaces of a line crack that is approximated by a narrow ellipse with $a/b = 10$ in a rectangular region $\mathcal{R}$ with dimensions $3c \times 4c$ with $c = 2$ in. This is illustrated in Figure 6.13. The material properties correspond to those for the T300/5208 composite as determined in [3]. Determination of dW/dV in equation (6.18) can be made from the parameters in Table 6.15. In addition, the physical parameters D^*, $\mathcal{D}^*$, λ^*, ν^* and N_c^* in equations (6.20) and (6.21) must be specified. They are given in Table 6.16 for the case of plane strain, depending on whether the moisture or temperature boundary conditions are used. In what follows, numerical results of dW/dV in equation (6.18) will be obtained for three applied stress levels of $\sigma_0 = 0.145$, 1.450 and 14.500 ksi in combination with the ΔT and ΔC

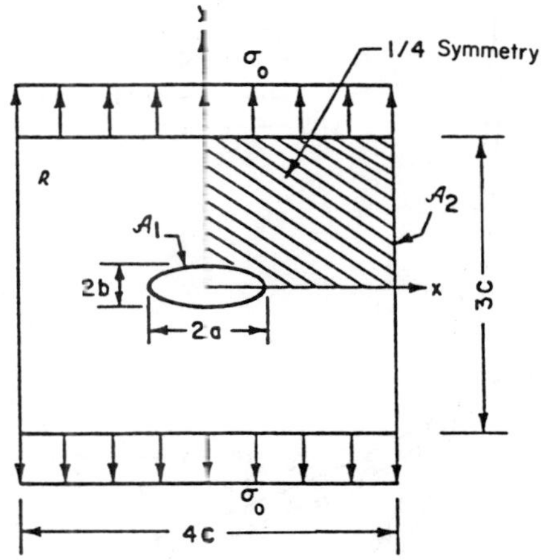

Figure 6.13. Rectangular region containing a crack approximated as narrow ellipse.

Table 6.15. Thermal/moisture/mechanical properties of T300/5208 composite.

Thermal expansion coefficient α (in./in K)	Moisture expansion coefficient β (in./in./wt% H_2O)	Poisson's ratio ν	Young's modulus E (ksi)
4.16×10^{-5}	3.56×10^{-3}	0.493	10.463

Table 6.16. Plane strain coupling coefficients for T300/5208 laminate subjected to moisture change at 21°C and temperature change at 75% RH.

Physical parameters	Moisture change (0% to 75% RH)	Temperature change (21°C to 61°C)
D^* (ft^2/h)	8.988×10^{-6}	30.802
D^* (ft^2/h)	1.586×10^{-4}	483.674
λ^* (lb_m/ft^3 K)	7.058×10^{-3}	6.058×10^{-3}
ν^* (ft^3 K/lb_m)	31.141	33.671
N_c^* (ft^2 K/lb_m)	1.121×10^{-8}	1.264×10^{-8}

boundary conditions. Moreover, the term

$$f(\Delta T, \Delta C) = \frac{3G(1+\nu)}{1-2\nu}[\alpha(T-T_0)+\beta(C-C_0)]^2 \tag{6.23}$$

that tends to be dominated by ΔT and ΔC will be evaluated such that a comparison between the influence of mechanical work and diffusion can be made.

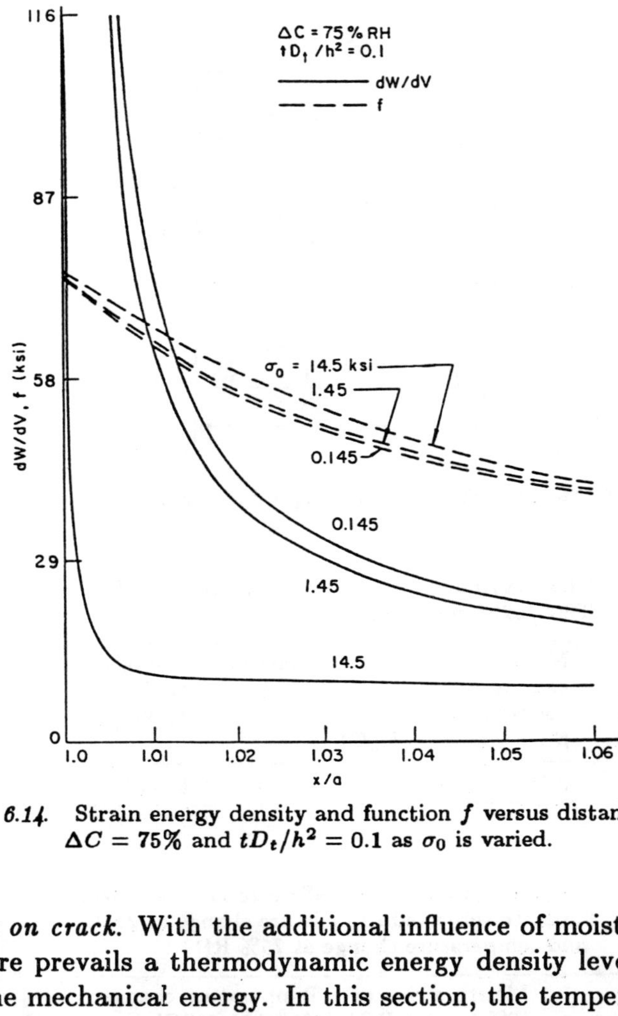

Figure 6.14. Strain energy density and function f versus distance for $\Delta C = 75\%$ and $tD_t/h^2 = 0.1$ as σ_0 is varied.

Moisture rise on crack. With the additional influence of moisture and temperature, there prevails a thermodynamic energy density level in addition to that for the mechanical energy. In this section, the temperature on the crack surface $\mathcal{A}_1$ and specimen boundary $\mathcal{A}_2$ in Figure 6.13 is maintained at 21°C. The moisture level on $\mathcal{A}_2$ is always kept at its initial value, say 0%, RH, being the same as that on $\mathcal{A}_1$. Time dependency is monitored by the normalized time parameter $\tau^* = tD_t/h^2$ in equations (6.20). The scale factor $D_t = 10^{-5}$ cm^2/h. is chosen for problems where moisture boundary conditions are applied.

Plotted in Figure 6.14 are the decay of dW/dV with the dimensionless ratio x/a for a crack surface moisture change of 75% RH and $tD_t/h^2 = 0.1$. Near the crack front $x/a = 1.0$ for $\sigma_0 = 14.5$ ksi, the strain energy density function dW/dV undergoes extremely large changes for a slight perturbation of the distance. This is indicative of the singular behavior of dW/dV near the crack tip. In contrast to intuition, the energy densities for $\sigma_0 = 0.145$ and 1.450 ksi attained higher amplitudes than those for $\sigma_0 = 14.5$ ksi at a

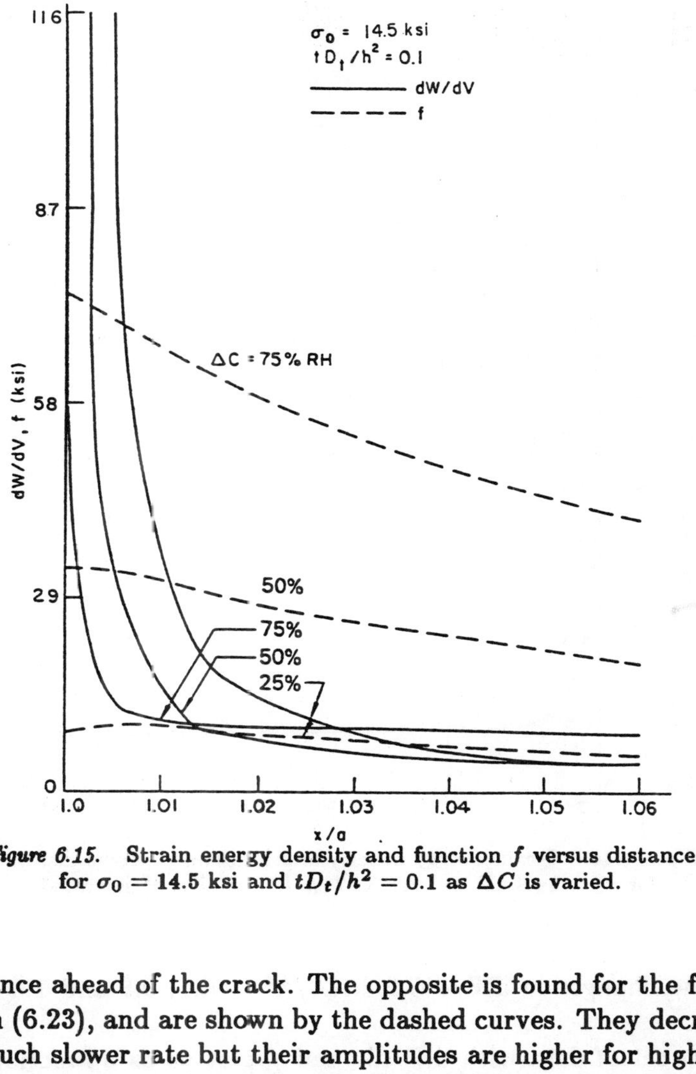

Figure 6.15. Strain energy density and function f versus distance for $\sigma_0 = 14.5$ ksi and $tD_t/h^2 = 0.1$ as ΔC is varied.

given distance ahead of the crack. The opposite is found for the function f in equation (6.23), and are shown by the dashed curves. They decrease with x/a at a much slower rate but their amplitudes are higher for higher σ_0. As the dashed curves are closely spaced, this implies that f is not sensitive to changes in mechanical loading or it is predominantly a diffusion-controlled function.

Let us suppose that the load σ_0 is kept at 14.5 ksi while the crack surface moisture change takes the values of 25%, 50% and 75% RH, all other conditions remaining the same as before. This large variation in diffusion is immediately reflected by the dashed curves representing f in Figure 6.15.

Note that its energy level changes significantly with the specified crack surface moisture. The variations of solid curves for dW/dV in Figure 6.15 are not nearly as large compared to those in Figure 6.14 when σ_0 is varied. With $\sigma_0 = 14.5$ ksi and a crack surface moisture rise of $\Delta C = 75\%$ RH, Figure 6.16 shows that diffusion is more sensitive to time change than deformation. This is evidenced by the larger variations of the function f as tD_t/h^2 is

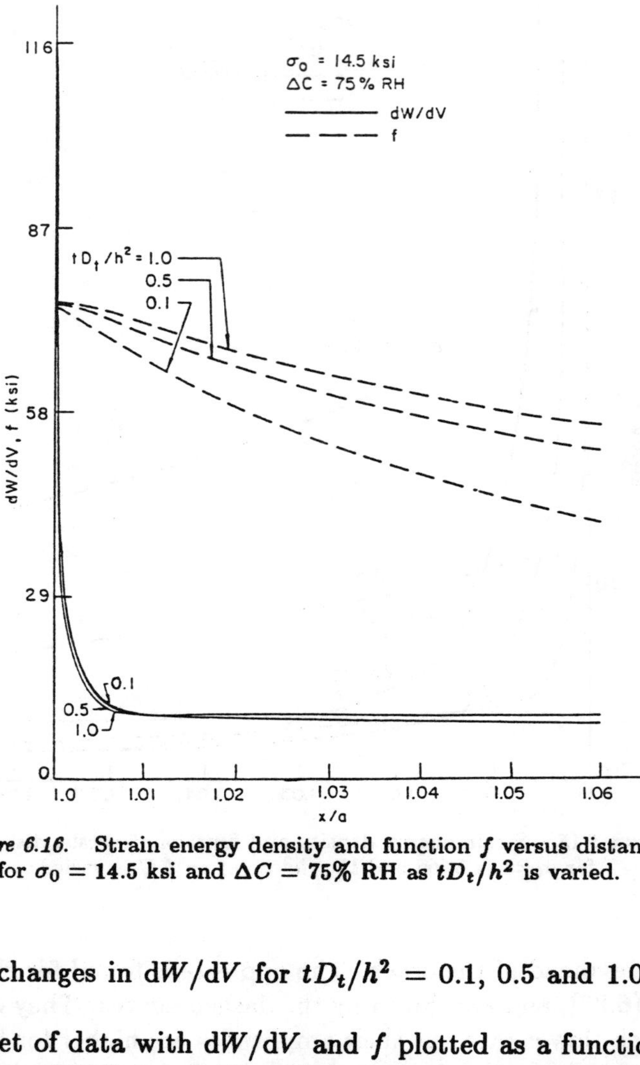

Figure 6.16. **Strain energy density and function f versus distance for $\sigma_0 = 14.5$ ksi and $\Delta C = 75\%$ RH as tD_t/h^2 is varied.**

varied. The changes in dW/dV for $tD_t/h^2 = 0.1$, 0.5 and 1.0 are hardly noticeable.

A similar set of data with dW/dV and f plotted as a function of x/a is given in Figure 6.17 for a crack surface moisture change of 50% RH and $tD_t/h^2 = 0.1$. The trends are nearly the same as those in Figure 6.14. A change in tD_t/h^2 from 0.1 to 0.5 did not alter the results quantitatively. This can be seen from Figures 6.17 and 6.18. Figure 6.19 displays the change of dW/dV and f with x/a with $\Delta C = 50\%$ RH $\sigma_0 = 10^8$ N/m^2 as tD_t/h^2 is varied. The data are not significantly different from those in Figure 6.16 for $\Delta C = 75\%$ RH.

Temperature rise on crack. **The condition of sudden temperature rise on the crack will be considered such that a constant 75% RH moisture level is kept on $\mathcal{A}_1$ and $\mathcal{A}_2$ at all time. At time $t = 0$, the temperatures on $\mathcal{A}_1$ and $\mathcal{A}_2$ are both at 21°C. Only the temperature on $\mathcal{A}_1$ will be altered. The value of 21°C will be maintained on $\mathcal{A}_2$ for all t. Plotted will be dW/dV and f as**

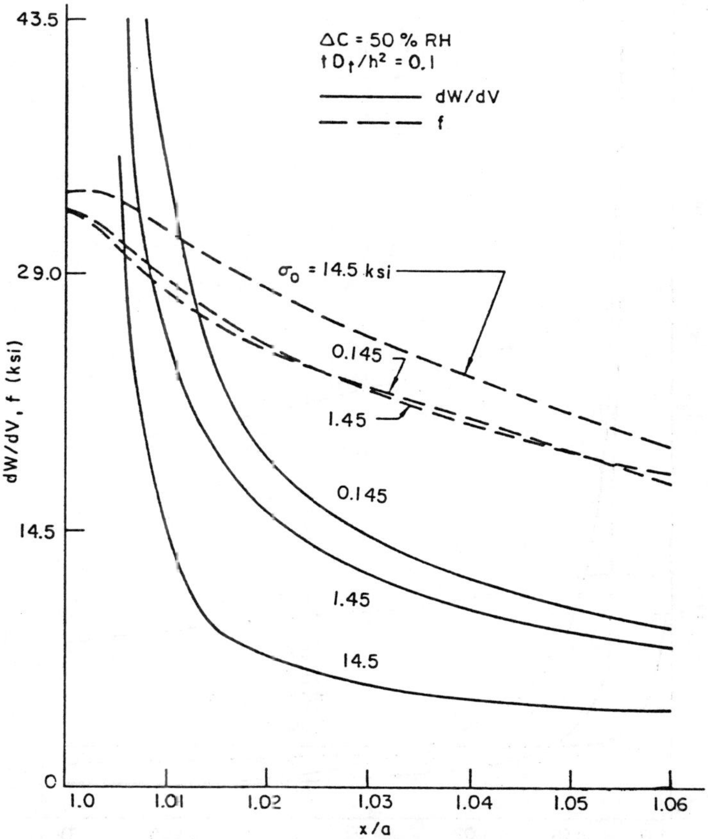

Figure 6.17. Strain energy density and function f versus distance for $\Delta C = 50\%$ and $tD_t/h^2 = 0.1$ as σ_0 is varied.

the time, applied stress σ_0 and ΔT on $\mathcal{A}_1$ take different values.

For a sudden temperature rise of 40°C on the crack and at $tD_t/h^2 = 0.1$, the variations of dW/dV and f with distance ahead of the crack are shown graphically in Figure 6.20 for $\sigma_0 = 0.145$, 1.45 and 14.5 ksi, the way with which σ_0 affects dW/dV, and f is opposite to the trend in Figure 6.14 for the case of sudden moisture change. The magnitude of the diffusion controlled function f decreased with measuring σ_0, whereas dW/dV rose higher for larger σ_0 at a given x. As the change of temperature ΔT and $\mathcal{A}_1$ is varied, the proportion between dW/dV and f will be affected accordingly. Figure 6.21 gives the results for σ_0 and tD_t/h^2 fixed, respectively, at 10^8 N/m^2 and 0.1 while three sets of data for $\Delta T = 20$, 30 and 40°C are obtained.

Larger ΔT on $\mathcal{A}_1$ gave a higher f but lower dW/dV. The wide spacing between the dashed curves again shows that f changes more readily as diffusion conditions are varied. The time parameter tD_t/h^2 has little influence on dW/dV and f as it takes on the values of 0.1, 0.5 and 1.0 for $\sigma_0 = 14.5$ ksi and $\Delta T = 40$°C on $\mathcal{A}_1$. Refer to Figure 6.22.

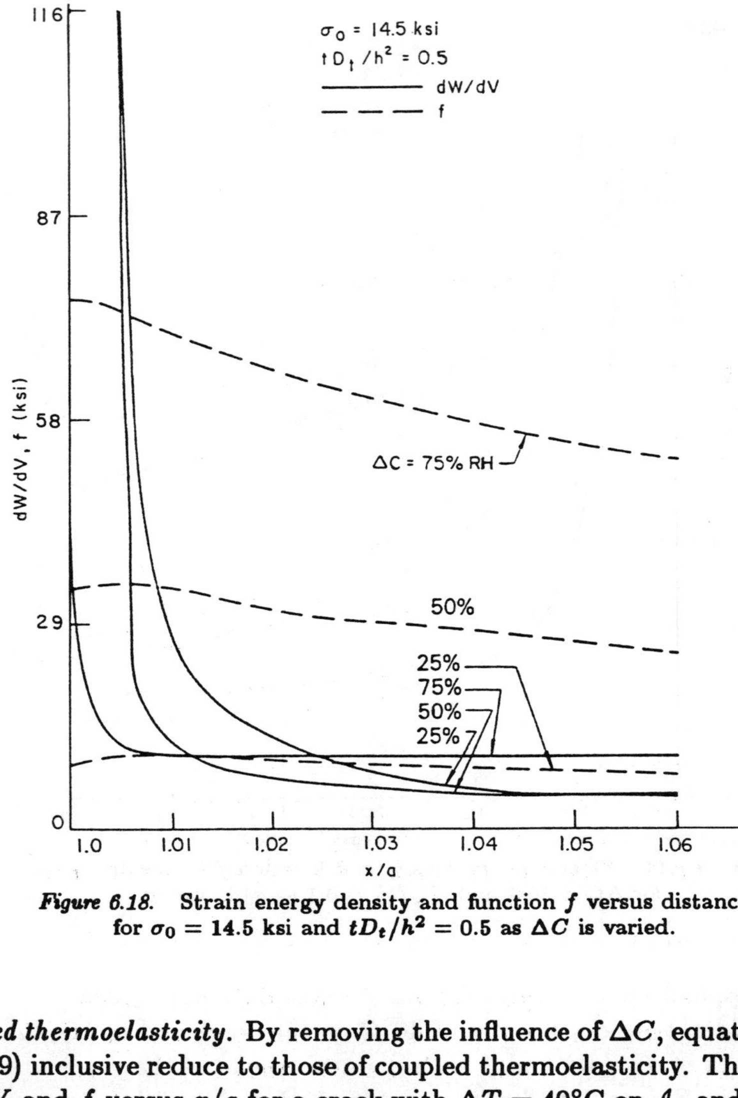

Figure 6.18. **Strain energy density and function f versus distance for $\sigma_0 = 14.5$ ksi and $tD_t/h^2 = 0.5$ as ΔC is varied.**

Coupled thermoelasticity. By removing the influence of ΔC, equations (6.16) to (6.19) inclusive reduce to those of coupled thermoelasticity. The results of dW/dV and f versus x/a for a crack with $\Delta T = 40°$C on $\mathcal{A}_1$ and $tD_t/h^2 = 0.1$ are given in Figure 6.23. This is essentially a special case of the data in Figure 6.20 by decoupling the influence of moisture. Although the trends of the curves are nearly the same, there are some noticeable differences in dW/dV and f ahead of the crack. Moisture decoupling gave lower values of energy stored per unit volume of material for a given position ahead of the crack. This is because moisture tends to aggravate the local crack tip triaxiality. The resulting dilatation elevates the magnitude of dW/dV.

Although the formulation of coupled thermoelasticity [11] is well known, numerical results are scarce because the coupling coefficients are difficult to obtain. The results in Figure 6.23 for the T300/5208 material are not only new but they can readily be used to analyze failure as the load values and the physical parameters are varied.

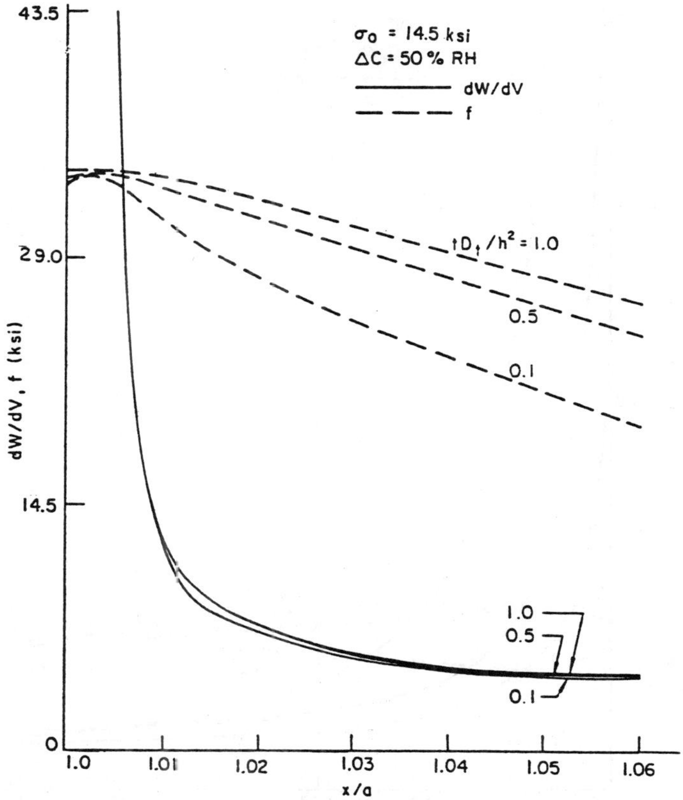

Figure 6.19. **Strain energy density and function f versus distance for $\sigma_0 = 14.5$ ksi and $\Delta C = 50\%$ RH as tD_t/h^2 is varied.**

6.8. Failure enhanced by temperature and moisture changes

Failure prediction deals with the forecast of conditions involving material damage that tends to decrease the capability of structural members to support load. With the added influence of moisture and temperature, the situation becomes more complicated because their interaction with mechanical stresses must also be considered. This concern arises since almost all materials will degrade to some extent when undergoing diffusion and/or mechanical deformation. The idea that material damage alters the apparent behavior of solids has been widely appreciated by those working in the area of material characterization and structure design.

The choice of a failure criterion is not always straightforward. Some of the more common candidates such as the maximum normal stress,* maximum normal strain, etc., may encounter contradictions which must then be

* For a running crack, the local maximum normal stress is directed parallel to the crack instead of normal to the site along which the free surface is being created. This is in contradiction to the original assumption of maximum normal stress condition.

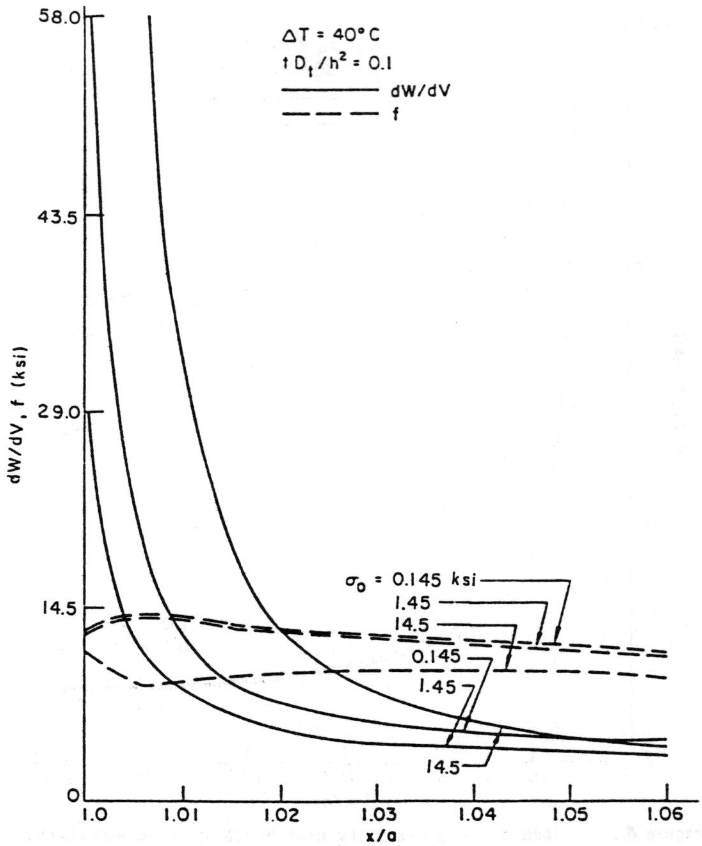

Figure 6.20. **Strain energy density and function f versus distance for $\Delta T = 40°\text{C}$ and $tD_t/h^2 = 0.1$ as σ_0 is varied.**

discarded at the start. The strain energy density criterion has been used successfully for predicting failure of solids due to yielding and/or fracture [9] and applies equally well even if moisture and temperature effects are taken into account. It assumes that failure coincides with locations where dW/dV attains relative maxima and minima. The existence of stationary values of dW/dV is guaranteed for all systems other than the trivial case of $dW/dV =$ const.* at every point in the system. Critical values of dW/dV are assigned to each material to characterize failure.

Stationary values. The strain energy density theory views the continuum as an assembly of small elements, each containing a unit volume of material and can store a finite amount of energy at a given instance of time. The energy

* The condition dW/dV = const. defines the resolution of the scale at which analysis is being made and is assumed to prevail in a continuum element. If all the elements within a system possess the same dW/dV value, the implication is that analysis must then be performed at a more refined scale level in order to discuss failure initiation within an element.

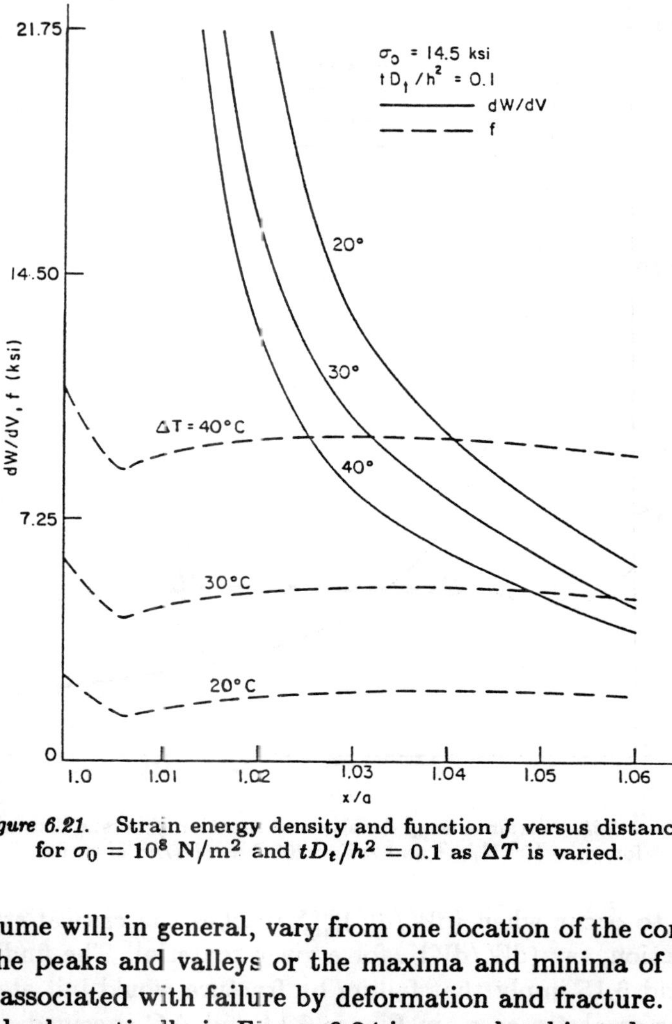

Figure 6.21. Strain energy density and function f versus distance for $\sigma_0 = 10^8$ N/m^2 and $tD_t/h^2 = 0.1$ as ΔT is varied.

per unit volume will, in general, vary from one location of the continuum to another. The peaks and valleys or the maxima and minima of the energy density are associated with failure by deformation and fracture.

Illustrated schematically in Figure 6.24 is a crack subjected to tension in the y-direction. Among many of the minima of dW/dV, the maximum of $(dW/dV)_{min}$ is assumed to coincide with macrocrack growth and the largest of $(dW/dV)_{max}$ with the direction of maximum permanent deformation. The former can be associated with excessive dilatation and the latter with excessive distortion. This is consistent with the intuition that large volume change leads to fracture and shape change to permanent deformation.*

Failure interpretation. The numerical results for dW/dV presented in Figures 6.14 to 6.23 inclusive represent those at $\theta = 0°$, i.e., the plane of expected crack growth. They are the relative minima. Failure by fracture

* The separation of dW/dV into its dilatational and distortional components, of course, can be carried out only by assuming a linear relationship between stress and strain. No such separation can be made *a priori* when the response becomes nonlinear.

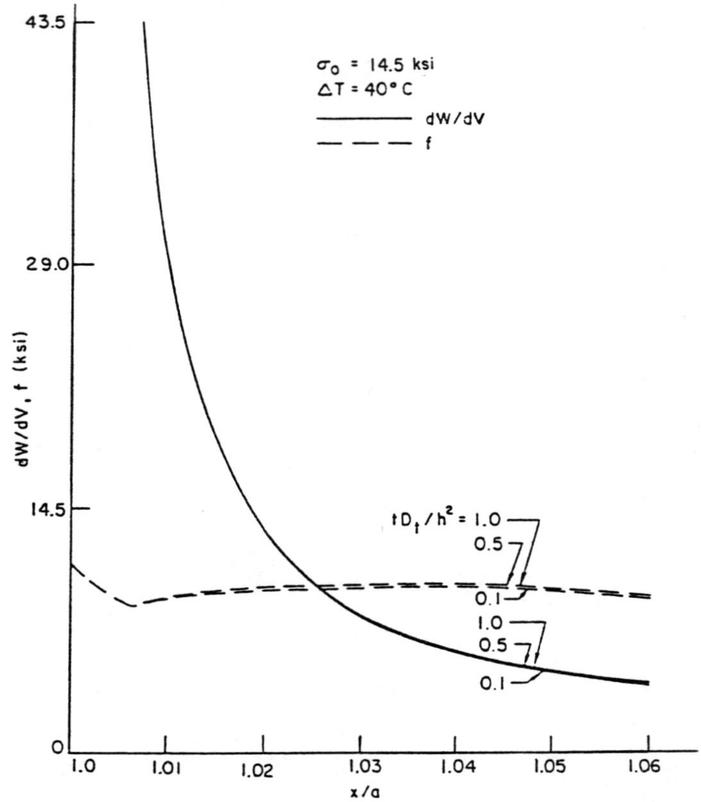

Figure 6.22. **Strain energy density and function f versus distance for $\sigma_0 = 10^8$ N/m^2 and $\Delta T = 40$°C as tD_t/h^2 is varied.**

is assumed to occur when $(dW/dV)^{\max}_{\min}$ reaches the critical strain energy density function, say $(dW/dV)_c$, for a given material. The findings in Figures 6.14 and 6.15 imply that failure by fracture would initiate closer and closer to the crack tip as σ_0 in Figure 6.14 and ΔC specified on the crack in Figure 6.15 are increased. When the crack surface temperature is raised, a higher mechanical load would tend to cause fracture in a region further ahead of the crack tip, Figure 6.20. For a fixed applied stress, failure by fracture tends to be more localized as ΔT on the crack is increased.

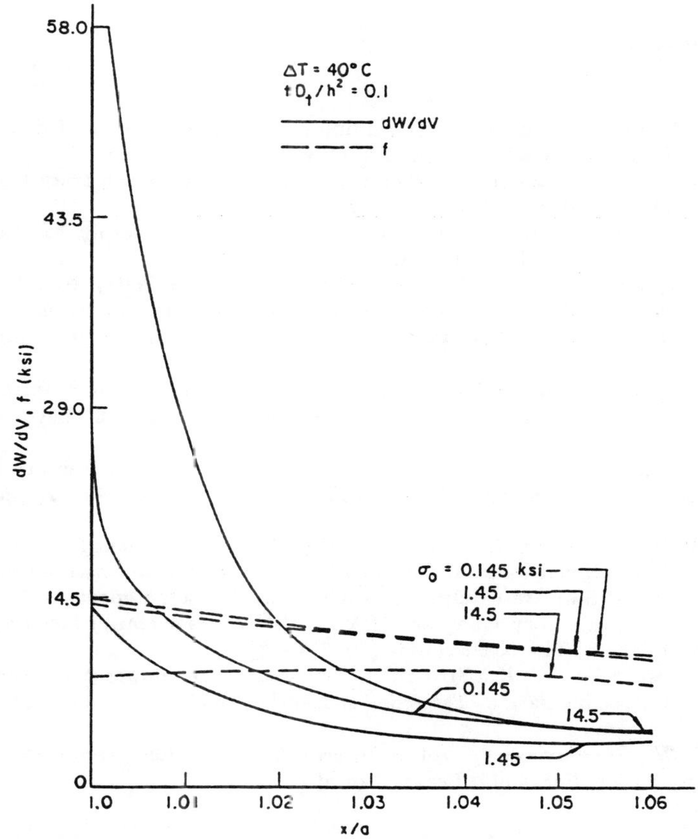

Figure 6.23. **Strain energy density and function f versus distance for $\Delta T = 40°C$ and $tD_t/h^2 = 0.1$ as σ_0 is varied by neglecting the moisture effect.**

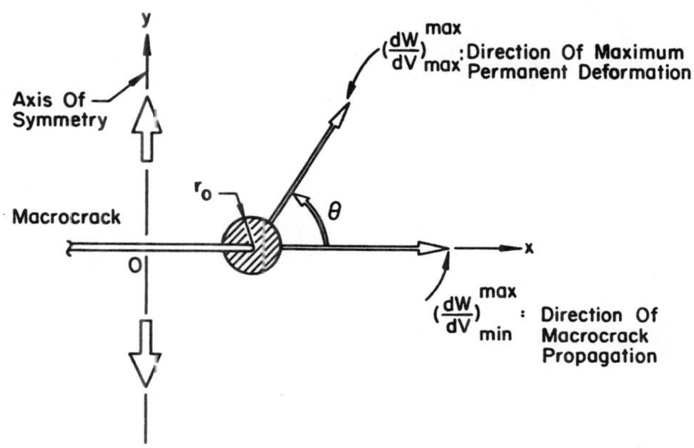

Figure 6.24. **Stationary values of strain energy density function associated with failure by deformation and fracture.**

6.9. References

[1] Sih, G. C., Heat conduction in the infinite medium with lines of discontinuities, *Journal of Heat Transfer* **87**, pp. 283–298 (1965).

[2] Sih, G. C., On the singular character of thermal stresses near a crack tip, *Journal of Applied Mechanics* **29**, pp. 287–589 (1962).

[3] Sih, G. C., Michopoulos, J. and Chou, S. C., *Hygrothermoelasticity*, Martinus Nijhoff Publishing, The Netherlands (1986).

[4] Boley B. A. and Weiner, J. H., *Theory of Thermal Stresses*, Wiley, New York (1960).

[5] Kipp, M. E. and Sih, G. C., The strain energy density failure criterion applied to notched elastic solids, *International Journal of Solids and Structures* **2**, pp. 153–173 (1975).

[6] Sih, G. C., Strain energy density and surface layer energy for blunted cracks or notches, in G. C. Sih, ed., *Mechanics of Fracture V*, Noordhoff, Leyden, XIII-CX (1978).

[7] Sih, G. C. and Bolton, R. C., Prediction of thermally induced fracture from notch and crack front, *Theoretical and Applied Fracture Mechanics* **5**, No. 2, pp. 125–132 (1986).

[8] Hilton, P. D., Gifford Jr., L. N. and Lomacky, O., Finite element fracture mechanics analysis of two-dimensional and axisymmetric elastic and elastic-plastic cracked structure, Naval Ship Research and Development Center, Report 4493 (1975).

[9] Sih, G. C., Introductory Chapters of Vols. I-VII, *Mechanics of Fracture*, Martinus Nijhoff Publishing, The Netherlands (1972–1982).

[10] Sih, G. C. and Chen, C. K., Growth of crack caused by temperature gradients with change in surface insulation, *Theoretical and Applied Fracture Mechanics* **5**, pp. 101–107 (1986).

[11] Nowacki, W., *Thermoelasticity*, Vol. 3, Interm. Series of Monographs on Aeronautics and Astronautics, Pergamon Press, Oxford (1962).

7

Failure of composites as predicted by the strain energy density theory

7.1. Preliminary remarks

If composite materials are to be used effectively in high performance structural applications, it is necessary to develop accurate reliable predictive methods for describing their failure behavior. The various failure modes observed at the laboratory scale level are fiber breakage, matrix separation and interface debonding. Should the load be increased indefinitely, the existing defects will grow while new ones may be created and damage of the composite will eventually reach catastrophic proportions. The progressive failure process in a composite is extremely complex and it would be impossible to analytically describe each successive stage in detail. This would not only be impractical but would also shed little light on engineering applications. On the other hand, methods of analysis procedures are needed to bridge the gap between composite testing and design. Only in this way can reliable estimates of the allowable stresses expected in service be made from strength and fracture toughness data obtained from small-scale laboratory tests. This capability still does not generally exist because of the complexity of cracks interacting with material inhomogeneity and anisotropy, effects that cannot be ignored in fiber-reinforced composites. To treat this subject, the discipline of fracture mechanics,* which is concerned with failure by crack initiation and propagation, is a natural tool to use for this purpose.

Prior to the application of fracture mechanics, however, a few remarks on scaling the continuum element relative to defect size and microstructure of the composite is in order. Loosely speaking, there are two schools of thought

* The term fracture mechanics acquires a much broader meaning in this communication than the conventional G_{1c} or k_{1c} concept. It addresses the initiation, stable growth and/or unstable propagation of cracks.

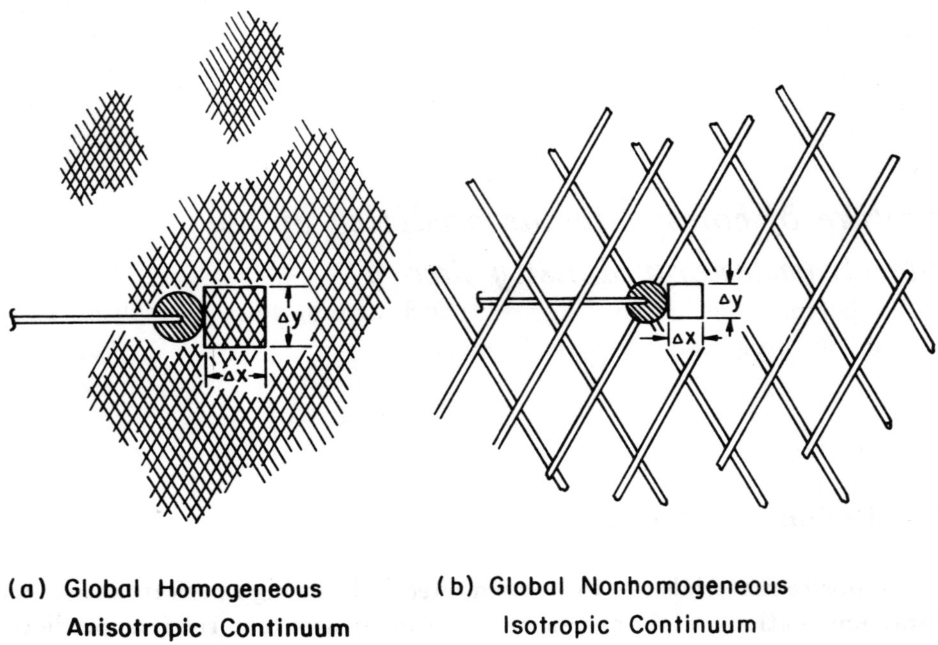

Figure 7.1. Scaling of the continuum element with reference to crack size and microstructure of the composite.

in the literature. The first assumes a global homogeneous anisotropic* continuum such that crack extension takes place in an idealized material with the gross combined properties of the constituents. This approach excludes detailed consideration of local damages such as broken fibers, cracks in matrix, etc. The second is to consider a global nonhomogeneous anisotropic medium where the crack tip with a known radius of curvature can exist only in one phase of the composite material at a given instance. The tip must either lie in the fiber, matrix or at the interface between two media. Figures 7.1(a) and 7.1(b) illustrate the scaling of the continuum element with reference to the crack size and microstructure of the composite. The validity of these approaches depends of course on how the specimen dimensions are scaled with respect to the allowable stress level and size of the cracks or damage zone. The reason is that failure modes are dictated by the rate and amount of energy dissipated which depends on specimen size and the

* The homogeneous anisotropic assumption has also been applied to continuum models using the concept of "micromechanical" elements whose mechanical properties are determined from laboratory specimens that are several orders of magnitude larger in size. The degree to which the mechanical behavior of composites is dependent on damage due to the size variance between the specimen and continuum element is not at all clear. Homogeneity at the global scale level may not hold in the local region where material is being damaged.

kind and rate of loading. The sequential order of discrete failure modes can also greatly influence the life of composite laminates. This is known as the stacking sequence effect.

The experimental procedure for measuring the effective moduli of a composite should be distinguished from that of strength determination. The former assumes the generic stress/strain relations and geometries to remain unchanged under load and material damage is kept to a minimum or is assumed to be negligible. Strength prediction implies the monitoring of progressive failure under increasing load up to the final stage of fracture. It is the specimen size and loading rate that determine whether analysis should account for progressive material damage.*

7.2. Strain energy density criterion

The main concern in fracture analysis consists in characterizing the action or phenomenon that causes a structural member to fail by slow and/or rapid crack extension. This necessitates the selection of a quantity that can most consistently relate the observed fracture mode with such factors as material properties, type of loading, shape of member, etc. The strain energy density function, dW/dV, as proposed by Beltrami, Haigh and Ludwik at the beginning of the 20th century possesses the unique feature that it can be expressed in terms of strains or stresses:

$$\frac{dW}{dV} = \frac{1+\nu}{2E}\left[\sigma_1^2 + \sigma_2^2 + \sigma_3^2 - \frac{\nu}{1+\nu}(\sigma_1 + \sigma_2 + \sigma_3)^2\right] \tag{7.1}$$

in which σ_1, σ_2 and σ_3 are the principal stresses, ν is Poisson's ratio and E is Young's modulus. In the linear theory of elasticity, dW/dV may be regarded as the sum of two energy components, $(dW/dV)_d$ and $(dW/dV)_v$. The former

$$\left(\frac{dW}{dV}\right)_d = \frac{1+\nu}{6E}[(\sigma_1 - \sigma_2)^2 + (\sigma_2 - \sigma_3)^2 + (\sigma_3 - \sigma_1)^2] \tag{7.2}$$

can be identified with shape change and is, in fact, the Huber–von Mises–Hencky yield condition while the latter

$$\left(\frac{dW}{dV}\right)_v = \frac{1-2\nu}{6E}(\sigma_1 + \sigma_2 + \sigma_3)^2 \tag{7.3}$$

is associated with volume change and condition for fracture. The value of dW/dV at fracture, say $(dW/dV)_c$, can be evaluated from a simple tension

* This is analogous to the typical phenomenon of metal fracture which is always preceded by slow crack growth.

test by calculating the area under the true stress/true strain curve as

$$\left(\frac{dW}{dV}\right)_c = \int_0^{\epsilon_c} \sigma \, d\epsilon \tag{7.4}$$

where ϵ_c is the critical strain. Both Matthaes [1] and Gillemot [2] have discussed the association of $(dW/dV)_c$ with the failure of tensile bar specimens. Experimental values of $(dW/dV)_c$ were also reported in [4].

The way in which $(dW/dV)_c$ is related to the condition of crack instability was established by Sih [3, 4] through the critical strain energy density factor

$$S_c = r_c \left(\frac{dW}{dV}\right)_c \tag{7.5}$$

with r_c being the radius of a core region surrounding the crack tip.* The size of this region is characteristic of the material [5, 6]. An important advancement of the strain energy density criterion [4] is that the energy absorbed in changing volume and shape should be considered simultaneously when searching for possible sites of failure. A possible means of weighing $(dW/dV)_d$ in equation (7.2) and $(dW/dV)_v$ in equation (7.3) is to consider the stationary values of the total strain energy density function with respect to a set of polar coordinates measured from the crack front. The locations of $(dW/dV)_{max}$ correspond to yielding and $(dW/dV)_{min}$ to fracture or crack extension. The critical value of equation (7.1), $(dW/dV)_c$, is assumed to remain constant along the crack front, a condition that has been used to determine crack profiles in elastic [7] and elastic–plastic materials [8].

Direction of crack initiation. Failure modes in composite materials are more complex than those in more conventional materials. Because of nonhomogeneity and anisotropy, self-similar crack growth is not likely to occur in composites even in simple tension. The classical concept of energy release rate G_{1c} may no longer be easily applied. A more convenient approach is to apply the strain energy density criterion. For the onset of rapid fracture, this criterion may be stated as follows:

(1) Crack initiation takes place in a direction determined by the relative minimum of the strain energy density factor, i.e.,

$$\frac{\partial S}{\partial \theta} = 0 \quad \left(\frac{\partial^2 S}{\partial \theta^2} > 0\right) \quad \text{at} \quad \theta = \theta_0. \tag{7.6}$$

(2) Rapid crack growth occurs when the minimum strain energy density factor reaches a critical value:

$$S_{min} = S_c \quad \text{at} \quad \theta = \theta_0. \tag{7.7}$$

* In ductile fracture [8], r_c represents the last ligament of slow crack growth just prior to the onset of rapid fracture.

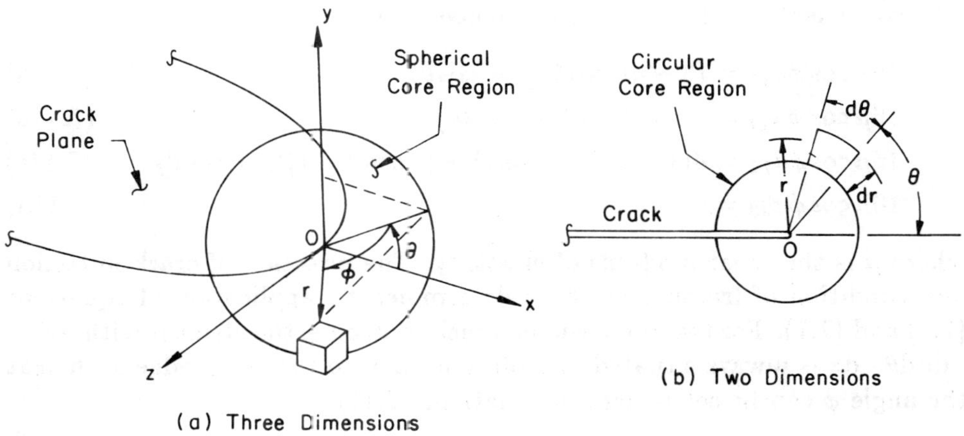

Figure 7.2. Core region surrounding the crack front.

The S_c-criterion is particularly useful for analyzing crack systems involving complex loadings where the direction of crack initiation is not known *a priori*. The parameter S_c is, in fact, directly related to K_{1c} by the relation*

$$S_c = \frac{(1+\nu)(1-2\nu)}{2\pi E} K_{1c}. \tag{7.8}$$

This implies that the S_c values for metal alloys can be found in the ASTM standard fracture toughness testing procedures. An alternate way of evaluating S_c has been presented in [10, 11].

Since the material in the immediate vicinity of the crack front is highly stressed and its behavior is not exactly known, the continuum mechanics analysis must necessarily be restricted to regions outside of a small zone surrounding the crack edge. For three-dimensional cracks, the shape of this core may be conveniently taken as the interior of all spheres like that of Figure 7.2(a), the three-dimensional crack, Figure 7.2(a), and circular for the two-dimensional crack, Figure 7.2(b). From the linear theory of isotropic and homogeneous elasticity, the strain energy density function is found to possess a singularity of the order** $1/r$ near the crack front and can be expressed in terms of three stress intensity factors k_j $(j = 1,2,3)$:

$$\frac{dW}{dV} = \frac{S}{r} + \cdots \tag{7.9}$$

where

$$S = \frac{1}{r}(a_{11}k_1^2 + 2a_{12}k_1k_2 + a_{22}k_2^2 + a_{33}k_3^2). \tag{7.10}$$

* The stress intensity factor K_1 and k_1 are related by $K_1 = \sqrt{\pi}\, k_1$.

** The order of strain energy density singularity remains the same for finite deformation [9].

The coefficients a_{ij} $(i, j = 1, 2, 3)$ are given by

$$16\mu \cos\phi a_{11} = (3 - 4\nu \cos\theta)(1 + \cos\theta) \tag{7.11a}$$

$$16\mu \cos\phi a_{12} = 2\sin\theta(\cos\theta - 1 + 2\nu) \tag{7.11b}$$

$$16\mu \cos\phi a_{22} = 4(1 - \nu)(1 - \cos\theta) + (3\cos\theta - 1)(1 + \cos\theta) \tag{7.11c}$$

$$16\mu \cos\phi a_{33} = 4 \tag{7.11d}$$

where μ is the shear modulus of elasticity. The direction of crack initiation and condition of fracture are then determined by application of equations (7.6) and (7.7). For two-dimensional crack problems, the element with $dV = r\,dr\,d\theta \cdot dz$ is always situated in a place normal to the crack edge such that the angle ϕ can be set to zero in equations (7.11).

Incremental crack growth. In order to achieve a predictive capability for composite fracture, the sequence of failure events, whereby a crack extends stably under an increasing load up to rapid growth, must be determined from the composite configurations and applied loads. The rate at which the crack grows incrementally can be associated with the amount of energy absorbed and subsequently released by each element ahead of the advancing crack. Suppose that the $(dW/dV)_c$ value for a material is known, then the condition

$$\left(\frac{dW}{dV}\right)_c = \frac{S_1}{r_1} = \frac{S_2}{r_2} = \ldots = \frac{S_i}{r_i} = \ldots = \frac{S_c}{r_c} = \text{const.} \tag{7.12}$$

prevails for stable crack growth as well as the onset of catastrophic fracture, equation (7.5).

If the fracture process due to increasing load is unstable, then each increment of crack growth will increase monotonically, i.e.,

$$r_1 < r_2 < \cdots < r_i < \cdots < r_c. \tag{7.13}$$

The corresponding strain energy density factors will also increase accordingly

$$S_1 < S_2 < \cdots < S_i < \cdots < S_c \tag{7.14}$$

such that the ratio S_j/r_j $(j = 1, 2, \ldots, n)$ is a constant as assumed by equation (7.12). The last ligament of growth r_c corresponds to instability as illustrated in Figure 7.3. A stable fracture process corresponds to decreasing increments of crack growth:

$$r_1 > r_2 > \cdots > r_i > \cdots > r_0, \tag{7.15}$$

where r_0 represents the last increment of growth before crack arrest. The strain energy density factors $S_1, S_2, \ldots, S_0$ also decrease according to

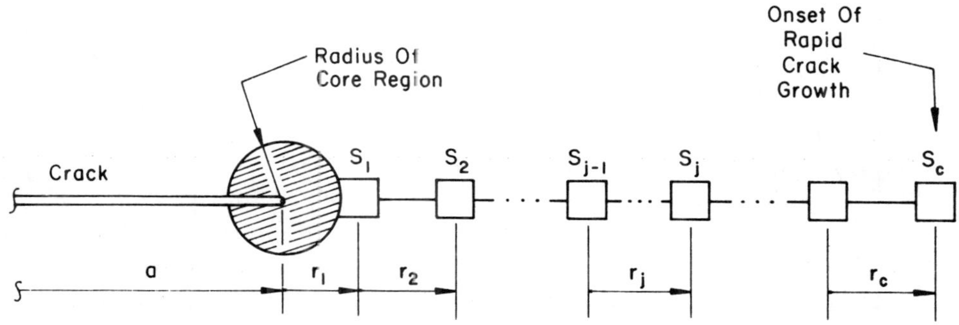

Figure 7.3. Incremental stable crack growth leading up to rapid fracture.

$S_j/r_j = \text{const.}$:

$$S_1 > S_2 > \cdots > S_i > \cdots > S_0. \tag{7.16}$$

The factor

$$S_0 = r_0 \left(\frac{dW}{dV} \right)_0 \tag{7.17}$$

is below the critical value S_c.

In a composite structure, a combination of the conditions described by equations (7.13) and (7.15) can exist. That is, the increments of crack growth may either increase or decrease depending on several factors such as the prevailing mechanical constraint, the material properties, nature of loading, etc.

7.3. Failure of fiber, matrix or interface

As mentioned earlier, real failure modes in fiber-reinforced composite systems are very complex as they may involve fiber breakage, matrix cracking, fiber/matrix debonding, and laminae delamination during a period of gradually increasing load. Although a precise quantitative assessment of all the failure modes is beyond the scope of this communication, reasonably accurate predictions of the allowable stress can be made if any one of the aforementioned failure modes tends to dominate at the onset of rapid fracture. The local damage done prior to global instability simply alters the stiffness and/or available energy to cause final fracture. Therefore, it is useful to have analytical representation of the individual failure modes and to understand how the physical properties of the constituents affect the fracture behavior of composite systems.

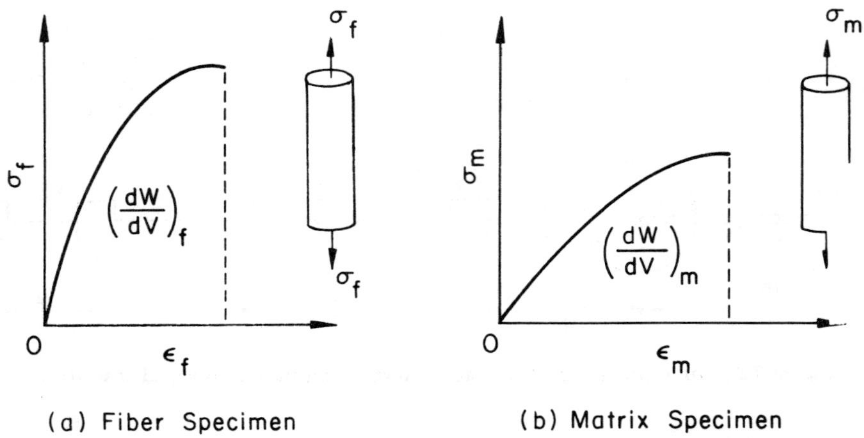

Figure 7.4. **Measurements of the strain energy density function at fracture.**

Critical strain energy density function for fiber and matrix. **Needless to say, data from fracture tests on isolated fibers and on matrix materials are prerequisite to the development of predictive procedures. Standard simple tension tests as shown in Figures 7.4(a) and 7.4(b) may be performed to obtain the values of $(dW/dV)_f$ and $(dW/dV)_m$ at fracture in accordance with equation (7.4). The specimen size and loading rate should be selected such that the influence of preexisting defects will not have to be considered in the strength calculation. Special consideration must also be given to the dependency of dW/dV on specimen size at fracture. The way in which $(dW/dV)_f$ and $(dW/dV)_m$ are utilized in estimating the fracture resistance of larger size composite members, in which existing defects or cracks must be accounted for, will be examined by application of the strain energy density criterion.**

Crack initiation in fiber, matrix or interface. **The commonly observed failure modes of fiber breaking and/or matrix cracking are generally determined by the combination of loading, location of defect and material nonhomogeneity. The fracture mechanics viewpoint is that all failure initiates from a preexisting defect. Figures 7.5(a) and 7.5(b) illustrate three sites of possible failure initiating from a crack of length $2a$ situated near the interface. The first possibility is matrix cracking if the loading angle β is sufficiently large since the shear modulus of the fiber μ_f is much greater than that of the matrix μ_m. As β is decreased towards zero, more load is transmitted to the fiber and there is more likelihood of fiber breakage. Failure within the interface is the third possibility.**

Assuming that the defect does not grow along the interface, the cohesive failure model in Figure 7.5(a) may be adopted. The upper half-plane is

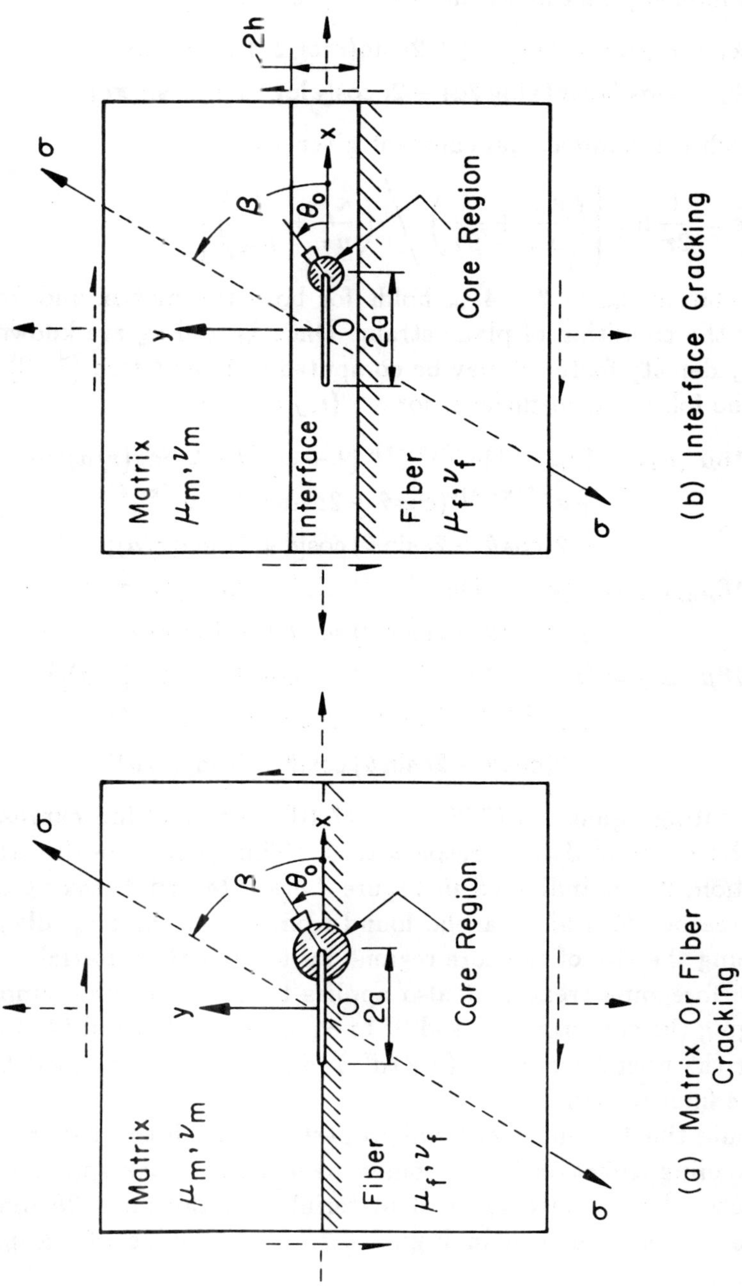

Figure 7.5. **Failure modes caused by defect near interface.**

occupied by the matrix material with elastic properties μ_m, ν_m and the lower half-plane by the fiber with elastic properties μ_f, ν_f. For this configuration, asymmetry prevails in both loading and material, and hence two types of stress intensity factors k_1 and k_2 are present [12]

$$k_1 = \sigma\sqrt{a}[\cos(\epsilon \log 2a) + 2\epsilon \sin(\epsilon \log 2a)]/\cosh(\pi\epsilon) \tag{7.18a}$$

$$k_2 = -\sigma\sqrt{a}[\sin(\epsilon \log 2a) - 2\epsilon \cos(\epsilon \log 2a)]/\cosh(\pi\epsilon) \tag{7.18b}$$

in which ϵ is a bimaterial constant given by

$$\epsilon = \frac{1}{2\pi} \log \left[\left(\frac{\kappa_m}{\mu_m} + \frac{1}{\mu_f} \right) \Big/ \left(\frac{\kappa_f}{\mu_f} + \frac{1}{\mu_m} \right) \right]. \tag{7.19}$$

The relation $\kappa_m = 3 - 4\nu_m$ holds for both the matrix and fiber material under the condition of plane strain. Once k_1 and k_2 are known, the strain energy density factor S may be computed from equation (7.10) with $k_3 = 0$ and the following coefficients for a_{ij} $(i, j = 1, 2)$:

$$\begin{aligned} 16\mu_m a_{11} = {} & (\kappa_m - 1)e^{-2(\pi-\theta)\epsilon}\{1 + \cos[\theta + 2\epsilon \log(r/a)]\} + \\ & + e^{-2(\pi-\theta)\epsilon}(\cos\theta + 2\epsilon \sin\theta)^2 + e^{2(\pi-\theta)\epsilon} - \\ & - 2(\cos\theta + 2\epsilon\sin\theta)\cos[\theta + 2\epsilon\log(r/a)] \end{aligned} \tag{7.20a}$$

$$\begin{aligned} 16\mu_m a_{12} = {} & -(\kappa_m - 1)e^{-2(\pi-\theta)\epsilon}\sin[\theta + 2\epsilon\log(r/a)] + \\ & + 2(\cos\theta + 2\epsilon\sin\theta)\sin[\theta + 2\epsilon\log(r/a)] \end{aligned} \tag{7.20b}$$

$$\begin{aligned} 16\mu_m a_{22} = {} & (\kappa_m - 1)e^{-2(\pi-\theta)\epsilon}\{1 - \cos[\theta + 2\epsilon \log(r/a)]\} + \\ & + e^{-2(\pi-\theta)\epsilon}(\cos\theta + 2\epsilon \sin\theta)^2 + e^{2(\pi-\theta)\epsilon} + \\ & + 2(\cos\theta + 2\epsilon\sin\theta)\cos[\theta + 2\epsilon\log(r/a)]. \end{aligned} \tag{7.20c}$$

Substituting equations (7.20) into (7.10), the resulting expression S may then be minimized with respect to θ. This yields the direction of crack initiation, θ_0, as indicated in Figure 7.5(a). Matrix breaking occurs when $S_{\min}$ reaches S_c which can be found from the product $r_m(dW/dV)_m$ with r_m being the size of the core region* for the matrix material.

The foregoing treatment also applies to fiber breaking simply by interchanging the subscript 'm' and 'f' in equations (7.19) and (7.20). The value S_c for the fiber becomes $r_f(dW/dV)_f$ in which case r_f must be evaluated for the fiber material.

Should the dissimilar material properties of the fiber, matrix and interface and loading direction be combined to enhance crack spreading along the interface, then a third layer of material with thickness $2h$ may be added to the model as shown in Figure 7.5(b). Modeling of the properties of

---

* The size of the core region around a crack tip for a given material has received much attention [6, 13]. It can be interpreted as the zone outside of which failure is assumed to initiate.

this interface layer, however, is open to uncertainty. Sih and Moyer [14] have shown that the difference between an interface with gradual modulus variation and constant average modulus can be significant depending on the boundary conditions specified remotely away from the interface. The results suggest that even though the interface may be thin in comparison with the dimensions of the two joining materials, it can exert appreciable influence on the gross mechanical behavior of the composite. A possible way of circumventing this problem is to optimize the properties of the constituents with the local stress/strain condition such that the interface will not be overly sensitive to the response of the composite.

If the original defect, say a crack, is located either in the matrix or fiber, then the material surrounding the crack may be regarded as homogeneous and equation (7.10) may be used. Stress intensity factor solutions for a variety of configurations and loadings involving matrix or fiber weakened by defects can be found in several chapters of this volume.

7.4. Unidirectional fiber/matrix composite

The unidirectional fibrous composite is the simplest two-phase material constructed by embedding a system of parallel fibers in a matrix. Two basic continuum mechanics approaches will be considered:

(1) a crack propagating in a layer of isotropic matrix material sandwiched between two anisotropic solids, and
(2) a crack extending in a homogeneous and anisotropic material having the gross mechanical properties of the fiber and matrix or composite.

Critical stresses for these two models will be obtained and compared with available experimental data.

Matrix cracking model. Experimental work [15] on Scotchply 1002 (trade name by the Minnesota Mining and Manufacturing Company) has shown that unidirectional composites fail by crack propagation between the fibers starting from an initial defect such as voids or air bubbles in the composite. On the basis of this observed failure mode, the analytical model in Figure 7.6(a) is chosen with a crack of length $2a$ in a layer of matrix material having elastic properties E_m, ν_m or μ_m, ν_m. The material surrounding this layer of thickness $2h$ is taken to be orthotropic with the elastic properties E_1, E_2, ν_{12}, ν_{21} and μ_{12}. The subscripts 1 and 2 designate the directions parallel and perpendicular to the fiber as illustrated in Figure 7.6(a). On the average, the spacing $2h$ between the fibers may be estimated from the fiber volume fraction V_f by assuming that all the fibers having radius R, are uniformly

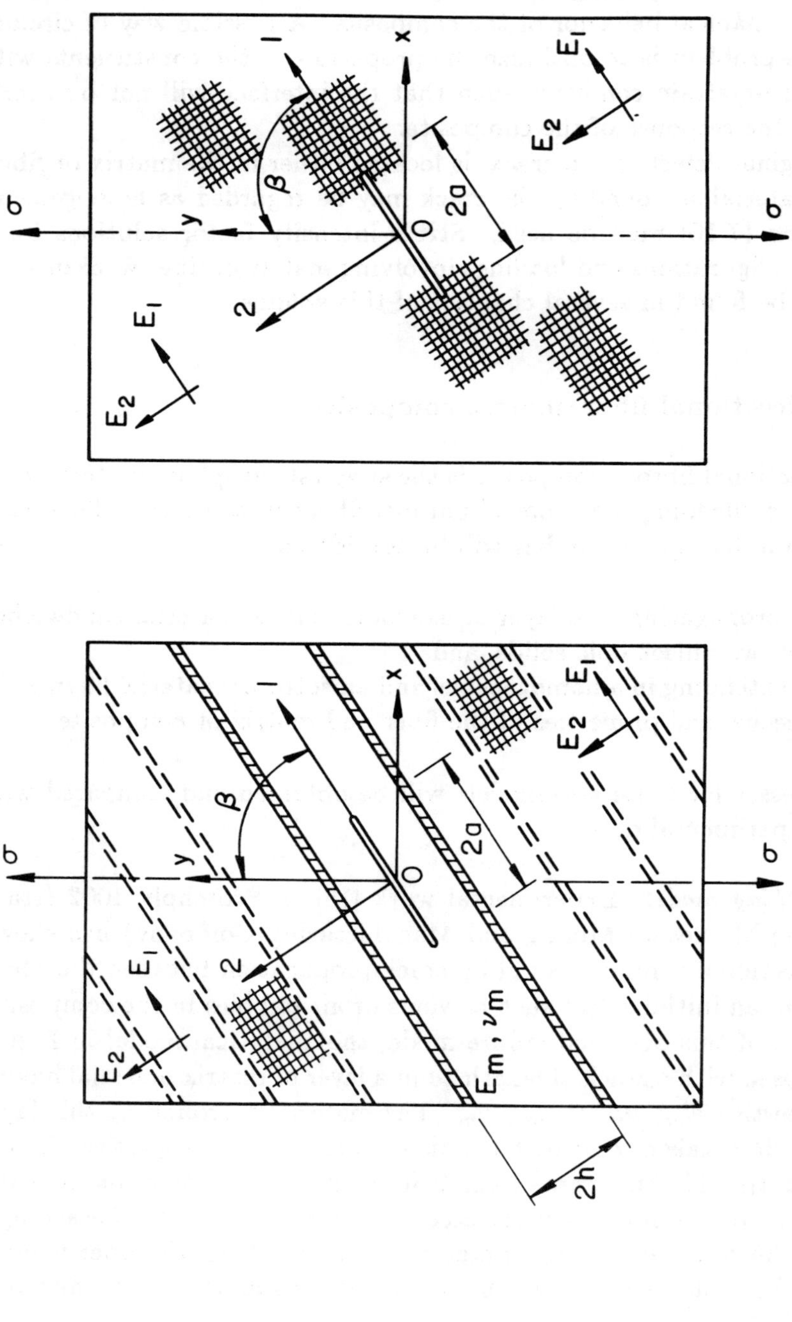

(a) Matrix Cracking

(b) Crack in Homogeneous and Anisotropic Medium

Figure 7.6. **Modeling of unidirectional fibrous composites under angle loading.**

distributed throughout the composite. The resulting expression is

$$2h = R\left[\sqrt{\pi/V_f} - 2\right]. \tag{7.21}$$

For the configuration shown in Figure 7.6(a), the applied stress σ makes an angle β with the crack plane and the intensity of the crack tip stress field is described by the following k_j $(j = 1,2)$ factors:

$$k_1 = \Phi(1)\sigma\sqrt{a}\sin^2\beta \tag{7.22a}$$
$$k_2 = \Psi(1)\sigma\sqrt{a}\sin\beta\cos\beta \tag{7.22b}$$

in which the functions $\Phi(\xi)$ and $\Psi(\xi)$ depend on the elastic constants and geometric parameters of the composite. They can be computed numerically from a system of Fredholm integral equations [16]. The gross elastic properties for the Scotchply 1002 composite with a fiber volume fraction of 56.5% can be obtained from equations (7.72) in Appendix 7.1 and they are given by

$$E_1 = 6.13 \times 10^6 \text{ psi}, \quad \nu_{12} = 0.27 \tag{7.23a}$$
$$E_2 = 1.42 \times 10^6 \text{ psi}, \quad \nu_{23} = 0.34 \tag{7.23b}$$
$$\mu_{12} = 0.53 \tag{7.23c}$$

where $\nu_{12}/E_1 = \nu_{21}/E_2$. The matrix is made of epoxy resin with $E_m = 4.5 \times 10^5$ psi and $\nu_m = 0.35$ and fibers with $E_f = E_{f_1} = E_{f_2} = 10.5 \times 10^6$ psi and $\nu_f = 0.20$. When the layer thickness is taken to be very small in comparison with the crack length $2a$, the values of $\Phi(\xi)$ and $\Psi(\xi)$ evaluated at $\xi = 1$ are

$$\Phi(1) = 0.31, \quad \Psi(1) = 0.16. \tag{7.24}$$

In general, both $\Phi(1)$ and $\Psi(1)$ decrease with V_f which means that fiber reinforcement tends to lower the intensification of stresses near the crack tip.

Before application of the S_c-criterion* to the unidirectional fibrous composite, it is essential to emphasize the failure mode of Scotchply 1002. First of all, the final fracture is known to occur in a brittle fashion in that there exists a distinct point of instability on the load versus deflection curve. Most of the stored energy is assumed to be released at the moment of incipient fracture where the crack begins to propagate. The remaining energy dissipated prior to global instability is sufficiently small so that the specimen geometry and stiffness are not affected appreciably as the critical load point is approached. To be recognized also is that failure initiates at an angle θ_0

---

* Stable crack growth is assumed to be governed by the condition of $(dW/dV)_c$ = const. as indicated in equation (7.12).

Table 7.1. **Fracture angles and normalized strain energy density factors for the matrix cracking model (Scotchply 1002 with a 56.5% fiber volume fraction).**

β	$-\theta_0$	$S_{\min}/\sigma^2 a$	β	$-\theta_0$	$S_{\min}/\sigma^2 a$
5.00	77.700	0.00012	50.00	35.400	0.01841
10.00	71.400	0.00052	55.00	31.600	0.02257
15.00	65.600	0.00123	60.00	27.700	0.02688
20.00	60.300	0.00231	65.00	23.600	0.03112
25.00	55.500	0.00380	70.00	19.300	0.03508
30.00	51.000	0.00575	75.00	14.800	0.03849
35.00	46.900	0.00819	80.00	10.000	0.04113
40.00	42.900	0.01113	85.00	5.100	0.04280
45.00	39.100	0.01456	90.00	0.000	0.04337

with reference to the crack plane although the final crack path may follow along the direction of the fiber. The values of θ_0 for each β may be found by first computing S in equation (7.10) with the aid of equations (7.22) and (7.24), and then applying the conditions specified by equation (7.6). The fracture angles θ_0 are found to be negative as listed in Table 7.1 together with the corresponding values of $S_{\min}/\sigma^2 a$ obtained from

$$S_{\min}=\sigma^2 a(0.096a_{11}\sin^4\beta+0.050a_{12}\sin^3\beta\cos\beta+0.26a_{22}\cos^4\beta). \quad (7.25)$$

It should be reiterated that the foregoing model assumes that energy is released only in the matrix. Incipient fracture is assumed to occur when $S_{\min}$ reaches S_c for the epoxy resin. Hence, only a knowledge of S_c for the matrix material and a valid stress analysis that accounts for a reduction in load transmitted to the matrix owing to fiber reinforcement are needed. When local nonhomogeneity of the composite is considered, it is no longer meaningful to discuss the global fracture toughness of the composite system since all the free surfaces are assumed to be created in the matrix, and none in the fibers nor the fiber-to-matrix interfaces.

Homogeneous anisotropic model. Inquiries have been made into the possibility of defining a fracture toughness value for the entire composite system in much the same way as for a single-phase homogeneous material such as metal. Sih *et al.* [17] first derived an energy release rate expression for a line crack in a solid possessing rectilinear anisotropy. Conceptually speaking, it is not clear how the creation of free surfaces in fracturing fibers, matrix and fiber-matrix interfaces can be combined in such a way as to use an average release of energy through a segment of equivalent homogeneous and anisotropic material at incipient fracture. The validity of this idealization will obviously depend on how the discrete nature of the composite affects the failure modes and will be assessed quantitatively by using the S_c-criterion.

Referring to Figure 7.6(b), a crack of length $2a$ is situated in a homogeneous anisotropic medium with the principal axes of elastic symmetry denoted by 1 and 2. A uniform tensile stress of magnitude σ is applied in the y-direction which makes an angle β with the crack plane. The stresses σ_1, σ_2 and τ_{12} in the neighborhood of the crack tip are given by [17]

$$\sigma_1 = \frac{k_1}{\sqrt{2r}}\mathrm{Re}\left[\frac{s_1 s_2}{s_1 - s_2}\left(\frac{s_2}{z_2} - \frac{s_1}{z_1}\right)\right] + \\ + \frac{k_2}{\sqrt{2r}}\mathrm{Re}\left[\frac{1}{s_1 - s_2}\left(\frac{s_2^2}{z_2} - \frac{s_1^2}{z_1}\right)\right] + \cdots \tag{7.26a}$$

$$\sigma_2 = \frac{k_1}{\sqrt{2r}}\mathrm{Re}\left[\frac{1}{s_1 - s_2}\left(\frac{s_1}{z_2} - \frac{s_2}{z_1}\right)\right] + \\ + \frac{k_2}{\sqrt{2r}}\mathrm{Re}\left[\frac{1}{s_1 - s_2}\left(\frac{1}{z_2} - \frac{1}{z_1}\right)\right] + \cdots \tag{7.26b}$$

$$\tau_{12} = \frac{k_1}{\sqrt{2r}}\mathrm{Re}\left[\frac{s_1 s_2}{s_1 - s_2}\left(\frac{1}{z_1} - \frac{1}{z_2}\right)\right] + \\ + \frac{k_2}{\sqrt{2r}}\mathrm{Re}\left[\frac{1}{s_1 - s_2}\left(\frac{s_1}{z_1} - \frac{s_2}{z_1}\right)\right] + \cdots \tag{7.26c}$$

where Re[ ] denotes the real part of the complex function in the bracket. The stress intensity factors* k_1 and k_2 take the same forms as those shown in equations (7.22) with $\Phi(1) = \Psi(1) = 1$. The complex variables z_1 and z_2 are:

$$z_1 = (\cos\theta + s_1 \sin\theta)^{1/2}, \quad z_2 = (\cos\theta + s_2 \sin\theta)^{1/2}. \tag{7.27}$$

The polar coordinates r and θ are measured from the right hand side crack tip. The angular distribution of the stresses is clearly affected by the anisotropy of the material through the parameters s_1 and s_2 which are roots of the fourth order characteristic equation**

$$b_{11}s^4 + 2b_{16}s^3 + (2b_{12} + b_{66})s^2 - 2b_{26}s + b_{22} = 0. \tag{7.28}$$

The constants b_{ij} correspond to those appearing in the strain-stress relationship:

$$\begin{bmatrix} \epsilon_1 \\ \epsilon_2 \\ \gamma_{12} \end{bmatrix} = \begin{bmatrix} b_{11} & b_{12} & b_{16} \\ b_{21} & b_{22} & b_{26} \\ b_{61} & b_{62} & b_{66} \end{bmatrix} \begin{bmatrix} \sigma_1 \\ \sigma_2 \\ \tau_{12} \end{bmatrix}. \tag{7.29}$$

---

* The stress intensity factors [17] are defined so that for cracks in an infinite medium under self-equilibrating loads, they are independent of the elastic constants and the same as those for the isotropic material.

** By an energy consideration, it can be shown that the roots of equation (7.28) are either complex or pure imaginary and cannot be real. Thus, the four roots separate into two sets of distinct complex conjugates. The parameters s_1 and s_2 correspond to those with positive imaginary parts.

For the condition of plane strain, b_{ij} can be related to s_{ij} as follows:

$$b_{11} = \frac{s_{11}s_{33} - s_{13}^2}{s_{33}}, \quad b_{12} = b_{21} = \frac{s_{12}s_{33} - s_{13}s_{23}}{s_{33}} \tag{7.30a}$$

$$b_{22} = \frac{s_{22}s_{33} - s_{23}^2}{s_{33}}, \quad b_{16} = b_{61} = \frac{s_{16}s_{33} - s_{13}s_{36}}{s_{33}} \tag{7.30b}$$

$$b_{66} = \frac{s_{66}s_{33} - s_{36}^2}{s_{33}}, \quad b_{26} = b_{62} = \frac{s_{26}s_{33} - s_{23}s_{36}}{s_{33}}. \tag{7.30c}$$

For unidirectional composites, there exist only three principal directions that are mutually perpendicular to one another and only nine independent elastic constants are nonzero. Since s_{44} and s_{55} do not appear in equations (7.30), only seven of the nine quantities of s_{ij} need to be identified with the engineering constants E_1, E_2, etc. They are:

$$s_{11} = \frac{1}{E_1}, \quad s_{22} = \frac{1}{E_2}, \quad s_{33} = \frac{1}{E_3}, \quad s_{66} = \frac{1}{\mu_{12}} \tag{7.31a}$$

$$s_{12} = s_{21} = -\frac{\nu_{21}}{E_2} = -\frac{\nu_{12}}{E_1} \tag{7.31b}$$

$$s_{13} = s_{31} = -\frac{\nu_{31}}{E_3} = -\frac{\nu_{13}}{E_1} \tag{7.31c}$$

$$s_{23} = s_{32} = -\frac{\nu_{32}}{E_3} = -\frac{\nu_{23}}{E_2} \tag{7.31d}$$

As in the case of isotropic elasticity, the stresses in equations (7.26) may be inserted into the strain energy density function

$$\frac{dW}{dV} = \tfrac{1}{2}(\sigma_1\epsilon_1 + \sigma_2\epsilon_2 + \tau_{12}\gamma_{12}) \tag{7.32}$$

to yield an expression involving $1/r$ the coefficient of which is the strain energy density factor

$$S = A_{11}k_1^2 + 2A_{12}k_1k_2 + A_{22}k_2^2. \tag{7.33}$$

The quantities A_{ij} $(i,j = 1,2)$ are complicated functions of the anisotropic elastic constants and depend on the angle θ through z_1 and z_2. They are different from the a_{ij} $(i,j = 1,2)$ in equations (7.11) for an isotropic material

$$A_{11} = \tfrac{1}{4}[b_{11}A^2 + b_{22}C^2 + b_{66}E^2 + 2b_{12}AC + 2b_{16}AE + 2b_{26}CE] \tag{7.34a}$$

$$A_{12} = \tfrac{1}{4}[b_{11}AB + b_{22}CD + b_{66}EF + b_{12}(AD + BC) + b_{16}(AF + BE) + b_{26}(CF + DE)] \tag{7.34b}$$

$$A_{22} = \tfrac{1}{4}[b_{11}B^2 + b_{22}D^2 + b_{66}F^2 + 2b_{12}BD + 2b_{16}BF + 2b_{26}DF]. \tag{7.34c}$$

The following contractions have been made:

$$A = \mathrm{Re}\left[\frac{s_1s_2}{s_1 - s_2}\left(\frac{s_2}{z_2} - \frac{s_1}{z_1}\right)\right], \quad B = \mathrm{Re}\left[\frac{1}{s_1 - s_2}\left(\frac{s_2^2}{z_2} - \frac{s_1^2}{z_1}\right)\right] \tag{7.35a}$$

$$C = \mathrm{Re}\left[\frac{1}{s_1 - s_2}\left(\frac{s_1}{z_2} - \frac{s_2}{z_1}\right)\right], \quad D = \mathrm{Re}\left[\frac{1}{s_1 - s_2}\left(\frac{1}{z_2} - \frac{1}{z_1}\right)\right] \tag{7.35b}$$

$$E = \mathrm{Re}\left[\frac{s_1 s_2}{s_1 - s_2}\left(\frac{1}{z_1} - \frac{1}{z_2}\right)\right], \quad F = \mathrm{Re}\left[\frac{1}{s_1 - s_2}\left(\frac{s_1}{z_1} - \frac{s_2}{z_2}\right)\right]. \tag{7.35c}$$

Suppose that the unidirectional composite is made by embedding Modulite II graphite fibers in an epoxy resin matrix. The elastic properties of these two constituents are

$$E_f = 35 \times 10^6 \text{ psi}, \quad \nu_f = 0.3 \tag{7.36a}$$

$$E_m = 0.5 \times 10^6 \text{ psi}, \quad \nu_m = 0.4. \tag{7.36b}$$

The fibers are assumed to be isotropic so that $E_f = E_{f_1} = E_{f_2}$ and the fiber volume fracture for the Modulite II 5206 composite is approximately 0.65. The gross mechanical properties of the composite may be calculated from equations (7.72) in Appendix 7.1. The results are

$$E_1 = 22.92 \times 10^6 \text{ psi}, \quad E_2 = 2.22 \times 10^6 \text{ psi} \tag{7.37a}$$

$$\mu_{12} = 0.8 \times 10^6 \text{ psi}, \quad \nu_{12} = 0.34, \quad \nu_{23} = 0.43. \tag{7.37b}$$

The constants $s_{11}, s_{12}, \ldots, s_{66}$ can then be obtained from equations (7.31) in units of $(\text{psi})^{-1}$:

$$s_{11} = 0.04 \times 10^{-6}, \quad s_{12} = s_{13} = -0.02 \times 10^{-6} \tag{7.38a}$$

$$s_{22} = 0.45 \times 10^{-6}, \quad s_{23} = -0.19 \times 10^{-6}, \quad s_{66} = 1.25 \times 10^{-6}. \tag{7.38b}$$

Consequently, $b_{11}, b_{12}, \ldots, b_{66}$ follow from equations (7.30) having the same units as s_{ij}:

$$b_{11} = 0.04 \times 10^{-6}, \quad b_{12} = -0.02 \times 10^{-6} \tag{7.39a}$$

$$b_{22} = 0.37 \times 10^{-6}, \quad b_{66} = 1.25 \times 10^{-6} \tag{7.39b}$$

where $b_{16} = b_{26} = 0$. Once the coefficients b_{ij} in equation (7.28) are known, it is found that

$$s_1 = 0.636i, \quad s_2 = 5.261i. \tag{7.40}$$

With k_1 and k_2 in equation (7.33) being the same as an inclined tensile crack in an isotropic material, the S factor is obtained:

$$S = \sigma^2 a(A_{11} \sin^4 \beta + 2A_{12} \sin^3 \beta \cos \beta + A_{22} \sin^2 \beta \cos^2 \beta). \tag{7.41}$$

Differentiating equation (7.41) with respect to θ and setting the result equal to zero, the values of θ_0 corresponding to $S_{\min}$ are found and tabulated in Table 7.2 for β increasing from 5° to 90° in increments of 5°. Although the negative fracture angle θ_0 decreases as β is increased, the normalized strain energy density factor $S_{\min}/\sigma^2 a$ tends to increase reaching a maximum at $\beta = 90°$. Unlike the isotropic material solution, the fracture angle θ_0 is

Table 7.2. Fracture angles and normalized strain energy density factors for the homogeneous anisotropic model (Modulite II 5206 graphite-epoxy composite).

β	$-\theta_0$	$S_{min}/\sigma^2 a$	β	$-\theta_0$	$S_{min}/\sigma^2 a$
5.00	75.200	0.00041	50.00	40.900	0.06078
10.00	69.500	0.00182	55.00	38.500	0.07176
15.00	64.500	0.00447	60.00	36.300	0.08232
20.00	60.100	0.00853	65.00	34.400	0.09206
25.00	56.100	0.01411	70.00	32.800	0.10062
30.00	52.500	0.02119	75.00	31.700	0.10774
35.00	49.300	0.02967	80.00	31.200	0.11321
40.00	46.200	0.03932	85.00	31.400	0.11682
45.00	43.500	0.04982	90.00	32.400	0.11839

different from zero even when the applied stress is normal to the crack plane. In addition to $\theta_0 = -32.4°$ shown in Table 7.1, $\theta_0 = +32.4°$ is also a possible solution which implies the tendency for the crack to bifurcate when the modulus of elasticity in the direction parallel to the crack is larger than that normal to it, i.e., $E_1 > E_2$.

7.5. Comparison of results obtained from anisotropic and matrix cracking model

Two possible analytical models have been presented in this section for describing the brittle fracture of unidirectional fiber/matrix composites under angle loading. Results obtained from these two models will now be compared theoretically for the Modulite II 5206 graphite-epoxy composite. Experimental data on the fracture of E-glass fiber-reinforced Scotchply 1002 composite will also be discussed in connection with the theoretical prediction.

Modulite II 5206 graphite-epoxy. The mechanical properties for this material are given in equations (7.36) and (7.37). For the matrix cracking model, $\Phi(1)$ and $\Psi(1)$ in equations (7.22) take the values

$$\Phi(1) = 0.22, \quad \Psi(1) = 0.07. \tag{7.42}$$

The strain energy density factor becomes

$$S = \sigma^2 a(0.0484 a_{11} \sin^4\beta + 0.0308 a_{12} \sin^3\beta\cos\beta + \\ + 0.0049 a_{22} \sin^2\beta\cos^2\beta) \tag{7.43}$$

where a_{ij} $(i, j = 1, 2)$ are given by equations (7.11) with $\phi = 0$ and $\nu = \nu_m$. Making use of the conditions in equations (7.6) and (7.43), the fracture angles θ_0 and $S_{min}/\sigma^2 a$ are obtained (Table 7.3). By assuming that fracture occurs when $S_{min} = S_c$, the ratio $(\sigma_c)_\beta/(\sigma_c)_{\beta=\pi/2}$ is obtained and plotted as a function of β in Figure 7.7 by using equation (7.43) for the matrix

Table 7.3. Fracture angles and normalized strain energy density factors for the matrix cracking model (Modulite II 5206 graphite–epoxy composite).

β	$-\theta_0$	$S_{min}/\sigma^2 a$	β	$-\theta_0$	$S_{min}/\sigma^2 a$
5.00	76.600	0.00002	50.00	27.700	0.00502
10.00	67.900	0.00010	55.00	24.200	0.00635
15.00	60.500	0.00024	60.00	20.600	0.00775
20.00	54.100	0.00046	65.00	17.200	0.00915
25.00	48.600	0.00081	70.00	13.700	0.01045
30.00	43.700	0.00219	75.00	10.300	0.01158
35.00	39.300	0.00195	80.00	6.900	0.01245
40.00	35.300	0.00279	85.00	3.400	0.01300
45.00	31.400	0.00382	90.00	0.000	0.01318

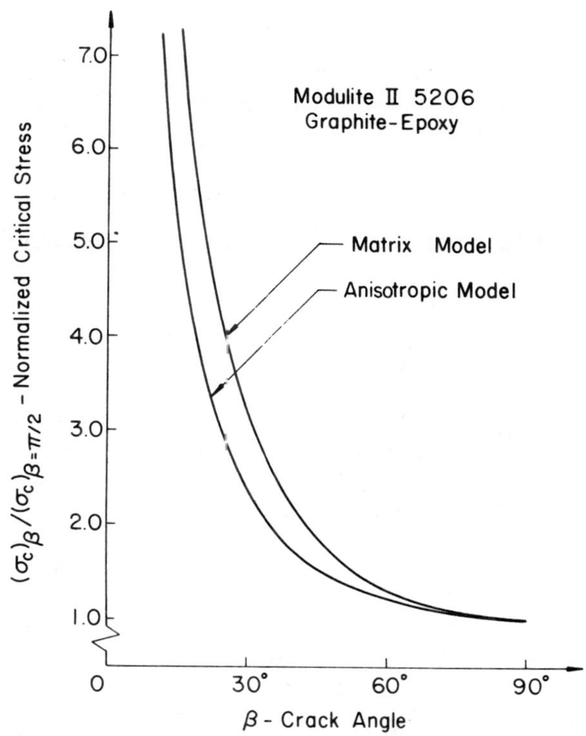

Figure 7.7. Normalized critical stress versus crack angle β for Modulite II 5206 graphite epoxy.

cracking model and equation (7.41) for the homogeneous anisotropic model. The two curves are normalized at $\beta = 90°$ and their deviations begin to increase significantly when β is reduced. The discrepancy can be attributed to the difference in the values for θ_0 and $S_{min}/\sigma^2 a$ given in Tables 7.2 and 7.3. Another point to be noted is that the matrix crack propagation model predicts a fracture angle $\theta_0 = 0$ at $\beta = 90°$ as opposed to $\theta_0 \neq 0$ for the anisotropic model.

Table 7.4. Fracture angles and normalized strain energy density factors for the homogeneous anisotropic model (Scotchply 1002 with a 56.5% fiber volume fraction).

β	$-\theta_0$	$S_{\min}/\sigma^2 a$	β	$-\theta_0$	$S_{\min}/\sigma^2 a$
5.00	75.200	0.00093	50.00	44.100	0.11599
10.00	70.600	0.00403	55.00	41.700	0.13528
15.00	66.400	0.00968	60.00	39.500	0.15358
20.00	62.500	0.01807	65.00	37.600	0.17026
25.00	58.900	0.02927	70.00	36.000	0.18485
30.00	55.500	0.04313	75.00	35.000	0.19697
35.00	52.400	0.05929	80.00	34.600	0.20634
40.00	49.500	0.07727	85.00	35.200	0.21264
45.00	46.700	0.09641	90.00	37.200	0.21546

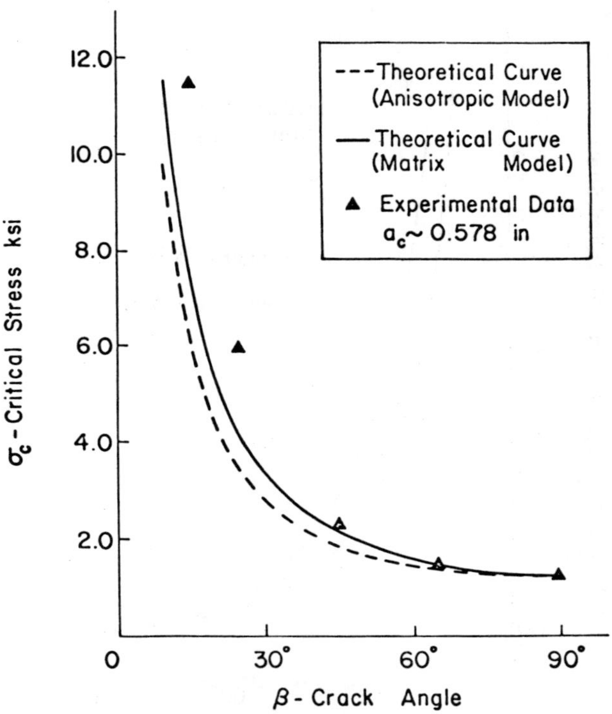

Figure 7.8. Critical stress versus crack angle for Scotchply 1002 with $a_c = 0.578$ in.

E-glass–epoxy resin composite. A comparison of the results presented in Table 7.1 for Scotchply 1002 will now be made with those for the homogeneous anisotropic model. Referring to the elastic constants in equations (7.23) and (7.30), the quantities b_{ij} can be computed. Equation (7.28) may then be solved to yield

$$s_1 = 0.611i, \quad s_2 = 3.222i. \tag{7.44}$$

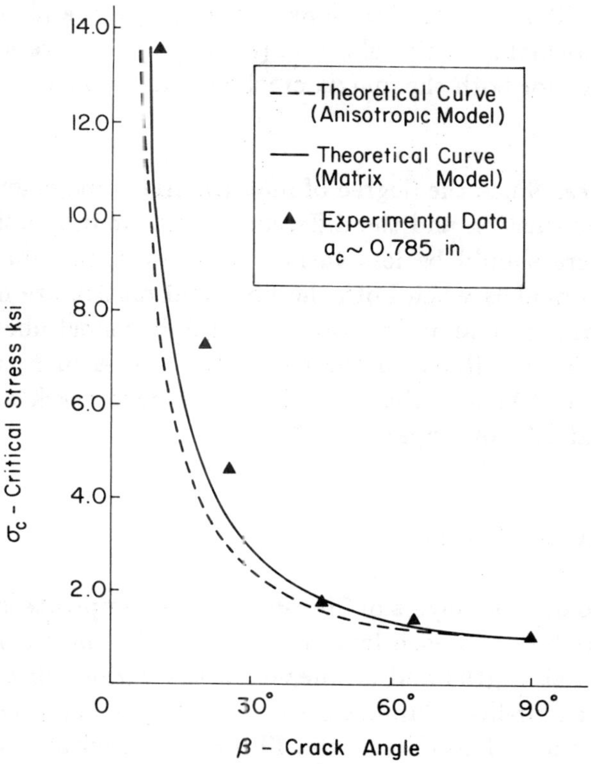

Figure 7.9. Critical stress versus crack angle for Scotchply 1002 with $a_c = 0.785$ in.

The strain energy density factor S takes the same form as equation (7.41) except the quantities A_{ij} are obtained from equations (7.34) and (7.44). The first condition of the strain energy density criterion given by equation (7.7) leads to the results in Table 7.4.

As β is varied, Tables 7.1 and 7.4 show that θ_0 and $S_{\min}/\sigma^2 a$ differ significantly for the two models. A graphical comparison of the critical stress as a function of β with experimental data [18] is displayed in Figures 7.8 and 7.9 for the critical half crack lengths of 0.578″ and 0.785″, respectively. All the curves are normalized with respect to the value of σ_c at $\beta = 90°$. For $45° \leq \beta \leq 90°$, the theoretical curves (solid for the matrix model and dotted for the anisotropic model) agreed reasonably well with the experimental data. The discrepancies become more noticeable as the angle β is reduced. This is not surprising, since at low crack angles, more and more fibers tend to be aligned with the tensile axis and to be broken and/or debonded from the matrix. These energy dissipative mechanisms were not included in the analysis. The solid curves for the matrix cracking model in Figures 7.8 and 7.9 appear to be in better agreement with the experimental data [18].

Similar results are reported in Appendix 7.2 for E-glass fiber–epoxy resin composite with fiber volume fractions of 0.1, 0.2, 0.5 and 0.6. Graphical display of the normalized critical stress $(\sigma_c)_\beta/(\sigma_c)_{\beta=\pi/2}$ versus crack angle β is also provided for both the matrix cracking and homogeneous anisotropic model.

Metal composites. Since the degree of material nonhomogeneity depends on the fiber volume fraction and the difference of elastic properties of the fiber and matrix, there should be less variations between the analytical predictions of the two models when both the fiber and matrix are made of metal. Results on aluminum matrix reinforced by stainless steel fibers indeed bear out this expectation. Refer to the calculated values of θ_0 and $S_{\min}/\sigma^2 a$ in Tables 7.15 to 7.17 and the critical stress versus crack angle curves in Figures 7.32 and 7.33 of Appendix 7.3.

7.6. Angle-ply laminates

A system of two or more layers of fiber-reinforced composite is referred to as a laminate. The fibers in each layer may be oriented in different directions to give different strengths and stiffnesses in the various directions. Figure 7.10(a) shows a four-layer laminate with each layer being considered as a plate element made of unidirectional fibrous composites. The plates are stacked and bonded together with various angles between the fibers in one plate to the next indicated by the angles $\pm\beta$ (angle-ply) as illustrated in Figures 7.11. Upon loading, the entire assembly acts as a single unit in that the load will be transferred from layer to layer through the interfaces. Moreover, the laminate is said to be balanced when the layers are stacked symmetrically with reference to the mid-plane of the composite plate, Figures 7.10 and 7.11.

It was observed in [19] that balanced laminates, when subjected to uniaxial loading, failed by a combination of thru-lamina and interlaminar (delamination) cracking. These two modes of failure tend to trade off with one another as the fiber angles are varied. For the Scotchply 1002E glass–epoxy composite, a large amount of delamination occurred at low fiber angles corresponding to $15° \leq |\beta| \leq 30°$ while thru-lamina cracking dominated at the higher fiber angles. Thru-lamina cracking is obviously more damaging as it leads to the eventual fracture of the composite into separate pieces. The two aforementioned failure modes always interact with each other near the free edges and their individual contributions cannot be easily accounted for.

Stress distribution. The interlaminar cracking between the plies, or delamination, at low fiber angles has been observed [19] to take place within a thin

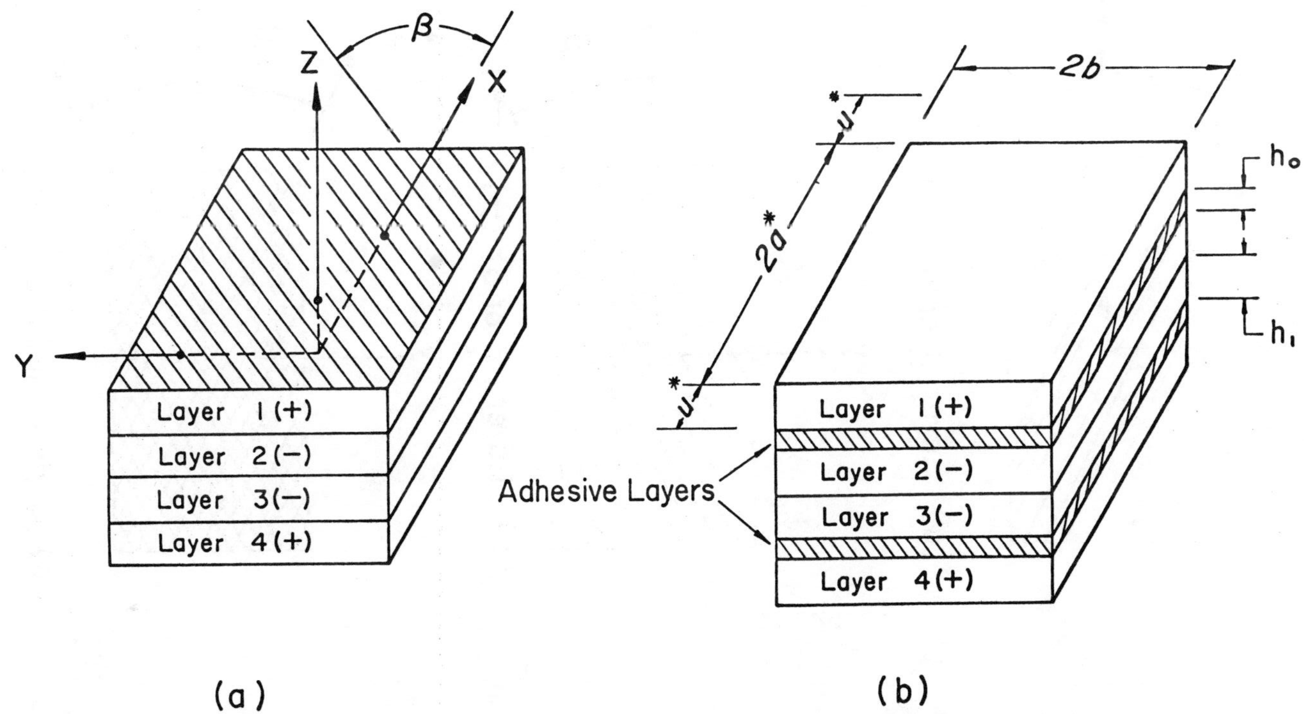

Figure 7.10. **Four-layer balanced laminate: (a) idealized model, (b) with adhesive layer.**

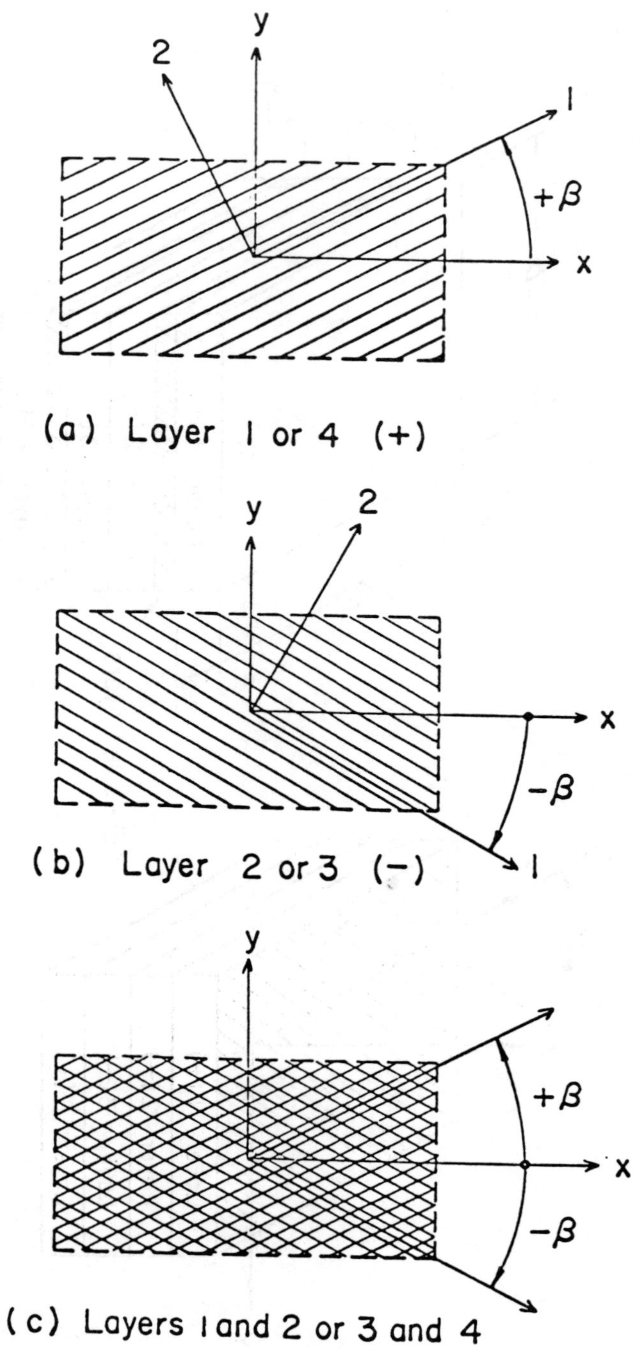

Figure 7.11. **Fiber orientations: (a) layer 1 or 4 (+), (b) layer 2 or 3 (−), (c) layers 1 and 2 or 3 and 4.**

thin layer of the fibrous material. Instead of taking this layer as an idealized interface of zero thickness* as in Figure 7.10(a), an additional layer of material representing the adhesive between the plies is introduced. Delamination is then assumed to occur by breaking the adhesive layer or the neighboring material. Figure 7.10(b) shows the analytical model of a four-layer laminate containing the additional adhesive layers with thickness h_0 which is small in comparison with the plate element with thickness h_1. Each plate element is assumed to possess three mutually perpendicular directions of material symmetry and will be described by the homogeneous, orthotropic theory of elasticity. The adhesive layers are assumed to be isotropic in nature.

Puppo and Evensen [21] have provided a solution to the problem in Figure 7.8(b) where the laminate is $2b$ in width and $2a^*$ in length and is extended uniformly by an amount $\pm u^*$ along the x-axis. The in-plane stresses in the anisotropic layers labelled by the subscripts 1 and 2 in Figure 7.12 are given by**

$$(\sigma_x)_1 = (\sigma_x)_2 = \sigma_x, \quad (\sigma_y)_1 = (\sigma_y)_2 = 0, \tag{7.45a}$$

$$(\tau_{xy})_1 = -(\tau_{xy})_2 = \tau_{xy} \tag{7.45b}$$

in which

$$\sigma_x = \frac{u^*}{a^*}\left[\overline{Q}_{11} - \frac{\overline{Q}_{12}^2}{\overline{Q}_{22}} - \frac{(\overline{Q}_{26}\overline{Q}_{12} - \overline{Q}_{22}\overline{Q}_{16})^2 \cosh(\sqrt{\omega}y)}{\overline{Q}_{22}(\overline{Q}_{22}\overline{Q}_{66} - \overline{Q}_{26}^2)\cosh(\sqrt{\omega}b)}\right] \tag{7.46a}$$

$$\tau_{xy} = \frac{u^*}{a^*}\left[\overline{Q}_{16} - \frac{\overline{Q}_{12}\overline{Q}_{26}}{\overline{Q}_{22}} + \frac{\overline{Q}_{26}\overline{Q}_{12} - \overline{Q}_{22}\overline{Q}_{16}}{\overline{Q}_{22}} \frac{\cosh(\sqrt{\omega}y)}{\cosh(\sqrt{\omega}b)}\right]. \tag{7.46b}$$

The interlaminar shear stresses in the adhesive layers are given by

$$\tau_{zx} = \frac{u^*}{a^*}\frac{2\mu^*}{h_0\sqrt{\omega}}\left(\frac{\overline{Q}_{26}\overline{Q}_{12} - \overline{Q}_{22}\overline{Q}_{16}}{\overline{Q}_{22}\overline{Q}_{66} - \overline{Q}_{26}^2}\right)\frac{\sinh(\sqrt{\omega}y)}{\cosh(\sqrt{\omega}b)}, \quad \tau_{zy} = 0 \tag{7.47}$$

where

$$\omega = \frac{2\mu^*}{h_1 h_0}\frac{\overline{Q}_{22}}{\overline{Q}_{22}\overline{Q}_{66} - \overline{Q}_{26}^2} \tag{7.48}$$

---

* By definition, delamination cannot take place at the idealized interface where perfect bonding is assumed by requiring the stresses and displacements to be continuous. The choice of cohesive or adhesive failure will depend on the composite system under consideration [20].

** Although the classical lamination theory in [22] and [23] gives a reasonably good estimate of the stress solution in the interior region of the lamimate, it does not account for the effect of interlaminar shears close to the edges of the composite plate. In general, interlaminar normal stress can also come into play depending upon the stacking sequence of the layers. Refer to Pipes and Pagano [24] for a discussion on this subject.

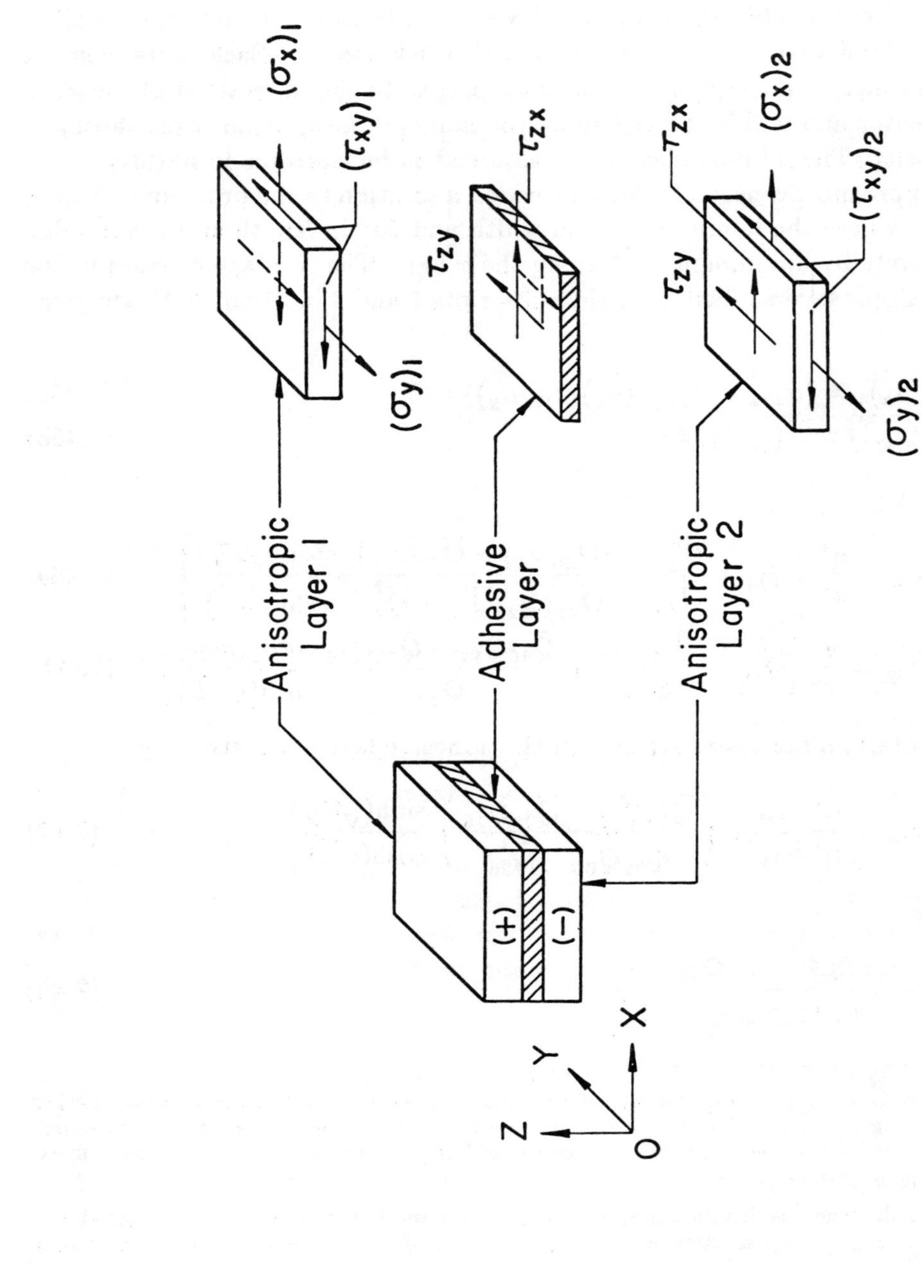

Figure 7.12. **Stress components on individual layers.**

with μ^* being the shear modulus of the adhesive. The quantities $\overline{Q}_{ij}$ are the stiffness coefficients referred to the axes x and y in Figure 7.11. They are identified with the generalized plane stress–strain relationship as follows:

$$\begin{bmatrix} \sigma_x \\ \sigma_y \\ \tau_{xy} \end{bmatrix} = \begin{bmatrix} \overline{Q}_{11}, & \overline{Q}_{12}, & 2\overline{Q}_{16} \\ \overline{Q}_{12}, & \overline{Q}_{22}, & 2\overline{Q}_{26} \\ \overline{Q}_{16}, & \overline{Q}_{26}, & 2\overline{Q}_{66} \end{bmatrix} \begin{bmatrix} \epsilon_x \\ \epsilon_y \\ \frac{1}{2}\gamma_{xy} \end{bmatrix}. \tag{7.49}$$

Since each of the plate elements is assumed to be in a state of generalized plane stress, this model does not account for interlaminar normal stress. The results in equations (7.46) can also be expressed in terms of the applied uniaxial tensile stress σ in the direction of the displacement u^*, i.e.,

$$\frac{u^*}{a^*} = \sigma \left[\overline{Q}_{11} - \frac{\overline{Q}_{12}^2}{\overline{Q}_{22}} - \frac{(\overline{Q}_{26}\overline{Q}_{12} - \overline{Q}_{22}\overline{Q}_{16})^2}{\overline{Q}_{22}(\overline{Q}_{22}\overline{Q}_{66} - \overline{Q}_{26}^2)} \frac{1}{\cosh(\sqrt{\omega}b)} \right]^{-1}. \tag{7.50}$$

Hence, the quantity u^*/a^* in equations (7.46) and (7.47) can be eliminated and σ_x, τ_{xy}, τ_{xz} can be expressed directly in terms of σ.

Of special interest is the interlaminar shear stress, equation (7.47), developed in the adhesive layer. For illustrating this, equation (7.47) is calculated numerically using the elastic constants for Scotchply 1002 glass–epoxy laminate and the results are plotted in Figure 7.13. Note that the shear stress τ_{xz} increases with the fiber angle β and reaches a maximum at $\beta \simeq 35°$. From there on, τ_{xz} decreases and becomes negative. The cases when $\beta = 0°$ and $90°$ correspond to the unidirectional composites with fiber parallel and perpendicular to the loading axis.

It should be mentioned that the existing lamination theories [21] through [24] do not account for pre-existing mechanical imperfections which normally act as stress raisers that lead to crack extension. Sih and Hilton [25] have used a variational principle and have developed an approximate three-dimensional isotropic lamination theory. The theory has been applied to solve the problem of a crack through each of the laminae with boundary layers constructed near the free surfaces and interfaces. The problem of a through crack in anisotropic laminates can also be solved in a similar manner.

Thru-laminar cracking. For laminates with fiber angles $45° \leq |\beta| \leq 90°$, the dominant fracture mode is that of cracking through the individual laminae with little or no delamination. Failure of the balanced four-layer Scotchply 1002 is attributed to crack propagation initiated from crack-like imperfections inherent in the cured laminate. It is assumed that the entire laminate fractures when any one lamina reaches the point of unstable rapid crack propagation. Thus, the S_c-criterion of fracture will be applied to a single lamina as in the case of the unidirectional composites.

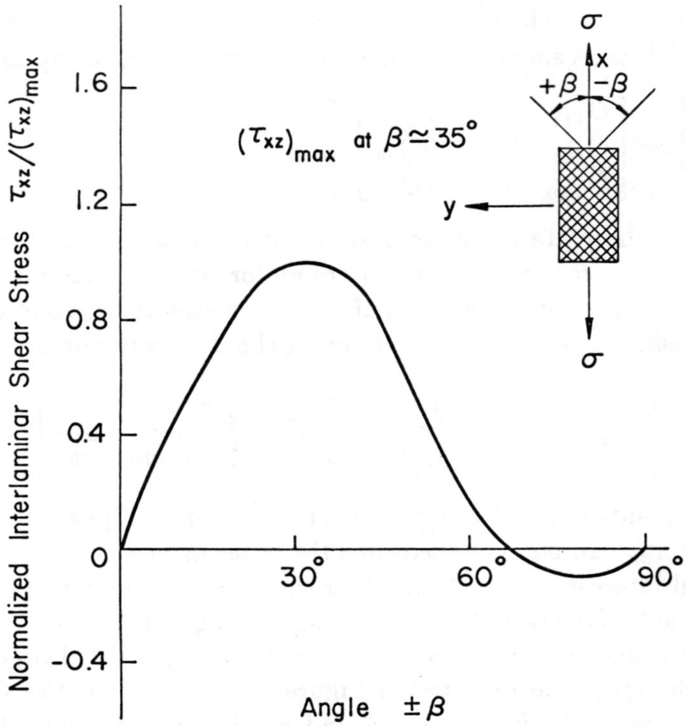

Figure 7.13. **Interlaminar shear stress versus fiber orientation.**

Since the crack is generally aligned along the fibers making an angle with the tensile axis, the initial direction of crack extension will not be collinear with the main crack. In this case, both stress intensity factors are present and they are

$$k_1 = 0.290\sigma_2\sqrt{a}, \quad k_2 = 0.170\tau_{12}\sqrt{a} \tag{7.51}$$

which correspond to the problem of a single crack in a laminae with 56.5% E-glass fiber volume fraction. The stresses σ_2 and τ_{12} are referred to the lamina principal axes as shown in Figure 7.11 and they can be obtained from

$$\begin{bmatrix} \sigma_1 \\ \sigma_2 \\ \tau_{12} \end{bmatrix} = \begin{bmatrix} \cos^2\beta, & \sin^2\beta, & \sin 2\beta \\ \sin^2\beta, & \cos^2\beta, & -\sin 2\beta \\ -\sin\beta\cos\beta, & \sin\beta\cos\beta, & \cos 2\beta \end{bmatrix} \begin{bmatrix} \sigma_x \\ \sigma_y \\ \tau_{xy} \end{bmatrix} \tag{7.52}$$

in which σ_2 and τ_{12} may be alternatively written as

$$\sigma_2 = \sigma_x \sin^2\beta + \sigma_y \cos^2\beta - \tau_{xy}\sin 2\beta \tag{7.53a}$$

$$\tau_{12} = (\sigma_y - \sigma_x)\sin\beta\cos\beta + \tau_{xy}\cos 2\beta. \tag{7.53b}$$

The stresses σ_x and τ_{xy} are given by equations (7.46) with $\sigma_y = 0$. Using

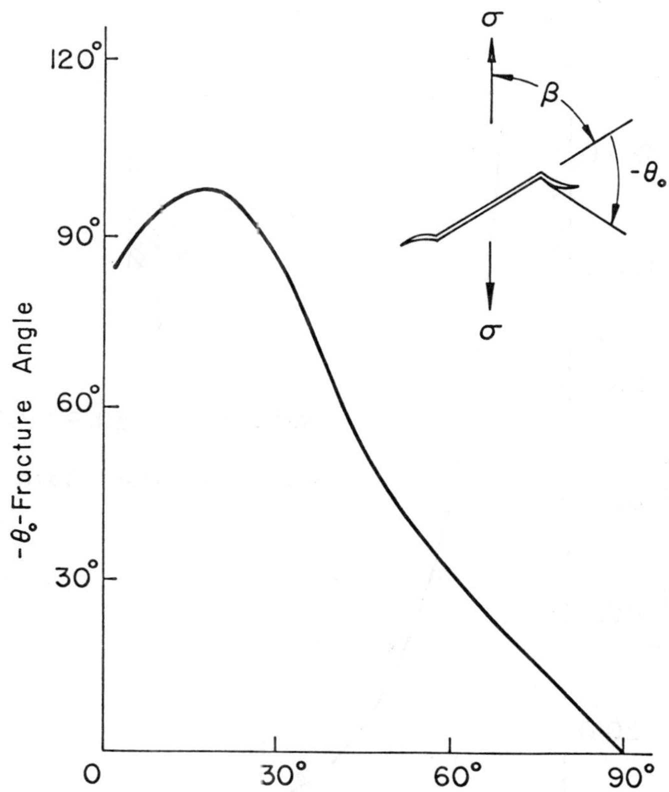

Figure 7.14. Direction of crack initiation as a function of fiber angle for angle-ply laminates.

equations (7.51) and (7.53), the strain energy density factor S in equation (7.10) may be computed:

$$S = (0.084a_{11}\sigma_2^2 + 0.049a_{12}\sigma_2\tau_{12} + 0.029a_{22}\tau_{12}^2)a. \tag{7.54}$$

Now, differentiating S in equation (7.54) with respect to θ and setting the result equal to zero, a relation between the direction of crack initiation, measured by the fracture angle θ_0, and the fiber angle β is obtained. The results, presented graphically in Figure 7.14, can be inserted into equation (7.54) to yield $S_{\min}$.

A comparison between theory and experiment will now be made. In [19], balanced four-layer Scotchply laminate specimens 10 in. long and 0.05 in. thick (0.01 in. per ply) with fiber orientations of ±15°, ±30°, ±45°, ±60°, and ±75° were tested by loading in the zero direction. The tensile loads at incipient fracture were measured and they are given as round dots in Figure 7.15 plotted as a function of the fiber angle β. The half crack length in these specimens is estimated to be approximately 0.03 in. The theoretical

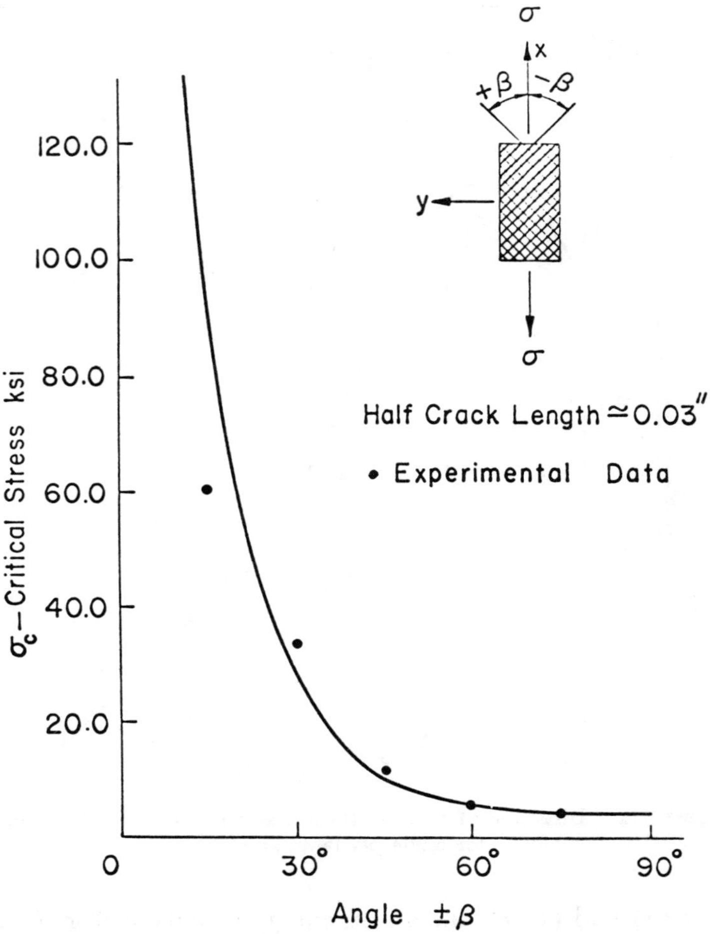

Figure 7.15. Critical stress variations with fiber orientation.

results represented by the solid curve are obtained from equation (7.54) by holding S_c at $\beta = 90°$ to be constant. Note that the agreement is good for $30° \leq |\beta| \leq 90°$. At $\beta = \pm 15°$, the theoretical prediction is somewhat higher than the experimental result. This is because a substantial amount of delamination has occurred at this fiber orientation which is not accounted for in the thru-laminar crack model.

Delamination. At low fiber orientations $15° \leq |\beta| \leq 30°$, the specimens exhibited interlaminar cracking (i.e., cracking between the "+" and "−" layers in Figure 7.10) prior to complete thru-laminar failure. The region of interlaminar cracking or delamination is triangular in shape as illustrated in Figure 7.16(a). This scissoring effect was not evident in the high angle fiber orientations. Experiments for angles below $\beta = \pm 15°$ are not available.

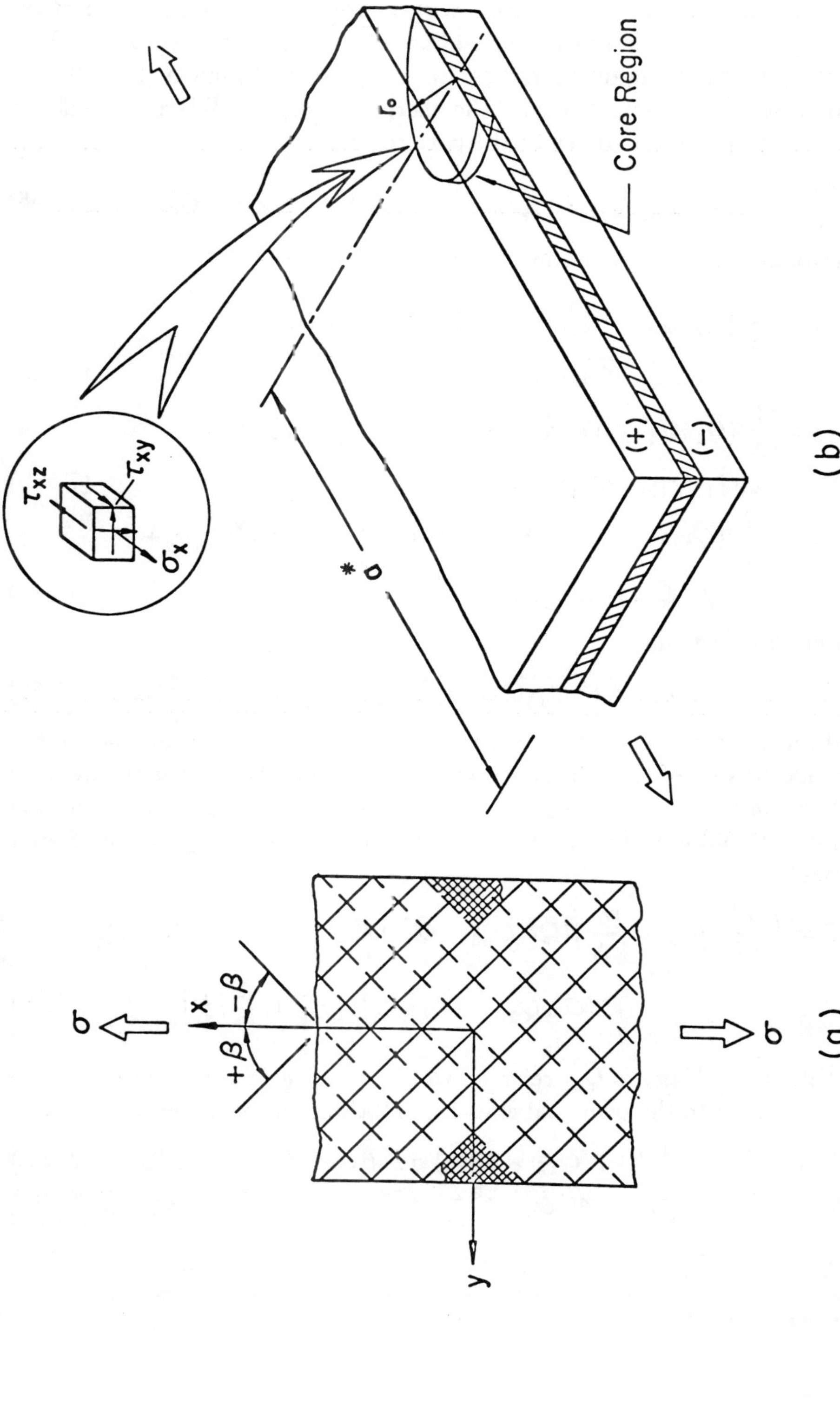

Figure 7.16. **Interlaminar cracking: (a) plane view, (b) core region.**

It is anticipated that the failure mode tends to that of the unidirectional composite for small values of β.

The possible sites of delamination will be predicted by application of the S_c-theory without assuming an initial flaw size. Referring to Figure 7.16(b), an interfacial core region of radius r_0 is centered around a point where delamination may start. Consider the energy stored in a differential volume $\mathrm{d}V$ situated on the boundary of the core region as

$$\frac{\mathrm{d}W}{\mathrm{d}V} = \tfrac{1}{2}(\sigma_x\epsilon_x + \sigma_y\epsilon_y + \tau_{xy}\gamma_{xy} + \tau_{xz}\gamma_{xz} + \tau_{yz}\gamma_{yz} + \sigma_z\epsilon_z). \tag{7.55}$$

The in-plane strains in equation (7.55) are given by

$$\epsilon_x = \frac{1}{\Delta}[(\overline{Q}_{22}\overline{Q}_{66} - \overline{Q}^2_{26})\sigma_x + (\overline{Q}_{16}\overline{Q}_{26} - \overline{Q}_{12}\overline{Q}_{66})\sigma_y + (\overline{Q}_{12}\overline{Q}_{26} - \overline{Q}_{16}\overline{Q}_{22})\tau_{xy}] \tag{7.56a}$$

$$\epsilon_y = \frac{1}{\Delta}[(\overline{Q}_{16}\overline{Q}_{26} - \overline{Q}_{12}\overline{Q}_{66})\sigma_x + (\overline{Q}_{11}\overline{Q}_{66} - \overline{Q}^2_{16})\sigma_y + (\overline{Q}_{12}\overline{Q}_{16} - \overline{Q}_{11}\overline{Q}_{26})\tau_{xy}] \tag{7.56b}$$

$$\gamma_{xy} = \frac{1}{\Delta}[(\overline{Q}_{12}\overline{Q}_{26} - \overline{Q}_{16}\overline{Q}_{22})\sigma_x + (\overline{Q}_{12}\overline{Q}_{16} - \overline{Q}_{11}\overline{Q}_{26})\sigma_y + (\overline{Q}_{11}\overline{Q}_{22} - \overline{Q}^2_{12})\tau_{xy}] \tag{7.56c}$$

in which Δ stands for

$$\Delta = \overline{Q}_{11}\overline{Q}_{22}\overline{Q}_{66} + 2\overline{Q}_{12}\overline{Q}_{16}\overline{Q}_{26} - \overline{Q}^2_{16}\overline{Q}_{22} - \overline{Q}_{11}\overline{Q}^2_{26} - \overline{Q}^2_{12}\overline{Q}_{66}. \tag{7.57}$$

The shearing strain γ_{xz} can be expressed in terms of τ_{xz} by calculating the shear modulus μ^* of the adhesive layer while the last two terms in equation (7.55) vanish because $\tau_{yz} = \sigma_z = 0$ for the present model. Inserting equation (7.56) into (7.55) and simplifying the result, the strain energy density factor becomes*

$$S = \left(\frac{\mathrm{d}W}{\mathrm{d}V}\right) r_0 = \frac{r_0}{2\Delta}\Big[(\overline{Q}_{22}\overline{Q}_{66} - \overline{Q}^2_{26})\sigma_x^2 + (\overline{Q}_{11}\overline{Q}_{22} - \overline{Q}^2_{12})\tau_{xy}^2 + 2(\overline{Q}_{12}\overline{Q}_{26} - \overline{Q}_{16}\overline{Q}_{22})\sigma_x\tau_{xy} + \frac{\Delta}{\mu^*}\tau_{xz}^2\Big]. \tag{7.58}$$

The stiffness coefficients $\overline{Q}_{ij}$ referred to the x and y axes can be related to the Q_{ij} referred to the principal axes 1 and 2 through the angle β as follows:

$$\overline{Q}_{11} = Q_{11}\cos^4\beta + 2(Q_{12} + 2Q_{66})\sin^2\beta\cos^2\beta + Q_{22}\sin^4\beta \tag{7.59a}$$

$$\overline{Q}_{12} = (Q_{11} + Q_{22} - 4Q_{66})\sin^2\beta\cos^2\beta + Q_{12}(\sin^4\beta + \cos^4\beta) \tag{7.59b}$$

---

* Although the exact shape of the initial defect which initiates delamination within the core region is not known, it is assumed that it reflects a $1/r$ singularity to the energy density field outside the core, i.e., $\mathrm{d}W/\mathrm{d}V = S/r$ for $r \to r_0$.

$$\overline{Q}_{16} = (Q_{11} - Q_{12} - 2Q_{66}) \sin\beta \cos^3\beta + (Q_{12} - Q_{22} + 2Q_{66}) \sin^3\beta \cos\beta \tag{7.59c}$$

$$\overline{Q}_{22} = Q_{11} \sin^4\beta + 2(Q_{12} + 2Q_{66}) \sin^2\beta \cos^2\beta + Q_{22} \cos^4\beta \tag{7.59d}$$

$$\overline{Q}_{26} = (Q_{11} - Q_{12} - 2Q_{66}) \sin^3\beta \cos\beta + (Q_{12} - Q_{22} + 2Q_{66}) \sin\beta \cos^3\beta \tag{7.59e}$$

$$\overline{Q}_{66} = (Q_{11} + Q_{22} - 2Q_{12} - 2Q_{66}) \sin^2\beta \cos^2\beta + Q_{66}(\sin^4\beta + \cos^4\beta). \tag{7.59f}$$

From equations (7.46) to (7.47), it is seen that the stresses in equation (7.47) depend only on the variable y. For the delamination problem, the S_c-theory has the following interpretation:

(1) Interlaminar cracking occurs at a location determined by the relative minimum value of the strain energy density factor S, i.e.,

$$\frac{\partial S}{\partial y} = 0 \quad \left(\frac{\partial^2 S}{\partial y^2} > 0\right) \quad \text{at} \quad y = y_0. \tag{7.60}$$

(2) Delamination occurs when the minimum strain energy density factor S reaches a critical value:

$$S_{\min} = S_c \quad \text{for} \quad y = y_0. \tag{7.61}$$

Figure 7.17 gives a plot of the normalized S-value in equation (7.58) versus the distance y, the path along which delamination may occur. Indeed, the strain energy density factor possesses a minimum value for each value of β. Failure is assumed to occur outside the core region with an average radius of $r_0 \simeq 0.02$ in. Experimental data on delamination is lacking. It must be cautioned that the measured values of the critical load or σ_c for $\beta = \pm 15°$ and $30°$ in [19] at which delamination was observed also include the effect of thru-laminar cracking. More refined experiments must be designed such that the quantitative effects of thru-laminar and interlaminar cracking may be separated.

7.7. Impact and cracking of unidirectional composites

When a fibrous composite is loaded suddenly, waves are reflected and refracted through the fibers and matrix in a complex pattern. Additional disturbances result from defects or cracks that are inherent in the composite. Because of the complexity of wave scattering in the presence of material nonhomogeneity and cracks, only a few simple cases of dynamic failure analyses have been treated [26–28]. To be considered in this section are the impact response of a unidirectional fibrous composite with an initial stationary crack and a running crack in the matrix.

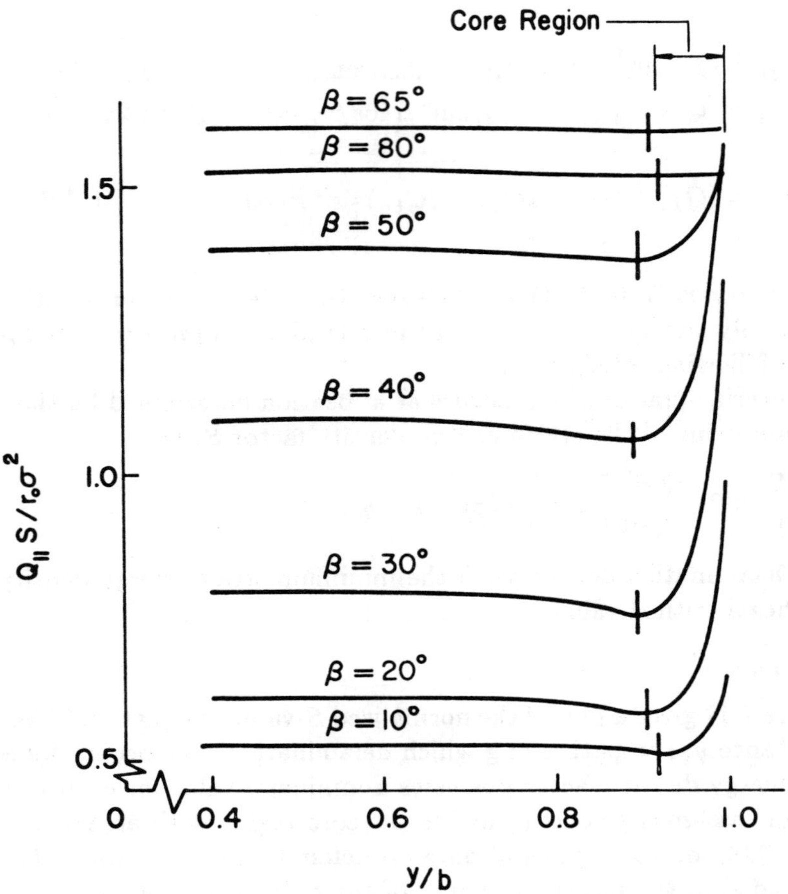

Figure 7.17. Variations of normalized S-value along the y-direction.

Impact loading. Without loss of generality, the unidirectional fibrous composite with a crack in the matrix will be modeled by a layer of cracked material with elastic properties μ_1, ν_1, ρ_1 sandwiched between two media with properties μ_2, ν_2, ρ_2. The number of layers surrounding the cracked layer is sufficiently large such that the average shear modulus μ_2, Poisson's ratio ν_2 and mass density ρ_2 can be used. Refer to Figure 7.18. The composite is subjected to an impact load oriented in a direction that makes an angle β with the crack plane. The crack surfaces experience both normal and shear impact with magnitude σ_0 and τ_0, respectively. This gives rise to both stress intensity factors $k_1(t)$ and $k_2(t)$ which in the dynamic case will vary as functions of time. The functional relationship of the dynamic stresses in terms of the local polar coordinates r and θ as indicated in Figure 7.18 is the same as that obtained from the static equations of elasticity. Suppose that the shear modulus μ_2 is much larger than μ_1 such that $\mu_2/\mu_1 = 10.0$

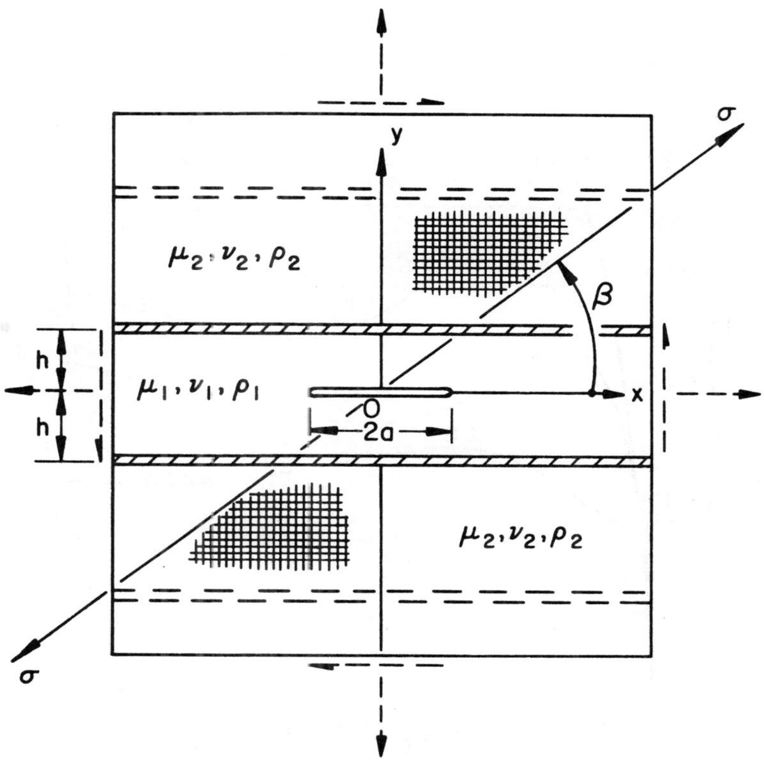

Figure 7.18. Unidirectional composite under normal and shear impact.

and that the same mass density and Poisson's ratio are used for the matrix and the surrounding composite material, i.e., $\rho_1 = \rho_2$ and $\nu_1 = \nu_2 = 0.29$. Figures 7.19 and 7.20 give the numerical values of the normalized dynamic stress intensity factors $k_1(t)/\sigma_0\sqrt{a}$ and $k_2(t)/\tau_0\sqrt{a}$ as functions of the dimensionless time variable $c_{21}t/a$ where $c_{21} = (\mu_1/\rho_1)^{1/2}$ is the shear wave velocity in the matrix material. All the curves tend to rise to a peak first and then oscillate wich decreasing amplitude. These peaks decrease in magnitude as the crack length $2a$ is increased relative to the height $2h$ of the matrix layer. A second peak is observed on the $k_1(t)/\sigma_0\sqrt{a}$ curve for small time t when $a/h = 2.0$ corresponding to a long crack in a narrow layer. This local disturbance corresponds to the arrival of the reflected waves from the interfaces that are in close proximity of the crack tip. In what follows, only the case of $a/h = 1.0$ will be discussed.

Once $k_1(t)$ and $k_2(t)$ are known, the dynamic strain energy density factor S can be calculated* from equation (7.10). In the dynamic problem, the

* Since the qualitative features of the stress field around a stationary crack for both static and dynamic loadings are the same, the coefficients a_{ij} are the same as those shown by equations (7.11). The only difference lies in the stress intensity factors.

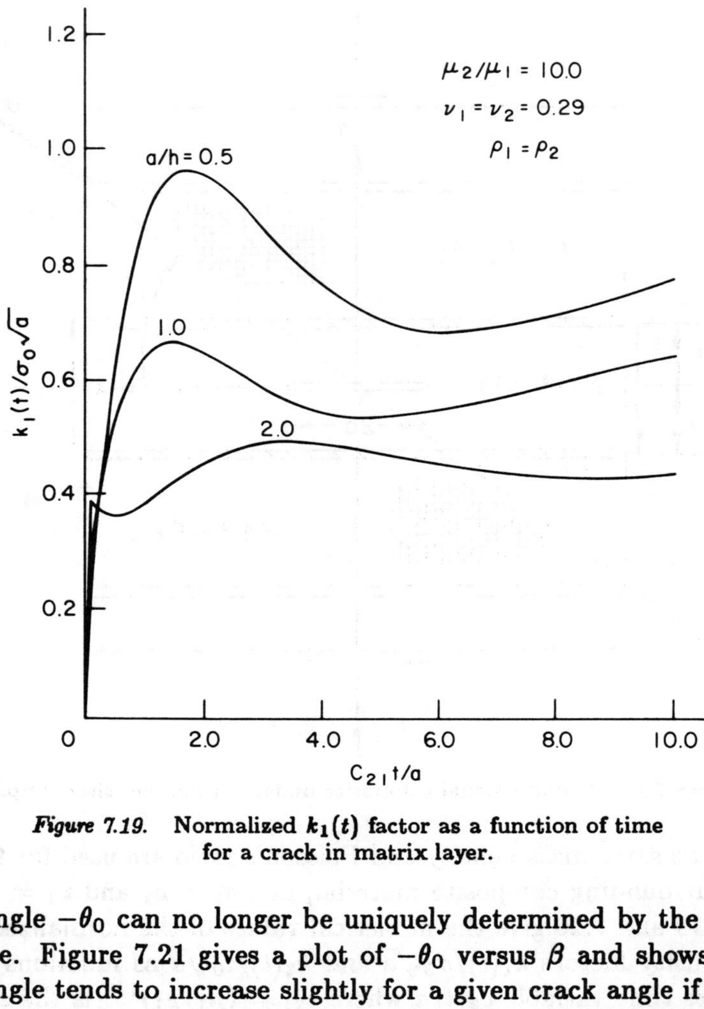

Figure 7.19. **Normalized $k_1(t)$ factor as a function of time for a crack in matrix layer.**

fracture angle $-\theta_0$ can no longer be uniquely determined by the crack angle β alone. Figure 7.21 gives a plot of $-\theta_0$ versus β and shows that the fracture angle tends to increase slightly for a given crack angle if the crack initiation occurs at the early stage of loading. This is better illustrated in Figure 7.22 which shows the variations for $-\theta_0$ with $c_{21}t/a$ for angles of $\beta = 15°, 30°, \ldots, 90°$. Note that the fracture angle is time dependent only for $c_{21}t/a < 1.0$. The minimum strain energy density factor $S_{\min}$ may now be obtained by inserting the values $-\theta_0$ from Figure 7.22 into the expression for S. The normalized strain energy density factor $16\mu_1 S_{\min}/\sigma^2 a$ is plotted against $c_{21}t/a$ for different crack angle β as shown in Figure 7.23. The quantities σ_0 and τ_0 in Figures 7.19 and 7.20 are related to σ as

$$\sigma_0 = \sigma \sin^2 \beta, \quad \tau_0 = \sigma \sin \beta \cos \beta. \tag{7.62}$$

All the curves in Figure 7.23 possess a maximum at a time t which depends on β. This maximum first increases with β from 30° to approximately 60° and then decreases as β approaches 90°.

In order to find the critical stress at which fracture occurs for a fixed

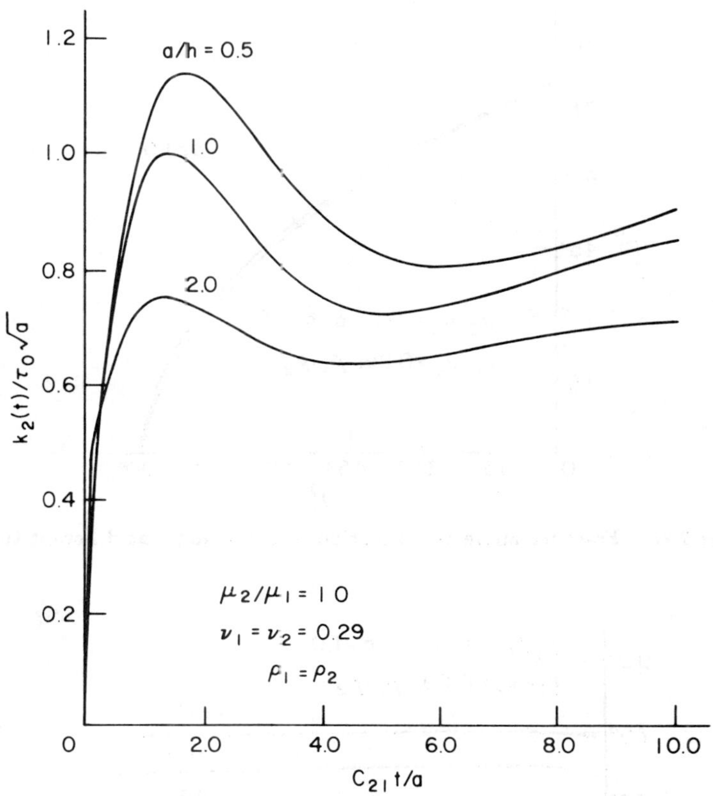

Figure 7.20. Normalized $k_2(t)$ factor as a function of time for a crack in matrix layer.

crack position β, the elapsed time at which $S_{\min}$ first reaches S_c must be known. To this end, it would be simpler to compare the ratio of the dynamic and static failure stress denoted by σ_m and σ_s, respectively. Let $16\mu_1 S_c = 4(1-2\nu_1)\sigma_s^2 a$ be given such that the same S_c governs unstable fracture regardless of whether the load is static or dynamic. Then the dynamic failure stress may be estimated from

$$\frac{\sigma_m}{\sigma_s} = \sqrt{\frac{4(1-2\nu_1)}{16\mu_1 S_{\min}/\sigma^2 a}}. \tag{7.63}$$

The ratio σ_m/σ_s plotted as a function of $c_{21}t/a$ is shown in Figure 7.24. The lowest failure stress σ_m occurs at $c_{21}t/a \simeq 1.5$ and $\beta \simeq 60°$. It is seen that the most vulnerable position of the crack is not at $\beta = 90°$ but when the crack is tilted at an angle to the direction of impact. The difference will depend on the material and geometric parameters describing the unidirectional composite.

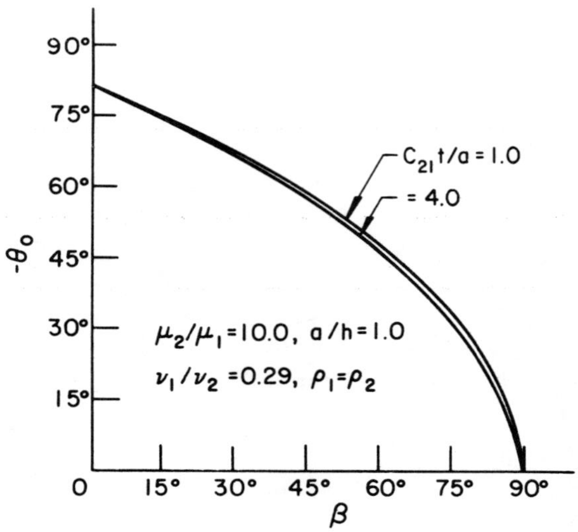

Figure 7.21. **Fracture angle as a function of crack angle at different time.**

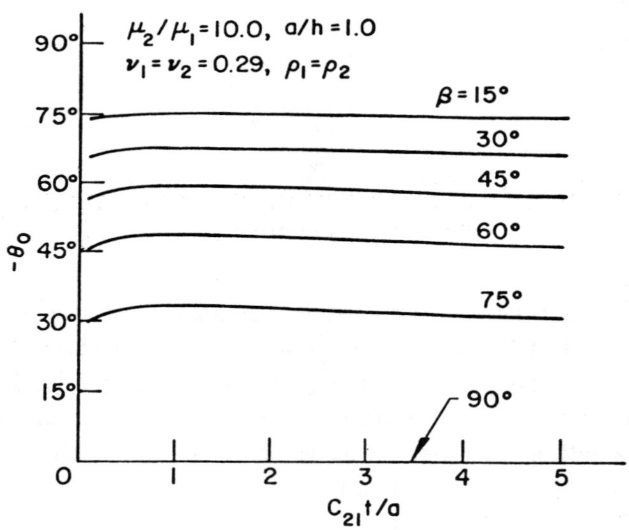

Figure 7.22. **Fracture angle as a function of time for different crack angle.**

Crack propagation. When the stress σ_m exceeds the critical level, failure of the composite can take place by a process of crack propagation which involves the cracking of the matrix at relatively high velocities. The process of fast fracture in a nonhomogeneous material is little understood because of the multitude of parameters that can affect crack propagation. Analytical modeling of composite failure by crack propagation is of great interest, since information on how the combination of fiber and matrix can influ-

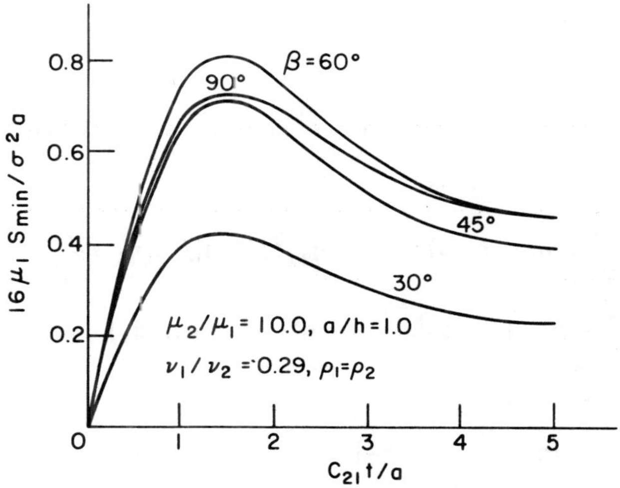

Figure 7.23. **Variations of normalized minimum strain energy density factor with time for different β.**

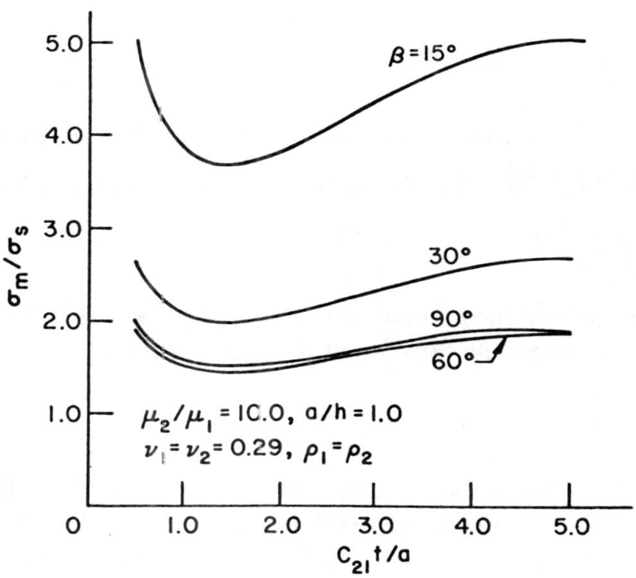

Figure 7.24. **Ratio of dynamic and static failure stress versus time for different β.**

ence the initiation and arrest of cracks is scarce and will be most useful for determining the resistance of fibrous composites to running cracks.

Referring to Figure 7.18, if the crack in the matrix layer begins to propagate at a velocity v under a uniform applied stress σ on the surrounding material for $\beta = 90°$, then the crack tip stress field when referenced to a

set of local polar coordinates r and θ will differ from that of the stationary crack. For a moving crack, the local stresses in a rectangular Cartesian coordinate system are [29]

$$\sigma_x = \frac{k_1(v)}{\sqrt{2r}}[(1+\eta_2^2)(2\eta_1^2+1-\eta_2^2)f(\eta_1) - 4\eta_1\eta_2 f(\eta_2)]/m(\eta_1,\eta_2) \tag{7.64a}$$

$$\sigma_y = \frac{k_1(v)}{\sqrt{2r}}[4\eta_1\eta_2 f(\eta_2) - (1+\eta_2^2)^2 f(\eta_1)]/m(\eta_1,\eta_2) \tag{7.64b}$$

$$\tau_{xy} = \frac{k_1(v)}{\sqrt{2r}}2\eta_1(1+\eta_2^2)[g(\eta_1) - g(\eta_2)]/m(\eta_1,\eta_2) \tag{7.64c}$$

in which $m(\eta_1,\eta_2)$ stands for

$$m(\eta_1,\eta_2) = 4\eta_1\eta_2 - (1+\eta_2^2)^2. \tag{7.65}$$

The functions $f(\eta_j)$ and $g(\eta_j)$ for $j = 1,2$ are defined as

$$f^2(\eta_j) = \tfrac{1}{2}\sec\theta[(1+\eta_j^2\tan^2\theta)^{-1/2} + (1+\eta_j^2\tan^2\theta)^{-1}] \tag{7.66a}$$

$$g^2(\eta_j) = \tfrac{1}{2}\sec\theta[(1+\eta_j^2\tan^2\theta)^{-1/2} - (1+\eta_j^2\tan^2\theta)^{-1}] \tag{7.66b}$$

where

$$\eta_j = [1-(v/c_{j1})^2]^{1/2} \tag{7.67}$$

while the dilatational and shear wave speeds designated by c_{11} and c_{21}, respectively, are given by

$$c_{11} = \left[\frac{\lambda_1+2\mu_1}{\rho_1}\right], \quad c_{21} = \left[\frac{\mu_1}{\rho_1}\right]^{1/2}. \tag{7.68}$$

Since the crack is assumed to run straight, only the stress intensity factor $k_1(v)$ appears in equations (7.64) and it takes the form

$$k_1(v) = F\left(\frac{v}{c_{21}}\right)\sigma\sqrt{a}. \tag{7.69}$$

If the material properties at a finite distance, h, away from the crack are different from those in which the crack is propagating, then the function F in equation (7.69) must depend on additional constants that describe the changes of the local environment, i.e.,

$$F = F\left(\frac{a}{h}, \frac{v}{c_{21}}, \frac{\mu_2}{\mu_1}, \frac{\nu_2}{\nu_1}, \frac{\rho_2}{\rho_1}\right). \tag{7.70}$$

Equation (7.70) shows the multiplicity of parameters that can influence the stress intensity factor $k_1(v)$. Their individual contributions can be studied analytically to gain a better understanding of the interaction of crack propagation with material nonhomogeneity.

Table 7.5. Bifurcation predicted by the strain energy density criterion for $\nu_1 = 0.25$.

v/c_{21}	0.1	0.2	0.3	0.4	0.5	0.6	0.8
θ_0	0°	0°	0°	0°	±15.1°	±48.6°	±65.2°
$8\mu_1 S_{\min}/\sigma^2 a$	1.008	1.036	1.090	1.185	1.361	1.631	3.892

The strain energy density criterion applies equally well to moving cracks without any major alteration except for the physical interpretation of the results. Substituting equations (7.64) into the strain energy density function expression for dW/dV, S is obtained since $S = r(dW/dV)$:

$$\frac{8\mu_1 S}{\sigma^2 a} = \left[\frac{F\left(\frac{a}{h}, \frac{v}{c_{21}}, \frac{\mu_2}{\mu_1}\right)^2}{m(\eta_1\eta_2)}\right]^2 \{(1+\eta_2^2)^2[2(1-\nu_1)(1+\eta_2^2)^2 -$$

$$-(1-2\nu_1)(2\eta_1^2+1-\eta_2^2)(1+\eta_2^2)]f^2(\eta_1) + 32\eta_1^2\eta_2^2 f^2(\eta_2) -$$

$$-16\eta_1\eta_2(1+\eta_1^2)(1+\eta_2^2)f(\eta_1)f(\eta_2) +$$

$$+8\eta_1^2(1+\eta_2^2)^2[g(\eta_1)-g(\eta_2)]^2\}. \tag{7.71}$$

In equation (7.71), Poisson's ratio and the mass density of the matrix layer 1 and surrounding material 2 have been assumed equal, i.e., $\rho_1 = \rho_2$ and $\nu_1 = \nu_2 = 0.25$.

In [29], it was shown that the ratio S_v/S_d for a moving crack remains nearly constant in the low crack velocity range and starts to decrease rapidly for v in the range of $0.4c_{21}$ and $0.5c_{21}$. The small volume change in an element moving ahead of the crack tends to suggest crack bifurcation* so that energy can be dissipated more quickly by material separation on both sides of the main crack. More specifically, $S_v > S_d$ at the locations of the minimum values of S or $S_{\min}$ and can be determined from $\partial S/\partial\theta = 0$ for $\theta = \theta_0$. For $\nu_1 = 0.25$, values of $S_{\min}$ were found at both sides of the crack corresponding to $\theta_0 = \pm 15.1°$ and occurred at a crack velocity of approximately $0.5c_{21}$. Table 7.5 shows the values of $8\mu_1 S_{\min}/\sigma^2 a$ and θ_0 for different crack velocities v. Bifurcation at the higher crack velocities is rare but may occur in materials that are relatively free of obstacles and hence it is less likely for the crack to encounter an obstacle and change direction. Consequently, it can be driven to a higher speed. This limitation also applies to crack propagation in a composite since the crack can only propagate in one phase of the composite at a given time. The occurence of bifurcation,

* The mathematical model of crack bifurcation should be interpreted with caution when compared to experiments where crack branching or curving is highly sensitive to the presence of obstacles such as inclusions or voids within the material.

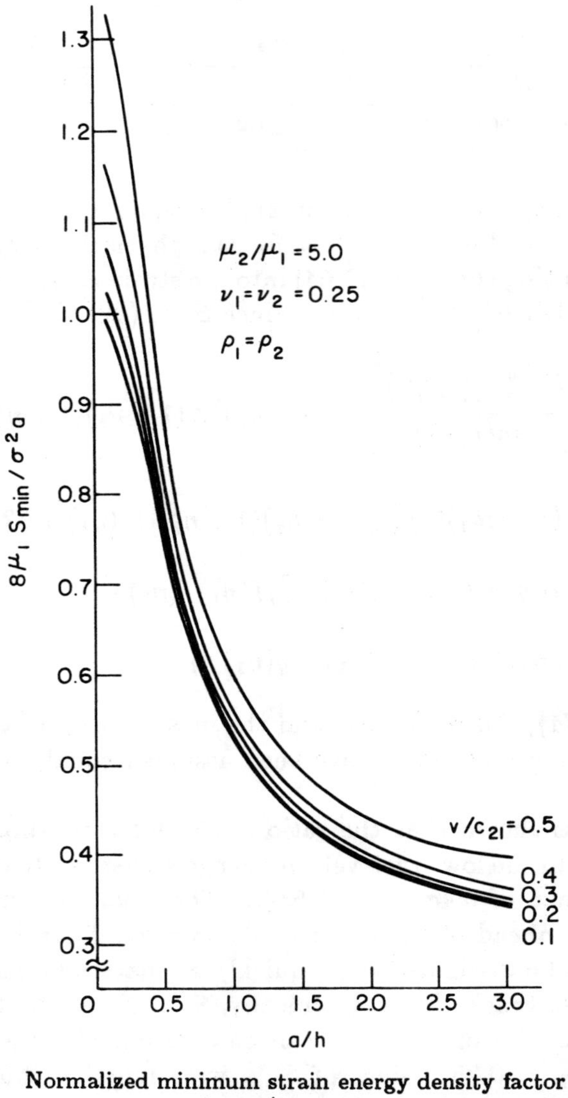

Figure 7.25. Normalized minimum strain energy density factor versus a/h for $\mu_2/\mu_1 = 5.0$.

however, will be affected by the nonhomogeneous nature of the composite material.

Now that the preliminary discussion on crack propagation in the homogeneous material has been completed, the effect of material nonhomogeneity will be analyzed. The minimum S or S_{min} in equation (7.71) for different values of a/h, v/c_{21} and μ_2/μ_1 is calculated for a moving crack. When $\mu_2/\mu_1 > 1$, the crack is in the matrix material that is relatively soft in comparison with the surrounding material located at a distance h from each side of the crack.

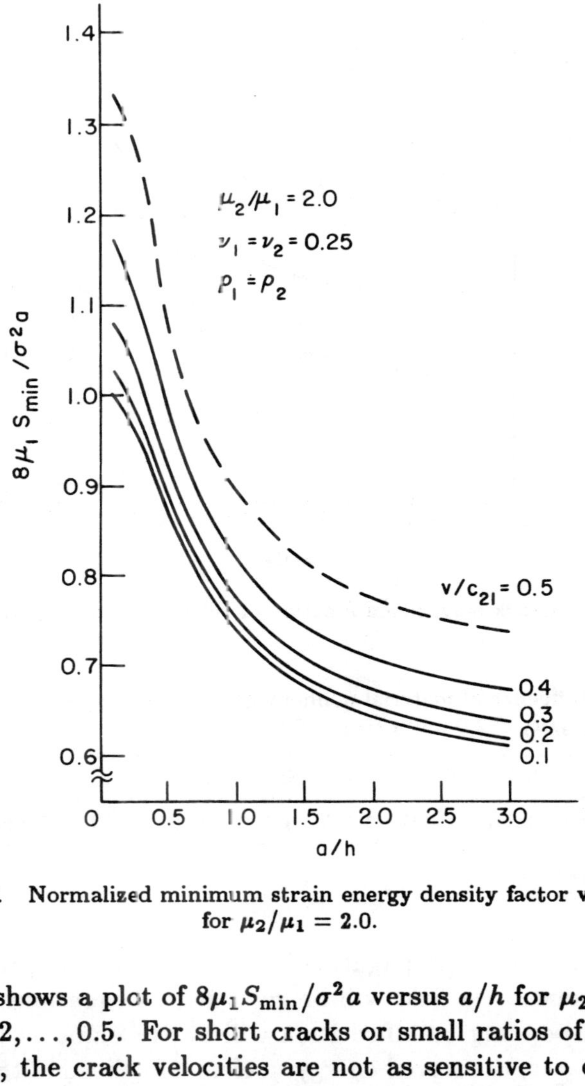

Figure 7.26. Normalized minimum strain energy density factor versus a/h for $\mu_2/\mu_1 = 2.0$.

Figure 7.25 shows a plot of $8\mu_1 S_{\min}/\sigma^2 a$ versus a/h for $\mu_2/\mu_1 = 5.0$ and $v/c_{21} = 0.1, 0.2, \ldots, 0.5$. For short cracks or small ratios of a/h assuming that h is fixed, the crack velocities are not as sensitive to changes in the applied stress or σ as compared with the longer cracks where a small change in σ can lead to larger variations of v. In general, $S_{\min}$ tends to decrease with increasing crack length. If the degree of nonhomogeneity is not as severe, i.e., for $\mu_2/\mu_1 = 2.0$, then Figure 7.26 suggests that the variations in σ for a given material or $S_{\min} = \text{const}$ seem to alter the crack velocities by approximately the same amount regardless of the crack size.

The influence of the relative stiffness or μ_2/μ_1 ratio on crack length is exhibited in Figure 7.27 in which $8\mu_1 S_{\min}/\sigma^2 a = 0.9$ is taken. As the crack travels at $v = 0.3c_{21}$, there is a tendency for it to increase its length should the ratio μ_2/μ_1 be suddenly altered, say decreased from 5.0 to 2.0. The crack length is observed to increase from $a = 0.35h$ to $a = 0.57h$, i.e., by a

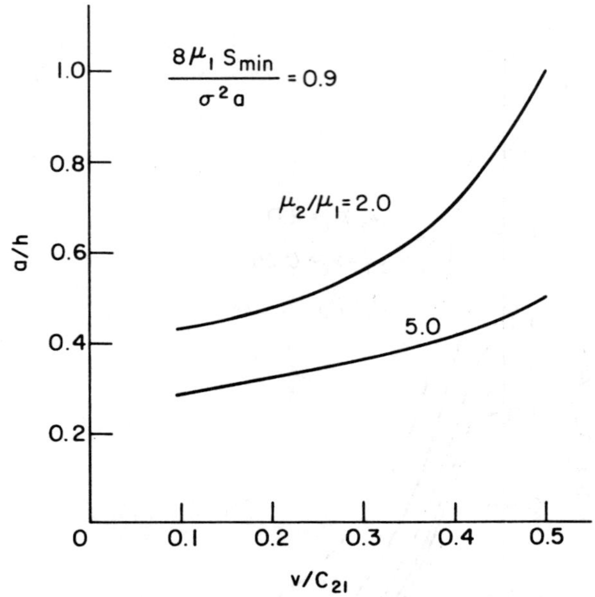

Figure 7.27. Crack length as a function of velocity for $8\mu_1 S_{min}/\sigma^2 a = 0.9$.

Table 7.6. Influence of material nonhomogeneity on S_{min} for $\mu_2/\mu_1 = 2.0$ and $v/c_{21} = 0.5$.

a/h	0.1	0.2	0.4	0.5	1.0	2.0	3.0
$8\mu_1 S_{min}/\sigma^2 a$	1.345	1.301	1.172	1.108	0.901	0.775	0.736

factor of 1.63.

As long as the crack is propagating in the material with properties μ_1, ν_1, ρ_1, the crack front stress field will be governed by equations (7.64) and the velocities corresponding to bifurcation will be the same as those shown in Table 7.5. However, the stored energy in elements ahead of the crack will be affected by μ_2/μ_1 and a/h. By letting $\mu_2/\mu_1 = 2.0$ and $v/c_{21} = 0.5$, Table 7.6 shows that $8\mu_1 S_{min}/\sigma^2 a$ tends to decrease with increasing a/h.

7.8. Concluding remarks

The strain energy density criterion has been applied to analyze the fracture behavior of unidirectional fiber-reinforced composites. Matrix cracking initiated from an inherent defect in the form of a crack is assumed to be the dominant mode of fracture. Depending on the fiber volume fraction and the material properties of the constituents, the composite system was modelled either by invoking the assumption of homogeneous anisotropy or nonhomogeneous isotropy. The former assumes fracture to occur in an idealized material with the combined anisotropic elastic properties of the fiber and matrix while the latter assumes cracking to take place in the isotropic matrix located in between the fibers. The agreement between theory and experiment for the E-glass epoxy resin with 56.5% fiber volume fraction was good for fiber orientations from 30° to 90° with respect to the direction of loading. As the fiber direction approaches the loading axis, there is the tendency for more fibers to break, a condition which was not included in the analysis. Theoretical predictions of Modulite II 5206 graphite, E-glass epoxy with different fiber volume fractions and stainless steel–aluminum composites were also presented.

The impact response of unidirectional composites is treated by assuming fracture to initiate from a finite crack in the matrix. Load transmission through the composite takes place in the form of wave propagation. The reflective waves from the composite interfaces and one crack tip to the other cause intensification of the local stress field. The results can be expressed in terms of the critical dynamic stress to initiate fracture. Unlike the behavior of a homogeneous material, the lowest critical stress corresponds to impact loading directed at an angle with respect to the crack plane rather than normal to it.

The failure of angle-ply laminates made of layers of unidirectional fiber-reinforced material is also considered. Thru-laminar and interlaminar (or delamination) failure modes are discussed separately. Delamination becomes significant when the orientation of the angle-ply with respect to the applied load is reduced. Failure of the laminate is assumed to coincide with unstable thru-laminar cracking of one of the laminae. The simultaneous occurrence of delamination and thru-cracking has yet to be investigated analytically.

7.9. Appendix 7.1: Calculation of gross mechanical properties of unidirectional composites

The gross mechanical properties of fiber-reinforced unidirectional composites may be calculated from the following formulas [16]

$$E_1 = E_{f_1}V_f + E_mV_m \tag{7.72a}$$

$$E_2 = \left[\frac{E_{f_2} + E_m + (E_{f_2} - E_m)V_f}{E_{f_2} + E_m - (E_{f_2} - E_m)V_f}\right] E_m \tag{7.72b}$$

$$\nu_{12} = \nu_f V_f + \nu_m V_m \tag{7.72c}$$

$$\nu_{23} = \nu_f V_f + \left[\frac{1 + \nu_m - \nu_{12}E_m/E_1}{1 - \nu_m^2 + \nu_m\nu_{12}E_m/E_1}\right] \nu_m V_m \tag{7.72d}$$

$$\mu_{12} = \left[\frac{\mu_f + \mu_m + (\mu_f - \mu_m)V_f}{\mu_f + \mu_m - (\mu_f - \mu_m)V_f} \right] \mu_m \tag{7.72e}$$

in which 1, 2 and 3 correspond to the principal axes of elastic symmetry. The axes 1 and 2 are in the plane of the composite sheet specimen while 3 is directed through the thickness. The symbols in equations (7.72) stand for

V_f = fiber volume fraction
V_m = matrix volume fraction $= 1 - V_f$
E_{f_1} = fiber longitudinal Young's modulus
E_{f_2} = fiber transverse Young's modulus
E_m = matrix Young's modulus
ν_f = fiber Poisson's ratio
ν_m = matrix Poisson's ratio
μ_f = fiber shear modulus
μ_m = matrix shear modulus.

7.10. Appendix 7.2: E-glass–epoxy resin composites

This section provides additional information on the fracture angle θ_0, normalized strain energy density factor $S_{\min}/\sigma^2 a$ and critical stress $(\sigma_c)_\beta/(\sigma_c)_{\beta=\pi/2}$ for the unidirectional E-glass epoxy resin composite modelled by cracking in the isotropic matrix and anisotropic homogeneous medium. Four different fiber volume fractions are considered. They are $V_f = 0.1, 0.2, 0.5$ and 0.6. The elastic properties of the E-glass fiber are

$$E_f = E_{f_1} = E_{f_2} = 10.50 \times 10^6 \text{ psi} \tag{7.73a}$$

$$\mu_f = 4.40 \times 10^6 \text{ psi}, \quad \nu_f = 0.20. \tag{7.73b}$$

The elastic properties of the epoxy resin are given by

$$E_m = 0.50 \times 10^6 \text{ psi} \tag{7.74a}$$

$$\mu_m = 0.19 \times 10^6 \text{ psi}, \quad \nu_m = 0.35. \tag{7.74b}$$

Fiber volume fraction 0.1. The gross mechanical response of the composite will be affected by the fiber volume fraction. For $V_f = 0.1$, equations (7.72) in Appendix 7.1 may be used to obtain

$$E_1 = 1.50 \times 10^6 \text{ psi}, \quad \nu_{12} = 0.34 \tag{7.75a}$$

$$E_2 = 0.62 \times 10^6 \text{ psi}, \quad \nu_{23} = 0.45 \tag{7.75b}$$

$$\mu_{12} = 0.27 \times 10^6 \text{ psi}. \tag{7.75c}$$

This leads to the characteristic roots

$$S_1 = 0.634i, \quad S_2 = 2.361i \tag{7.76}$$

that satisfy equation (7.28). The functions $\Phi(1)$ and $\Psi(1)$ in the stress intensity factor equations (7.22) for the matrix cracking model are

$$\Phi(1) = 0.71, \quad \Psi(1) = 0.50. \tag{7.77}$$

The final results for the anisotropic homogeneous and matrix cracking model are given in Tables 7.7 and 7.8. A plot of the normalized critical stress versus crack angle given in Figure 7.28 shows that there is little difference between the two models.

Fiber volume fraction 0.2. As before, the gross elastic properties of the composite for $V_f = 0.2$ are found as

$$E_1 = 2.50 \times 10^6 \text{ psi}, \quad \nu_{12} = 0.32 \tag{7.78a}$$
$$E_2 = 0.72 \times 10^6 \text{ psi}, \quad \nu_{23} = 0.44 \tag{7.78b}$$
$$\mu_{12} = 0.27 \times 10^6 \text{ psi}. \tag{7.78c}$$

The roots S_1 and S_2 take the values

$$S_1 = 0.615i, \quad S_2 = 2.851i. \tag{7.79}$$

For the matrix cracking model, it is found that

$$\Phi(1) = 0.65, \quad \Psi(1) = 0.38. \tag{7.80}$$

Refer to Tables 7.9 and 7.10 for the results on θ_0 and $S_{\min}/\sigma^2 a$. The curves in Figure 7.29 reveal a slightly larger deviation between the two models when the fiber volume fraction is increased from 0.1 to 0.2.

Fiber volume fraction 0.5. The elastic constants for the composite with $v_f = 0.5$ give rise to the following modulus of elasticity for the composite:

$$E_1 = 5.50 \times 10^6 \text{ psi}, \quad \nu_{12} = 0.28 \tag{7.81a}$$
$$E_2 = 1.33 \times 10^6 \text{ psi}, \quad \nu_{23} = 0.36 \tag{7.81b}$$
$$\mu_{12} = 0.50 \times 10^6 \text{ psi}. \tag{7.81c}$$

Equations (7.81) can then be used to obtain

$$S_1 = 0.620i, \quad S_2 = 3.140i. \tag{7.82}$$

The required values of $\Phi(1)$ and $\Psi(1)$ for k_1 and k_2 are

$$\Phi(1) = 0.37, \quad \Psi(1) = 0.19. \tag{7.83}$$

The necessary information in Tables 7.11 and 7.12 can be found by making use of equations (7.82) and (7.83). Figure 7.30 shows that as the crack angle

Table 7.7. **Fracture angle and normalized strain energy density factor for homogeneous anisotropic model (E-glass epoxy resin with $V_f = 0.2$).**

β	$-\theta_0$	$S_{min}/\sigma^2 a$	β	$-\theta_0$	$S_{min}/\sigma^2 a$
5.00	76.500	0.00272	50.00	45.700	0.27568
10.00	72.400	0.01149	55.00	42.900	0.31609
15.00	68.500	0.02690	60.00	40.200	0.35359
20.00	64.800	0.04903	65.00	37.600	0.38738
25.00	61.300	0.07752	70.00	35.200	0.41704
30.00	57.900	0.11154	75.00	33.200	0.44249
35.00	54.700	0.14987	80.00	31.700	0.46386
40.00	51.600	0.19107	85.00	31.100	0.48127
45.00	48.600	0.23353	90.00	32.400	0.49431

Table 7.8. **Fracture angle and normalized strain energy density factor for matrix cracking model (E-glass epoxy resin with $V_f = 0.2$).**

β	$-\theta_0$	$S_{min}/\sigma^2 a$	β	$-\theta_0$	$S_{min}/\sigma^2 a$
5.00	79.500	0.00104	50.00	42.400	0.10167
10.00	74.800	0.00431	55.00	38.400	0.11889
15.00	70.300	0.00993	60.00	34.300	0.13602
20.00	65.900	0.01785	65.00	29.900	0.15252
25.00	61.700	0.02793	70.00	25.900	0.16773
30.00	57.700	0.03994	75.00	19.600	0.18089
35.00	53.800	0.06867	80.00	13.600	0.19117
40.00	50.000	0.08480	85.00	7.000	0.19776
45.00	46.200	0.08480	90.00	0.000	0.20003

Table 7.9. **Fracture angle and normalized strain energy density factor for homogeneous anisotropic model (E-glass–epoxy resin with $V_f = 0.2$).**

β	$-\theta_0$	$S_{min}/\sigma^2 a$	β	$-\theta_0$	$S_{min}/\sigma^2 a$
5.00	75.800	0.00199	50.00	44.700	0.22672
10.00	71.400	0.00853	55.00	42.200	0.26254
15.00	67.300	0.02024	60.00	39.800	0.29622
20.00	63.500	0.03740	65.00	37.600	0.32679
25.00	59.900	0.05993	70.00	35.700	0.35358
30.00	56.500	0.08739	75.00	34.300	0.37614
35.00	53.400	0.11897	80.00	33.600	0.39421
40.00	50.300	0.15358	85.00	33.900	0.40744
45.00	47.500	0.18996	90.00	35.800	0.41511

Table 7.10. **Fracture angle and normalized strain energy density factor for matrix cracking model (E-glass–epoxy resin with $V_f = 0.2$).**

β	$-\theta_0$	$S_{min}/\sigma^2 a$	β	$-\theta_0$	$S_{min}/\sigma^2 a$
5.00	78.600	0.00062	50.00	38.500	0.07609
10.00	73.100	0.00260	55.00	34.700	0.09148
15.00	67.900	0.00609	60.00	30.600	0.10720
20.00	63.000	0.01118	65.00	26.400	0.12260
25.00	58.400	0.01793	70.00	21.800	0.13690
30.00	54.100	0.02638	75.00	16.900	0.14925
35.00	50.100	0.03651	80.00	11.500	0.15883
40.00	46.200	0.04829	85.00	5.900	0.16491
45.00	42.300	0.06157	90.00	0.000	0.16700

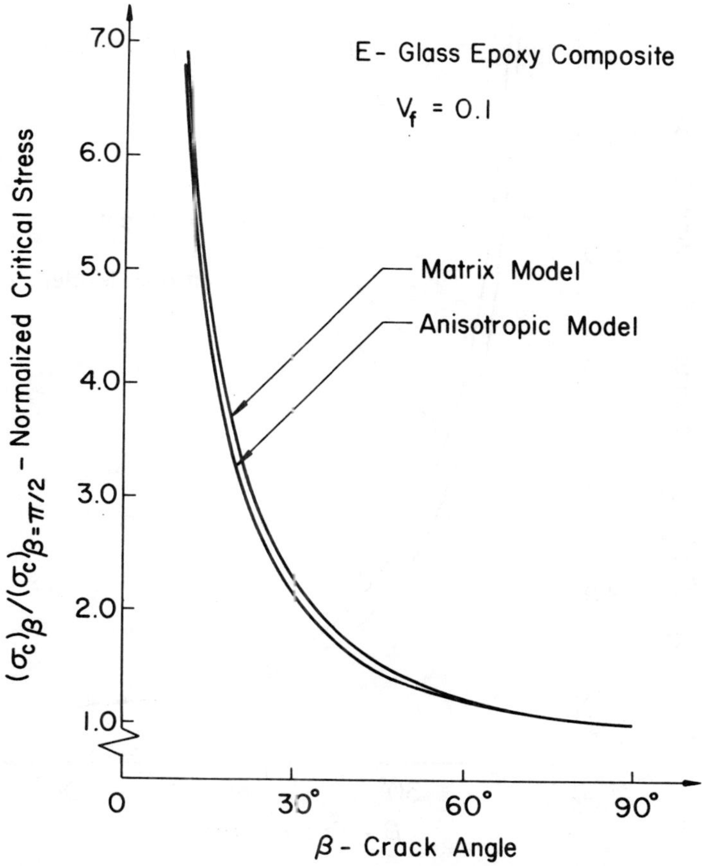

Figure 7.28. **Normalized σ_c versus β curve for E-glass epoxy composite ($V_f = 0.1$).**

β decreases, the difference between the critical stresses predicted by the two curves can differ by an appreciable amount.

Fiber volume fraction **0.6. The elastic constants for the composite with a 60% fiber volume fraction are**

$$E_1 = 6.50 \times 10^6 \text{ psi}, \quad \nu_{12} = 0.26 \tag{7.84a}$$

$$E_2 = 1.70 \times 10^6 \text{ psi}, \quad \nu_{23} = 0.33 \tag{7.84b}$$

$$\mu_{12} = 0.64 \times 10^6 \text{ psi}. \tag{7.84c}$$

based upon which S_1 and S_2 are calculated:

$$S_1 = 0.627i, \quad S_2 = 3.009i. \tag{7.85}$$

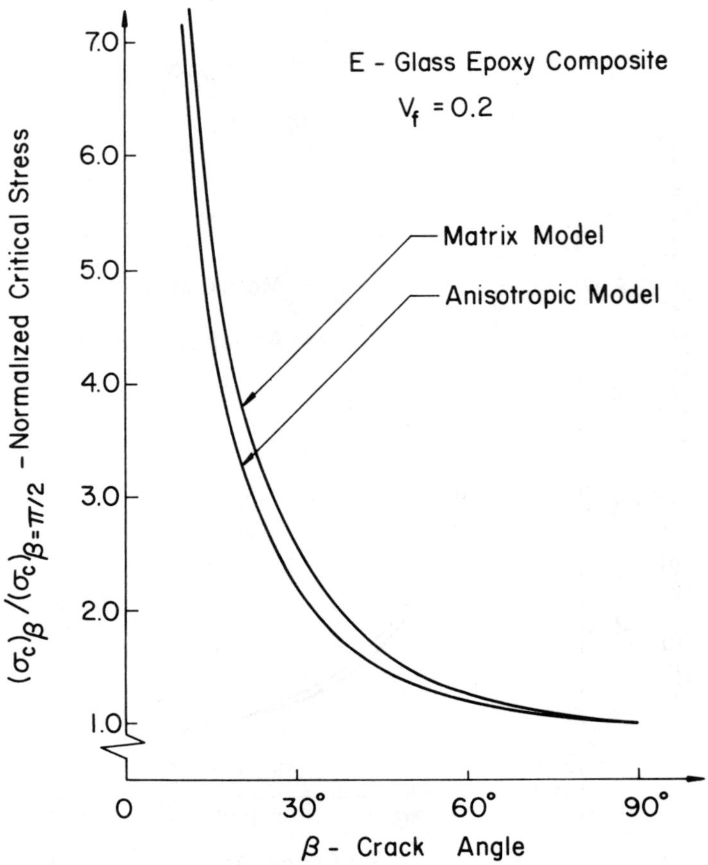

Figure 7.29. Normalized σ_c versus β curve for E-glass–epoxy composite ($V_f = 0.2$).

The k_1 and k_2 factors in equations (7.22) are determined as

$$\Phi(1) = 0.29, \quad \Psi(1) = 0.16. \tag{7.86}$$

Tables 7.13 and 7.14 summarize, respectively, the results for the anisotropic and matrix model. Comparing the curves in Figures 7.30 and 7.31, there appears to be negligible differences in the critical stresses for the two models when V_f is increased from 0.5 to 0.6.

7.11. Appendix 7.3: Stainless steel–aluminum composite

Consider an aluminum matrix whose elastic properties are given by

$$E_m = 10.50 \times 10^6 \text{ psi} \tag{7.87a}$$

$$\mu_m = 3.94 \times 10^6 \text{ psi}, \quad \nu_m = 0.33. \tag{7.87b}$$

Table 7.11. **Fracture angle and strain energy density factor for homogeneous anisotropic model (E-glass–epoxy resin with $V_f = 0.5$).**

β	$-\theta_0$	$S_{min}/\sigma^2 a$	β	$-\theta_0$	$S_{min}/\sigma^2 a$
5.00	74.900	0.00102	50.00	44.300	0.12845
10.00	70.400	0.00442	55.00	41.900	0.14995
15.00	66.200	0.01062	60.00	39.800	0.17037
20.00	62.400	0.01987	65.00	38.000	0.18899
25.00	58.800	0.03222	70.00	36.600	0.20524
30.00	55.500	0.04753	75.00	35.600	0.21866
35.00	52.400	0.06544	80.00	35.400	0.22891
40.00	49.500	0.08538	85.00	36.200	0.23559
45.00	46.800	0.10665	90.00	38.400	0.23820

Table 7.12. **Fracture angle and strain energy density factor for matrix cracking model (E-glass–epoxy resin with $V_f = 0.5$).**

β	$-\theta_0$	$S_{min}/\sigma^2 a$	β	$-\theta_0$	$S_{min}/\sigma^2 a$
5.00	77.900	0.00016	50.00	35.900	0.02293
10.00	71.700	0.00067	55.00	32.100	0.02801
15.00	66.100	0.00159	60.00	28.200	0.03327
20.00	60.800	0.00296	65.00	24.100	0.03844
25.00	56.000	0.00485	70.00	19.800	0.04326
30.00	51.600	0.00730	75.00	15.200	0.04741
35.00	47.500	0.01034	80.00	10.300	0.05063
40.00	43.500	0.01399	85.00	5.200	0.05266
45.00	39.700	0.01821	90.00	0.00	0.05336

Table 7.13. **Fracture angle and normalized strain energy density factor for homogeneous anisotropic model (E-glass–epoxy resin with $V_f = 0.6$).**

β	$-\theta_0$	$S_{min}/\sigma^2 a$	β	$-\theta_0$	$S_{min}/\sigma^2 a$
5.00	74.800	0.00083	50.00	44.500	0.10380
10.00	70.300	0.00360	55.00	42.200	0.12110
15.00	66.200	0.00863	60.00	40.100	0.13752
20.00	62.400	0.01613	65.00	38.300	0.15247
25.00	58.900	0.02614	70.00	36.900	0.16551
30.00	55.600	0.03852	75.00	36.000	0.17625
35.00	52.600	0.05299	80.00	35.900	0.18441
40.00	49.700	0.06908	85.00	36.800	0.18967
45.00	47.000	0.08623	90.00	39.200	0.19161

Table 7.14. **Fracture angle and normalized strain energy density factor for matrix cracking model (*E*-glass–epoxy resin with $V_f = 0.6$).**

β	$-\theta_0$	$S_{min}/\sigma^2 a$	β	$-\theta_0$	$S_{min}/\sigma^2 a$
5.00	78.000	0.00010	50.00	36.600	0.01476
10.00	72.100	0.00045	55.00	32.700	0.01796
15.00	66.500	0.00106	60.00	28.800	0.02127
20.00	61.300	0.00197	65.00	24.600	0.02453
25.00	56.600	0.00320	70.00	20.200	0.02755
30.00	52.200	0.00479	75.00	15.500	0.03017
35.00	48.100	0.00675	80.00	10.600	0.03219
40.00	44.290	0.00908	85.00	5.400	0.03347
45.00	40.300	0.01177	90.00	0.000	0.03391

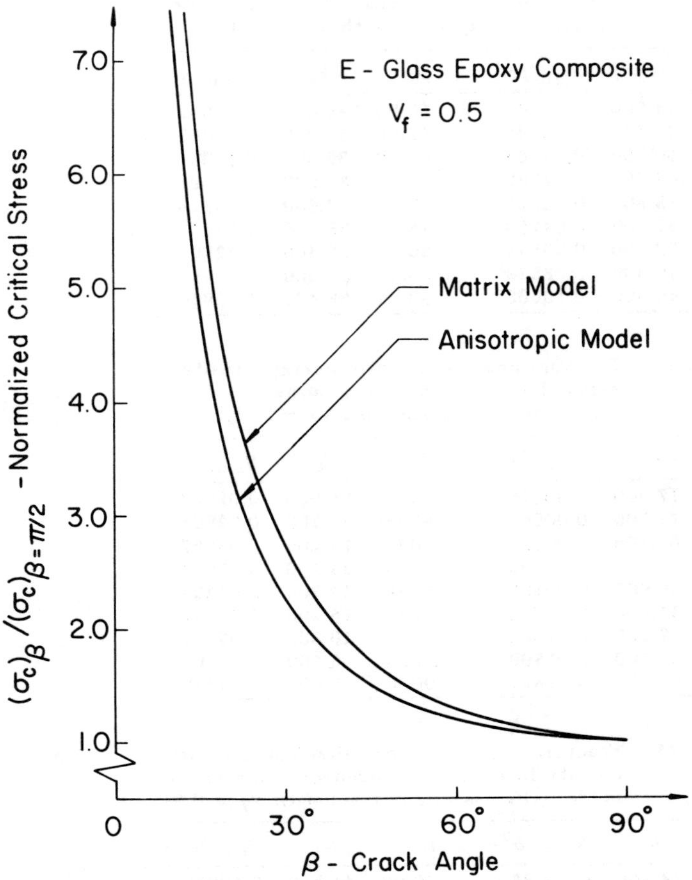

Figure 7.30. **Normalized σ_c versus β curve for E-glass–epoxy composite ($V_f = 0.5$).**

Stainless steel fibers with elastic properties

$$E_f = E_{f_1} = E_{f_2} = 30.0 \times 10^6 \text{ psi} \tag{7.88a}$$

$$\mu_f = 11.54 \times 10^6 \text{ psi}, \nu_f = 0.30 \tag{7.88b}$$

are embedded in the aluminum matrix to form a unidirectional composite. To be considered are two different fiber volume fractions, namely $V_f = 0.1$ and 0.4.

Fiber volume fraction 0.1. Application of the formulas in equations (7.72) yields the following anisotropic constants for the stainless steel–aluminum composite with $V_f = 0.1$:

$$E_1 = 12.45 \times 10^6 \text{ psi}, \quad \nu_{12} = 0.33 \tag{7.89a}$$

$$E_2 = 11.56 \times 10^6 \text{ psi}, \quad \nu_{23} = 0.35 \tag{7.89b}$$

$$\mu_{12} = 4.35 \times 10^6 \text{ psi}. \tag{7.89c}$$

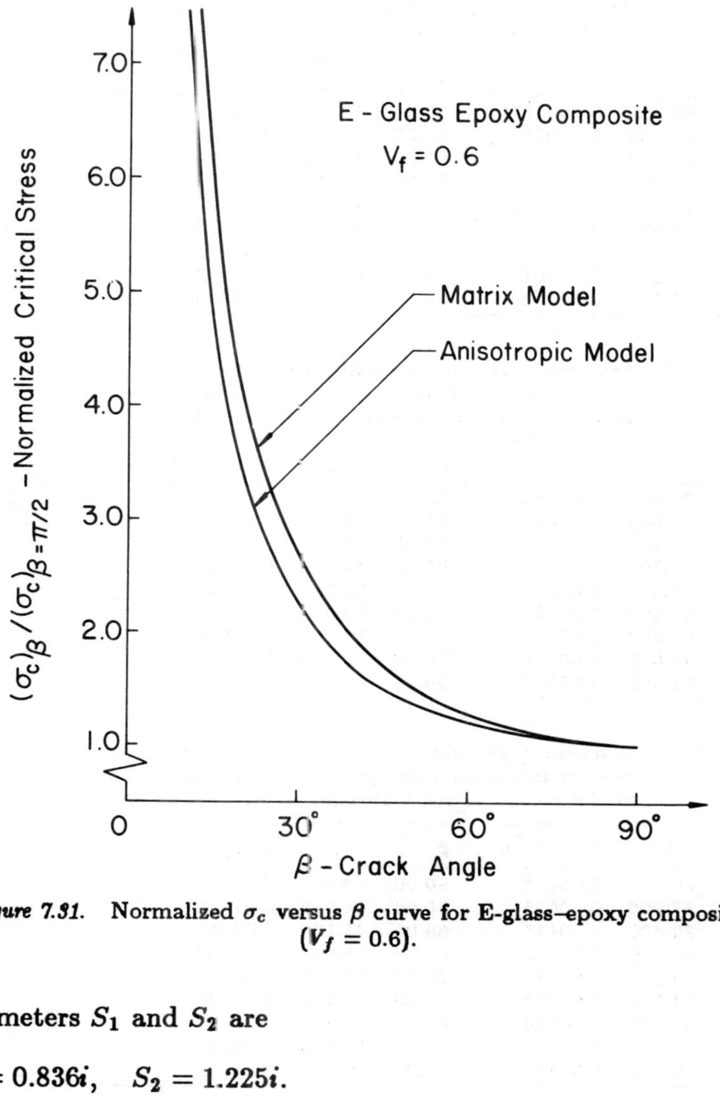

Figure 7.31. **Normalized σ_c versus β curve for E-glass–epoxy composite ($V_f = 0.6$).**

The parameters S_1 and S_2 are

$$S_1 = 0.836i, \quad S_2 = 1.225i. \tag{7.90}$$

For the matrix cracking model, it is necessary to know

$$\Phi(1) = 0.90, \quad \Psi(1) = 0.88 \tag{7.91}$$

for finding the crack tip stress intensity factors. Without going into details, $-\theta_0$ and $S_{\min}/\sigma^2 a$ for both of the models can be found with the aid of equations (7.90) and (7.91). Refer to Tables 7.15 and 7.16. It can be seen from Figure 7.32 that both models give nearly the same results.

Fiber volume fraction 0.4. Let the fiber volume function of the stainless steel–aluminum composite be increased to 0.4. This will result in a corre-

Table 7.15. **Fracture angle and normalized strain energy density factor for homogeneous anisotropic models (stainless steel–aluminum with $V_f = 0.1$).**

β	$-\theta_0$	$S_{min}/\sigma^2 a$	β	$-\theta_0$	$S_{min}/\sigma^2 a$
5.00	79.900	0.00018	50.00	49.100	0.01322
10.00	76.500	0.00075	55.00	45.300	0.01463
15.00	73.100	0.00169	60.00	41.300	0.01586
20.00	69.700	0.00297	65.00	37.000	0.01691
25.00	66.400	0.00453	70.00	32.100	0.01781
30.00	63.000	0.00627	75.00	26.300	0.01858
35.00	59.600	0.00809	80.00	19.300	0.01921
40.00	56.200	0.00991	85.00	10.500	0.01966
45.00	52.700	0.01164	90.00	0.000	0.01983

Table 7.16. **Fracture angle and normalized strain energy density factor for matrix cracking model (stainless steel–aluminum with $V_f = 0.1$).**

β	$-\theta_0$	$S_{min}/\sigma^2 a$	β	$-\theta_0$	$S_{min}/\sigma^2 a$
5.00	80.000	0.00016	50.00	48.800	0.01149
10.00	76.600	0.00065	55.00	45.000	0.01275
15.00	73.100	0.00146	60.00	40.900	0.01385
20.00	69.700	0.00257	65.00	36.400	0.01481
25.00	66.300	0.00392	70.00	31.200	0.01564
30.00	62.900	0.00542	75.00	25.300	0.01634
35.00	59.500	0.00700	80.00	18.100	0.01690
40.00	56.000	0.00858	85.00	9.600	0.01728
45.00	52.500	0.01009	90.00	0.000	0.01742

Table 7.17. **Fracture angle and normalized strain energy density factor for homogeneous anisotropic model (stainless steel–aluminum with $V_f = 0.4$).**

β	$-\theta_0$	$S_{min}/\sigma^2 a$	β	$-\theta_0$	$S_{min}/\sigma^2 a$
5.00	79.300	0.00013	50.00	48.600	0.01010
10.00	77.900	0.00055	55.00	44.900	0.01123
15.00	72.400	0.00125	60.00	41.100	0.01222
20.00	69.000	0.00220	65.00	36.900	0.01308
25.00	65.600	0.00337	70.00	32.400	0.01383
30.00	62.300	0.00468	75.00	27.100	0.01447
35.00	58.900	0.00608	80.00	20.500	0.01501
40.00	55.500	0.00748	85.00	11.800	0.01541
45.00	52.100	0.00884	90.00	0.000	0.01557

Table 7.18. **Fracture angle and normalized strain energy density factor for matrix cracking model (stainless steel–aluminum with $V_f = 0.4$).**

β	$-\theta_0$	$S_{min}/\sigma^2 a$	β	$-\theta_0$	$S_{min}/\sigma^2 a$
5.00	79.900	0.00008	50.00	48.100	0.00613
10.00	76.300	0.00034	55.00	44.300	0.00684
15.00	72.800	0.00076	60.00	40.200	0.00748
20.00	69.300	0.00134	65.00	35.700	0.00805
25.00	65.800	0.00204	70.00	30.600	0.00855
30.00	62.400	0.00284	75.00	24.600	0.00897
35.00	58.900	0.00368	80.00	17.600	0.00930
40.00	55.400	0.00453	85.00	9.250	0.00953
45.00	51.800	0.00536	90.00	0.000	0.00961

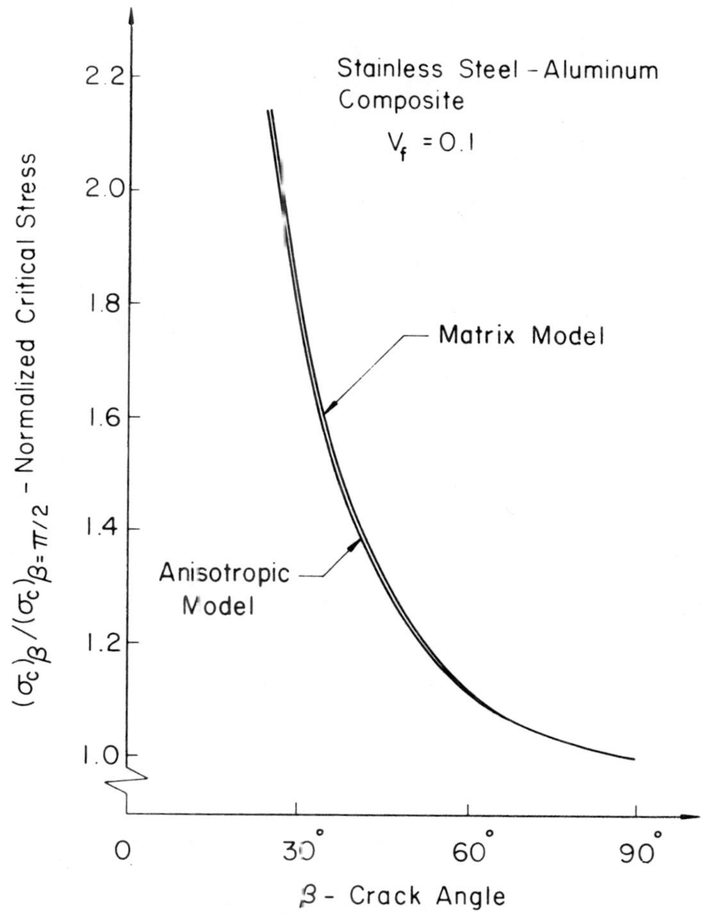

Figure 7.32. Normalized σ_c versus β curve for steel–aluminum composite ($V_f = 0.1$).

sponding increase of the elastic moduli:

$$E_1 = 18.30 \times 10^6 \text{ psi}, \quad \nu_{12} = 0.32 \tag{7.92a}$$

$$E_2 = 15.51 \times 10^6 \text{ psi}, \quad \nu_{23} = 0.36 \tag{7.92b}$$

$$\mu_{12} = 5.87 \times 10^6 \text{ psi}. \tag{7.92c}$$

which may be used to render

$$S_1 = 0.777i, \quad S_2 = 1.365i. \tag{7.93}$$

The quantities $\Phi(1)$ and $\Psi(1)$ take the values

$$\Phi(1) = 0.67, \quad \Psi(1) = 0.63 \tag{7.94}$$

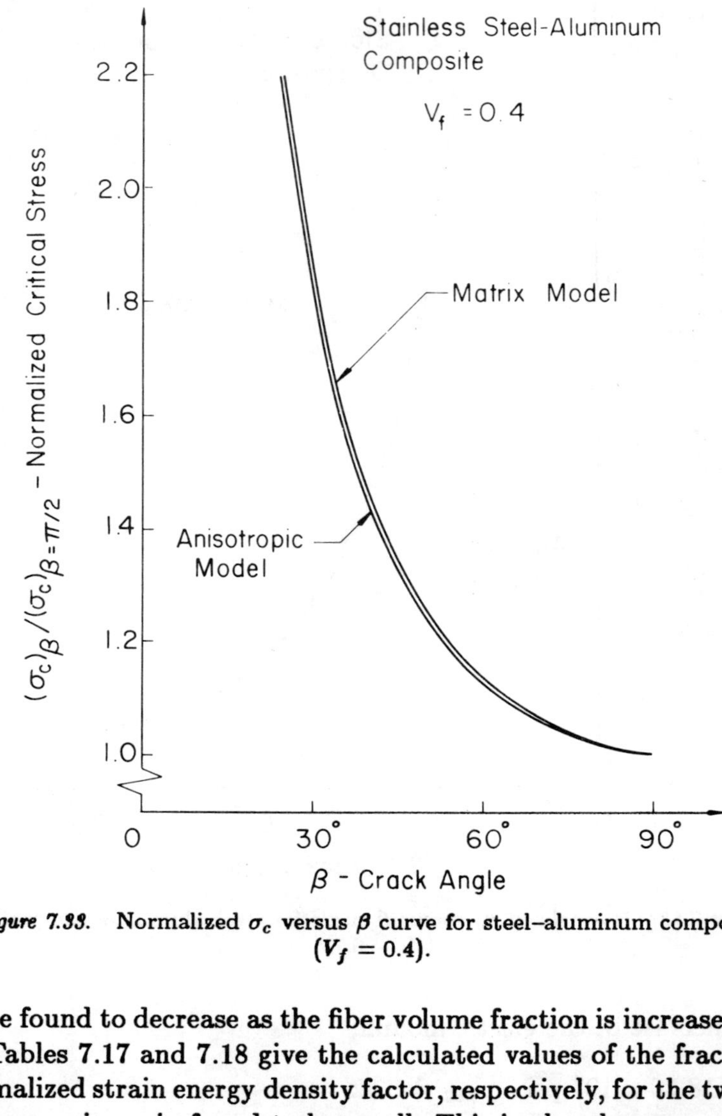

Figure 7.33. **Normalized σ_c versus β curve for steel–aluminum composite ($V_f = 0.4$).**

which are found to decrease as the fiber volume fraction is increased from 0.1 to 0.4. Tables 7.17 and 7.18 give the calculated values of the fracture angle and normalized strain energy density factor, respectively, for the two models. The difference is again found to be small. This is also shown graphically in Figure 7.33.

Equations (7.89) and (7.92) show that the elastic moduli in the principal directions of material symmetry do not vary appreciably. The prevailing nonhomogenity has a negligible influence on the way with which the metal composite is modelled. This observation can also be seen from the curves in Figures 7.32 and 7.33.

7.12. References

[1] Matthaes, K., Betrachtungen zur Theorie der Werkstoff-Festigkeit, ***Zeitschrift für Metallkunde*** **43**, pp. 11–19 and 90–95 (1952).

[2] Gillemot, L. F., Criterion of crack initiation and spreading, ***International Journal of Engineering Fracture Mechanics*** **8**, pp. 239–253 (1976).

[3] Sih, G. C., A special theory of crack propagation, ***Methods of Analysis and Solutions of Crack Problems, Mechanics of Fracture I***, edited by G. C. Sih, Noordhoff International Publishing, Leyden, pp. 21–45 (1973).

[4] Sih, G. C., A three-dimensional strain energy density factor theory of crack propagation, ***Three-Dimensional Crack Problems, Mechanics of Fracture II***, edited by G. C. Sih, Noordhoff International Publishing, Leyden, pp. XV–LIII (1975).

[5] Sih, G. C., Strain energy density and surface layer energy for blunted cracks or notches, ***Stress Analysis of Notch Problems, Mechanics of Fracture V***, edited by G. C. Sih, Noordhoff International Publishing, Alphen aan den Rijn, pp. XIII-CX (1978).

[6] Theocaris, P. S., Experimental determination of the core region in mixed-mode fracture, ***Mixed Mode Crack Propagation***, edited by G. C. Sih and P. S. Theocaris, Sijthoff and Noordhoff International Publishers, Alphen aan den Rijn, pp. 21–36 (1981).

[7] Sih, G. C. and Kiefer, B. V., Stable growth of surface cracks, ***Journal of the Engineering Mechanics Division***, ASCE **106**, No. EM2, pp. 245–253 (1980).

[8] Sih, G. C. and Kiefer, B. V., Nonlinear response of solids due to crack growth and plastic deformation, Nonlinear and Dynamic Fracture Mechanics, edited by N. Perrone and S. N. Atluri, ***The American Society of Mechanical Engineers***, AMD **35**, pp. 135–156 (1979).

[9] Stephenson, R. A., The equilibrium field near the tip of a crack for finite plane strain of incompressible elastic materials, ***California Institute of Technology***, Report No. 44 (1980).

[10] Ivanova, V. S. and Ginsztler, J., The examination of fracture toughness with the aid of Gillemot's absorbed specific energy till fracture, ***Absorbed Specific Energy and/or Strain Energy Density Criterion***, edited by G. C. Sih, E. Czoboly and F. Gillemot, Sijthoff and Noordhoff International Publishers, Alphen aan den Rijn, pp. 101–106 (1981).

[11] Wnuk, M. P. and Mazur, A., Experimental verification of the strain energy density criterion for the stable phase of tensile fracture, ***Absorbed Specific Energy and/or Strain Energy Density Criterion***, edited by G. C. Sih, E. Czoboly and F. Gillemot, Sijthoff and Noordhoff International Publishers, Alphen aan den Rijn, pp. 101–106 (1981).

[12] Rice, J. R. and Sih, G. C., Plane problems of cracks in dissimilar media, ***Journal of Applied Mechanics*** **32**, pp. 418–423 (1965).

[13] Kipp, M. E. and Sih, G. C., The strain energy density failure criterion applied to notched elastic solids, ***International Journal of Solids and Structures*** **2**, pp. 153–173 (1975).

[14] Sih, G. C. and Moyer, Jr., E. T., Influence of interface on composite failure, ***Proceedings of Advanced Composites***, Technology Conferences, El Segundo, California, pp. 283–312 (1979).

[15] Lauraitis, K., Tensile strength of off-axis unidirectional composites, ***University of Illinois TAM Report*** **344**, (1971).

[16] Sih, G. C. and Chen, E. P., Fracture analysis of unidirectional composites, ***Journal of Composite Materials*** **7**, pp. 230–244 (1973).

[17] Sih, G. C., Paris, P. C. and Irwin, G. R., On cracks in rectilinearly anisotropic bodies, ***International Journal of Fracture Mechanics*** **1**, pp. 189–203 (1965).

[18] Wu, E. M. and Reuter, R. C., Crack extension in fiber-glass reinforced plastics, ***University of Illinois TAM Report*** **275** (1965).

[19] Lauraitis, K., Failure modes and strength of angle-ply laminates, ***University of Illinois TAM Report*** **345** (1971).

[20] Sih, G. C., Fracture mechanics of adhesive joints, ***Journal of Polymer Engineering and Science*** **20**, No. 14, pp. 977–981 (1980).

[21] Puppo, A. H. and Evensen, H. A., Interlaminar shear in laminated composites under generalized plane stress, *Journal of Composite Materials* **4**, pp. 204–220 (1970).

[22] Reissner, E. and Stavsky, Y., Bending and stretching of certain types of heterogeneous aeolotropic elastic plates, *Journal of Applied Mechanics* **28**, pp. 402–408 (1961).

[23] Stavsky, Y., Bending and stretching of laminated aeolotropic plates. *Proceedings American Society of Civil Engineering, Journal of the Engineering Mechanics Division* **87**, pp. 31–56 (1961).

[24] Pipes, R. B. and Pagano, N. J., Interlaminar stresses in composite laminates under uniform axial extension, *Journal of Composite Materials* **4**, pp. 538–548 (1970).

[25] Sih, G. C. and Hilton, P. D., Three-dimensional analysis of laminar composites with through cracks, *American Society of Testing Materials*, STP 539, pp. 3–35 (1975).

[26] Sih, G. C. and Chen, E. P., Normal and shear impact of layered composite with a crack: dynamic stress intensification, *Journal of Applied Mechanics* **47**, pp. 351–358 (1980).

[27] Sih, G. C. and Chen, E. P., Effect of material nonhomogeneity on crack propagation characteristics, *International Journal of Engineering Fracture Mechanics* **13**, No. 3, pp. 431–438 (1980).

[28] Sih, G. C., Influence of interface modeling on composite failure, *Proceedings of the Third International Symposium on Continuum Models of Discrete Systems*, University of Waterloo, pp. 837–864 (1980).

[29] Sih, G. C., Dynamic crack problems: strain energy density fracture theory, *Elastodynamic Crack Problems, Mechanics of Fracture IV*, edited by G. C. Sih, Noordhoff International Publishing, Leyden, pp. XVII–XLVII (1977).

8

Experimental fracture mechanics: strain energy density criterion

8.1. Preliminary remarks

The objective of this chapter is to impart a better understanding of the mutual dependence of theory and experiment in fracture mechanics. An experiment cannot be considered successful unless the results are interpreted within the framework of some preconceived concept and/or theoretical prediction. Otherwise the test results are of no more value than to predict the results of other tests conducted under exactly the same conditions. Interpreting experimental results requires a sound knowledge of the fracture process and governing physical principles. In addition, theoretical speculations on these processes must be guided by experimental observations. The interplay between theory and experiment is essential to the development of both theories and experiments.

Linear elastic fracture mechanics [1] has not only been receiving worldwide attention from the academic community in recent years but it has also been increasingly employed as a useful engineering approach to optimizing material selection and engineering design. The wide acceptance of this relatively new discipline is largely due to the ease with which the theory can be applied quantitatively for making design decisions to prevent sudden fracture of structural components. As research investigators devoted more effort to ductile fracture, their viewpoints began to diverge; instead of a discipline with unifying concepts and methods, diverse parameters, methods of testing, and jargons emerged. The proliferation of organizations and committees within professional societies had led to the endorsement of a variety of test procedures and specimens. The variety of different specimens now being used in laboratories throughout the world make comparison of results obtained in different laboratories increasingly difficult. In addition, there is a widespread feeling that neither modern electronic computers nor laboratory equipment are being used effectively. Instead of serving to provide insight into the physical fracture process and to establish order within the confusing abundance of data, they seem to have become self-serving and generated an exhaustive amount of numerical results and unsupported data. The close relationship which should exist between methods of analysis and perceptual concepts of the physics of fracture is not being maintained in current research. Analytical work in fracture mechanics is often solely concerned with refinement of computations and fails to grasp the connection of the problem with reality. Experimental work on the other hand often becomes merely the collection of data even after the physical phenomena are better understood. Undoubtedly, these tendencies are responsible for the paucity of progress toward discovering the physical laws that govern material failure. Whenever possible, the separate efforts and multifarious facts should be united by clearly identifying their internal connections.

It is not the intention of this communication to propose a unified theory for fracture mechanics, rather the aim is to call for an increase in the degree of consistency and rigor in research. Too much emphasis cannot be placed on the development of theoretical concepts as an adjunct to experimentation so that different forms of failure may be treated not as unrelated independent events but rather as different aspects of a single, more fundamental phenomenon.

8.2. Mechanical properties of materials

The study of the mechanical properties of material is concerned with two kinds of quantities: those such as displacement, force, etc., which are directly measurable, and those such as stress, energy, etc., which are defined in terms of measurable quantities by means of mathematical relations. Both types are used to describe the response of materials to external mechanical influences, and are involved in predicting the ability of materials to sustain reversible and irreversible deformations and to resist fracture. The commonly known properties such as modulus of elasticity, yield strength, ultimate strength, etc., have values which measured on a certain scale, depend on rate of loading, size and shape of specimen, environmental conditions such as temperature, moisture, etc. The material response can vary widely from long-time stable failure by excessive deformation to short-time unstable fracture by sudden separation of material.

Physical hypotheses. To bring some order into this wide range of material behavior, failures should be related to experimentally measurable or mathematically defined quantities through physical hypotheses. Among the more common quantities used as criteria of failure are the maximum normal stress, maximum shear stress, maximum strain, total strain energy density, energy of distortion and octahedral shearing stress. It should be made clear at the start that the validity of a hypothesis cannot be proved mathematically from

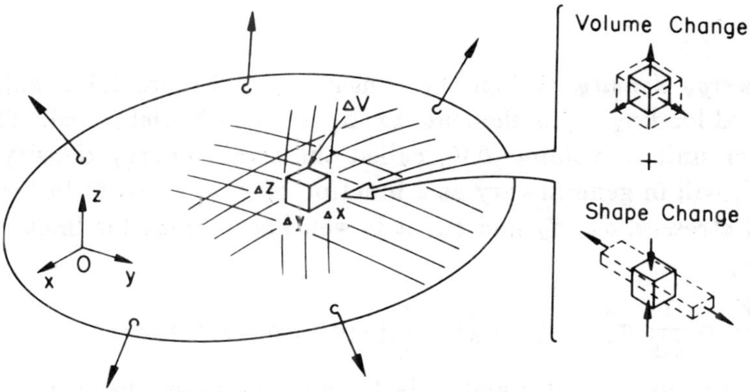

Figure 8.1. A continuum divided into finite number of macroscopic elements undergoing volume and shape change.

any relations previously obtained by definition and physical hypothesis. A case in point is the energy release rate quantity defined in linear elastic fracture mechanics to represent the energy dissipated by the unit extension of a crack.*

Mathematics cannot prove or disprove any hypothesis and should be regarded strictly as a tool to obtain verifiable consequences for testing physical hypotheses and through new experiments. The test of any hypothesis of a physical nature rests on the *consistency* of its derived consequences with measurements and its usefulness in forecasting physical events. Passing such a test, the hypothesis is satisfactory and useful.

Continuum scale. All hypotheses advanced in the study of material behavior presuppose that the physical behavior of systems is decribed using a predetermined scale. The continuum concept that a material can be repeatedly subdivided into smaller elements with dimensions approaching zero must, of course, be modified when examining real materials. The element size (volume $\Delta V = \Delta x\ \Delta y\ \Delta z$ in Figure 8.1) must obviously be small enough for numerical accuracy, yet not so small that the continuum model loses its validity. The conflicting needs of numerical analysis and experimentation must be assessed and made compatible with one another. The continuum level, visible to an unaided eye, is normally referred to as the macroscopic scale. Nonhomogeneity of a real material at the microscopic level that prevails within the basic element is neglected.** The validity of the continuum

---

* The calculation of the energy release rate from a stress field involves all of the assumptions used to obtain the stresses. The hypothesis that the crack grows when this equals some critical value is an additional, separate hypothesis.

** This simply means that the continuum model usually is applied to situations where the microstructure of the material is neglected unless otherwise accounted for.

approximation is tested through comparison of theory and experiment.

Strain energy density. When the continuum in Figure 8.1 is subjected to mechanical loading, each element $\Delta x\ \Delta y\ \Delta z$ will be deformed. The energy stored per unit of volume ΔV, called the strain energy density function $\Delta W/\Delta V$, will in general vary as a function of x, y, z and t. In terms of the principal stresses, σ_1, σ_2 and σ_3, it is well known that for linear, isotropic materials,

$$\frac{\Delta W}{\Delta V} = \frac{1}{2E}(\sigma_1^2 + \sigma_2^2 + \sigma_3^2) - \frac{\nu}{E}(\sigma_1\sigma_2 + \sigma_1\sigma_3 + \sigma_2\sigma_3) \tag{8.1}$$

where ν is Poisson's ratio and E is Young's modulus. Equation (8.1) may be written as the sum of a part due to change of volume,

$$\left(\frac{\Delta W}{\Delta V}\right)_v = \frac{1-2\nu}{6E}(\sigma_1 + \sigma_2 + \sigma_3)^2 \tag{8.2}$$

and of a part due to change of shape,

$$\left(\frac{\Delta W}{\Delta V}\right)_d = \frac{1+\nu}{6E}[(\sigma_1 - \sigma_2)^2 + (\sigma_2 - \sigma_3)^2 + (\sigma_3 - \sigma_1)^2]. \tag{8.3}$$

Hence,

$$\frac{\Delta W}{\Delta V} = \left(\frac{\Delta W}{\Delta V}\right)_v + \left(\frac{\Delta W}{\Delta V}\right)_d. \tag{8.4}$$

The expressions $(\Delta W/\Delta V)_v$ and $(\Delta W/\Delta V)_d$ are useful as their individual contribution when weighed appropriately may be associated with the cause of failure resulting from fracture and yielding.

Test of failure criterion. A typical failure criterion states that failure takes place in a material when some quantity, say $\Delta W/\Delta V$ in equation (8.1) as proposed by Beltrami and Haigh, reaches a certain critical value $(\Delta W/\Delta V)_c$ characteristic of the material. A simple test of the criterion is to compare the results of two simple tests, namely a bar loaded to failure in tension or torsion as shown in Figures 8.2(a) and 8.2(b). For a linear elastic material, $\Delta W/\Delta V$ in simple tension can be found from equation (8.1) by letting $\sigma_1 = \sigma$ and $\sigma_2 = \sigma_3 = 0$. Then

$$\left(\frac{\Delta W}{\Delta V}\right)_c = \frac{1}{2E}\sigma^2. \tag{8.5}$$

Setting $\sigma_1 = \tau$, $\sigma_2 = -\tau$ and $\sigma_3 = 0$ for case of torsion, equation (8.1) yields

$$\left(\frac{\Delta W}{\Delta V}\right)_c = \frac{1+\nu}{E}\tau^2. \tag{8.6}$$

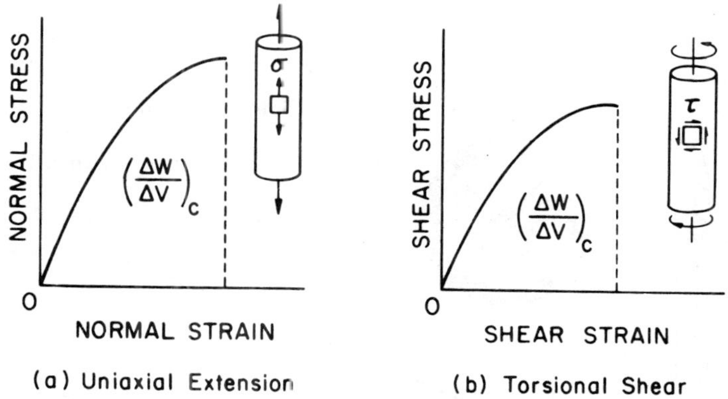

Figure 8.2. Bar specimen in tension and torsion.

at the instant of failure. If the bars are made of the same material, the normal stress σ in equation (8.5) and shear stress τ in equation (8.6) corresponding to failure should be related as

$$\frac{\tau}{\sigma} = \frac{1}{\sqrt{2(1+\nu)}} \tag{8.7}$$

if the failure criterion is applicable. As ν changes from 0.0 to 0.5, the ratio τ/σ varies in the range 0.707 to 0.577.

The strain energy density criterion is not limited to linear elastic materials but may be applied also to materials with nonlinear behavior as well. The value $(\Delta W/\Delta V)_c$ represents the area under the true stress and strain curve and can be computed directly from experimental data.* Measured values of $(\Delta W/\Delta V)_c$ for a variety of engineering materials can be found in [3, 4].

Specimen size effect. The tensile strength of smooth round bar specimens that are metallurgically and geometrically** similar has been shown to decrease as the bar diameter is increased. This is known as the size effect. The broken surfaces of the specimen show three distinct features of fracture development labelled as I, II and III in Figure 8.3. Failure by stable crack growth starts from a state of hydrostatic tension inside the region I. This is followed by rapid or unstable crack propagation denoted by region II. Final separation of the material occurs by slant fracture terminating at the specimen boundary referred to as region III. A plot of the lineal dimensions of these regions versus an order of magnitude increase in the specimen diam-

---

* A procedure for determining $(\Delta W/\Delta V)_c$ is presented in an English translation of the Hungarian Standard MSz 4929-76 [2].

** Round bars of different sizes with the same radius to length ratio, say r/L, are said to be geometrically similar. However, even for geometrically similar bars, such quantities as volume to surface ratio will be different for different sizes.

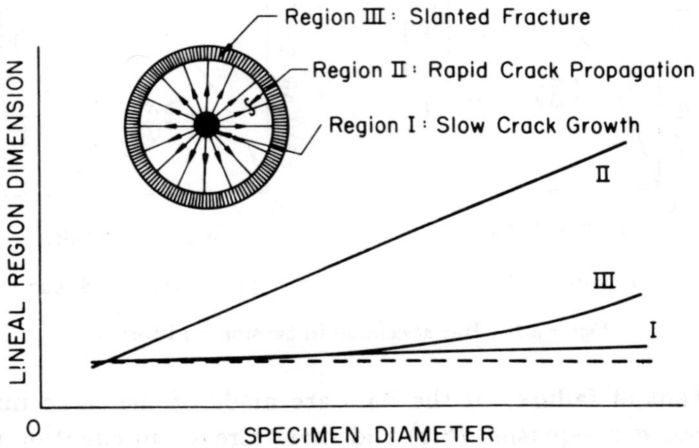

Figure 8.3. **Fracture pattern of metal specimen.**

eter is displayed schematically in Figure 8.3. Note that region II increases almost directly with specimen size while regions I and III are affected only slightly [5]. The larger specimen would tend to be more brittle because of the larger zone of rapid crack propagation. The lower tensile strength for larger specimens is a direct result of the definition of $\sigma = P/A$ where the rate of increase of the specimen cross-sectional area A exceeds that of the failure load P. The variation of tensile strength with specimen size would be greatly reduced if σ were defined in terms of the area occupied by region I or III instead of the total of all three regions, A. The three regions shown in Figure 8.3 are typical of aluminum, steel and titanium metal alloys [6] and are relatively independent of the shape of the cross-section.

Size sensitivity of specimens has frequently been attributed to statistical considerations and the weakest-link theory because of inability to understand the fracture development process and to incorporate it into analysis. Since the fracture pattern for many engineering materials is clearly deterministic, the apparent size effect can be resolved by judicious application of a fracture criterion that can methodically account for the physical fracture process. The analysis should distinguish and relate crack initiation, stable and unstable crack growth. These distinct stages of the total fracture process cannot be assumed to coincide nor be treated separately even though their interval of occurrence can be very short for some materials. In fact, the trade-off between specimen size and strength depends intimately on the interaction of the various intervening stages of crack growth. A quantitative assessment of the stability and instability of the fracture process can be

made from the strain energy* content of the system whose rate of release** dictates the different stages of fracture. The creation of free surfaces originating from defects is thus an inherent part of the size effect phenomenon associated with strength.

8.3. Fracture mechanics discipline

Fracture mechanics, like the other branches of mechanics, is concerned with predicting the response of a system to external disturbances. The idealized system is assumed to contain a large aggregate of continuous elements whose mechanical properties can be determined from ***material testing***. The specimens and test conditions must be carefully selected to simulate the failure mode of the smaller element as closely as possible. When defects or cracks are present, their significance must be assessed through analysis and ***fracture testing*** based on certain physical hypotheses. The energy content of specimens for material and fracture testing should be proportioned to control the degree of material damage at different scale levels and monitored as material behavior by the load versus displacement relationships. The connection between observed data and underlying theories must be clearly and consistently established for material damaged at the microscopic and macroscopic scale level. To reiterate, the primary objective of fracture mechanics is to develop procedures so that ***data collected on laboratory specimens will be useful in the design of larger size structural components***. The final test of predictive capability lies in expressing results in terms of directly measurable quantities such as allowable load and dimensions of structural components.

Classification of fracture. Several types of fractures in various metal alloys have been observed and classified qualitatively according to the test conditions under which they were created. Two of the most common types of fracture are known as cleavage and shear. They are differentiated by the orientation of the fracture surface with reference to the maximum value of a neighboring stress component. For instance, cleavage fracture would be associated with a surface normal to the largest tensile stress and shear fracture with a surface parallel to the direction of maximum shear stress. This approach can be subjective and lead to ambiguous interpretation. Funda-

---

* The events of slow and rapid crack growth can be described from the strain energy density criterion, $(\Delta W/\Delta V)_c = S_c/r_c$, where the critical ligament size r_c determines the termination of slow crack growth and the beginning of rapid crack propagation. It is characteristic of the material and cannot be taken as zero. The condition $(\Delta W/\Delta V)_c =$ const. corresponds to the failure of local elements while S_c to the onset of global instability.

** The rate of energy release is determined by conditions around the defect that can either be stress rate or strain rate controlled.

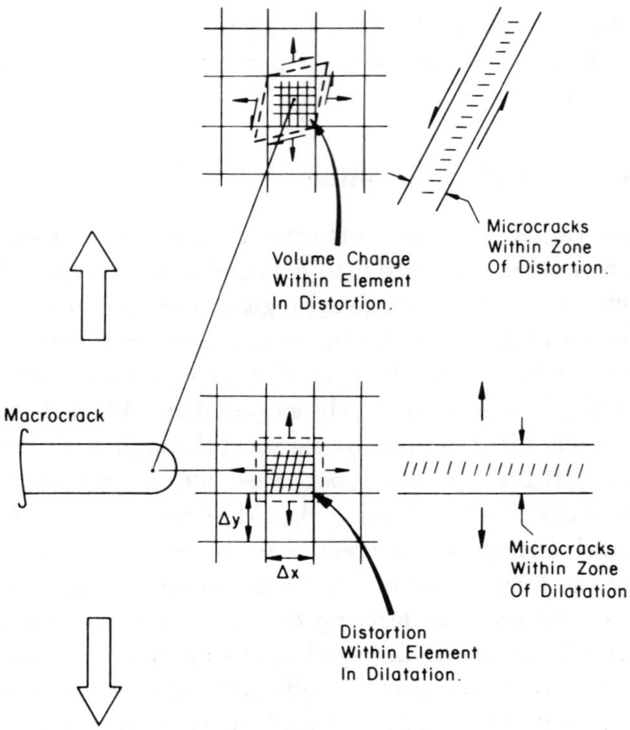

Figure 8.4. **Zones of distortion and dilatation ahead of a macrocrack in tension.**

mental difficulties are immediately disclosed when the fracture surface is examined more closely at the microscopic level; an apparent cleavage fracture may consist of minute shear planes and an apparent shear fracture may contain minute cleavage planes. This is illustrated by the macrocrack under tension shown in Figure 8.4. The elastic zone ahead of the macrocrack contains shear planes and the plastic region to the side of the macrocrack contains cleavage planes. Additional confusion arises in attempting to relate brittle fracture to cleavage and ductile fracture to shear. Descriptive terminologies based on phenomenological studies alone are not a proper substitute for basic understanding of the fracture of a real material. Although there are no theories at present that can account for the detailed mechanism of fracture, still the problem is not being approached in a more systematic and intelligent manner.

The assumption of material homogeneity is usually invoked in a theory for the determination of stresses and deformations in a solid because of the enormous complexity required to specify the various forms of irregularities in a real material. For this obvious reason, experimental observation would tend to scatter about the orderly behavior predicted by an idealized theory of homogeneous bodies. The mechanical conditions which cause solids to de-

Table 8.1. Values of $(\Delta W/\Delta V)_v/(\Delta W/\Delta V)_d$ for $\nu = 0.3$.

σ_1/σ_2	1	2	3	4	5
$(\Delta W/\Delta V)_v/(\Delta W/\Delta V)_d$	6.500	0.839	1.143	2.108	0.696

form permanently or to fracture can be related to shape change (distortion) and volume change (dilatation). In general, a material is subjected to both distortion and dilatation and the corresponding energies can be computed from equations (8.2) and (8.3) depending on the state of stress. Suppose that an element $\Delta x \Delta y$ ahead of the crack in Figure 8.4 is under the condition

$$\sigma_3 = \nu(\sigma_1 + \sigma_2) \tag{8.8}$$

known as plane strain. Let σ_1 be directed normal to the crack and σ_2 along the crack. Making use of equation (8.8), $(\Delta W/\Delta V)_v$ becomes

$$\left(\frac{\Delta W}{\Delta V}\right)_v = \frac{1-2\nu}{E}[\sigma_1 + \sigma_2 + \nu(\sigma_1 + \sigma_2)]^2 \tag{8.9}$$

and $(\Delta W/\Delta V)_d$ takes the form

$$\left(\frac{\Delta W}{\Delta V}\right)_d = \frac{1+\nu}{3E}[\sigma_1 + \sigma_2 + \nu(\sigma_1 + \sigma_2)]^2 - \frac{1+\nu}{E}[\sigma_1\sigma_2 + \nu(\sigma_1 + \sigma_2)^2]. \tag{8.10}$$

The ratio of equations (8.9) and (8.10) can be expressed in terms of σ_1/σ_2 as follows:

$$\frac{(\Delta W/\Delta V)_v}{(\Delta W/\Delta V)_d} = \frac{(1-2\nu)(1+\nu)[(\sigma_1/\sigma_2)+1]^2}{2(1+\nu)^2[(\sigma_1/\sigma_2)+1]^2 - 6(\sigma_1/\sigma_2) - 6\nu[(\sigma_1/\sigma_2+1)]}. \tag{8.11}$$

A simple calculation for $\nu = 0.3$ shows that $(\Delta W/\Delta V)_v$ is 6.5 times larger than $(\Delta W/\Delta V)_d$ in the case $\sigma_1 = \sigma_2$. This explains why failure is most likely to occur from the crack tip region where the material undergoes a large change in volume. The ratio $(\Delta W/\Delta V)_v/(\Delta W/\Delta V)_d$ decreases rapidly when σ_1/σ_2 increases. The strain energy associated with shape change becomes higher than the strain energy associated with volume change when $\sigma_1/\sigma_2 = 3.36$. Equation (8.11) is solved numerically for $\nu = 0.3$ and the results are given in Table 8.1.

In general, $(\Delta W/\Delta V)_v$ and $(\Delta W/\Delta V)_d$ are of the same order of magnitude and hence both quantities should be weighed* as a sum according to equation (8.4). The proportion between $(\Delta W/\Delta V)_v$ and $(\Delta W/\Delta V)_d$ corresponding to a nonuniform state of stress will be different for each element within the continuum and used to explain failure by deformation and fracture. Referring to the macro element $\Delta x\ \Delta y$ ahead of the crack in Figure 8.4, the principal stresses σ_1 and σ_2 are equal. Even though $(\Delta W/\Delta V)_v$

* The yield criterion of von Mises only accounts for $(\Delta W/\Delta V)_d$ in equation (8.3) and neglects the energy component due to volume change.

is 2.24 times larger than $(\Delta W/\Delta V)_d$, the distortional component of the strain energy density is not negligible and is assumed to be responsible for the creation of slip planes or microcracks within a macroscopic layer where dilatation dominates. Similarly, maximum distortion of the continuum element occurs to the side of the macrocrack. By the same argument, the dilatational component of the strain energy density is assumed to be associated with the creation of cleavage planes that are perpendicular to the direction of tension. It appears that material damage at the microscopic and macroscopic scale can be connected by referring to the proportion of $(\Delta W/\Delta V)_v$ and $(\Delta W/\Delta V)_d$ within the continuum element. For a macrocrack under tension, the model in Figure 8.4 suggests macrofracture to coincide with the location where $(\Delta W/\Delta V)_v > (\Delta W/\Delta V)_d$ and macroyielding with the direction where $(\Delta W/\Delta V)_d > (\Delta W/\Delta V)_v$. These locations may be determined by appealing to physical hypotheses.

Yielding and fracture. Since microyielding may attribute to macrofracture and microfracture to macroyielding, the two processes of yielding and fracture are inseparable. They are unique features of material damage and should be treated simultaneously by a single failure criterion, if possible. To this end, the strain energy density function $\Delta W/\Delta V$ will be used in the form

$$\frac{\Delta W}{\Delta V} = \frac{S}{r} \tag{8.12}$$

where S is the strain energy density factor* [7, 8] and r the radial distance measured from the site of possible failure initiation,** say the crack tip. The singular dependency $1/r$ is a fundamental character† of the Newtonian potential and is independent of the constitutive relation. The response of a particular material is reflected through the factor S. There are three basic hypotheses of the strain energy density criterion:

Hypothesis (1): ***The location of fracture coincides with the location of minimum strain energy density $(\Delta W/\Delta V)_{\min}$, and yielding with maximum strain energy density $(\Delta W/\Delta V)_{\max}$.***

---

* The factor S can be computed from the stress intensity factors k_1, k_2 and k_3 only for predicting the initiation of crack propagation in a linear elastic material. The S-criterion is applied through a given stress and strain relation, but is not restricted by it.

** The site of failure can be a reentrant corner, inclusion, void, etc., and is not restricted to the crack tip as in the conventional theory of linear fracture mechanics.

† Failure criteria based on the amplitude of the asymptotic stresses would tend to be restrictive since the singular character of the stresses depends on the constitutive relation of the material. Moreover, the order of singularity for each stress component may be different as in the case of finite deformation theory.

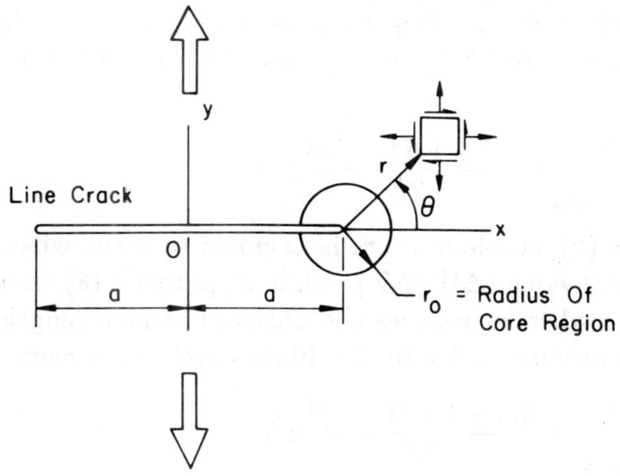

Figure 8.5. Line crack in tension.

Hypothesis (2): *Failure by stable fracture or yielding occurs when* $(\Delta W/\Delta V)_{\min}$ *or* $(\Delta W/\Delta V)_{\max}$ *reach their respective critical values.*

Hypothesis (3): *The amount of incremental growth* $r_1, r_2, \ldots, r_j, \ldots, r_c$ *is governed by*

$$\left(\frac{\Delta W}{\Delta V}\right)_c = \frac{S_1}{r_1} = \frac{S_2}{r_2} = \cdots = \frac{S_j}{r_j} = \cdots = \frac{S_c}{r_c} = \text{const.} \tag{8.13}$$

There is unstable fracture or yielding when the critical ligament size r_c *is reached.*

The intent of these proposed hypotheses is to provide a means for correlating the data obtained from material and fracture testing. By this procedure, a means of transferring results from specimen to structure may be gained. Such information is not only valuable to a better understanding of size effect but also opens the door to design in the inelastic range.

Line crack in tension. In order to be more specific, consider a line crack of length $2a$ in a material under uniform tension. Let attention be focused on the variation of the strain energy density $\Delta W/\Delta V$ with the angle θ in Figure 8.5 for a fixed value of r. The value, r_0 is taken to be sufficiently small that the expression [7]

$$r_0\left(\frac{\Delta W}{\Delta V}\right) = \frac{(1+\nu)}{8E}(3 - 4\nu - \cos\theta)(1 + \cos\theta)\sigma^2 a \tag{8.14}$$

derived from the asymptotic stresses in a linear elastic material is valid. The magnitude of the remote uniform stress is denoted by σ.

The conditions specified by Hypothesis (1) may be satisfied by requiring that $\partial(\Delta W/\Delta V)/\partial\theta = 0$. This gives two solutions for θ. The first solution, $\theta_0 = 0$, corresponds to $(\Delta W/\Delta V)_{\min}$ and determines the direction of crack propagation:

$$r_0\left(\frac{\Delta W}{\Delta V}\right)_{\min} = \frac{(1+\nu)(1-2\nu)}{2E}\sigma^2 a. \tag{8.15}$$

By Hypothesis (2), stable fracture is assumed to occur when $(\Delta W/\Delta V)_{\min}$ takes the critical value $(\Delta W/\Delta V)_c$ while Hypothesis (3) assumes that rapid fracture begins when r_0 reaches the critical ligament length, r_c. Equation (8.15) may be decomposed into the dilatational component

$$r_0\left(\frac{\Delta W}{\Delta V}\right)_v = \frac{2(1-2\nu)(1+\nu)^2}{3E}\sigma^2 a \tag{8.16}$$

and distortional component

$$r_0\left(\frac{\Delta W}{\Delta V}\right)_d = \frac{(1+\nu)(1-2\nu)}{2E}\sigma^2 a. \tag{8.17}$$

A comparison of equations (8.16) and (8.17) reveals that the strain energy density for volume change in an element along the path of expected crack extension is greater than the strain energy density for shape change. The maximum value of $\Delta W/\Delta V$ corresponds to $\theta_0 = \pm\cos^{-1}(1-2\nu)$ and is given by

$$r_0\left(\frac{\Delta W}{\Delta V}\right)_{\max} = \frac{(1+\nu)(1-\nu)^2}{2E}\sigma^2 a. \tag{8.18}$$

Based on the definitions of $(\Delta W/\Delta V)_v$ and $(\Delta W/\Delta V)_d$ in equations (8.2) and (8.3), equation (8.18) may be separated into

$$r_0\left(\frac{\Delta W}{\Delta V}\right)_v = \frac{(1+\nu)^2(1-\nu)(1-2\nu)}{3E}\sigma^2 a \tag{8.19}$$

for dilatation and

$$r_0\left(\frac{\Delta W}{\Delta V}\right)_d = \frac{(1-\nu^2)(1-\nu+4\nu^2)}{6E}\sigma^2 a \tag{8.20}$$

for distortion. Since $(\Delta W/\Delta V)_d > (\Delta W/\Delta V)_v$ as given by equations (8.19) and (8.20) corresponds to yielding in the areas to the sides of the crack at $\theta_0 = \pm\cos^{-1}(1-2\nu)$.

Additional light may be shed on yielding and fracture by studying the variation of $(\Delta W/\Delta V)_v/(\Delta W/\Delta V)_d$* in the vicinity of the crack tip. Figure

---

* The gradient of this ratio is related to the determination of flat and slant fracture [9].

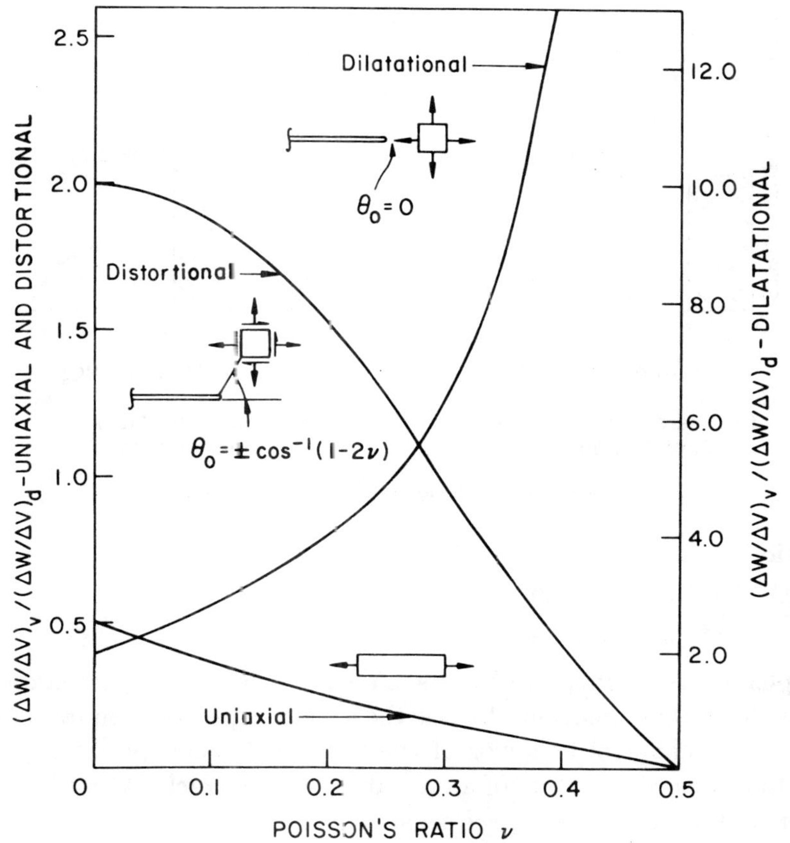

Figure 8.6. Variations of strain energy density ratio with Poisson's ratio.

8.6 gives a plot of this quantity with Poisson's ratio ν. For an element situated at $\theta_0 = 0$, equations (8.16) and (8.17) yield

$$\frac{(\Delta W/\Delta V)_v}{(\Delta W/\Delta V)_d} = \frac{2(1+\nu)}{1-2\nu} \tag{8.21}$$

which takes a minimum value of 2.0 at $\nu = 0.0$ and increases rapidly as ν increases. Along the directions of yielding, however, the ratio

$$\frac{(\Delta W/\Delta V)_v}{(\Delta W/\Delta V)_d} = \frac{2(1+\nu)(1-2\nu)}{1-\nu+4\nu^2} \tag{8.22}$$

acquires a maximum value of 2.0 at $\nu = 0.0$ and decreases to zero at $\nu = 0.5$. The deviation between the curves becomes appreciable as ν is raised beyond 0.2.

For a uniaxial tensile specimen with no crack, it can be shown that

$$\left(\frac{\Delta W}{\Delta V}\right)_v = \frac{1-2\nu}{6E}\sigma^2, \quad \left(\frac{\Delta W}{\Delta V}\right)_d = \frac{1+\nu}{3E}\sigma^2. \tag{8.23}$$

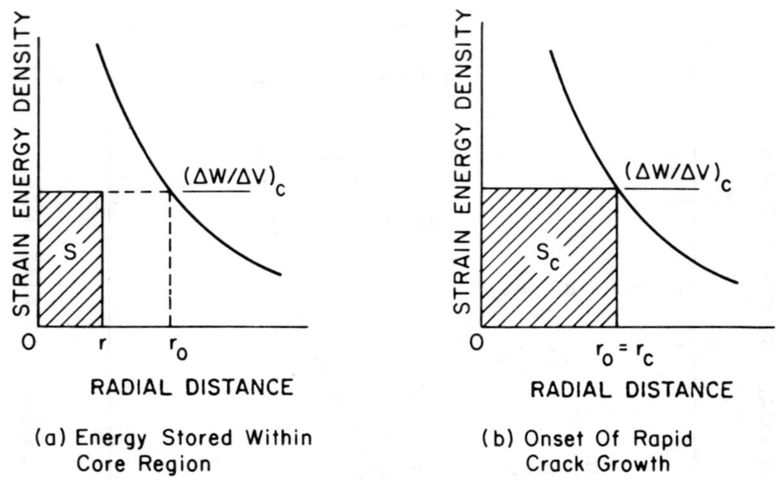

Figure 8.7. Critical strain energy density.

The ratio

$$\frac{(\Delta W/\Delta V)_v}{(\Delta W/\Delta V)_d} = \frac{1-2\nu}{2(1+\nu)} \tag{8.24}$$

is also plotted as a function of Poisson's ratio in Figure 8.6 for comparison purpose. Note that equation (8.24) is the exact reciprocal of equation (8.21). Hence, the mechanical behavior of the uniaxial tensile specimen does not appear to represent the state of affairs ahead of the crack. There is the need for improved design in test specimens.

Critical ligament. The circle of radius r_0 in Figure 8.5 has been referred to as the core region. Depending on the local stress state, r_0 may or may not coincide with the critical ligament, say r_c, that corresponds to the onset of unstable crack extension. For simplicity sake, let $r_0 = r_c$ as illustrated in Figure 8.7. Material failure is initiated in the core region as soon as the strain energy density level reaches $(\Delta W/\Delta V)_c$, Figure 8.7(a). However, rapid crack growth will not begin until the area under the $\Delta W/\Delta V$ versus r curve reaches S_c, Figure 8.7(b). For fracture specimens with load applied perpendicular to the crack, it can be easily shown that

$$S_c = r_c\left(\frac{\Delta W}{\Delta V}\right)_c = \frac{(1+\nu)(1-2\nu)K_{1c}^2}{2\pi E}. \tag{8.25}$$

An approximate estimate of r_c can be made by relating the fracture test result in equation (8.25) to the tensile yield strength σ_{ys}. The distortional strain energy density reduces to

$$\left(\frac{\Delta W}{\Delta V}\right)_d = \frac{1+\nu}{3E}\sigma_{ys}^2. \tag{8.26}$$

Table 8.2. Material parameters for 4140 steel.

σ_{ys} (ksi)	K_{1c} (ksi$\sqrt{\text{in.}}$)	S_c (lb/in.)	r_c (in.)
220	25	3.379	0.00065
205	52	7.459	0.00164
180	80	17.655	0.00510
165	120	39.724	0.01345

Substituting equations (8.25) and (8.26) into (8.4) and rearranging the result, an expression for r_c is derived:

$$r_c = \frac{3(1-2\nu)}{2\pi}\left(\frac{K_{1c}}{\sigma_{ys}}\right)^2\left[1+\frac{(\Delta W/\Delta V)_v}{(\Delta W/\Delta V)_d}\right]^{-1}. \tag{8.27}$$

Equation (8.27) shows that r_c depends not only on K_{1c} and σ_{ys} but also on the local stress state through the ratio $(\Delta W/\Delta V)_v/(\Delta W/\Delta V)_d$. Several values of r_c for the 4140 steel with $\nu = 0.3$ are calculated and tabulated in Table 8.2 together with other material parameters. A value of $(\Delta W/\Delta V)_v/(\Delta W/\Delta V)_d$ equal to 6.5 is used and is obtained from Figure 8.6 for $\nu = 0.3$.

The quantity $(\Delta W/\Delta V)_c$ may be obtained independently from the area under the true stress and strain curve or calculated from S_c and r_c by means of equation (8.12).

Data interpretation. The fundamental physical principles that govern material failure have not been well-established over the years. The value of material and fracture testing is uncertain where physical interpretation of data cannot be made possible. More emphasis must be placed on a combination of theory and experimentation. Otherwise, no confidence can be placed on judging the reliability of new results.

A frequent problem of interest is the failure of specimens under combined loading where both normal stress σ and shear stress τ are present. According to the maximum principal stress criterion, the value of σ and τ at failure, when normalized by the ultimate strength σ_u, can be written as

$$\left(\frac{\sigma}{\sigma_u}\right)^2 + \left(\frac{\tau}{\sigma_u}\right)^2 = 1 \tag{8.28}$$

while the maximum total strain energy density criterion renders

$$\left(\frac{\sigma}{\sigma_u}\right)^2 + 2(1+\nu)\left(\frac{\tau}{\sigma_u}\right)^2 = 1. \tag{8.29}$$

at failure. The choice of selecting equations (8.28) and (8.29) depends on data which may often require fitting an empirical equation of the form

$$A\left(\frac{\sigma}{\sigma_u}\right)^m + B\left(\frac{\tau}{\sigma_u}\right)^n = 1. \tag{8.30}$$

The parameters A, B, m and n for a particular material are determined from experimental data. It is desirable to consider these parameters as material constants and to attribute apparent changes in material behavior* to the presence of defects or differences in loading and/or specimen geometry.

Internal defect. Suppose that the loading and specimen geometry in a given test are fixed but the material may contain initial defects. The specific case of a cylindrical bar subjected to combined tension and torsion will be considered for evaluating the effect of initial defects on the parameters in equation (8.30).

Let the cylindrical bar of radius b in Figure 8.8 contain a flat circular defect of radius a. A force P and torque T are applied such that the bar experiences both a nominal normal stress $\sigma = P/\pi b^2$ and a nominal shear stress $\tau = 2Ta/\pi b^4$. The corresponding stress intensity factors are** [10]

$$k_1 = \frac{\sigma}{1-(a/b)^2}\sqrt{\frac{ac}{b}}\, f_1\left(\frac{a}{b}\right) \tag{8.31a}$$

$$k_3 = \frac{\tau}{1-(a/b)^4}\sqrt{\frac{ac}{b}}\, f_3\left(\frac{a}{b}\right) \tag{8.31b}$$

in which the functions $f_1(a/b)$ and $f_3(a/b)$ stand for

$$f_1\left(\frac{a}{b}\right) = \frac{2}{\pi}\left(1+\frac{1}{2}\frac{a}{b}-\frac{5}{8}\frac{a^2}{b^2}+0.268\frac{a^3}{b^3}\right) \tag{8.32a}$$

$$f_3\left(\frac{a}{b}\right) = \frac{4}{\pi}\left(1+\frac{1}{2}\frac{a}{b}+\frac{3}{8}\frac{a^2}{b^2}+\frac{5}{16}\frac{a^3}{b^3}-\frac{93}{128}\frac{a^4}{b^4}+0.038\frac{a^5}{b^5}\right). \tag{8.32b}$$

Referring to [7], the strain energy density factor S for mixed mode crack extension with $k_2 = 0$ is given by

$$S = a_{11}k_1^2 + a_{33}k_3^2 \tag{8.33}$$

where the coefficients are given by

$$a_{11} = \frac{1+\nu}{8E}(3-4\nu-\cos\theta)(1+\cos\theta) \tag{8.34a}$$

$$a_{33} = \frac{1+\nu}{2E}. \tag{8.34b}$$

---

* Parameters associated with mechanical behavior should be clearly distinguished from those material properties governed by the chemical composition and microstructure.

** The critical value k_{1c} differs from the conventional fracture toughness value K_{1c} by a factor $\sqrt{\pi}$, i.e., $K_{1c} = \sqrt{\pi}\, k_{1c}$.

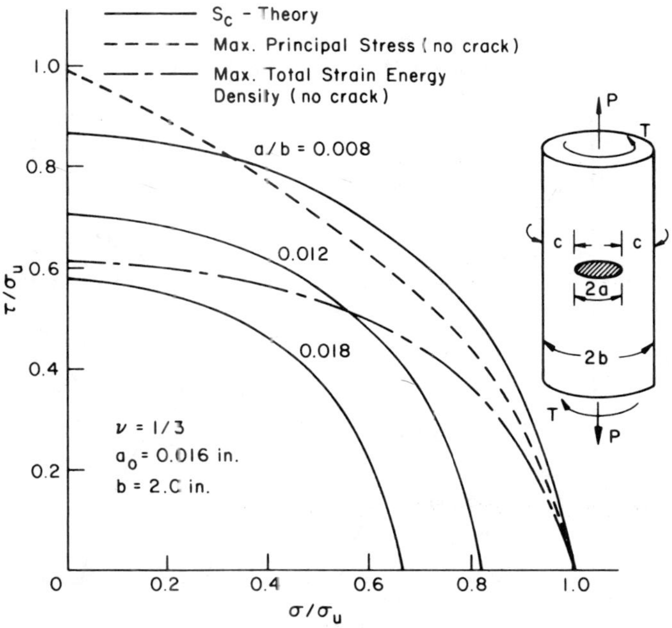

Figure 8.8. Cylindrical bar with an internal crack subjected to tension and torsion.

The angle θ that makes S a minimum is found by taking $\partial S/\partial\theta = 0$. Letting $S_{\min} = S_c$ and using equation (8.31), it is found that

$$(1-2\nu)f_1^2\left(\frac{a}{b}\right)\sigma^2 + \frac{f_3^2(a/b)}{[1+(a/b)^2]^2}\tau^2 = \frac{2Eb[1-(a/b)^2]^2 S_c}{ac(1+\nu)}. \tag{8.35}$$

In view of equation (8.25), the fracture toughness S_c can be determined from a K_{1c} fracture test:

$$K_{1c} = \frac{\sigma_u}{1-(a_0/b)^2}\sqrt{\frac{\pi a_0 c}{b}}\, f_1\left(\frac{a_0}{b}\right) \tag{8.36}$$

in which a_0 is the crack radius at instability. With the aid of equations (8.25) and (8.36), equation (8.35) becomes

$$\frac{f_1^2(a/b)}{f_1^2(a_0/b)}\left(\frac{\sigma}{\sigma_u}\right)^2 + \frac{[1+(a/b)^2]^{-2}}{1-2\nu}\,\frac{f_3^2(a/b)}{f_3^2(a_0/b)}\left(\frac{\tau}{\sigma_u}\right)^2 = \frac{a_0 c_0[1-(a/b)^2]^2}{ac[1-(a_0/b)^2]^2}. \tag{8.37}$$

Equation (8.37) may be rearranged and cast into the form shown by equation (8.30) with $m = n = 2$. The parameter A is a complicated function of defect size and cylinder geometry and does not depend on material constants. Poisson's ratio ν is the only material constant in the expression for B.

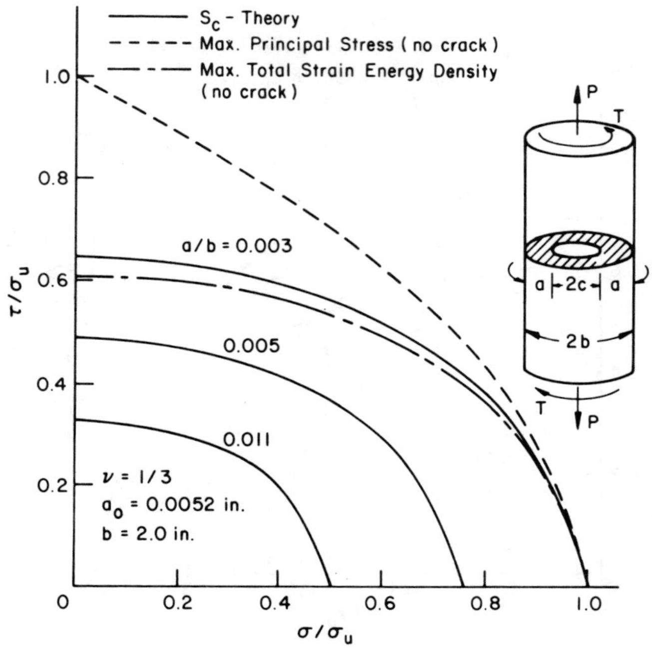

Figure 8.9. Cylindrical bar with an external crack subjected to tension and torsion.

A graphical display of τ/σ_u versus σ/σ_u is given by Figure 8.8 for $a_0 = 0.016$ in., $b = 2.0$ in. and a material with $\nu = 1/3$. The three solid curves labelled $a/b = 0.008$, 0.012 and 0.018 are calculated from equation (8.37) and they are plotted next to the dotted curves found from equations (8.28) and (8.29) with no consideration given to the presence of initial defects. The relation between τ/σ_u and σ/σ_u is seen to depend sensitively on the size of defects. Therefore defects may contribute significantly to data scattering if they are not accounted for in the analysis.

Surface defect. A cylindrical bar with a surface defect in the form of a circumferential crack may be treated by the same procedure. The configuration of the cracked bar is shown in Figure 8.9. Under combined tension and torsion, there arise the stress intensity factors [10]

$$k_1 = \sigma\left(\frac{b}{c}\right)^2\sqrt{\frac{ac}{b}}\,g_1\left(\frac{c}{b}\right) \tag{8.38a}$$

$$k_3 = \tau\left(\frac{b}{c}\right)^3\sqrt{\frac{ac}{b}}\,g_3\left(\frac{c}{b}\right). \tag{8.38b}$$

The functions $g_1(c/b)$ and $g_3(c/b)$ take the forms

$$g_1\left(\frac{c}{b}\right) = \frac{1}{2}\left[1 + \frac{1}{2}\frac{c}{b} + \frac{3}{8}\frac{c^2}{b^2} - 0.363\frac{c^3}{b^3} + 0.731\frac{c^4}{b^4}\right] \tag{8.39a}$$

$$g_3\left(\frac{c}{b}\right) = \frac{3}{8}\left[1 + \frac{1}{2}\frac{c}{b} + \frac{3}{8}\frac{c^2}{b^2} + \frac{5}{16}\frac{c^3}{b^3} + \frac{25}{128}\frac{c^4}{b^4} + 0.208\frac{c^5}{b^5}\right]. \tag{8.39b}$$

As in the previous example, the S factor obtained from equation (8.33) by means of equations (8.38) may be minimized with respect to θ in the coefficients a_{ij} $(i,j = 1,3)$. The resulting expression is

$$(1-2\nu)g_1^2\left(\frac{c}{b}\right)\sigma^2 + \left(\frac{b}{c}\right)^2 g_3^2\left(\frac{c}{b}\right)\tau^2 = \frac{2Eac^3S_c}{b^3(1+\nu)}. \tag{8.40}$$

With K_{1c} known as

$$K_{1c} = \sigma_u\left(\frac{b}{c_0}\right)^2\sqrt{\frac{\pi a_0 c_0}{b}}\,g_1\left(\frac{c_0}{b}\right) \tag{8.41}$$

S_c in equation (8.40) may be eliminated using equation (8.25) to yield the failure envelope,

$$\frac{g_1^2(c/b)}{g_1^2(c_0/b)}\left(\frac{\sigma}{\sigma_u}\right)^2 + \frac{(b/c)^2}{1-2\nu}\frac{g_3^2(c/b)}{g_1^2(c_0/b)}\left(\frac{\tau}{\sigma_u}\right)^2 = \frac{a_0c^3}{ac_0^3}. \tag{8.42}$$

Equation (8.42) is evaluated numerically for $a_0 = 0.0052$ in., $b = 2.0$ in. and $\nu = 1/3$. The results are displayed in Figure 8.9 for $a/b = 0.003$, 0.005 and 0.011. The curves in Figure 8.9 for a surface defect are quite different from those in Figure 8.8 for an internal defect. Both the size and type of defect can affect the combination of σ and τ leading to fracture. This may explain why conventional failure criteria cannot consistently correlate fracture data.

8.4. Incremental crack growth

Crack growth as a continuous motion is thinkable only in mathematical terms. It is not a precise description because of the discrete character of nature. Physically speaking, what is perceived as a continuous motion is actually a series of ordered states of rest. The interval between these states is not necessarily constant. Depending on the rate of energy release, the increment of length in each successive stage of crack growth can either increase or decrease. The former may lead to unstable crack propagation and the latter to crack arrest. Another commonly observed phenomenon of ductile fracture is the transition from stable to unstable cracking. The quantitative assessment of such behavior requires not only a knowledge of the availability of energy for creation of free surface but also a criterion that determines

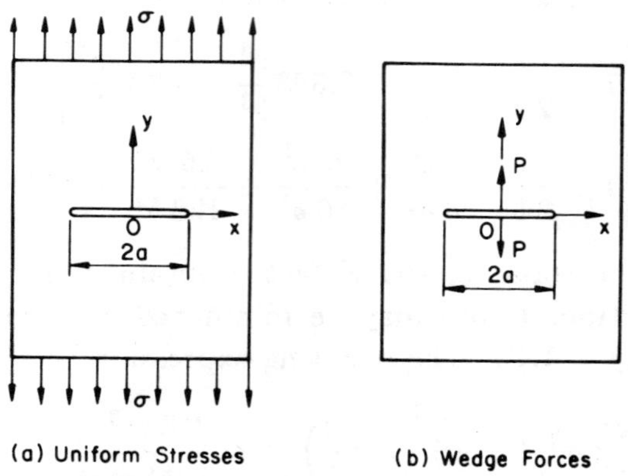

Figure 8.10. Cracked specimens for (a) unstable crack growth and (b) stable crack growth.

the rate of incremental crack growth and shape of crack profiles which are needed for experimental verification.

For illustration, the cracking behavior of the two specimens in Figure 8.10 will be considered. When a crack is completely engulfed in a uniform stress field as shown in Figure 8.10(a), the energy available to create new free surfaces increases as the crack expands. For the case of a pair of wedge forces applied to the crack surfaces in Figure 8.10(b), cracking once initiated tends to slow down because the energy available to drive the crack decreases.

Monotonically increasing crack growth. Let the crack motion of the specimen in Figure 8.11(a) be regarded as a series of finite segments r_1, r_2, etc., of crack growth. The strain energy density factor S is proportional to $\sigma_c^2 a$, where σ_c is the critical stress level* maintained after crack initiation. Hence, the strain energy density factors S_1, S_2, etc., will increase with each increment of growth r_1, r_2, etc., i.e.,

$$S_1 < S_2 < \cdots < S_j < \cdots < S_c \tag{8.43a}$$

$$r_1 < r_2 < \cdots < r_j < \cdots < r_c. \tag{8.43b}$$

It follows from the growth criterion in equation (8.13) that

$$\frac{1}{r_1} = \frac{a + r_1}{r_2} = \frac{a + r_1 + r_2}{r_3} = \cdots = \text{const.} \tag{8.44}$$

If crack initiation starts when $r_1 = r_0$, then a recursion relation for incre-

* An overload is inherent in the process of fracture initiation where the crack tip tends to rest against a stronger grain which causes an abrupt increase in energy demand. This is followed by a sudden excess of energy available for release when the local material inhomogeneity is overcome.

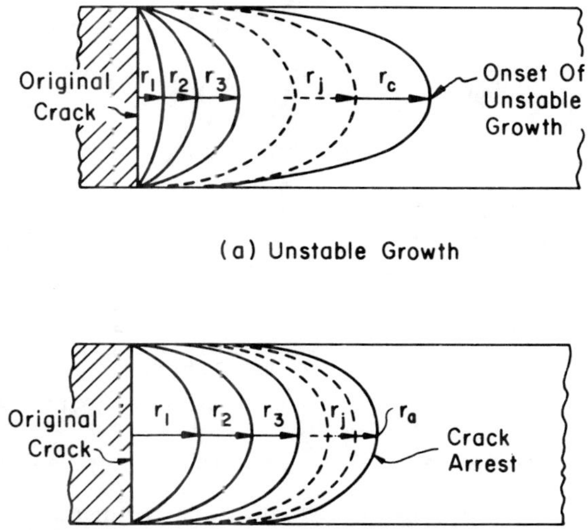

(a) Unstable Growth

(b) Stable Growth

Figure 8.11. Incremental crack growth leading to (a) instability and (b) crack arrest.

mental crack advancement can be obtained:

$$r_{n+1} = \left(1 + \frac{r_0}{a}\right) r_n, \quad n \geq 1. \tag{8.45}$$

Each consecutive segment of growth is seen to increase and unstable fracture occurs when r_j reaches the critical ligament r_c. In reality, the crack front acquires a curvature across the specimen thickness with the maximum extension taking place in the mid-plane as illustrated in Figures 8.11. The growth increments $r_1, r_2, \cdots, r_j, \cdots, r_c$ in equation (8.45) correspond to the maximum values on each crack profile, Figure 8.11(a).

Monotonically decreasing crack growth. Refer to the crack loaded by wedge forces in Figure 8.10(b). The crack begins to move by segments r_1, r_2, etc., with decreasing strain energy density S_1, S_2, etc., because S is inversely proportional to the half crack length a, i.e., $S \sim P_c^2/a$ with P_c being the critical load. The conditions

$$S_1 > S_2 > \cdots > S_j > \cdots > S_a \tag{8.46a}$$

$$r_1 > r_2 > \cdots > r_j > \cdots > r_a \tag{8.46b}$$

imply that the fracture process leads to eventual crack arrest as r_j approaches r_a. Since $(\Delta W/\Delta V)_c = \text{const.}$ during the crack growth process, equation (8.13) requires that

$$\frac{1}{ar_1} = \frac{1}{(a+r_1)r_2} = \frac{1}{(a+r_1+r_2)r_3} = \cdots = \text{const.} \tag{8.47}$$

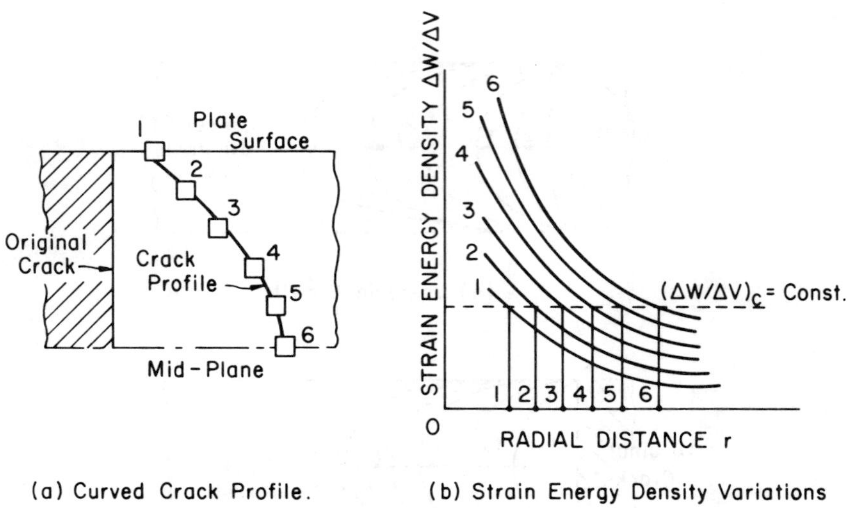

Figure 8.12. Crack growth development.

which for $r_1 = r_0$ gives the expression

$$r_{n+1} = \frac{r_n}{1 + (r_n/r_0)(r_n/a)}, \quad n \geq 1. \tag{8.48}$$

Figure 8.11(b) shows the curved crack profiles through the specimen thickness and the decreasing increments of maximum crack growth $r_1, r_2, \ldots, r_j, \ldots, r_a$ at the mid-plane.

Development of crack profile. It is useful to construct a procedure for developing crack profiles so that experiments can be devised to check the validity of the theory. In three dimensions, all the elements located at the prospective sites of the crack profile at a given instance are assumed to satisfy the condition of constant strain energy density $(\Delta W/\Delta V)_c$ as stated by equation (8.13). The growth condition can be written more precisely as

$$\left(\frac{\Delta W}{\Delta V}\right)_c = \frac{S_1^{(i)}}{r_1^{(i)}} = \frac{S_2^{(i)}}{r_2^{(i)}} = \cdots = \frac{S_j^{(i)}}{r_j^{(i)}} = \cdots = \text{const.} \tag{8.49}$$

where $i = 1, 2, \ldots, m$ represents the number of points or elements on a crack profile and $j = 1, 2, \ldots, n$ represents the number of crack profiles required to simulate the fracture process.

The procedure for applying the constant strain energy density criterion can be illustrated by referring to a crack located in a tensile plate specimen as shown in Figure 8.12(a). Because of symmetry, consider only one-half of the plate thickness and a typical crack profile constructed by six points or elements labelled as $1, 2, \ldots, 6$. The precise locations of these elements $r_1, r_2, \ldots, r_6$ are fixed by the condition of $(\Delta W/\Delta V)_c = \text{const}$. The variations of $\Delta W/\Delta V$ with the distance measured from the crack front may be found analytically or experimentally. Figure 8.12 (b) shows the $\Delta W/\Delta V$ versus r relation for each of the six elements in Figure 8.12(a). The intersections of the line $(\Delta W/\Delta V)_c$ = const. for a given material with the curves determines the values of $r_1, r_2, \ldots, r_6$ that form the crack profile. This step by step procedure can be repeated such that a series of crack profiles are developed to describe the total fracture process from initiation to termination.

Transition from stable to unstable fracture. Instantaneous crack instability is a mathematical abstraction that has no physical counterpart. All fracture processes may be regarded as transitions from stable* to unstable crack propagation. In particular, the commonly observed phenomenon of ductile fracture in structural components is characterized by slow and stable cracking prior to rapid crack propagation. This is normally associated with a nonlinear response on the load versus deformation curve and is attributed to nonself-similar crack growth and material deformation beyond the elastic limit. These two effects occur simultaneously under load. There are no known experimental methods that can separate them and measure their individual contributions. It is inaccurate and incorrect to simply identify the nonlinear behavior of metal alloys with plasticity. The presence of a slowly growing crack also contributes to nonlinearity. Nonlinearity is controlled by the rate of loading, specimen size and geometry, crack growth pattern and material properties and is not a well-understood subject for metals. It depends on the scale level, size of physical domain, and time interval on the basis of which the response is recorded. Yielding of the grains in a polycrystal does not necessarily correspond to yielding of polycrystalline specimens measured by a tensile test. Comparison of physically recorded results on the different scale levels should be made with the utmost care. Quantities such as microstress and macrostress, which are frequently interchanged in the literature, must be distinguished in the analysis.

Deformation and fracture are unique in that they occur in an orderly fashion dictated by the laws of nature. Therefore, it is only logical that man-made measurements should be identified with physical events. Otherwise, the results are neither meaningful nor helpful for gaining new experience. This is particularly true in conventional material and fracture tests relative to the initiation of yielding and fracture. It is not clear how these quantities could be used to predict the performance of the materials tested when operated under conditions that are different from those in the laboratory. It is fundamental knowledge that a nonlinear process is path dependent and

---

* The details of stable fracture initiation require a consideration of material damage at the microscopic level.

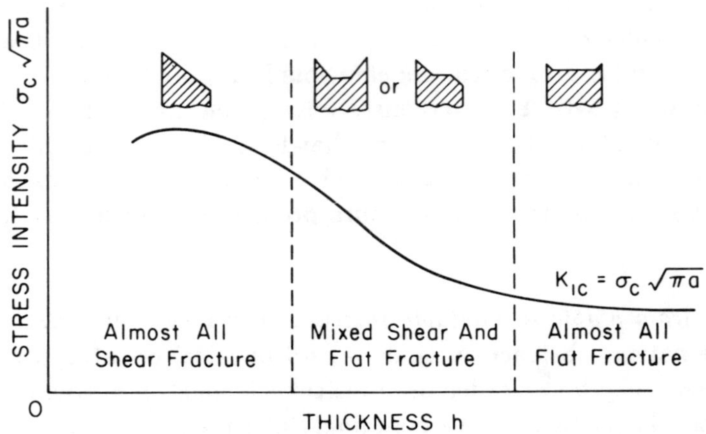

Figure 8.13. **Size effect for a cracked tensile specimen.**

so must be identified with material damage at each stage of loading on the load–deformation curve. Such capability must be established theoretically and verified experimentally to resolve the size effect* problem.

It is well known that the deformation and fracture pattern in a material can be controlled by the rate of loading, specimen size and temperature. However, it would be overwhelming to discuss failure modes corresponding to all possible combinations of the variables. To focus ideas, only the specimen thickness h will be altered. The specimen size effect is customarily exhibited by a plot of the critical stress σ_c times the square root of the half crack length a and π, i.e., $\sigma_c\sqrt{\pi a}$,** against the plate thickness h, Figure 8.13(a). Only when $\sigma_c\sqrt{\pi a}$ ceases to change for h sufficiently large is the quantity $K_{1c} = \sigma_c\sqrt{\pi a}$ a measure of the fracture toughness of the material. In this case, the excess available energy produces fracture surfaces that are predominantly flat. As h is decreased, $\sigma_c\sqrt{\pi a}$ loses its physical meaning and should no longer be referred to as fracture toughness or stress intensity factor. The fact that ductile fracture is accompanied by an increase in the load carrying capacity of the specimen should not be interpreted as a change in fracture toughness. First of all, yielding affects only the material behavior on a load deformation diagram and does not constitute a change in material failure properties. On physical grounds, if the fracture toughness is interpreted as characterizing the resistance of the material to fracture, it

---

* The conventional approach of measuring a so-called 'fracture toughness' parameter for each different specimen thickness is not sound and does not provide any predictive capacity. It only confirms that the measured parameter cannot be adequately used as a material constant.

** Because the stress intensity factor concept is restricted to flat fracture, the measurement of K_{1c} requires the specimen to be overly thick so that a large amount of energy can be stored and released to initiate the onset of rapid crack propagation.

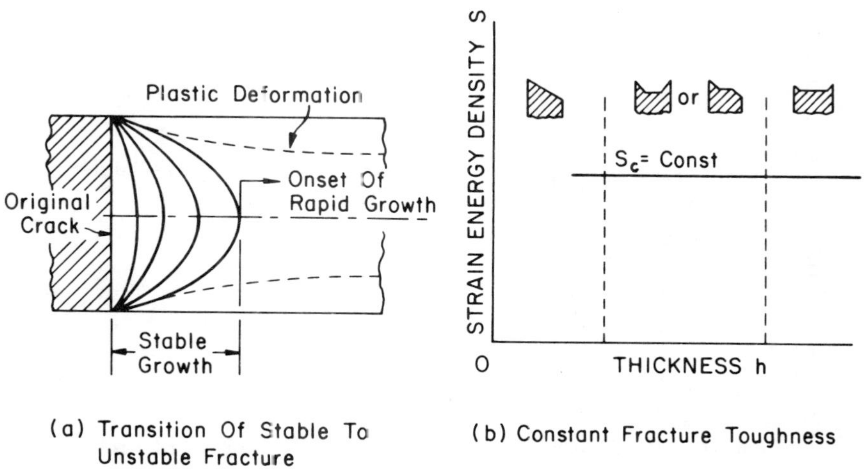

Figure 8.14. Ductile fracture in monotonically loaded tensile specimen.

should be a constant with or without yielding. Because K_{1c} is specimen size sensitive, it is more consistent to regard it simply as a measure of incipient rapid fracture rather than an inherent material parameter. Plastic deformation or yielding reduces the amount of energy available to cause unstable fracture. Ductile fracture can only be explained and resolved when the interaction between deformation and fracture are properly accounted for. The nature of this path dependent process must be understood before experiments can be designed. Reference can be made to [11, 12] for a more detailed discussion on this subject.

An inherent characteristic of ductile fracture is that both flat and slant fracture surfaces are produced. The strain energy density criterion [7, 8] has been shown in many previous problems to be well-suited for analyzing mixed mode fracture where the path of crack extension is not known as *a priori*. It is informative to visualize the deformation and fracture history in a tensile specimen with thickness h, Figure 8.10(a). A crack of length $2a$ with a straight edge lies in a plane normal to the applied stress σ. Figure 8.14(a) shows the crack profiles during the stage of slow and stable growth while the material near the plate surfaces is deformed beyond the yield point. The local plastic deformation constrains crack growth and delays the start of unstable crack propagation. This problem has been analyzed [13] for the ductile fracture of plates with different thicknesses. The critical strain energy density factor or fracture toughness S_c was held constant, Figure 8.14(b), for several plate thickness to crack length ratio. As the plate thickness is continuously decreased, deformation becomes the dominant mode of failure and S_c fails to be applicable. This is why the constant S_c line has not been extended beyond the value of h below which fracture instability no

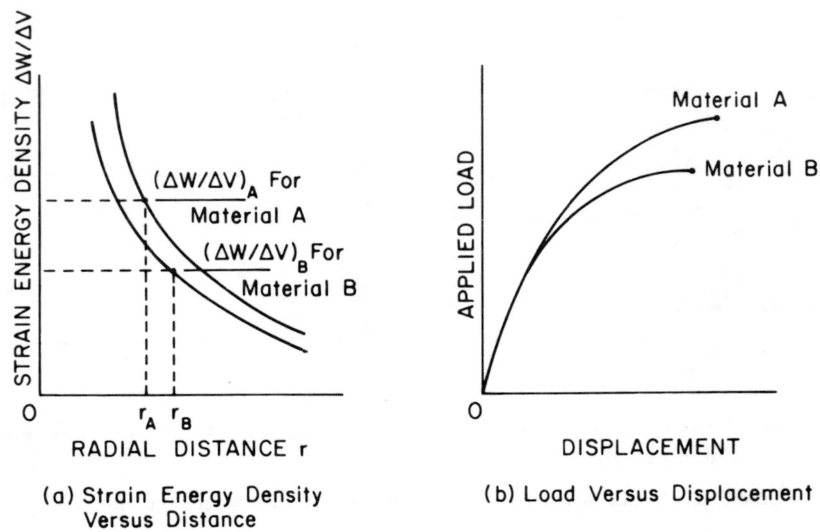

Figure 8.15. Relationship between toughness and critical load for two materials.

longer occurs.

The transition from stable to unstable fracture is controlled by the critical ligament size r_c and fracture toughness measured by S_c which is proportional to $(\Delta W/\Delta V)_c$. Figures 8.15 give a comparison of $(\Delta W/\Delta V)_c$, S_c and critical load for two different materials A and B. Material A is said to be tougher than material B, Figure 8.15(a), if

$$\left(\frac{\Delta W}{\Delta V}\right)_A > \left(\frac{\Delta W}{\Delta V}\right)_B, \quad \left(\frac{\Delta W}{\Delta V}\right)_A r_A > \left(\frac{\Delta W}{\Delta V}\right)_B r_B. \tag{8.50}$$

The tougher material may also sustain a higher critical load as shown in Figure 8.15(b).

The above procedure gives a complete description of deformation* and fracture, Figure 8.14(a), as the material is loaded incrementally. Each point on the load–deformation curve, Figure 8.15(b), can be identified with material damage such that the critical load corresponds to satisfying the condition $S_c = r_c(\Delta W/\Delta V)_c$. Allowable load and net section size, which are the basic information needed in engineering design, can thus be found for structural members undergoing ductile fracture. The method and analytical results in [13] serve as a guide to experimentally identify each point of the load–deformation diagram with material damage. Such analytical capability when verified by experiments will place increased confidence in translating laboratory data for use in the design of larger size structural members.

---

* In a continuum analysis, microscopic damage of the material enters into the problem through plastic deformation [13].

8.5. Fatigue crack growth: a path-dependent process

The behavior of metals in fatigue is a complex process which is not yet properly understood despite the enormous amount of data generated by testing. Admittedly, much has been done to examine fracture test specimens at the microscopic level by identifying cracks at the earliest possible stage and following their subsequent growth. Many of the results have unavoidably been couched in terms of microscopic entities and their interactions: e.g., dislocation, vacancies, grain inhomogeneity, inclusions, voids, etc. Although these studies offer a detailed description of the mechanisms fatigue failure, little insight will be gained unless the results are tied together by theory.

Fatigue is normally separated into initiation* and propagation. The necessity for addressing these two processes** individually arises because the theory is not able to bridge the gap between physical events observed at the microscopic and macroscopic scale level. Unless efforts are made to relate macroscopic variables with microscopic entities, the fatigue process will not be completely understood.

No attempt will be made to explain all qualitative observations, for this would lead to no useful results. The objective is to develop a procedure within the framework of continuum mechanics such that experimental data can be interpreted with order and consistency. As in the case of monotonic loading, there is a need to infuse fatigue crack growth data with results obtained from materials testing.

Fatigue specimen. Unlike monotonically loaded tensile specimens, the load amplitude is maintained at a level such that the specimen will not break on the first peak load but will eventually break after many cycles of loading. The implication is that damage is being accumulated in the material. A stress amplitude versus number of cycles to failure diagram is often obtained for determining the endurance limit† of a particular material. The loading usually has constant amplitude and frequency while the mean stress level can be changed to determine its effect on the fatigue life.

A basic assumption of continuum mechanics is that the mechanical properties of a material can be measured from a test specimen, Figure 8.16(a). This notion applies to cyclic loading as well where the material now experiences reversal of stress and strain forming hysteresis loops are shown in

---

* The mechanism of fatigue initiation has been attributed to the dislocation process of extrusion and intrusion and will not be discussed in this communication.

** Fatigue initiation and propagation are intimately related and cannot be justified as two separate processes on the basis of difficulties arising from theoretical treatment.

† The endurance limit is defined as the stress amplitude corresponding to that portion of the curve which no longer changes with the number of cycle of loading. Not all metals possess this property.

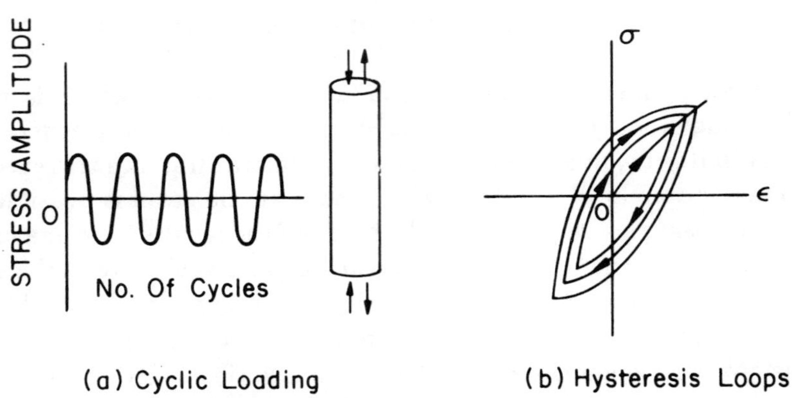

Figure 8.16. Material testing specimen in fatigue.

Figure 8.16(b). The object is to predict the global response of a solid by assuming that the material properties of all the local elements within the solid can be determined from material testing. As mentioned previously for materials subjected to monotonic loading, there are difficulties associated with this procedure because of the well-known size effect. The uniaxial test specimen may not accurately simulate the failure behavior ahead of a crack. This was discussed in terms of the strain energy density ratio $(\Delta W/\Delta V)_v/(\Delta W/\Delta V)_d$ which is a more sensitive measure of material failure due to deformation and fracture. The fundamental problem of specimen size effect must also overcome in fatigue.

Cracked specimen. Fracture testing considers a larger size specimen with an initial macrocrack. The case of a transverse crack* in a plate has been used extensively for collecting crack propagation data. The standard cyclic loading has constant amplitude and frequency, Figure 8.17. The uniaxial stress amplitude and number of cycles applied to the specimen at distances remote from the crack shall be denoted by Σ and N, respectively. An element of material ahead of the crack, however, is in a multiaxial stress state where the amplitude and frequency response of a typical stress component, say σ, may, in general, be different from the input. This is illustrated in Figure 8.18(a) with n being the number of cycles experienced by the local element. The influence of phase shift between the applied load spectrum (Σ, N) and material response (σ, n) should not be overlooked.** A knowl-

* When the material deforms plastically, the specimen can no longer be considered as two-dimensional as the most important aspects of material damage occur in the plate thickness direction.

** The problem of an elastic plate with an angle crack subjected to cyclic uniaxial loading has been solved [14]. The stress intensity factors k_1 and k_2 are found to depend on the frequency of loading.

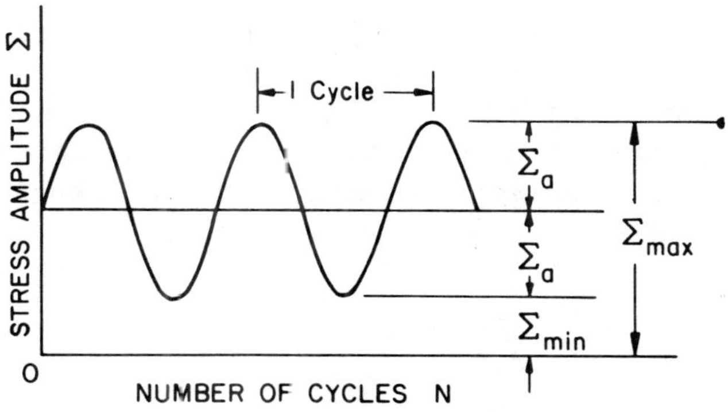

Figure 8.17. Cyclic loading applied to a crack specimen.

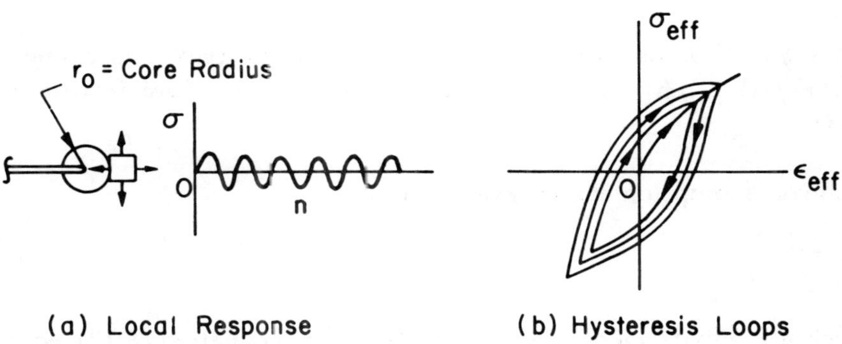

Figure 8.18. Response of element ahead of crack.

edge of the cyclic response on the local element is pertinent to testing the fatigue properties of metals. Fatigue, as a path-dependent process, depends on the load history. Hence, the applied cyclic loads on the test specimen in Figure 8.16(a) cannot be chosen at will.

Damage accumulation. Under fatigue, the material microstructure is damaged before any failure can be detected or described as a process of macrocrack propagation. It is in this sense that material damage is said to be accumulated.* A convenient measure of this accumulation can be expressed

* The terminology 'damage accumulation' should not imply a direct correspondence between fatigue life and number of load cycles. An occasional high positive overload is known to increase the fatigue life of cracked specimens. Nonlinear behavior of materials is not always intuitively obvious.

in terms of the effective* stress σ_{eff} and strain ϵ_{eff} from which the hysteresis energy density, $\Delta W/\Delta V$, can be computed, Figure 8.18(b).

Suppose that an estimate of the additional hysteresis energy density accumulation in the material element ahead of the crack can be made for each cycle, then a saturation point will eventually be reached at which the element is broken and the macrocrack advances. The amount of growth will depend on the rate of energy release which cannot be arbitrarily assumed. Let the total damage be represented as an average,

$$\sum_{j=1}^{n}\left(\frac{\Delta W}{\Delta V}\right)_j = \left(\frac{\Delta W}{\Delta V}\right)_{\text{ave}} \Delta n \tag{8.51}$$

where n is the number of cycles for the element to break. In equation (8.51), an average hysteresis energy density $(\Delta W/\Delta V)_{\text{ave}}$ is defined in terms of the interval number of cycles Δn. For simplicity, assume that the energy density datum line corresponds to zero. A fatigue crack growth hypothesis can be stated:

Crack growth is assumed to occur when the total hysteresis energy density $(\Delta W/\Delta V)_{\text{ave}}\Delta n$ reaches a critical value that is characteristic of the material.

The above assumption can be expressed as

$$\left(\frac{\Delta W}{\Delta V}\right)_{\text{ave}} \Delta n = A \tag{8.52}$$

in which A is a material constant. With the help of equation (8.12), equation (8.52) can be put into the form

$$\left(\frac{\Delta W}{\Delta V}\right)_{\text{ave}} = \frac{\Delta S}{\Delta r} = \frac{A}{\Delta n}. \tag{8.53}$$

If the load is applied symmetrically with respect to the crack plane, $\Delta r = \Delta a$ and an expression for $\Delta a/\Delta N$ is obtained:

$$\frac{\Delta a}{\Delta N}\left(\frac{\Delta N}{\Delta n}\right) = B(\Delta S). \tag{8.54}$$

---

* The effective stress and strain can be defined in terms of the principal stresses σ_1, σ_2 and σ_3 and the principal strains ϵ_1, ϵ_2 and ϵ_3, i.e.,

$$\sigma_{\text{eff}} = \tfrac{1}{3}[(\sigma_1-\sigma_2)^2+(\sigma_2-\sigma_3)^2+(\sigma_3-\sigma_1)^2]$$
$$\epsilon_{\text{eff}} = \tfrac{1}{3}[(\epsilon_1-\epsilon_2)^2+(\epsilon_2-\epsilon_3)^2+(\epsilon_3-\epsilon_1)^2].$$

Neither the constant $B = 1/A$ nor $\Delta N/\Delta n$ can be set to unity unless demonstrated by analysis or experimentation. The quantity ΔS is the change of strain energy density factor for a given increment of crack growth and must be computed for a material that dissipates energy through irreversible deformation and fracture. Unfortunately, there does not exist a suitable constitutive relation* that can realistically describe the response of metals undergoing fatigue. The theory of continuum plasticity leaves much to be desired since the yield conditions have yet to be established on a firm foundation. One of the major shortcomings of the conventional elastic–plastic crack analyses is that the yield criterion is not consistent with the criterion governing the propagation of the macrocrack. The nature of physics suggests that a single criterion should govern material damage at both the microscopic and macroscopic level.

Fracture surface appearance. Fatigue cracks grow under the conditions of a sustained mean stress $(\Sigma_{max} + \Sigma_{min})/2$ with a superimposed cyclic stress of amplitude Σ_a, Figure 8.17. The loads may also vary in amplitude depending on the service conditions. For many engineering structures, cracks are initiated very early so that most of the fatigue life is taken up with its growth.**

It has been observed in aluminum sheet specimens [15] that fatigue cracks tend to advance more rapidly in the mid-plane than at the surface. Figure 8.19 shows that fracture surfaces produced in fatigue have two distinct features, one appearing dull and the other bright. The dull regions are produced by sudden bursts of crack growth. This is dominated by the release of dilatational energy ahead of the crack and occurs relatively fast. It tends to spread more in the interior while the material at the trailing edge near the free surface is being deformed. In the bright textured areas, growth is controlled by distortion in the material near the surface† while much less growth (denoted in x_m in Figure 8.19) occurs at the mid-plane. This is because most of the energy is momentarily stored in the material near the surface layer and less in the central portion. When sufficient crack growth has taken place at the surface, say x_s, to allow an increase in energy at the center, another sudden burst of growth occurs in the interior. As growth

* Hopefully, nonlinear analyses can be performed to establish the range of validity of fatigue predictions based on results obtained from the linear theory of elasticity which is used to compute the change of stress intensity factors ΔK in many crack growth models. The conditions under which the linear analysis yields a good approximation should be clearly understood and established.

** The proportion of crack initiation and propagation life of a polycrystalline metal will depend on the microstructure of the material and surface treatment which can prolong the time before crack initiation.

† Fatigue crack growth has been related to the number of striations observed from fractured specimen surfaces.

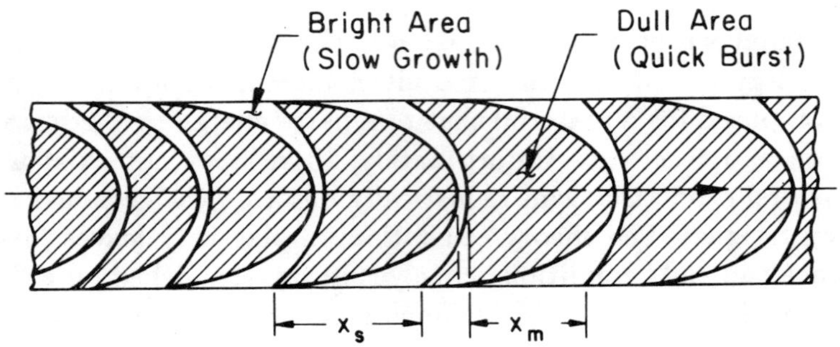

Figure 8.19. Fatigue fracture growth pattern.

continues, the crack length at the center exceeds than at the surface by increasing amounts until the specimen is completely broken.

Crack growth rate. The intermittent behavior of fatigue cracks with periodic changes from slow growth to quick bursts indicates that the shape of the crack front may be the controlling factor. The two-stage propagation process is basically the same as ductile fracture under monotonic loading characterized by slow stable crack growth prior to rapid fracture. The method for developing the crack profiles in Figure 8.19 follows the same procedure described earlier for ductile fracture.

The crack growth condition will be assumed to follow a state of critical strain energy density accumulation. The hysteresis energy density $(\Delta W/\Delta V)_j$, $j = 1, 2, \ldots, n$, ahead of the crack for each cycle of loading is counted and an increment of crack growth is assumed to occur when the critical condition in equation (8.52) is fulfilled. The quantity ΔS in equation (8.54) for each increment of crack growth is found from

$$\Delta S = S_{j+k} - S_j \tag{8.55}$$

which represents the area underneath the strain energy density versus radial distance curve, Figure 8.20. The corresponding segment of crack growth is Δr or Δa as the load is cycled from N_j to N_{j+k} with k being the number of cycles corresponding to an increment of growth Δa. The difference between the two curves in Figure 8.20 represents the amount of irreversibility within the system. An elasticity calculation will not exhibit such a character since the curves for N_j and N_{j+k} would be identical. This process is repeated and a new ΔS is calculated for each increment of crack growth. A crack growth relationship between a and N can then be formed with the initial crack length $2a_0$ being specified and the critical crack length $2a_c$ determined from the fracture toughness value K_{1c}. However, the crack path between a_0 and a_c will depend on the rate of crack growth and/or the spectrum of cyclic loading. The path-dependency nature of the fatigue crack growth process

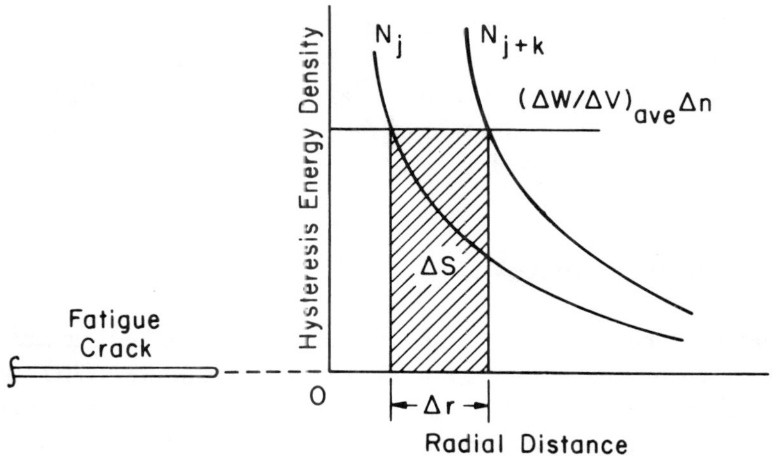

Figure 8.20. An increment of crack growth.

can be better understood by considering the following two cases:

Case (1): Determine the number of cycles for each constant increment of crack growth.

Case (2): Determine the non-uniform rate of crack growth increment for a fixed number of load cycles.

The conditions under which Cases (1) and (2) yield different or similar crack path can be examined for a variety of load spectrums and specimen geometries.

Material microstructure. Having discussed the interaction between cyclic loading and specimen geometry, a few remarks on the influence of material microstructure is in order. Needless to say, much uncertainty is encountered when attempting to analyze the conversion of available energy to irreversible material damage due to microcracks in the grains, grain boundary rumpling, creation of voids, etc. A more manageable approach would be to account for the microscopic entities through a crack tip radius of curvature which varies as the fatigue crack grows and encounters microstructure inhomogeneity. The change in the crack tip radius can be related to the grain size, local modulus and characteristic dimensions representing microdefects or inclusions. Special attention should be given to quantities that are sensitive to changes in scale such as microstress and macrostress.

Experimental data for materials with different microstructures can be collected with conventional and crack growth fatigue specimens from which the constants A, C and $\Delta a/\Delta N$ can be obtained. Once ΔS is found,

$\Delta N/\Delta n = B(\Delta S)/(\Delta a/\Delta N)$ becomes known and can give an indication of load transfer to cause damage in the material near the crack tip region and will reflect the influence of material microstructure. The success depends on a reliable determination of ΔS by theory or experiment which is by no means straightforward.

8.6. Concluding remarks

An effort has been made to define the objective of experimental fracture mechanics within the framework of continuum mechanics. Without a clear concept as to how the several types of fractures in metals can be explained and described categorically by theory, there is nothing else that remains other than to perform tests and study the conditions of fracture on phenomenological grounds. Such an approach has limited use in comparing different material behavior under the same operating conditions but does not establish predictive capability for estimating the remaining life of structural members containing defects or cracks.

Nondestructive testing methods have demonstrated the capability for locating and describing the size and shape of defects in structures. This information is incomplete unless a fracture criterion can be applied to determine the critical combinations of defect size, geometry of the structure and loading. Otherwise, no reliable inspection procedures could be developed. There is a definite need to generalize and refine the theory of fracture mechanics beyond the conventional approach of fracture testing. The discipline has the potential to resolve the fundamental problem of scaling or size effect in the field of solid mechanics.

This communication is intentionally confined to the basic aspects of deformation and fracture in metals leaving out the abundance of test data and fracture observations which often tend to prejudice the development of theory. Attention is focused on applying the strain energy density criterion, mainly because it offers a degree of quantitativeness and qualitativeness never before achieved from any single criterion. In particular, the $1/r$ dependence of the strain energy density $\Delta W/\Delta V$ offers a unique characterization of the crack tip field that is completely independent of the constitutive relations. The factor S, defined as $r(\Delta W/\Delta V)$ in equation (8.12), represents the local energy release for a segment of crack growth r. It can be related to the tearing energy [16] commonly used in measuring the rate at which energy is released by extension of a crack in rubber. Another salient feature of the criterion is that it can be most easily integrated with stress analysis for determining crack profiles and growth in three dimensions. Such capability is a prerequisite for analyzing ductile and fatigue fracture that are triaxial in nature even though the loading may be confined to a plane. The step-by-step procedures described in this chapter offer guidance to analytical and experimental work in fracture mechanics. Particular emphasis should be given to the measurement of crack front profiles in terms of parameters that control the rates of crack growth.

It can be concluded that the basic concepts of fracture mechanics are sound. Much of what remains to be done is concerned with finding the appropriate constitutive equations to account for the influence of material damage. This implies that changes in material behavior due to physical damage and alteration in microstructure and/or chemical composition should be distinguished in terms of measurable quantities. The varying degree of irreversibility needs to be quantified.

8.7. References

[1] *Linear Fracture Mechanics*, edited by G. C. Sih, R. P. Wei and F. Erdogan, Envo Publishing Co., Inc., Lehigh Valley, Pennsylvania (1976).

[2] *The Determination of Absorbed Specific Fracture Energy*, Hungarian Standard MSz 4929–76 (English Translation), Institute of Fracture and Solid Mechanics Publication, Lehigh University (1978).

[3] Gillemot, L. F., Criterion of crack initiation and spreading, *Journal of Engineering Fracture Mechanics* **8**, pp. 239–253 (1976).

[4] Czoboly, E., Havas, I. and Gillemot, F., The absorbed specific energy till fracture as a measure of the toughness of metals, *Proceedings of International Symposium on Absorbed Specific Energy and/or Strain Energy Density Criterion*, edited by G. C. Sih, E. Czoboly and F. Gillemot, Sijthoff and Noordhoff International Publishers, Alphen aan den Rijn, pp. 107–130 (1981).

[5] Sih, G. C., *Fracture toughness concept*, American Society of Testing Materials, STP **605**, pp. 3–15 (1976).

[6] De Sisto, T. S., Carr, F. L. and Larson, F. R., *Proceedings of American Society for Testing and Materials* **63**, pp. 768–779 (1963).

[7] Sih, G. C., A special theory of crack propagation: methods of analysis and solutions of crack problems, *Mechanics of Fracture I*, edited by G. C. Sih, Noordhoff International Publishing, Leyden, pp. 21–45 (1973).

[8] Sih, G. C., A three-dimensional stress energy density factor theory of crack propagation: three-dimensional crack problems, *Mechanics of Fracture II*, edited by G. C. Sih, Noordhoff International Publishing, Leyden, pp. 15–53 (1975).

[9] Sih, G. C., The analytical aspects of macrofracture mechanics, *Proceedings of International Conference on Analytical and Experimental Fracture Mechanics*, edited by G. C. Sih and M. Mirabile, Sijthoff and Noordhoff International Publishers, Alphen aan den Rijn, pp. 3–15 (1981).

[10] Sih, G. C., *Handbook of Stress Intensity Factors*, Institute of Fracture and Solid Mechanics Publication, Lehigh University, Bethlehem, Pennsylvania (1973).

[11] Sih, G. C., Elastic-plastic fracture mechanics, *Proceedings of International Conference of Prospects of Fracture Mechanics*, edited by G. C. Sih, H. C. van Elst and D. Broek, Noordhoff International Publishing, Leyden, pp. 613–621 (1974).

[12] Sih, G. C., Mechanics of ductile fracture, *Proceedings of International Conference in Fracture Mechanics and Technology*, edited by G. C. Sih and C. L. Chow, Sijthoff and Noordhoff International Publishers, Alphen aan den Rijn, pp. 767–784 (1977).

[13] Sih, G. C., Mechanics of crack growth: geometrical size effect in fracture, *Proceedings of International Conference on Fracture Mechanics in Engineering Application*, edited by

G. C. Sih and S. R. Valluri, Sijthoff and Noordhoff International Publishers, Alphen aan den Rijn, pp. 3–29 (1979).

[14] Hartranft, R. J. and Sih, G. C., Application of the strain energy density fracture criterion to dynamic crack problems, *Proceedings of International Conference on Prospects of Fracture Mechanics*, edited by G. C. Sih, H. C. van Elst and D. Broek, Noordhoff International Publishing, Leyden, pp. 281–297 (1974).

[15] Forsyth, P. J. E. and Ryder, D. A., Some results of examination of aluminum alloy specimen fracture surfaces, *Metallurgia* **63**, No. 377, pp. 117-124 (1961).

[16] Lake, G. J. and Thomas, A. G., Mechanics of fracture of rubber-like materials, *Proceedings of International Symposium on Absorbed Specific Energy and/or Strain Energy Density Criterion*, edited by G. C. Sih, E. Czoboly and F. Gillemot, Sijthoff and Noordhoff International Publishers, Alphen aan den Rijn, pp. 223–242 (1981).

9

Isoenergy density theory: exchange of surface and volume energy

9.1. Introduction

Fracture mechanics grew out of the need for a better understanding of how to control material and structure failure behavior after World War II. Notwithstanding the progress that has already been made in the past two decades, the fundamentals of the fracture process remain fragmented due to the lack of a general theory. The performance limits and reliability objectives dealing with the application of advanced materials and structures have altered considerably in recent times and call for a quantitative and consistent assessment of the material damage process that involves changes at the atomic, microscopic and macroscopic scale level. These changes though reflect a complex dependence on the material microstructure; they continue to be elusive if the combined interaction of space/time/temperature is not considered. Past methodologies [1–3] that relied on a single-parameter characterization would no longer be adequate for situations where the failure modes are more complex.

The inability of continuum mechanics theories to address failure by unexpected fracture has necessitated the invokement of fracture mechanics. Characterization of material required considering the trade-off between strength and fracture toughness. The latter was based on the concept of Linear Elastic Fracture Mechanics (LEFM), which is concerned primarily with determining the threshold condition of go and no-go. It considers a unique amount of energy being material-specific released for a small crack extension that triggers rapid fracture. The critical stress intensity factor K_{1c} or critical energy release rate G_{1c} has been commonly used [3] to distinguish the onset of unstable fracture. Although the crack opening displacement (COD) advanced by Wells [4] and J-integral developed by Rice and his co-workers [5, 6] have also been advocated as parameters for characterizing the fracture behavior, they in essence are based on the same underlying idea of energy released for a segment of crack extension. This is the specific surface energy or ***surface energy density**** dW/dA. In fact, J is equivalent to G for the elastic case where the material dissipates no energy other than that used in extending the crack. It is when crack extension is accompanied by other forms of energy dissipation that the parameters K, G, J or COD fail to apply as the failure process involves the simultaneous description of crack initiation, growth and onset of rapid fracture. Shih, DeLorenzi and Andrews [7] used dJ/da for characterizing subcritical crack growth, but this resulted in nonlinear relations.** Linearity on the crack growth resistance curves has been achieved using the ***volume*** energy density criterion. Confirmation was made by Sih and Tzou [11] for subcritical crack growth accompanied by plastic deformation in two dimensions while Sih and Chen [12] gave results on the three-dimensional crack problem.

For materials and/or specimen geometries that could deform substantially prior to each segment of crack growth, the total input energy would involve those consumed in plastic deformation and creating new crack surfaces. Simulation of the release of this total energy by crack extension alone would result in oversized specimens. Besides, the mode of failure would not be ductile. Smaller specimen tests under dynamic loads were thus advanced by empirical means. This motivated the Charpy V-Notch (CVN) impact test which has been used mostly for establishing the nil ductility temperature (NDT) with reference to a threshold energy level. There are conceptual difficulties associated with this approach because the rate effect is not specified on the CVN test. It has been shown by Sih and Tzou [13] that the dynamic fracture of the Charpy V-Notch specimen depends directly on the ***rate*** of input energy and not the energy level. Moreover, dynamic fracture is inherently a load-dependent process; it cannot be characterized by a single parameter such as K_{1D}. A more detailed discussion on this can be found in the works of Sih [14, 15].

Even less understood is the process of fatigue crack growth where the material fails by damage accumulation. What has prevailed over the years is the empirical '4th power law' of Paris [16] on relating the crack growth rate da/dN and change in Mode I stress intensity factor ΔK despite the obvious limitations of such an approach. It was Wei [17] who first showed that such a correlation failed to hold in the presence of moisture. Subsequent development of fatigue crack propagation laws followed the same footsteps and resorted in applying corrections to mean stress, specimen thickness, temperature, crack size, crack opening or closure effects. Variations of da/dN

---

* The notation G, rather than dW/dA, is usually used to denote energy release rate.

** The data of Bernstein [8] on precracked polycarbonate did not satisfy the $dJ/da =$ const. condition. This was pointed out and discussed by Sih and Tzou [9]. The strain energy density factor S, when plotted against crack growth, did yield a linear relationship such that specimen size and loading rate effects can then be easily resolved by interpolation.

versus ΔK should be interpreted with care because the same crack growth data would have different trends when plotted against other parameters. For example, thresholds in ΔK would disappear if ΔG were used. The influence of the mean stress on da/dN could reverse its trend for some metals when ΔK was replaced by ΔG. Crack size and geometry can also affect the scatter of the fatigue data. This was pointed out by Vecchio and Hertzberg [18] in association with the so-called 'small or short cracks'. The fundamental mechanisms that control crack retardation due to occasional overload in fatigue are yet to be found. Crack closure cannot be used as an explanation, for it can occur with and without retardation.

Fracture mechanics, being a supplement to the theories of continuum mechanics, cannot be regarded as a fundamental entity in itself. The unique correspondence between uniaxial and fracture data must be established; otherwise arbitrariness and inconsistency could result in the analysis as a result of oversupply of information. Verbal and/or elegant mathematical descriptions of already known and observed events cannot be considered as original research for they merely serve as a means of bookkeeping. *Predictive capability* is needed such that simple test data can be used to forecast the behavior of more complex systems regardless of whether they are loaded monotonically, repeatedly or dynamically. To this end, this chapter will provide an account of the importance of the exchange between surface and volume energy density which have historically been regarded as independent quantities.

9.2. Surface energy: micro- and macrocrack models

The process of fracture* can be referred to at the microscopic and macroscopic scale levels. It normally entails a detailed description of the material damage zone, analyzed experimentally and/or analytically. Quantitative assessment of the fracture process is, however, complex because it involves the combined interaction of loading rate and specimen geometry, in addition to the material microstructure. These effects influence the overall irreversible failure behavior of the material and their individual contributions cannot easily be separated.

Theories of continuum mechanics address material damage by permanent deformation and/or fracture. This process disturbs the homogeneity in a system and results in nonuniform fluctuation of the energy in a unit volume of material from one location to another. A most dominant exhibition of this inhomogeneity pertains to the state of affairs near a sharp crack tip

---

* *Fracture* pertains to material discontinuities at the microscopic and macroscopic levels. Imperfections at the atomic level should be referred to as vacancies, dislocations, etc.

where the energy density attains extremely high elevation. Depending on the load–time history, subcritical crack growth may or may not occur prior to rapid fracture. A distinction between local and global failure becomes necessary. In the simplest case of brittle fracture, local and global failure occur so rapidly after each other that they can be assumed to coincide. Although global instability can be predicted by considering either breaking a microelement or macroelement ahead of the crack, the specific fracture surface energy or energy required to break a microelement will differ from that of a macroelement. Both models, however, should yield the same critical load at fracture. Sih [19] showed that the correspondence between the micro- and macro-cracking model is related by a scale shifting factor. When yielding* occurs near a macrocrack tip, global instability no longer occurs immediately but is delayed by slow crack growth after initiation [20]. The damage process then becomes load history and specimen size dependent for a given material. The influence of microstructure becomes more and more important as the rate of energy input to the crack tip region is slowed down, such that the nonhomogeneous character of the grains begins to influence the load transfer characteristics. Creep and/or fatigue failure are cases in point. For the same microstructure, material can exhibit different failure behavior, ranging from brittle fracture to plastic collapse or excessive elongation.

The earlier attempts of Griffith [21] and Irwin [22] on characterizing fracture considered the energy released for creating a unit of new crack surface, say dW/dA. This concept was also associated with the amplitude of the local crack tip stress field, a parameter generally known as the stress intensity factor. Such an approach fails to hold if nonuniform yielding prevails around the crack. The nonhomogeneous nature of the elastic–plastic local stresses with varying field strength or stress singularity cannot be uniquely characterized by the so-called plastic stress intensity factor of Hilton and Hutchinson [23]. The model of uniform yielding by Hutchinson [24] in a local region near the crack also tends to oversimplify reality. The size of this region cannot be claimed to be microscopic in size since the analysis invokes the assumption of material homogeneity that can be justified only on the macroscopic scale level, where inhomogeneity caused by material microstructure can be neglected.

Scaling of free surface creation at instability. All solids are known to contain imperfections or defects whose size may be microscopic or macroscopic. Their dimensions can be made with reference to those of the crystal structure, e.g., the mean linear dimension α of the grains in a polycrystal. For

* Macroyielding reflects material damage at the microscopic level and the effects are smeared over a region in the continuum analysis referred to as the plastic enclaves.

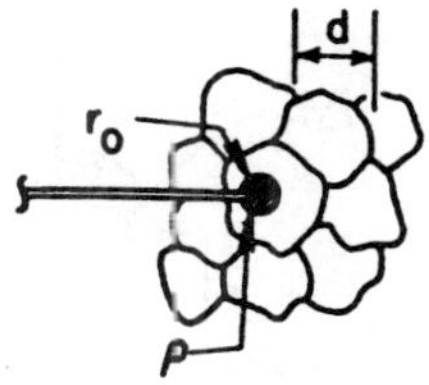

(a) Crack Tip In A Grain: $\rho << d$

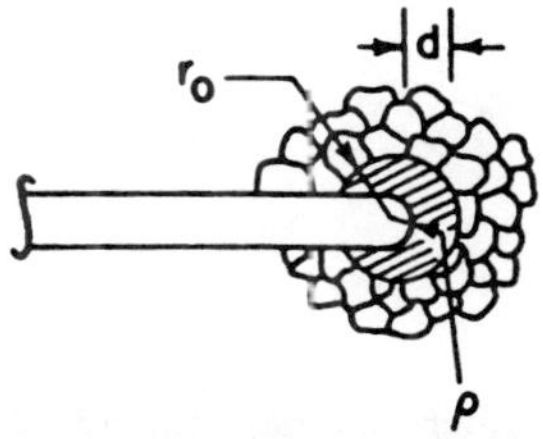

(b) Crack Tip Among Grains: $\rho \simeq d$

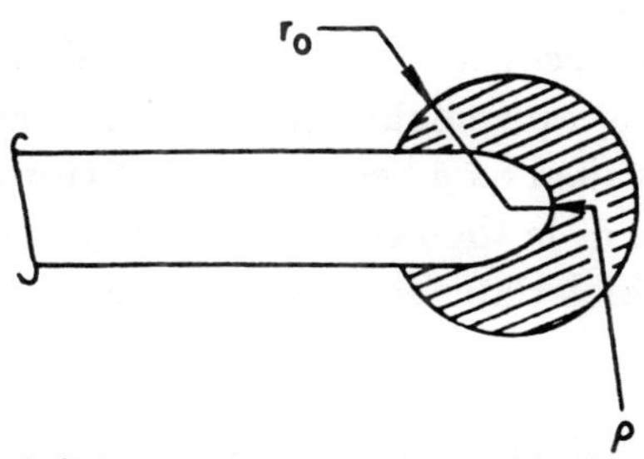

(c) Crack Tip In Homogeneous Material: $\rho >> d$

Figure 9.1. Size of crack tip compared with material microstructure.

a line defect, say a crack,* the mean radius of curvature ρ may be used to compare with d. Figures 9.1(a) to 9.1(c) illustrate three different cases where $\rho \ll d$, $\rho \simeq d$ and $\rho \gg d$. In the first situation, given by Figure 9.1(a),

---

* On physical grounds, it is necessary to consider a core region of radius r_0 (Figure 9.1) around the crack front which serves as a measure of the resolution of analysis within which failure will not be considered. Such an assumption can be justified even at the atomic level where a dislocation free zone prevails directly ahead of the crack tip with r_0 being of the order of an ångstrom. This has been observed experimentally by Ohr [25].

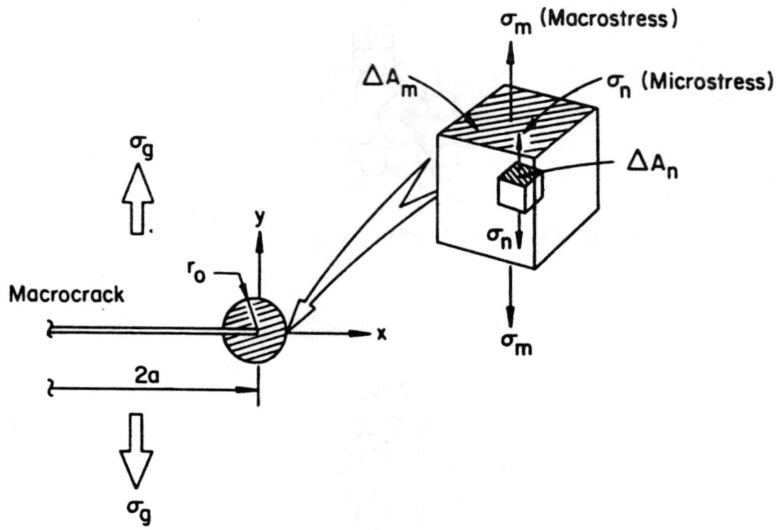

Figure 9.2. Macro- and micro-element ahead of crack.

the crack tip is located in a single grain and the basic element would be submicroscopic in size requiring information on the properties within a single grain. When $\rho \simeq d$ as indicated in Figure 9.1(b), the core region radius r_0 would have to be increased accordingly and a microelement consisting of the average properties of the grains should suffice. The assumption of material homogeneity can be justified as ρ becomes many times greater than d, Figure 9.1(c). The dimension r_0 is then comparable with the macroelement size. It is the scale level at which damage is being analyzed that determines the size of r_0.

Stress localization. When material homogeneity is disturbed by the presence of a pre-existing defect, there prevails a high elevation of local stress field. In other words, the distribution of force per unit area becomes nonuniform such that the magnitude increases as the region close to the defect is approached. A precise description of this localization depends on whether the observation is made over a unit of macro or micro area. Referring to Figure 9.2, the local macrostress σ_m defined over the macroarea ΔA_m should be distinguished from the microstress σ_n defined over the microarea ΔA_n. They, of course, are not equal. For the case of a remotely applied stress σ_g on a solid containing a through crack of length $2a$, the magnification of the local macrostress σ_m is given by

$$\frac{\sigma_m}{\sigma_g} = 1 + 2\sqrt{\frac{a}{\rho}} \tag{9.1}$$

which corresponds to the case of an elliptical slit of length $2a$ with ρ being the radius of curvature at the end of the major axis. If the slit is very narrow, then $a \gg \rho$ and equation (9.1) may be written as

$$\sigma_g\sqrt{a} = \tfrac{1}{2}\sigma_m\sqrt{\rho}. \tag{9.2}$$

An estimate of the macrocrack tip radius of curvature ρ depending on the ratio of σ_g/σ_m can thus be made:

$$\rho = 4a\left(\frac{\sigma_g}{\sigma_m}\right)^2. \tag{9.3}$$

The radius of curvature for a microcrack will depend on the ratio σ_g/σ_n and would be many times smaller than that given in equation (9.3). This is because σ_n would be much greater than σ_m.

Line crack model. Modelling of material failure near the crack tip region has been considered at both the microscopic and macroscopic levels. The distinction between the creation of a unit macrocrack and microcrack surface area is not only necessary, but must be assessed accordingly when computing the specific surface energy* dW/dA. In what follows, $(dW/dA)_m$ or $(dW/dA)_n$ will be used depending upon whether the surface creation process takes place at the macroscopic or microscopic scale level.

One of the simplest fracture models corresponds to the coincidence of local and global failure. This idealized situation assumes that the rate at which energy is dissipated to form a unit of free surface occurs very suddenly with little or no warning, and is known as the process of brittle fracture.** Returning to the example in Figure 9.2, the initiation of crack growth coincides with global instability as σ_g becomes critical†:

$$\sigma_g = \sqrt{\frac{E}{\pi a}}\sqrt{(dW/dA)_m} \tag{9.4}$$

where E is Young's modulus. Equations (9.1) and (9.4) may be combined to yield

$$\sigma_m = \sqrt{\frac{4E}{\pi\rho}}\sqrt{(dW/dA)_m}. \tag{9.5}$$

---

* The symbol γ with different subscripts is generally used to denote the energy per unit surface area.

** This is a process that excludes slow crack growth and energy dissipation over a unit volume of material, generally referred to as yielding.

† For simplicity, let it be understood that Young's modulus E in equation (9.4) stands for $E(1-\nu^2)$ with ν being Poisson's ratio such that the stress state ahead of the crack is plane strain.

The quantity $(\mathrm{d}W/\mathrm{d}A)_m$ represents the energy required to break a macroelement.

Alternatively, the same global failure stress σ_g can be predicted from the brittle fracture of a microelement. Suppose that $(\mathrm{d}W/\mathrm{d}A)_n$ represents the energy required to create the microcrack surface ΔA_n, as shown in Figure 9.2. The critical local microstress σ_n takes a form similar to σ_m in equation (9.5), i.e.,

$$\sigma_n = \sqrt{\frac{4e}{\pi d}}\sqrt{(\mathrm{d}W/\mathrm{d}A)_n} \tag{9.6}$$

except that e is the microscopic modulus and d the mean length of the microcracks. The macroscopic parameters σ_m and $(\mathrm{d}W/\mathrm{d}A)_m$ cannot be directly equated to the microscopic parameters σ_n and $(\mathrm{d}W/\mathrm{d}A)_n$. They can be related by invoking the assumption that the micro- or macroelement fails by the same critical *force*:

$$f_m = f_n \quad \text{or} \quad \sigma_m \cdot \Delta A_m = \sigma_n \cdot \Delta A_n. \tag{9.7}$$

The condition of critical *total energy** at fracture gives

$$\left(\frac{\mathrm{d}W}{\mathrm{d}A}\right)_m \sum \Delta A_m = \left(\frac{\mathrm{d}W}{\mathrm{d}A}\right)_n \sum \Delta A_n \tag{9.8}$$

which is the same as equation (9.7) because

$$\frac{\Delta A_m}{\Delta A_n} = \frac{\sigma_n}{\sigma_m} = \frac{(\mathrm{d}W/\mathrm{d}A)_m}{(\mathrm{d}W/\mathrm{d}A)_n} \simeq 10^3 \ \text{to} \ 10^4 \ \text{in}. \tag{9.9}$$

Force and total energy are unaltered by a change in scale and they can serve a useful purpose to bridge the gap between macroscopic and microscopic material damage models. The specific microsurface energy $(\mathrm{d}W/\mathrm{d}A)_n$ is therefore 10^3 to 10^4 times greater than the specific macrosurface energy $(\mathrm{d}W/\mathrm{d}A)_m$ while the material behavior remains brittle. It was first pointed out in [19] that $(\mathrm{d}W/\mathrm{d}A)_n$ cannot be claimed as plastic work** for the brittle fracture of metals.

Further clarification of the foregoing discussion can be obtained by expressing σ_g in terms of $(\mathrm{d}W/\mathrm{d}V)_n$. This can be achieved with the aid of

$$\frac{(\mathrm{d}W/\mathrm{d}A)_n}{(\mathrm{d}W/\mathrm{d}A)_m} = \left(\frac{e}{E}\right)\left(\frac{\rho}{d}\right) \tag{9.10}$$

---

* The ratio $\sum \Delta A_m / \sum \Delta A_n = \Delta A_m / \Delta A_n$ is the macro- and microelement and are assumed to be of the same shape.

** This removes the ambiguity raised in Orowan's work [26] for associating the brittle fracture of low carbon steel with the large amount of plastic work $(\mathrm{d}W/\mathrm{d}A)_n$ or γ_p that is many times larger than the elastic work $(\mathrm{d}W/\mathrm{d}A)_m$ or γ_g. The energy per unit of macrosurface area should be clearly distinguished from that per unit of microsurface area.

which is obtained by substituting equations (9.4) and (9.5) into (9.9). Putting equation (9.10) into (9.4), the result is

$$\sigma_g = \sqrt{\frac{E}{\pi a}}\sqrt{\left(\frac{E}{e}\right)\left(\frac{d}{\rho}\right)}\sqrt{(\mathrm{d}W/\mathrm{d}A)_n}. \tag{9.11}$$

The factor* $\sqrt{Ed/e\rho}$ provides the scale shift that relates the macro- and microquantities.

Expanding narrow ellipse. A free surface can be created either by cracking a solid, as discussed earlier, or by expanding the surface area of a solid, as in the case of liquid or viscous material. The energy associated with these two processes, however, would differ considerably. Such a distinction was not made clear by Griffith [21], which later contributed to misinterpretations.

The process of surface expansion can be most easily illustrated by considering the equilibrium of a spherical bubble of radius R in a uniform pressure field p. Application of the theorem of minimum potential energy gives

$$p\frac{\mathrm{d}V}{\mathrm{d}A} = \frac{\mathrm{d}W}{\mathrm{d}A} \tag{9.12}$$

in which the rate change of volume with surface $\mathrm{d}V/\mathrm{d}A$ is equal to $R/2$ and $\mathrm{d}W/\mathrm{d}A$ stands for the surface tension of the liquid, normally designated by the symbol γ. This leads to the familiar expression $pR = 2\gamma$.

Equation (9.12) applies to other cavity shapes provided that a state of uniform pressure prevails. For an elliptical hole with major and minor axes $2a$ and $2b$, respectively, in the xy-plane with unity in the z-direction, its volume V and area A are given by

$$V = \pi ab; \quad A = 4aE(k). \tag{9.13}$$

The quantity $E(k)$ is the complete elliptical integral of the second kind with argument k, i.e.,

$$E(k) = \int_0^{\pi/2} \sqrt{1 - k^2 \sin^2\phi} \cdot \mathrm{d}\phi, \quad k^2 = 1 - (b/a)^2. \tag{9.14}$$

Substituting equation (9.13) into (9.12) and carrying out the algebraic ma-

---

* This factor was taken as unity by Orowan [26]. It is not justified to arbitrarily add γ_p onto γ_g and then drop γ_g on the basis that $\gamma_p \gg \gamma_g$, i.e.,

$$\sigma_g = \sqrt{\frac{2E}{\pi a}}\sqrt{\gamma_g + \gamma_p} \simeq \sqrt{\frac{2E}{\pi a}}\sqrt{\gamma_p}.$$

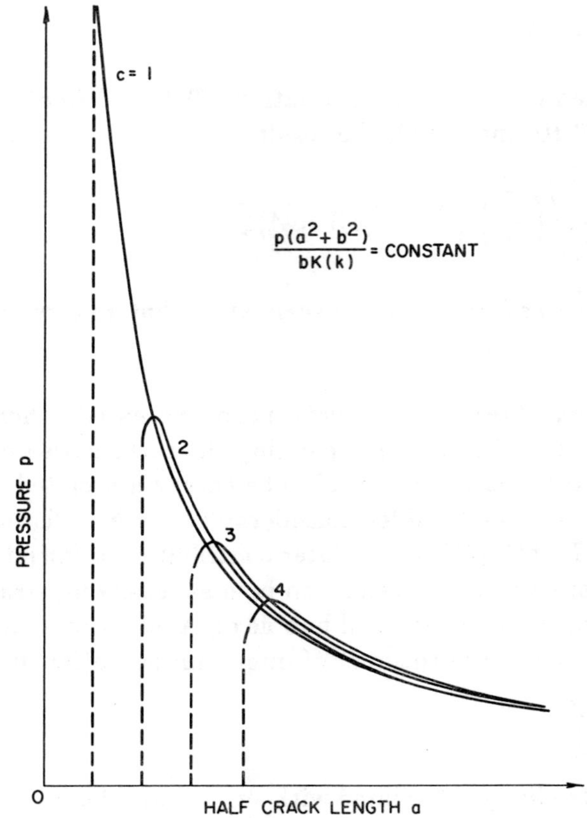

Figure 9.3. Pressure versus half crack length for the surface tension model.

nipulations, it can be shown that

$$\frac{\pi}{4}\frac{\sqrt{a/\rho}}{K(k)}pa = \frac{\mathrm{d}W}{\mathrm{d}A} \tag{9.15}$$

where $\rho = b^2/a$ is the radius of curvature at the apex of the major axis and $K(k)$ is the complete elliptical integral of the first kind. Suppose that a narrow elliptical cavity undergoes shape change while the focal distance $2c$ with $c = \sqrt{a^2 - b^2}$ remains unchanged, then a family of curves may be obtained to represent the variations of p with a for fixed $\mathrm{d}W/\mathrm{d}A$. A more general interpretation of equation (9.15) is given in Figure 9.4 which shows a three-dimensional plot of the p–a–b surface with different values of c. As p is increased, the elliptical cavity is assumed to grow confocally while a and b are varied from those for a circle ($a = b$) to that for a crack ($a \gg b$).

The physical process of free surface creation depends on the material behavior. For a liquid or viscous material, new surfaces may be increased by stretching the substance in a thin layer. Vacancies would be created and occupied by molecules from the bulk. The surface tension quantity is generally used for estimating the energy required to create a unit area of additional surface. This concept was adopted by Griffith [21] to compute

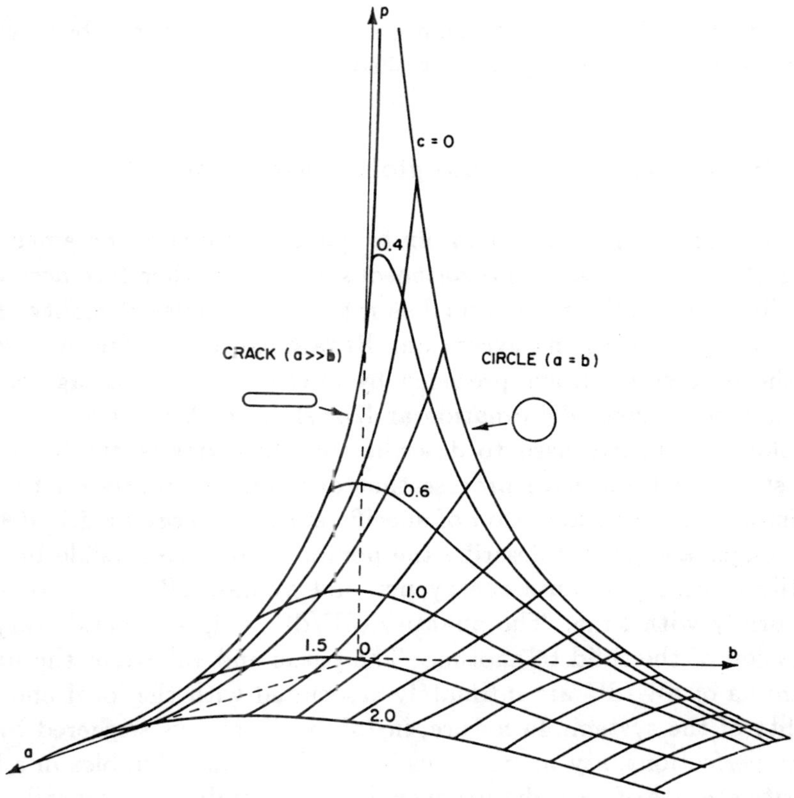

Figure 9.4. Variations of pressure with dimensions of elliptical cavity in equilibrium with surface tension.

the increase in surface area of a narrow ellipse or crack of length $2a$. In his energy balance consideration, he used $4a \times$ unit depth for the surface energy term instead of $4\Delta a \times$ unit depth with Δa being an additional free surface created by the breaking of cohesive bonds rather than the replacement of molecular vacancies. Although the results displayed in Figure 9.3 for $p^2 a =$ const. resemble those obtained by Griffith [21] for a line crack under uniform pressure, the underlying theoretical assumptions and physical reasoning differ.* This was first pointed out and discussed by Sih [19].

Surface energy alone, however, is not sufficient for a complete description of material damage in general. It is well known in crystal growth that the rate of nucleation is intimately associated with the exchange of surface and volume energy. They determine the boundary and size of grains in a polycrystal. As in all physical processes, it involves loss of irrecoverable

* Griffith [21] considered the creation of free surface around the periphery of a narrow elliptical boundary or crack rather than the breaking of cohesive bonds in the solid ahead of the crack. He made use of the solution of the theory of elasticity which was not needed in the surface tension model.

energy. A more detailed discussion on this will be made subsequently in relation to the strain energy density theory.

9.3. Volume energy: local and global instability

The discreteness of nature calls for finite space and/or time increment in describing physical processes. Instantaneous events are therefore accountable only as limits in mathematical models rather than physical reality. Failure behavior of materials is no exception. Regardless of how fast a solid fractures, the process is always preceded by initial material damage either in the form of permanent deformation and/or slow crack growth. A common terminology frequently used to describe such behavior is ductile fracture. It is a strain rate sensitive process that depends inherently on the load–time history. The consideration of specific surface energy or dW/dA alone can no longer adequately describe the nonlinear and irreversible behavior* of ductile fracture, because energy per unit volume dW/dV also changes nonuniformly with time. The quantity dW/dV will, in general, vary from one location of the solid to another. The peaks and valleys or the maxima and minima of dW/dV are intimately associated with the local and global instability of the system. In nature, instability is always triggered by some initial imperfections, say microcracks in a solid or microbubbles in a liquid. A subcritical state of local disturbance, however small, must prevail prior to reaching the critical condition at large. The physical significance of the local and global stationary values of dW/dV as applied to structural components can be found in the works of Sih [27] and Sih and Chu [28].

Stationary values of strain energy density function. When a solid is deformed, the energy per unit volume of material or the strain energy density function dW/dV changes** from one location to another. This function can be computed as

$$\frac{dW}{dV} = \int_0^{\epsilon_{ij}} \sigma_{ij}\, d\epsilon_{ij} \tag{9.16}$$

where σ_{ij} and ϵ_{ij} are the stress and strain components in a solid. Changes in temperature and/or moisture on the stress and strain states can also be included in computing dW/dV, as shown by Sih, Michopoulos and Chou [29]; given an instance of time, it can be employed to describe the progressive

* Invoked in the classical theory of plasticity is the von Mises yield condition that consists of the distortional component of the strain energy density function, $(dW/dV)_p$.

** In general, there would be many maxima and minima of dW/dV in a continuum. It is the maximum of the maxima and maximum of the minima denoted by $(dW/dV)^{\max}_{\max}$ and $(dW/dV)^{\max}_{\min}$ that are assumed to reach their respective critical values at failure.

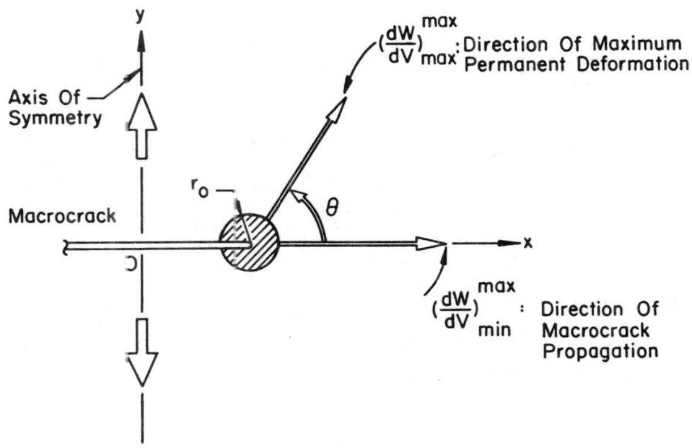

Figure 9.5. **Stationary values of volume energy density function associated with damage by deformation and/or fracture.**

damage of material. The underlying idea of the strain energy density theory of Sih [27] can be stated as follows:

> ***Progressive change in material states is assumed to be uniquely associated with the rate at which energy is dissipated in a unit volume of material.***

The criterion can be used to determine the direction of yield and fracture initiation in addition to subsequent stages of damage caused by yielding/fracture.

Distortion and dilatation. Excessive distortion and dilatation are two of the most common modes of failure in engineering structure. They can be associated with the maximum and minimum of dW/dV. Illustrated schematically in Figure 9.5 is a two-dimensional macrocrack subjected to tension in the y-direction. Among many of the minima of dW/dV, the maximum of $(dW/dV)_{min}$ is assumed to coincide with macrocrack growth and the largest of $(dW/dV)_{max}$ with the direction of maximum permanent deformation. The former can be associated with excessive dilatation while the latter with excessive distortion. This is consistent with the intuition that large volume change leads to fracture and shape change to permanent deformation.*

The physical interpretation of the stationary values of the strain energy density function can best be illustrated in Figure 9.6. According to the criterion, along the path of prospective macrocracking the macroelements experience macrodilatation and microdistortion. This accounts for the cre-

---

* The separation of dW/dV into its dilatational and distortional components, of course, can be carried out only by assuming a linear relationship between stress and strain. No such separation can be made *a priori* when the response becomes nonlinear.

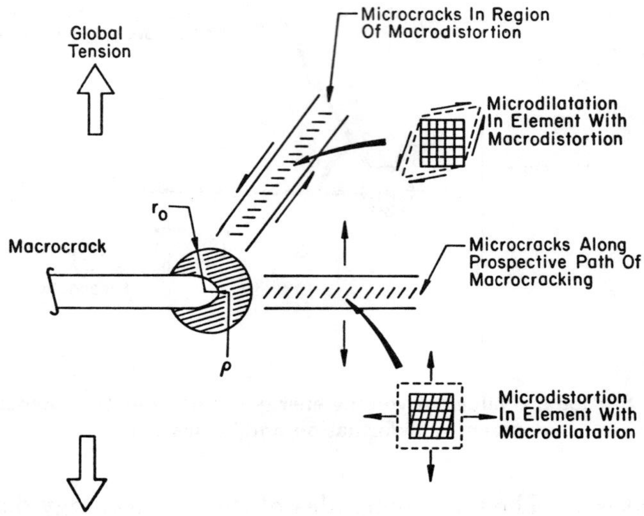

Figure 9.6. Schematic of macro- and microdamage in region ahead of crack tip.

ation of slanted microcracks prior to macrocracking along planes normal to the applied tension. In the same vein, the macroelements off to the sides are subjected to macrodistortion and microdilatation. They are responsible for permanent deformation and creation of microcracks.

The interplay between distortion and dilatation is an inherent part of material behavior. They vary in proportion depending on the load history and location due to nonuniformity in stress or energy distribution even for isotropic and homogeneous solids. The dilatational effect tends to dominate along the path of macrocracking while the distortional effect governs permanent macrodeformation. Moreover, macrocracking is initiated by microdistortion and permanent macrodeformation by microdilatation.*

Dissipated energy density. Permanent deformation prevails when a solid is stretched or compressed beyond the linear range as illustrated by the uniaxial stress and strain curve in Figure 9.7. The rate of change of area under the σ versus ϵ curve gives the change of dW/dV as a function of time. The shaded area *Oypq* denoted by $(dW/dV)_p$ represents the energy dissipated by permanent deformation while the area *qpg* given by**

---

* The von Mises yield condition considered only the distortional component as a linear component of the total strain energy density function. It not only neglects the dilatational contribution but invokes the principle of superposition that is invalid in nonlinear analysis.

** The incremental change of dW/dV with respect to that of $(dW/dV)_p$, i.e., $\Delta(dW/dV)/\Delta(dW/dV)_p$ represents a measure of temperature for a given strain state as in thermodynamics.

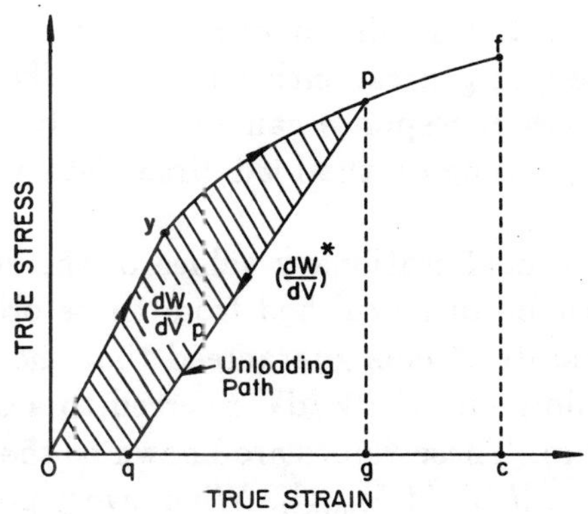

Figure 9.7. **Schematic of true stress and true strain.**

$$\left(\frac{dW}{dV}\right)^* = \left(\frac{dW}{dV}\right) - \left(\frac{dW}{dV}\right)_p \tag{9.17}$$

is the available energy density that is released when a unit of macrovolume fails as $(dW/dV)^*$ becomes critical. The unloading path may or may not be parallel with Oy, the initial path of loading. Its slope depends on the loading rate.

The dissipated energy density function $(dW/dV)_p$ in equation (9.17) is logically a measure of irreversibility* of physical systems. It is an increasing function of the state variables, i.e.,

$$\left(\frac{dW}{dV}\right)_p \geq 0 \tag{9.18}$$

regardless of the loading, geometry and material. The equality sign in equation (9.18) holds for the hypothetical reversible process. The concept of reversibility is needed for specifying the scale level of investigation and degree of irreversibility neglected in any physical measurements. An exact evaluation of irreversibility is obviously beyond the comprehension of the human mind for it is an inherent part of the evolution of physical systems. The inaccuracies, however, can always be suppressed to a lower scale level.

Theories of continuum mechanics as developed conventionally are inadequate for assessing the irreversible behavior of materials in a consistent fashion. The invokement of separate fracture criteria on the theory of plasticity is a case in point and has been known to introduce arbitrariness. Even more serious is the inability of the theories to translate irreversibility measured from the uniaxial data to multiaxial stress or energy states. The

---

* It serves the same purpose as the specific entropy quantity in thermodynamics.

arbitrary assumption in plasticity that the uniaxial data coincide with the effective stress and effective strain curve has limited range of application* and cannot be justified, in general, either theoretically or physically. Moreover, the stress and strain response can vary widely from one location to another in a solid, depending on the load–time history.

System stability. The local stationary values of the strain energy density function dW/dV must be distinguished from those obtained globally. For any nontrivial stress state,** it is guaranteed that there exists at least one maximum and one minimum of dW/dV referred to a coordinate system at each point in a medium. These values are known as the *local* stationary values $[(dW/dV)_{max}]_l$ and $[(dW/dV)_{min}]_l$. When every point in a system is referred to the same coordinate system, the resulting maxima and minima are known as the *global* stationary values $[(dW/dV)_{max}]_g$ and $[(dW/dV)_{min}]_g$. The relative distances between the local and global stationary values of dW/dV can serve as a measure of system stability that involves the combined effects of loading rate, structure geometry and material type.

In the case of fracture instability, it suffices to consider the distance l between $[(dW/dV)_{min}^{max}]_l$ and $[(dW/dV)_{min}^{max}]_g$ as illustrated in Figure 9.8 for a line crack under symmetrical loading. It can measure the degree of system stability or instability. The constant contours of $[(dW/dV)_{min}]_l$ can be calculated analytically and they are displayed in Figure 9.9 for a line crack of length $2a = 10$ in. in an infinite domain subjected to a remote uniform stress of 100 ksi. The intensity of the contours increases as the crack tip is approached with the maximum of $[(dW/dV)_{min}]_l$ or $[(dW/dV)_{min}^{max}]_l$ stopping short at the core region. Crack initiation is assumed to occur when $[(dW/dV)_{min}^{max}]$ reaches certain critical values for a given material. The maximum of the global minimum or $[(dW/dV)_{min}^{max}]_g$ is at infinity. This means that l in Figure 9.8 for a crack engulfed in a uniform stress field of infinite extent is infinite. Hence, unstable fracture† prevails regardless of the material. For any finite plate geometry which is necessarily the case in any

* At best, it applies to situations where distortion dominates such that failure takes place in the form of plastic collapse. In regions of high dilatation such as the crack tip, plasticity yields very poor results as shown by Sih and Tzou[30].

** Constant dW/dV corresponds to a uniform stress state. The criterion would then predict the simultaneous failure of all elements. Should experiment show otherwise, the fluctuation of dW/dV at a lower scale level, say microscopic, must then be analyzed for a detailed account of the initiation of local failure.

† In retrospect, the Griffith mathematical crack model is basically unsuited for studying stable crack growth. This point was raised many years ago by Barenblatt [31, 32] but has received very little attention by the fracture mechanics community. Slow crack growth is not exclusively associated with the ductility of the material. Gdoutos and Sih [33] have shown that slow crack growth can arise due to changes in load even without permanent deformation.

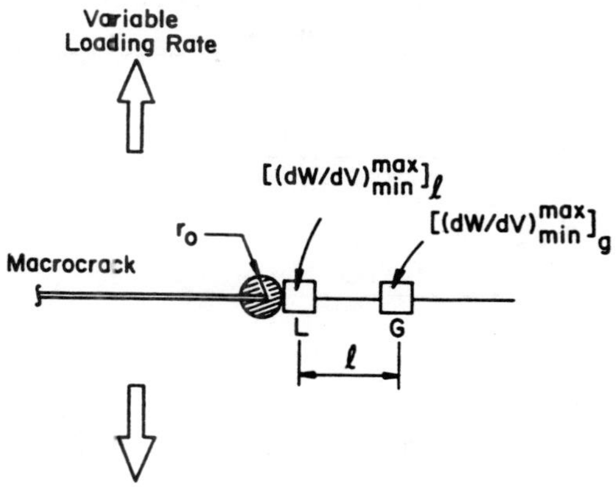

Figure 9.8. **Relative distance between local and global minimum of strain energy density function.**

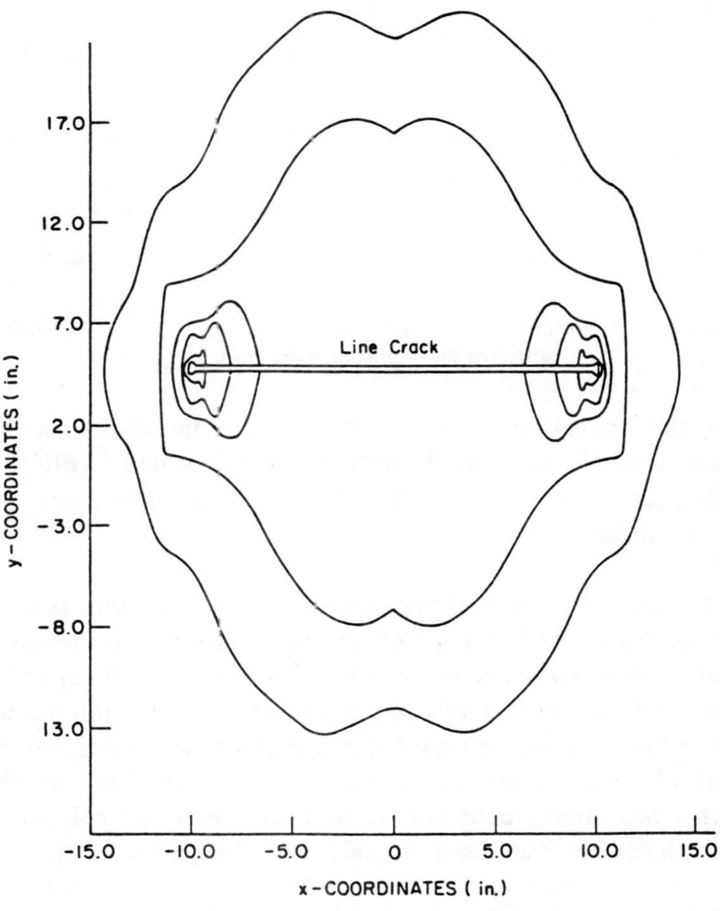

Figure 9.9. **Constant contours of local minimum strain energy density function around crack in uniform tension.**

Table 9.1. Variations of distance between $[(dW/dV)_{\min}^{\max}]_l$ and $[(dW/dV)_{\min}^{\max}]_g$ with initial crack length for uniform and concentrated load.

	l (in.)	
a (in.)	Uniform load	Concentrated load
1	4.00	3.00
2	3.00	2.00
3	1.53	1.30
4	0.60	0.60

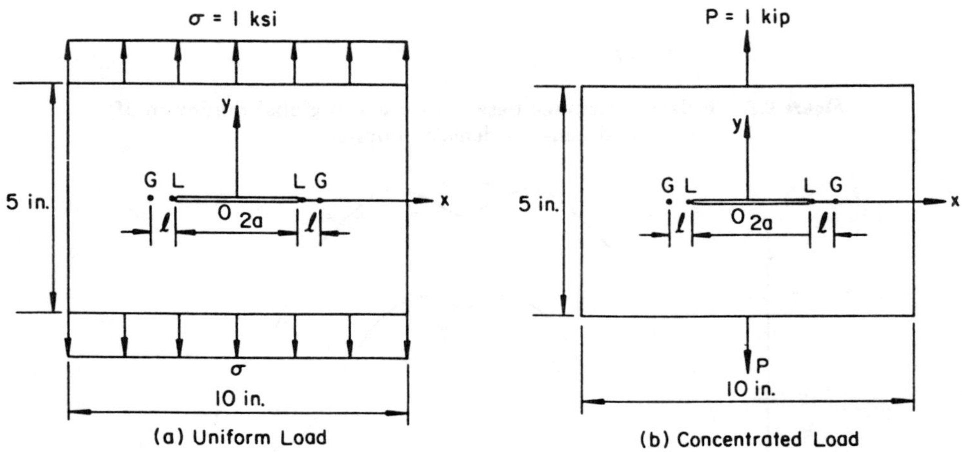

Figure 9.10. A center-cracked plate 5 in. × 10 in. × 1 in. subjected to uniform and concentrated load.

experiments, the location of $[(dW/dV)_{\min}^{\max}]_g$ would fall either on the specimen boundary or at the interior. A finite distance l between $[(dW/dV)_{\min}^{\max}]_l$ and $[(dW/dV)_{\min}^{\max}]_g$ can be found as an indication of the degree of system stability or instability.

Load type. The influence of load type on the parameter l can best be illustrated by considering the behavior of a center-cracked specimen subjected to uniform and concentrated load. Illustrated in Figures 9.10(a) and 9.10(b) are the cases of a 1 kip load distributed uniformly over the plate with dimensions 5 in. × 10 in. × 1 in. and concentrated along the y-axis, respectively. The locations of the local and global stationary values of the maximum of the minimum strain energy density functions are found by application of the theory of elasticity for four different half crack lengths $a = 1$, 2, 3 and 4 in. Table 9.1 gives the values of l. The most unstable situation corresponds to $a + l = 5$ in., i.e., when the location of $[(dW/dV)_{\min}^{\max}]_g$ or the point G coincides with the plate boundary. This means that once the crack starts to spread from L, the plate fractures completely. For the plate geometry with a height to width ratio of one-half, the concentrated load configuration is more stable than that of uniform load because of the smaller distance between L and G or l-value. The stabilities of the two loading systems approach one another as the crack length is made to approach the plate width. Except for $a = 1$ and 2 in., in the case of uniform load where $a + l = 5$ in., the other cases in Table 9.1 correspond to $a + l < 5$ in. This means that G is interior to the plate whose stability increases with decreasing l, i.e., failure by crack growth becomes more localized near the crack tip. Experimental measurement of l would not be possible as fracture initiation would always involve overload* that changes with the compliance of the specimens which, in turn, depends on plate dimension, crack size, loading rate and material. Moreover, since the above results are based on the theory of elasticity, which is inadequate for describing the state of affairs near the crack, the foregoing results should be interpreted and used only within the limitations of the theory.

Material inhomogeneity. The influence of material inhomogeneity on the system's stability can also be measured by l. To this end, consider the two-material system in Figure 9.11. Let E_1, ν_1 and E_2, ν_2 represent the elastic properties of the inclusion and plate with dimensions $2H \times 2B$, respectively. Loading consists of a uniform stress with stress σ (Type I) and a linearly varying load (Type II), which depends on the variable x, as $\sigma_0(1 - x/B)$, with σ_0 being constant. The inclusion is an elongated ellipse with semi-major axis a and semi-minor axis b. The total length of the inclusion is $2c > 2a$. Referring to Figure 9.11, the locations of $[(\mathrm{d}W/\mathrm{d}V)^{\max}_{\min}]_l$ and $[(\mathrm{d}W/\mathrm{d}V)^{\max}_{\min}]_g$, corresponding to the points L and G separated by l, always lie along the prospective path** of failure.

Based on the Type I loading in Figure 9.11, numerical values of l are found for $\sigma = 689.5$ MPa, $E_2 = 68.95$ GPa while E_1/E_2 is varied with $\nu_1 = \nu_2$. For $c = 12.70$ cm, $b/a = 0.5$, $B = 15.88$ cm, three different ratios of $H/B = 0.45$, 0.50 and 0.55 are considered. The results are given in Table 9.2 and plotted in Figure 9.12. The three solid lines are nearly straight except when they reach the upper limit $l = 3.18$ cm corresponding to G on the boundary. For $E_1/E_2 < 1$, l increases or decreases with E_1/E_2. The case of a notch is recovered in the limit as $E_1/E_2 \to 0$. Thus, reduction in

---

* The so-called 'critical load' referred to any fracture criteria corresponds to the meta-stable state of any system and cannot be realized experimentally unless there is overload to guarantee crack motion. Such an effect introduces scatter into the measurements of parameters associated with incipient fracture that, as a rule, are not reported in the literature.

** In the absence of load and/or crack symmetry, the path on which L and G are situated can be curved.

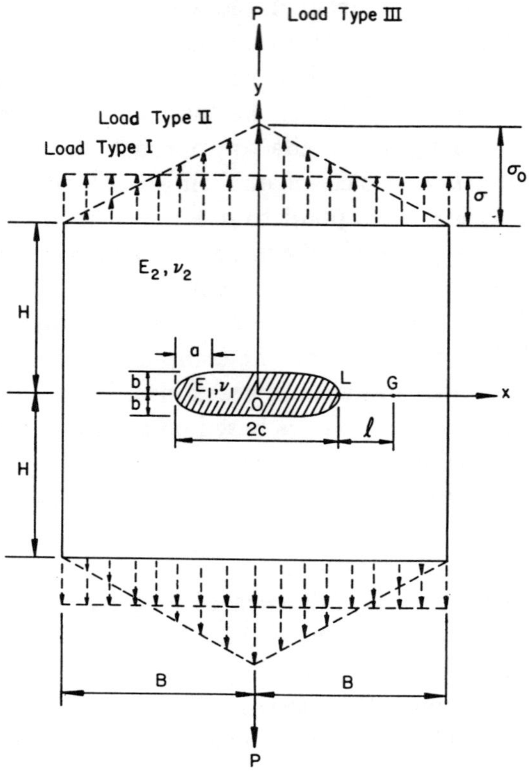

Figure 9.11. Plate with inclusion subjected to uniform and varying load.

E_1/E_2 and H/B result in more localized failure and enhances subcritical material damage. As $E_1/E_2 \to 1$, l approaches the entire ligament B–c and failure tends to be more catastrophic as a greater portion of the solid is involved in the damage process at a given instant.

Type II loading that decays linearly in Figure 9.11 is also analyzed with $\sigma_0 = 137.9$ MPa. The other geometric and material parameters are the same as those for Type I loading. A plot of l versus E_1/E_2 in Figure 9.13 reveals that l does not increase as rapidly with E_1/E_2 as in the case of uniform loading. The failure behavior is therefore more stable. This is intuitively obvious because the ligament ahead of the inclusion experiences a lower load magnitude. Table 9.3 gives the numerical values of l for different E_1/E_2 ratios. Note that the values of l in Table 9.3 are lower than those in Table 9.2. These two examples show that failure mode stability in a nonhomogeneous system depends sensitively on the interaction of load with geometry.

Subcritical yielding and fracture. A salient feature of the strain energy density function dW/dV near a defect or any other sites of possible failure

Table 9.2. Distance l between maximum of local and global minimum strain energy density functions for uniform loading.

	H/B		
E_1/E_2	0.45	0.50	0.55
0.10	2.725	2.778	2.858
0.11	2.802	2.858	2.934
0.12	2.883	2.937	2.996
0.13	2.946	3.016	3.068
0.14	3.018	3.089	3.150
0.15	3.096	3.149	3.175
0.16	3.150	3.175	–
0.17	3.175	–	–

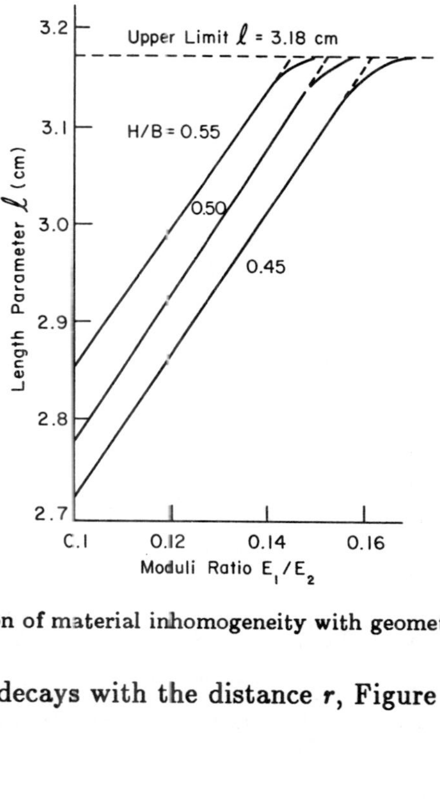

Figure 9.12. Interaction of material inhomogeneity with geometry under uniform load.

initiation is that it decays with the distance r, Figure 9.14. It follows that the relationship

$$\frac{dW}{dV} = \frac{S}{r} \tag{9.19}$$

can be defined with S being the strain energy density factor, representing the area under the dW/dV versus r curve at the location under consideration. The $1/r$ character holds for all geometries, loadings and material types. It is independent of the form of the assumed constitutive relations. The dW/dV criterion, therefore, remains valid as the loading is changed from monotonic to cyclic fatigue. Additional conditions need not be imposed

Table 9.3. Distance l between maximum of local and global minimum strain energy density functions for linearly decaying load.

	H/B			H/B	
E_1/E_2	0.50	0.55	E_1/E_2	0.50	0.55
0.1	1.640	1.693	0.6	2.249	2.302
0.2	1.693	1.746	0.7	2.434	2.514
0.3	1.799	1.826	0.8	2.646	2.725
0.4	1.985	1.958	0.9	2.910	2.995
0.5	2.064	2.117	0.95	3.043	3.149

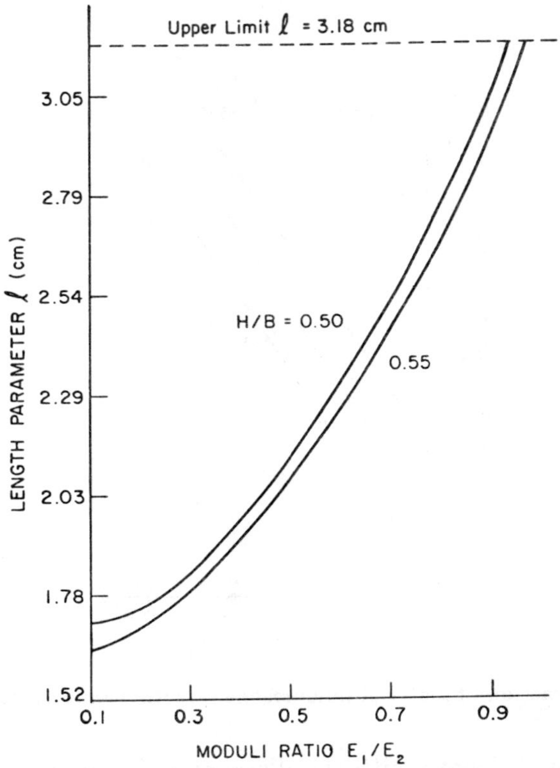

Figure 9.13. Interaction of material inhomogeneity with geometry under linearly varying load.

for materials that behave nonlinearly and undergo irreversible deformation. This, of course, does not apply to other failure criteria that involve stress or strain quantities. The singular behavior of the local stresses and strains, as determined from the theory of plasticity, is known to depend on the strain hardening character that changes with each increment of crack growth. The fact that stress and strain alter their character with loading makes them unsuitable for application in failure prediction.

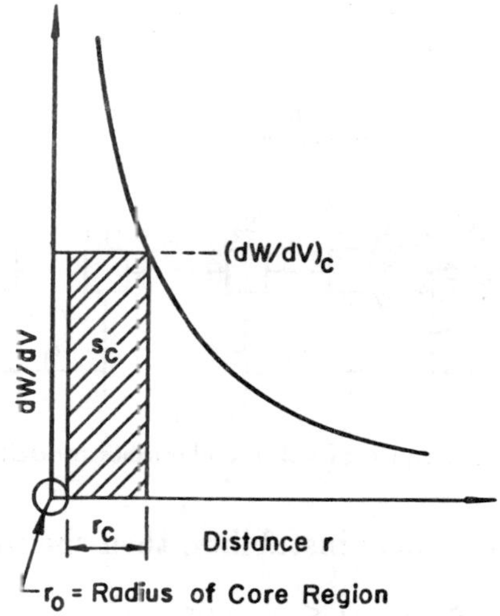

Figure 9.14. Variations of strain energy density function with distance near site of failure initiation.

Hypotheses. The foregoing concepts can be summarized in terms of applying dW/dV as a failure criterion. Stated below are three fundamental hypotheses.

Hypothesis (1): The relative local and global minima of dW/dV or $(dW/dV)_{min}$ and maxima, $(dW/dV)_{max}$, are assumed to coincide with the locations of fracture and yielding,* respectively.

Hypothesis (2): Yielding and fracture are assumed to occur when the maximum of $(dW/dV)_{max}$ or $(dW/dV)_{max}^{max}$ and the maximum of $(dW/dV)_{min}$ or $(dW/dV)_{min}^{max}$ reach their respective critical values.

Hypothesis (3): The rate of yielding and fracture are assumed to obey the relation

$$\left(\frac{dW}{dV}\right)_c \quad \text{or} \quad \left(\frac{dW}{dV}\right)_c^* = \frac{S_1}{r_1} = \frac{S_2}{r_2} = \cdots = \frac{S_j}{r_j} = \cdots = \frac{S_c}{r_c} \quad \text{or} \quad \frac{S_c^*}{r_c^*}. \tag{9.20}$$

* The term yielding is used to describe failure by excessive distortion and is not necessarily associated with the von Mises yield condition as defined in the classical theory of plasticity. The energy density dW/dV, which includes both distortion and dilatation, can be used as the condition for yield.

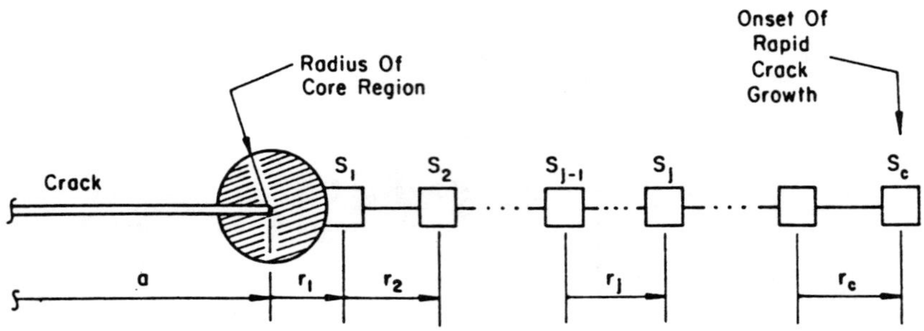

Figure 9.15. Schematic of volume elements modelling crack growth.

If the process leads to global instability, then the inequalities

$$\begin{matrix} S_1 < S_2 < \cdots < S_j < \cdots < S_c & \text{or} & S_c^* \\ r_1 < r_2 < \cdots < r_j < \cdots < r_c & \text{or} & r_c^* \end{matrix} \tag{9.21}$$

will hold. In situations where yielding and fracture come to arrest, the following shall apply:

$$\begin{matrix} S_1 > S_2 > \cdots > S_j > \cdots > S_0 & \text{or} & S_0^* \\ r_1 > r_2 > \cdots > r_j > \cdots > r_0 & \text{or} & r_0^* \end{matrix} \tag{9.22}$$

In equations (9.21) and (9.22), $S_c^* < S_c$ ($r_c^* < r_c$) and $S_0^* < S_0$ ($r_0^* < r_0$). The quantity with the asterisk corresponds to the portion associated with the creation of macrocrack surface, i.e., excluding dissipation due to yielding.

It should be kept in mind that subcritical crack growth need not be exclusively accompanied by plastic deformation. It can occur elastically if the applied loads vary in amplitude. This was shown by Gdoutos and Sih [33]. The conditions described by equations (9.21) are illustrated schematically in Figure 9.15 in which the crack path is modelled by loci of broken elements. Fracture is regarded as a discrete process in the volume energy density criterion. This is in contrast to the surface energy density or energy release approach where crack extension is assumed to be a continuous process occurring directly from the crack tip as implied by the limit process:

$$G = \frac{1}{2h} \lim_{\Delta a \to 0} \frac{W(a + \Delta a) - W(a)}{\Delta a} = 2\gamma \tag{9.23}$$

where G, being an energy release quantity, is understood to be negative. The plate thickness is h while γ is the specific surface energy or the equivalent of dW/dA.

Hypotheses (2) and (3) apply equally well to fatigue where the failure of a volume element would correspond to the summation of energy density

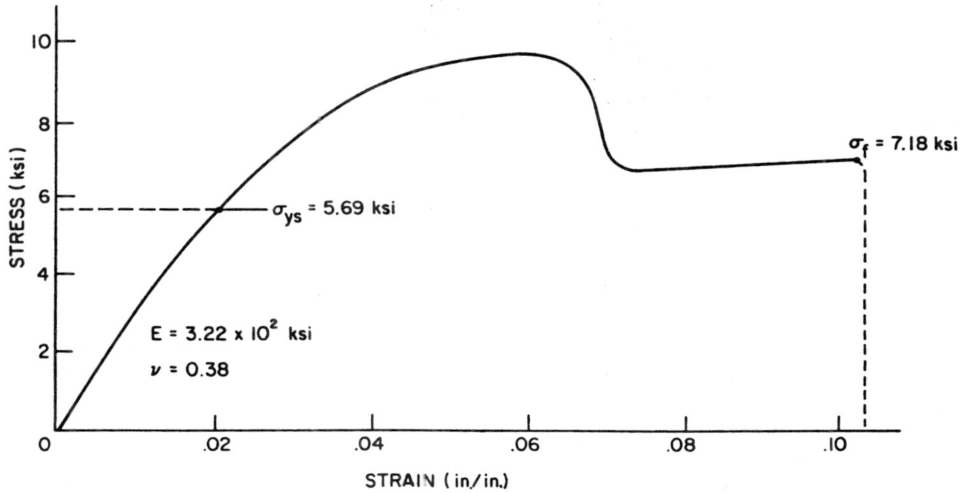

Figure 9.16. Stress and strain curve for polycarbonate material.

accumulation during each cycle, reaching the critical value obtained from uniaxial tests. Fatigue crack initiation and propagation have been treated in the same way by Sih and Chao [34, 35].

Slow crack growth. In practice, fracture never occurs instantaneously but is preceded by some subcritical material damage at the microscopic or macroscopic scale level. Slow crack growth followed by rapid crack propagation is a phenomenon commonly observed in ductile fracture. The load versus displacement response then becomes nonlinear and changes as specimen size and loading rate are altered. In the case of a precracked metal specimen, the crack length is frequently monitored with load for assessing the severity of damage. The data usually exhibit a high degreee of nonlinearity and offer little or no insight into the combined influence of those parameters affecting material damage. The objective is to linearize* the raw data such that specimen size and testing time effects could be scaled appropriately.

Polycarbonate material has received a great deal of attention because of its unique stress and strain response that exhibits both strain hardening and softening as illustrated in Figure 9.16. A host of slow crack growth data were collected by Bernstein [8] using the three-point bend specimen as shown in

* No useful purpose would have been served if the nonlinear load versus crack length data are reproduced nonlinearly by other parameters such as J [8] or COD [36].

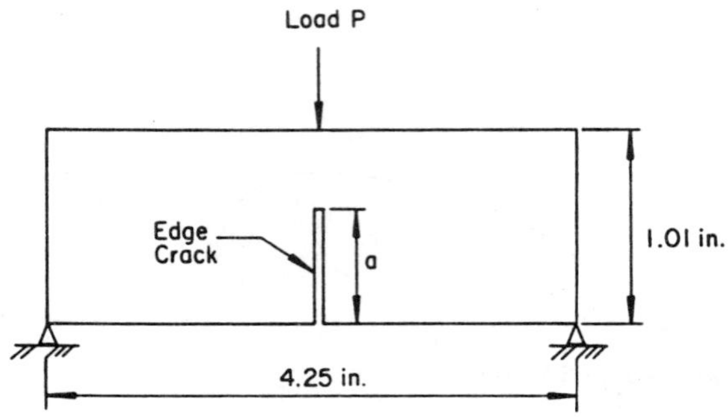

Figure 9.17. Edge crack in three-point bend specimen.

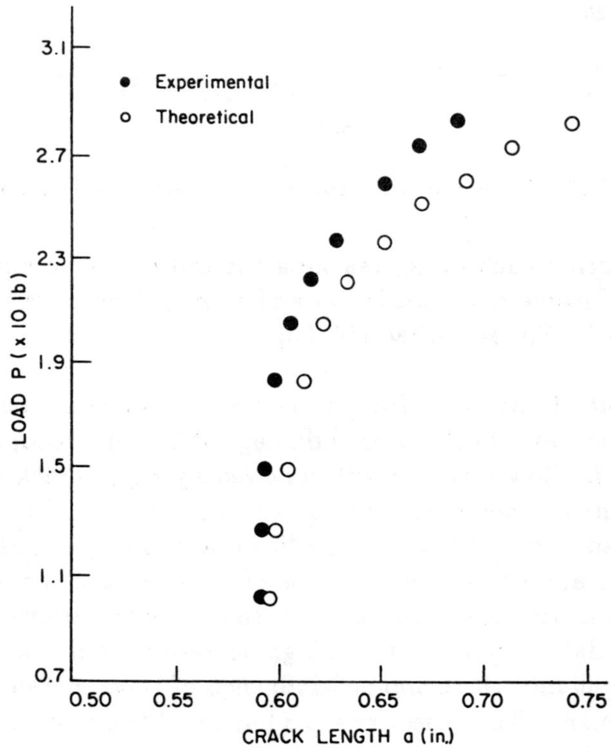

Figure 9.18. Variations of load with crack length for polycarbonate three-point bend specimen.

Figure 9.17 at the rate of 0.02 in./min. The strain energy density theory and incremental theory of plasticity assuming isotropic and kinematic hardening were used by Sih and Tzou [9] to analyze this problem theoretically. A critical strain energy density function of $(dW/dV)_c^* = 150.75$ psi was used

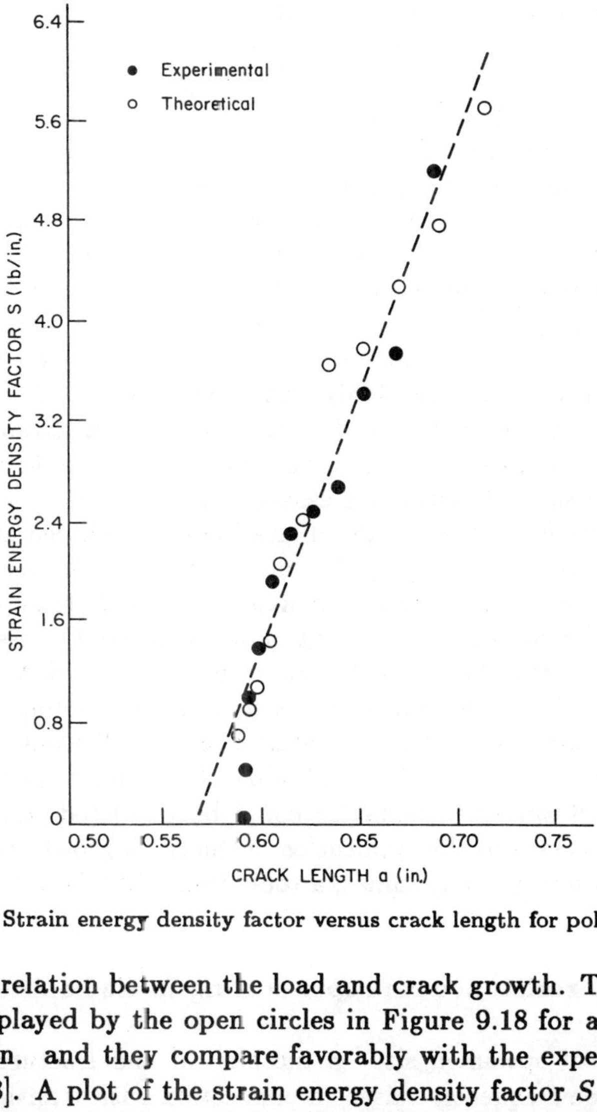

Figure 9.19. Strain energy density factor versus crack length for polycarbonate.

to predict the relation between the load and crack growth. The theoretical results are displayed by the open circles in Figure 9.18 for a starter crack of $a = 0.591$ in. and they compare favorably with the experimental data of Bernstein [8]. A plot of the strain energy density factor S against crack length can be found in Figure 9.19. The solid circles are obtained simply by multiplying $(dW/dV)_c^*$ and the experimentally measured crack growth increment Δa. Indeed, a straight line relation is obtained in the S versus a plot. Correlation based on the strain energy density theory gave a standard deviation* of 32.02% for the experimental data and 25.03% for the theoretical prediction. As mentioned earlier, the theory of plasticity cannot account

* The data of Bernstein [8] were interpreted in terms of the J parameter which varied nonlinearly with crack length and resulted in a standard deviation of 69.54% for this example. No theoretical prediction could be made from the J-integral because it does not apply to nonlinear material that experiences irreversible deformation.

for strain rate effects ahead of the crack and hence no general conclusion can be made at present. This, however, will not alter the qualitative character of the $dS/da = \text{const.}$ prediction.

Additional remarks. The rate at which energy is dissipated in a unit volume of material is intimately associated with permanent deformation. Yielding, as assumed in the theory of plasticity, addresses the extent of irreversibility of a solid when it is damaged by excessive distortion or shape change. The onset of this process is generally taken to coincide with the threshold of the distortional component of the strain energy density function known as the von Mises yield condition. The influence of the dilatational component, however, is not negligible in a region local to the crack tip where the strains and strain rates are highly nonhomogeneous. Moreover, it would not be possible to separate the distortional and dilatational effects when the response is nonlinear. It would be more appropriate to apply the total strain energy density function for assessing material damage such that the proportion between distortion and dilatation is not chosen arbitrarily.

The advantage of associating dW/dV with failure is because it can be easily obtained as area under the uniaxial stress and strain curve. Since it can also be calculated for an element under multiaxial stress and/or strain field, a means for translating the test data to the design of structural components can thus be established. This is essential for resolving the well-known specimen size and time effects in material testing. It would be desirable if large specimen behavior could be obtained from small specimen test data and long-period specimen behavior could be predicted from short-period test data. This leads to the application of the crack growth resistance curve or R-curves in fracture mechanics, a topic that will follow.

9.4. Damage resistance concept: scaling in size and time

The influence of specimen size on mechanical and fracture properties of metals has been frequently raised in connection with translating specimen test data to the design of structural components. Experiences have shown that the apparent strength of a larger sized specimen is lower than a small sized specimen of the same geometry and material. This so-called 'size effect' has been attributed to the presence of randomly distributed flaws in a given volume of material. Extreme value statistics have been applied by Weibull [37] to assess the variations of strength with specimen size. The simplest one-parameter relation is of the form

$$\frac{\sigma_1}{\sigma_2} = \left(\frac{V_2}{V_1}\right)^{1/\xi} \tag{9.24}$$

in which σ_1 and σ_2 are the average strengths measured in the tests of two specimen sizes for which the flaw distribution volumes are V_1 and V_2, respectively. Test results have been obtained for unprepared glass fibers whose surfaces are relatively rough. The data gave an ξ-value of 4 while $\xi = 12$ corresponded to fibers with coating that are protected from damage by handling. The test volume was changed in proportion to the fiber length. According to equation (9.24), the longer fiber with the larger volume of material damage has the lower strength. In these tests, the fibers snapped with little or no noticeable change in shape. For steels fractured in a brittle fashion by tension and bending, the ξ-value was found to be 25. When the toughness of the steel specimens was increased such that general yielding occurred prior to material separation, the data did not correlate well with equation (9.24) as ξ varied from 50 to 100. This implies that the volume change ratio V_2/V_1 is not sufficient for describing yielding. The surface area change ratio A_1/A_2 should also be considered and the process becomes sensitive to load–time history. A modified version of equation (9.24) can be written as

$$\frac{\sigma_1}{\sigma_2} = \left(\frac{V_2}{V_1}\right)^{1/\xi} \left(\frac{A_1}{A_2}\right)^{1/\eta} \tag{9.25}$$

with $\eta \to \infty$ for the case of brittle failure. Here, A_1/A_2 are proportional to the yield surface of two different sized specimens. Although the statistical approach to size effect may be useful in a practical way, it does not provide any in-depth understanding of the physical process of material damage.

The mechanical and fracture properties are also affected by stress and time or loading rate even if the specimen sizes are unchanged. As the material is stressed, say uniaxially, flaws coalesce and develop into cracks which extend until one crack dominates and reaches critical size, thus breaking the specimen. As a given crack extends, the local stress intensifies due to increasing crack size and the process feeds on all the smaller cracks and flaws nearby. The exact nature of the global response depends on the rate at which energy is transferred to different locations in the system. Interaction of loading rate with material damage is therefore essential and must be accounted for in any analytical model that attempts to explain size and time effects. In what follows, attention will be focused on the rate of energy dissipation in a unit volume of material for assessing the data of small and large specimens and slow and fast tests.

Dominant crack model. When the stress state in a specimen is dominated by the influence of a single macrocrack, it suffices to concentrate on the local region of prospective crack extension. The resistance of the material against complete separation of the specimen is known as toughness in fracture mechanics. An appropriate testing method has been proposed by ASTM [3] to measure this toughness so as to avoid variation in test results due to choice of specimen size or thickness. The relevant parameter is the crack tip stress intensity factor K_1. For a small crack of length $2a$ in a large plate subjected to uniform applied stress σ, the K_1 parameter increases with the square root of the half crack length*, i.e.,

$$K_1 = \sigma\sqrt{\pi a}. \tag{9.26}$$

A critical value of K_1 or K_{1c} can be associated with the initiation of crack growth. The amount of slow crack extension preceding rapid propagation depends on the specimen size and/or thickness.

Thickness effect. Test results on metals have shown that the quantity $\sigma_c\sqrt{\pi a}$ in equation (9.26) is a function of the specimen thickness and may not correspond to K_{1c}. For the thinner specimens, $\sigma_c\sqrt{\pi a}$ is relatively high and the fracture surface is slanted through the thickness. As the specimen thickness is increased, $\sigma_c\sqrt{\pi a}$ decreases while a portion of the fracture surface becomes flat or normal to the plate surface and the amount of slow crack growth is also reduced. This mid-portion of the curve shown in Figure 9.20 may be regarded as the fracture mode transition region from ductile to brittle. There prevails a specimen thickness beyond which $\sigma_c\sqrt{\pi a}$ no longer varies** while the fracture surface is completely flat with little or no observable slow crack growth. This is the ASTM valid K_{1c}-value used to rank the toughness of metal alloys.

Since the K-concept derives from the theory of linear elasticity, it cannot be applied to explain the behavior of ductile fracture where irrecoverable energy is dissipated during crack growth. The path-independent J-integral is also invalid because it is based on the assumption that the physical system must undergo reversible deformation. This excludes the inherent behavior of material damage. A detailed account of the limitations of linear elastic fracture mechanics, J-integral – and other failure criteria when applied to ductile fracture – can be found in the work of Sih [14].

As specimen size is increased, more energy can be stored in the system and released suddenly to initiate rapid fracture. The lower energy level in the

---

* There is no physical significance attached to the arbitrary insertion of the factor $\sqrt{\pi}$ other than to absorb π into K_1 in the energy release rate expression $G = K_1^2/E$ for plane stress. In retrospect, this choice of definition served no purpose as the stress ahead of the crack is always in a state of plane strain. The factor $(1 - \nu^2)$ is needed between G and K_1^2/E and defeats the original intention of simplifying the $G - K_1$ relationship.

** A plastic zone correction factor $K^2/2\pi(\sigma_{ys})^2$, with σ_{ys} being the yield strength, has been used to adjust the original crack length in order to eliminate the specimen size effect upon $\sigma_c\sqrt{\pi a}$. This empirical approach is based on the argument that a segment of material along the line of prospective crack extension is completely relaxed by plastic deformation. Macroscopic yielding occurring off to the sides of the crack, however, is more significant than that directly ahead of the crack.

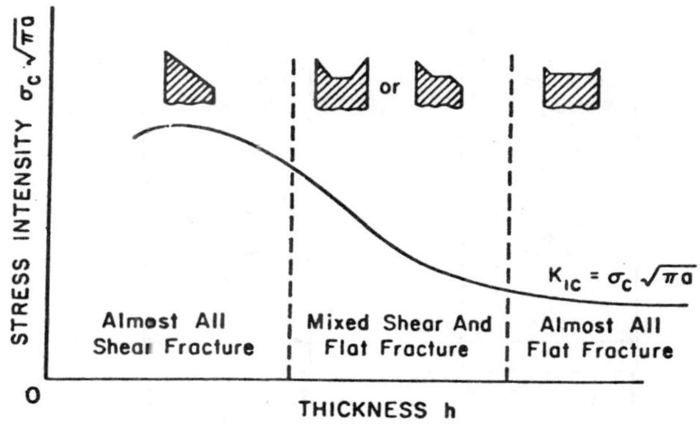

Figure 9.20. Variation of critical stress intensity factor with specimen plate thickness.

smaller specimens enhances slow energy dissipation by yielding and/or slow crack growth. This rate-dependent process is not governed by geometric similitude but rather by the rate at which energy is dissipated in a unit volume of material. Specimen size effect will not be understood unless the energies dissipated by yielding and fracture are accounted for simultaneously for each increment of loading.

Gross yielding. The stress intensity factor K_1 in equation (9.26) can be related to the strain energy density factor S given in equation (9.19) as

$$S = \frac{(1+\nu)(1-2\nu)K_1^2}{2\pi E}. \tag{9.27}$$

Equation (9.27), however, is valid only when fracture occurs suddenly, at which point a relation prevails between S_c and K_{1c}. There corresponds a critical ligament size r_c which triggers the onset of rapid fracture. Unlike K_1, the factor S defined in the strain energy density* criterion applies equally well to materials that exhibit a nonlinear failure behavior. It can be used to characterize the resistance of material against yielding and fracture when specimen size and loading rate are altered.

Depending on the combination of component size, loading rate, temperature and material, failure by crack growth can occur in several different ways. That is, the creation of macrocrack surface may not always take place in accordance with equation (9.27). Possible failure modes may be described by referring to the center-cracked specimen loaded in tension, as illustrated in Figures 9.21(a) to 9.21(d). Then yielding is confined to the immediate vicinity of the crack such that the energy released during crack growth is pre-

* For a nonlinear irreversible material, S is no longer related to the stress intensity factors and must be determined from appropriate nonlinear theories.

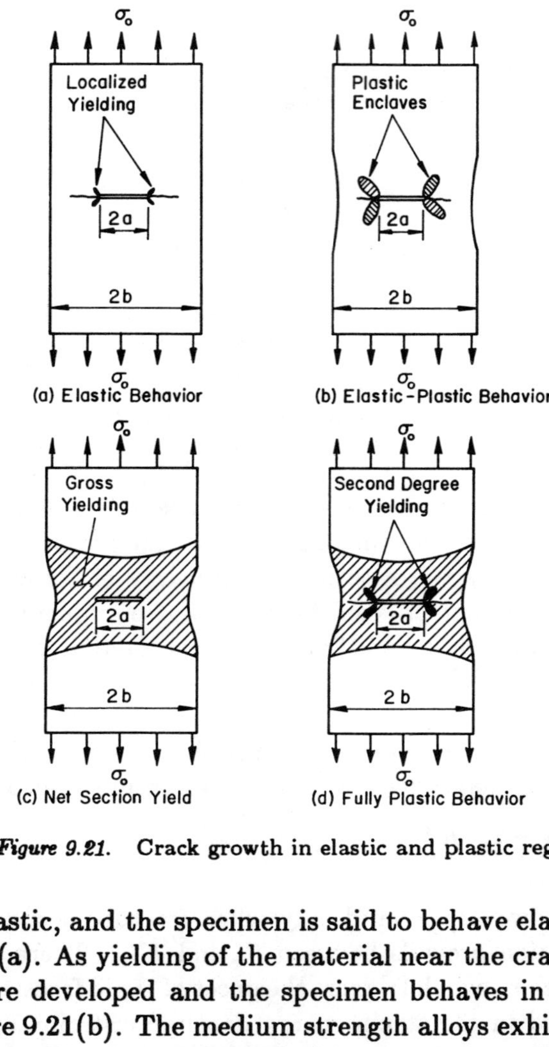

Figure 9.21. Crack growth in elastic and plastic regime.

dominantly elastic, and the specimen is said to behave elastically* as shown in Figure 9.21(a). As yielding of the material near the crack increases, plastic enclaves are developed and the specimen behaves in an elastic–plastic manner, Figure 9.21(b). The medium strength alloys exhibit this characteristic. Crack growth takes place in the elastic portion of the elastic–plastic material. The creation of plastic enclaves requires additional energy which is beneficial in that it reduces the available energy to drive the crack. This increases the loading capacity of the system if the same amount of input energy were transferred to the specimen in Figure 9.21(a), say, made of a brittle material and in Figure 9.21(b), say, made of a ductile material. As long as crack growth at instability releases elastic energy, even though material away from the crack path is yielded, the fracture toughness value is still governed by K_{1c} or S_c as related by equation (9.27). Any attempts

* Elasticity does not imply linearity. The global behavior of a specimen can be nonlinear even if the material remains within the elastic range. Nonlinearity could be attributed by nonuniform and/or nonself-similar crack growth [20].

made to introduce or define a different fracture toughness for the situation in Figure 9.21(b) would be fundamentally incorrect.

For a very ductile material and/or very slow loading rates, yielding can spread to the specimen boundary prior to crack growth. This situation is shown in Figure 9.21(c) and can be described as gross yielding. Failure can occur by plastic collapse without macrocracking, Figure 9.21(c), or by macrocracking, Figure 9.21(d). The presence of a crack causes the material near the tips to yield more intensely. This is referred to as *second degree* yielding in Figure 9.21(d). Since crack growth now takes place in the yielded material, the energy required to create the free crack surfaces would be different from that required if the material were elastic. The energy released for a segment of crack growth in the elastic material, Figure 9.21(a) or 9.21(b), will obviously differ from that for the material in the plastic state, Figure 9.21(c) or 9.21(d) even though global instability may be reached in both situations. Energy release is a rate-dependent process that is not uniquely associated with material type unless all the load and geometric parameters are held constant.

Crack growth resistance curves. Numerous attempts have been made in the past to seek a parameter that can uniquely characterize the failure behavior of materials whose load and displacement responses are highly nonlinear. For the dominant crack model, the crack growth rate serves as a measure of the degree of material damage. The objective is to establish a linear relationship between the fracture and damage parameter such as that illustrated in Figure 9.19 for the polycarbonate material.

Two-dimensional line crack model. The straight line relationship between the strain energy density factor S and half crack length a provides a simple means for evaluating the specimen size, loading rate and material type effects. Sih and Tzou [11] have considered the two-dimensional through-crack model in Figure 9.22. A slowly rising monotonic load in increments of $\Delta\sigma$ will be applied until σ_0 is reached while specimen size, loading step and material type are varied. The dimensions b, c and h in Table 9.4 are chosen such that their ratios always remain constant while the volume to surface ratio of the specimens is varied as $V/A = 0.435$, 1.154 and 2.693 in. The three loading steps are $\Delta\sigma = 35.0$, 27.5 and 22.5 ksi while three different materials are also selected, as shown in Table 9.5 and Figure 9.23. The quantity $(\mathrm{d}W/\mathrm{d}V)_c$ represents the total area under the true stress and true strain curve at fracture and $(\mathrm{d}W/\mathrm{d}V)_c^*$ pertains to that portion converted to create a free macrocrack surface. The difference is dissipated via plastic deformation and is not available during crack growth.

Figure 9.24 shows the crack growth resistance curves for the geometrically similar specimens where a constant load step of $\Delta\sigma = 35.0$ ksi is used.

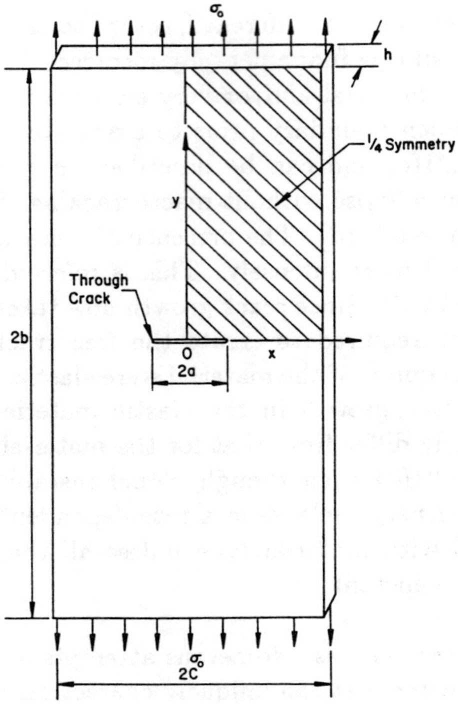

Figure 9.22. **Center-cracked specimen subjected to incremental tensile load.**

Table 9.4. **Geometrically similar specimens.**

Specimen type	V/A (in.)	Dimensions a (in.)	b (in.)	c (in.)	h (in.)
I	0.435	1.0	10.0	5.0	1.0
II	1.154	3.0	30.0	15.0	3.0
III	2.693	7.0	70.0	35.0	7.0

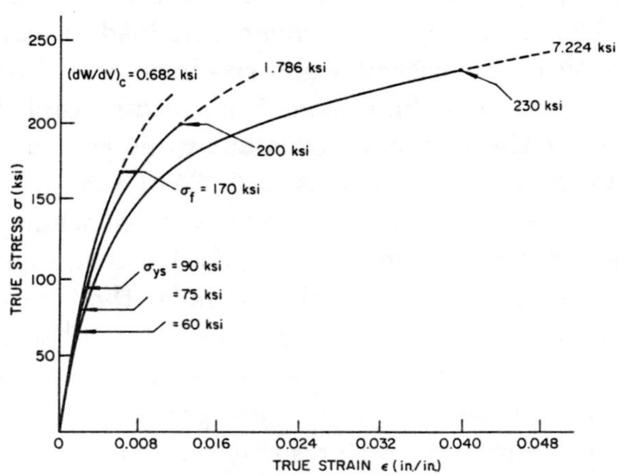

Figure 9.23. **Stress and strain curves for three different metal alloys.**

Table 9.5. Three different metal alloys.

Material Type	σ_{ys} (ksi)	σ_f (ksi)	$(dW/dV)_c$ (ksi)	$(dW/dV)_c^*$ (ksi)
I	90	170	0.682	0.481
II	75	200	1.786	0.667
III	60	230	7.224	0.882

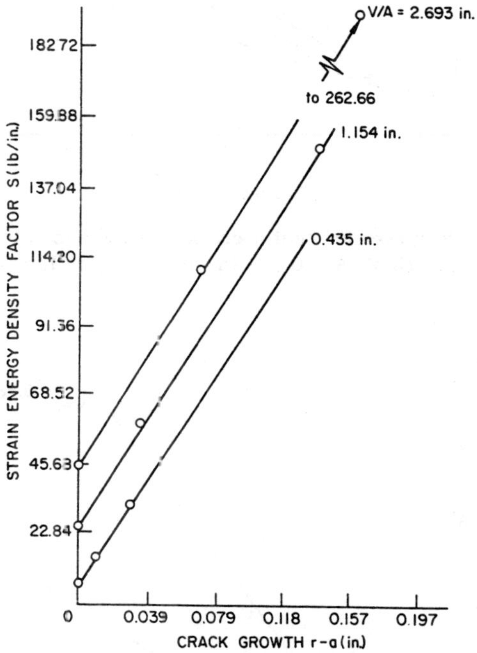

Figure 9.24. *R*-curves for geometrically similar center-cracked panel with $\Delta\sigma = 35.0$ ksi and $(dW/dV)_c^* = 0.667$ ksi.

The material has a $(dW/dV)_c$ value of 1.786 ksi. For a given material or constant S_c value, the smaller specimen with the lower V/A ratio sustains more subcritical crack growth. This explains the tendency for the thicker or larger specimen to behave in a more brittle fashion, i.e., failure with little or negligible amount of slow crack growth.

By fixing $V/A = 0.435$ in. for the $(dW/dV)_c = 1.786$ ksi material, the loading step $\Delta\sigma$ is varied from 22.5 to 35.0 ksi as illustrated in Figure 9.25. The S versus a lines are again straight but they now rotate in a counterclockwise direction as $\Delta\sigma$ is increased. This trend confirms the well-known experimental observation that small load steps enhance slow crack growth while large load steps lead to brittle-like fracture.

Material properties such as yield strength and fracture toughness are known to influence the crack growth characteristics. This is illustrated by the *R*-curves in Figure 9.26 for materials with $(dW/dV)_c^* = 0.481$, 0.667 and 0.882 ksi. Volume to surface ratio V/A and loading step $\Delta\sigma$ are fixed

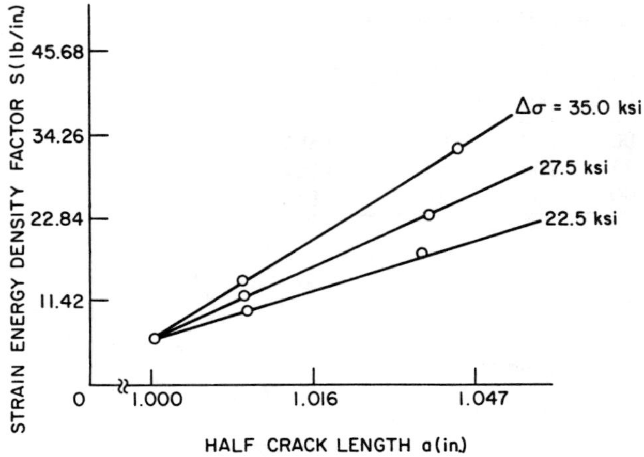

Figure 9.25. **R-curves for center-cracked panel subjected to different load steps with $V/A = 0.435$ in. and $(dW/dV)_c^* = 0.667$ ksi.**

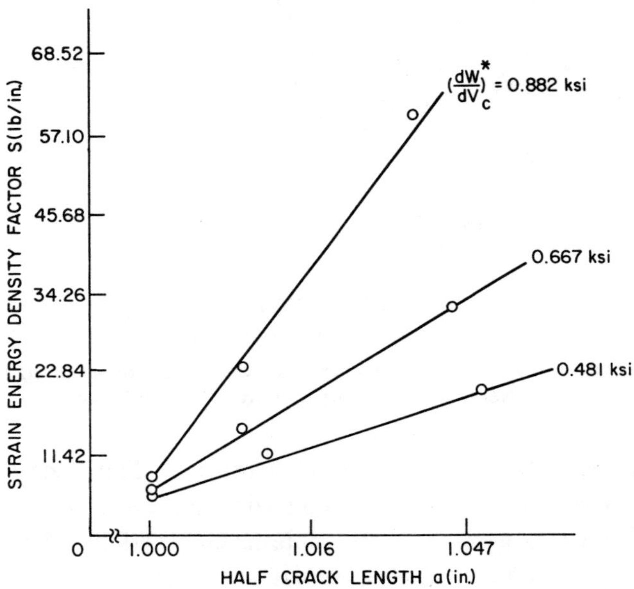

Figure 9.26. **R-curves for center-cracked panel of different materials with $V/A = 0.435$ in. and $\Delta\sigma = 35.0$ ksi.**

at 0.435 in. and 35.0 ksi, respectively. The S versus a line rotates counterclockwise as the material toughness is increased. Here, S_c^* will intersect each line at a different location with the tougher material offering better resistance to failure by slow crack growth.

In view of the results in Figures 9.24 to 9.26, the combined influence of specimen size, loading rate or step and material type can be summarized by the translation and rotation of the S–a lines as summarized in Figures 9.27(a) to 9.27(c). Only a limited number of fracture tests would be needed

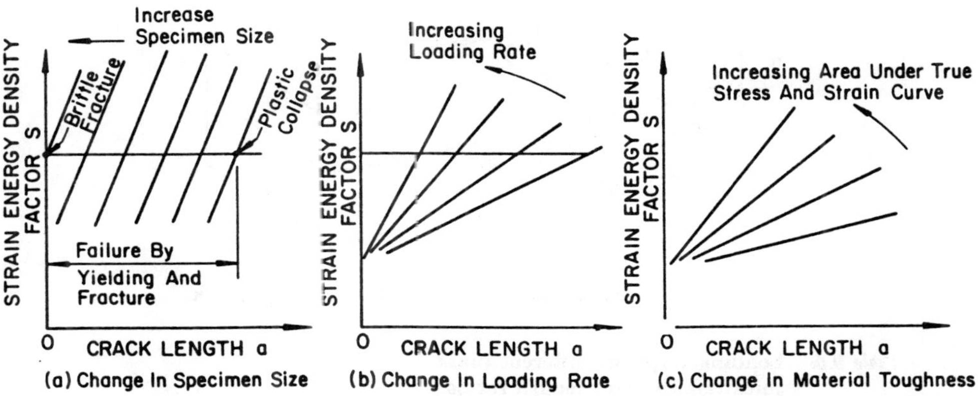

Figure 9.27. Schematic of resistance curves for changes in specimen size, loading rate and material toughness.

to obtain a complete description of failure modes from plastic collapse to brittle fracture for different loading rates and material types described by uniaxial specimen data. Small specimen tests involving slow crack growth can be used to predict the brittle fracture of larger specimens or the corresponding K_{1c} value. An order of magnitude prediction in specimen size was made by Carpinteri and Sih [39] for the three-point bend concrete specimen with an edge crack.

Similarly, by relating the data of specimens subjected to very small load increments that fail over a long period of time to those which failed more quickly when load step or rate is increased, as shown schematically in Figure 9.27(b), a considerable gain in testing time can be accomplished. This is particularly useful for establishing creep and fatigue* data in design.

The way uniaxial data influences subcritical crack growth can be analyzed from the R-curves in Figure 9.27(c). They are useful for pin-pointing the trade-off between yield strength and fracture toughness in a given application.

Three-dimensional through-crack model. Ductile fracture is inherently a three-dimensional material damage process [12] depending on the geometry of the specimen and crack configuration. Figure 9.28 refers to a through crack of length $2a$ in a finite thickness plate stretched uniformly in increments of $\Delta\sigma$ up to σ_0. The material next to the crack border is yielded nonuniformly through the thickness and attains a curve shape as growth

---

* R-curves similar to those in Figures 9.24 to 9.26 for monotonic loading were also found for the case of fatigue by Sih and Chao [35].

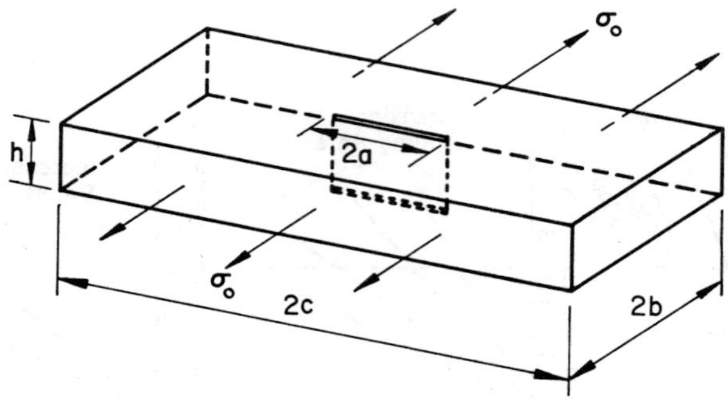

Figure 9.28. Three-dimensional model of through crack.

Table 9.6. Loading steps for different plate thickness to crack length ratios.

Load step $\Delta\sigma$ (ksi)	Thickness to crack length ratio $h/2a$
0.3	0.15
0.5	0.20
1.0	0.25

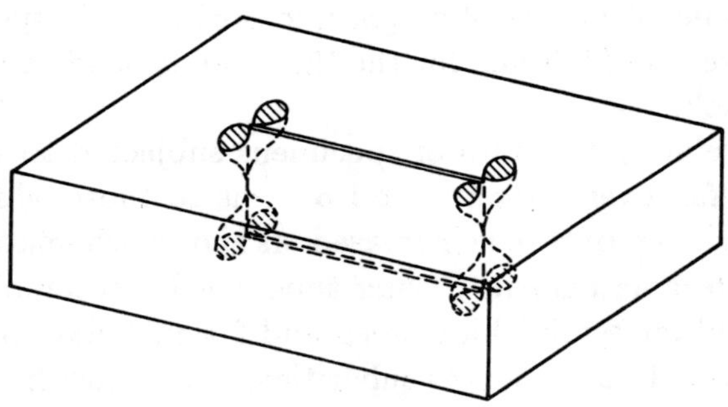

Figure 9.29. Yield zone along crack border.

progresses. A schematic of the yield zone around such a crack front is shown in Figure 9.29. Material elements near the plate surface are more readily distorted and therefore experience more yielding than those in the interior.

The original crack front in the plate is assumed to be straight and its length is kept at $a = 2.0$ in. with $b/a = 1.0$ and $c/a = 2.0$. Table 9.6 gives the three different loading steps in accordance with $h/2a$ that were analyzed in [12]. The first step represents the load at crack growth initiation. As the load is increased, the crack grows incrementally with distances r_1, r_2, etc., at the mid-point of the plate shown in Figure 9.30 and adopts a curved front in a nonself-similar manner. The profiles become more and more elongated

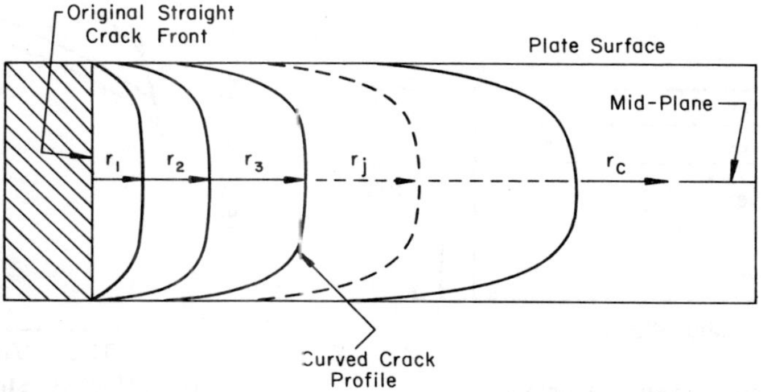

Figure 9.30. **Incremental crack growth leading up to global instability.**

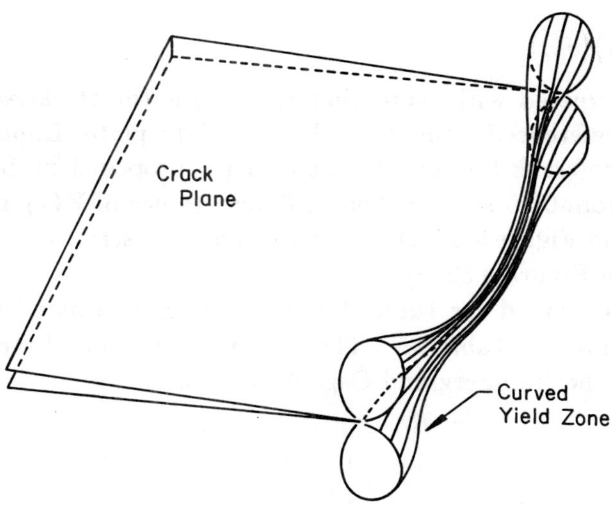

Figure 9.31. **Schematic of plastic zone along curved crack border.**

as the point of global instability is approached. This corresponds to the last increment of slow crack growth with $r = r_c$. Figure 9.30 displays some typical profiles that were predicted from the strain energy density theory in conjunction with the incremental theory of plasticity with modification made on the apparent yield strength through the plate thickness. The yield zone would also be curved along the crack front, as shown in Figure 9.31.

Because of the large variation in constraint on the elements near the curved crack front, the ASTM recommended [40] tensile yield strength $(\sigma_{ys})_0$

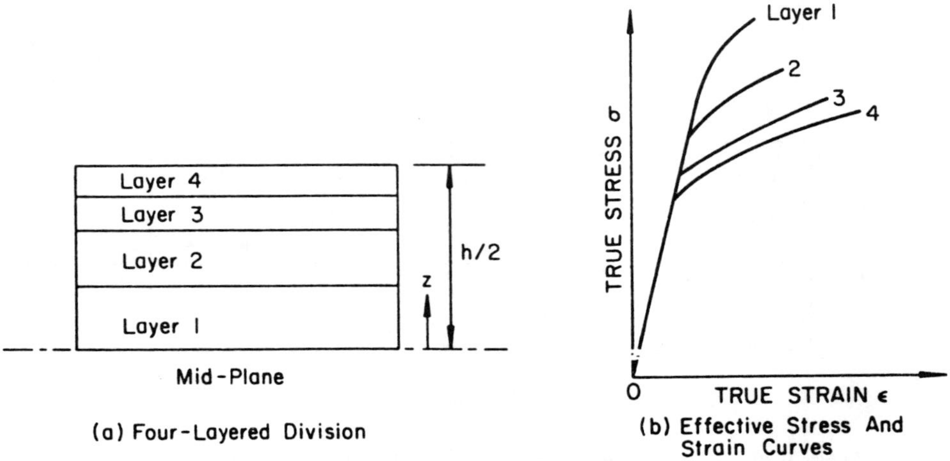

Figure 9.32. Stress and strain curves with modified yield strength.

is modified as

$$\sigma_{ys} = Y(z)\left(\sigma_{ys}\right)_0 \tag{9.28}$$

where $Y(z)$ decreased with increasing z, which is the thickness coordinate in Figure 9.28 referenced from the mid-plane of the plate. Depending on the curvature of the crack border, the factor $Y(z)$ proposed by Sih and Chen [12] can vary from 0.70 to 1.50. Four different values of $Y(z)$ are chosen for the four layers in Figure 9.32(a). The corresponding stress and strain curves can be found in Figure 9.32(b).

Results are obtained for three different material properties having the properties outlined in Table 9.7. Their uniaxial stress and strain behavior is modelled by the Ramberg and Osgood relation

$$\epsilon = \begin{cases} \dfrac{\sigma}{E}, & \sigma < \sigma_{ys} \\ \dfrac{1}{E}\left[\sigma + \phi\left\{\left(\dfrac{\sigma}{\sigma_{ys}}\right)^{\psi} - 1\right\}\sigma_{ys}\right], & \sigma > \sigma_{ys} \end{cases} \tag{9.29}$$

in which ϕ and ψ are the hardening coefficients given in Table 9.7. Young's modulus $E = 30 \times 10^3$ ksi and Poisson's ratio $\nu = 0.3$ are the same for all three materials.

Based on the assumption that the fracture toughness $(dW/dV)_c$ or S_c is independent of the specimen thickness and loading step, crack growth profiles can be developed for each of the materials in Table 9.7 as the specimen sizes and loading steps are varied in accordance with those in Table 9.6. This makes a total of 27 different sets of crack profiles which were calculated numerically in [12]. The onset of rapid crack propagation is then determined by matching the last ligament of crack growth r_c for each ma-

Table 9.7. Mechanical properties of materials for plate with through crack.

Material type	Yield strength $(\sigma_{ys})_0$ (ksi)	Critical strain energy density $(dW/dV)_c$ (ksi)	Hardening coefficients ϕ	ψ
I	80	0.15	0.2	2.0
II	70	0.20	0.5	5.0
III	58	0.25	0.5	5.0

Table 9.8. Calculated average critical strain energy density factor for different materials with $(dW/dV)_c$ given in Table 9.7.

Material type	r_c (in.)	$\overline{S}_c$ (lb/in.)
I	0.100	15
II	0.115	23
III	0.132	23

terial. Tabulated in Table 9.8 are the average values of r_c, from which S_c can be calculated from $r_c(dW/dV)_c$. Since ν and E for each material are already known, K_{1c} follows immediately from equation (9.27). Note that material I has the highest yield strength and the lowest K_{1c} value while the opposite holds for material III. This trade-off between σ_{ys} and K_{1c} is well known.

It is of interest to compare the results predicted from the three-dimensional model with those of ASTM based on the two-dimensional plane strain assumption [3]. To this end, an effective half crack length $\overline{a}$ will be defined by equating the area $\overline{A}$ of a rectangle to the actual fracture surface A with a curved crack front, as illustrated in Figure 9.33. Similarly, an effective strain energy density factor $\overline{S}$ can also be defined as

$$\overline{S} = \frac{\sum_{j=1}^{n} r_j (dW/dV)_c z_j}{\sum_{j=1}^{n} z_j} \tag{9.30}$$

in which r_j and z_j are the coordinates of a point on the crack border. Table 9.9 gives the values of $\overline{S}_c$ calculated from equation (9.30) from which K_{1c} can be found by means of equation (9.27). The minimum thickness $h_{\min}$ according to ASTM [3] is obtained from

$$h_{\min} \geq 2.5 \left[\frac{K_{1c}}{(\sigma_{ys})_0} \right]^2 . \tag{9.31}$$

It is also possible to impose limits on the specimen thickness for plane strain fracture directly from the strain energy density prediction. This is accomplished by constructing plots of $2\overline{a}/h$ versus $\overline{a}$ for a given material

Table 9.9. ASTM specimen thickness requirement for valid K_{1c} test.

Material type	S_c^* (lb/in.)	K_{1c} (ksi$\sqrt{\text{in.}}$)	$h_{min}/2$ (in.)
I	15	11.75	1.060
II	23	14.54	2.126
III	33	17.43	4.450

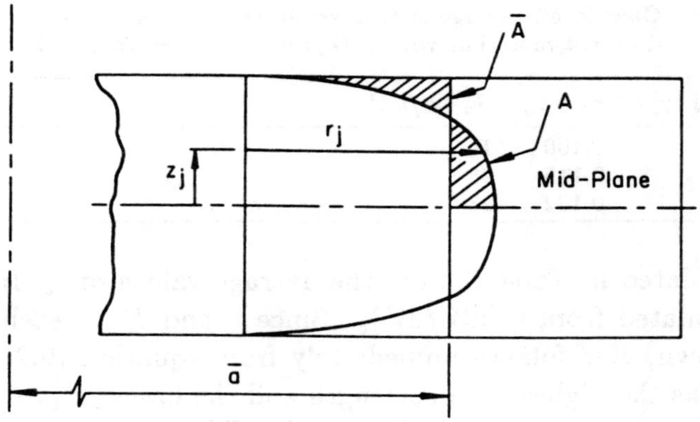

Figure 9.33. Effective crack length.

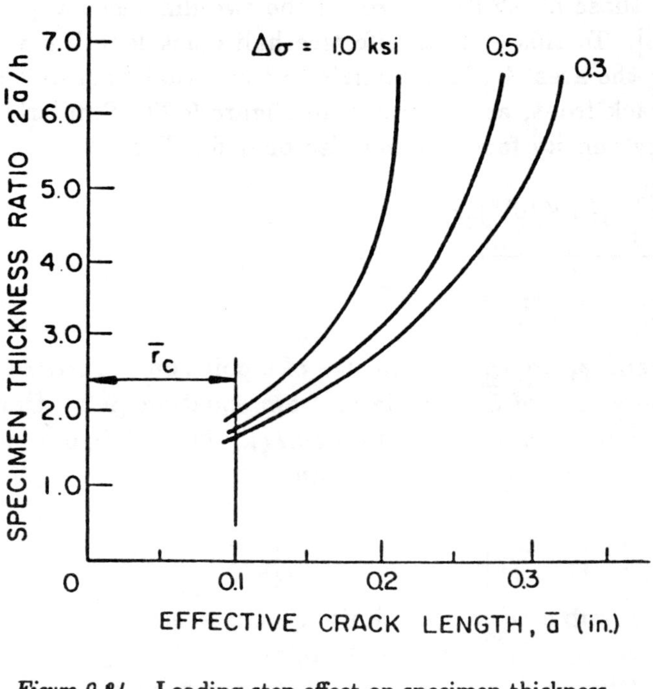

Figure 9.34. Loading step effect on specimen thickness.

Table 9.10. Influence of loading step on thickness requirement for brittle fracture.

Material type	Load step (ksi)	Thickness $h_{min}/2$ (in.)
I	0.3	1.189
	0.5	1.177
	1.0	1.051
II	0.3	3.115
	0.5	2.098
	1.0	1.598
III	0.3	5.000
	0.5	4.161
	1.0	2.768

and loading step. Figure 9.34 displays such a plot for material I undergoing loading steps of $\Delta\sigma = 0.3$, 0.5 and 1.0 ksi. All curves tend to converge to a common point. Their intersection with the effective critical ligament size* $\bar{r}_c$ gives the limit corresponding to brittle or sudden fracture with no slow crack growth, i.e., when $\bar{a}_c = \bar{r}_c$. A set of values for h_{min} can thus be obtained and are given in Table 9.10 for material type II and III as well. The minimum thickness h_{min} is seen to vary with loading step, an influence that has not been considered by ASTM, equation (9.31). Referring to Table 9.10, the average values of $h_{min}/2$ for materials I, II and III are, respectively, 1.139, 2.270 and 3.976 in. This trend is similar to the ASTM recommendations in Table 9.9. The variation of h_{min} with $\Delta\sigma$ is not significant for the stiffer material type I. However, $\Delta\sigma$ does play a role when the materials are more ductile. It is evidenced from Table 9.10 for materials II and III that h_{min} varies substantially with $\Delta\sigma$. For instance, the predicted minimum thickness for material III with $\Delta\sigma = 1.0$ ksi is about 45% less than that for $\Delta\sigma = 0.3$ ksi. The additional requirement on loading step for the K_{1c} test can be important.

The material properties in Table 9.7 refer to typical engineering metal alloys used in structural applications. It is, therefore, useful to construct *R*-curves such that specimen size, loading step and material type effects can be summarized graphically. For the sake of brevity, only a few of the 27 cases treated by Sih and Chen [12] will be discussed. Figure 9.35 gives a plot of $\bar{S}$ versus $\bar{a}$ for material I with $\bar{S}_c = 15$ lb/in. and loading step of $\Delta\sigma = 0.3$ ksi. Three parallel lines are obtained for $h/2a = 0.15$, 0.20 and 0.25. The interactions of the line $\bar{S}_c =$ const. with these lines give the effective critical crack lengths $\bar{a}_c$. It is apparent that the thinner specimens with the lower $h/2a$ ratio sustain more slow crack growth prior to the onset of global instability. As the loading step $\Delta\sigma$ is increased, the spacing between the parallel lines becomes narrower. This reduces slow crack growth and

* The effective size $\bar{r}_c$ is equal to the average size r_c in Table 9.8.

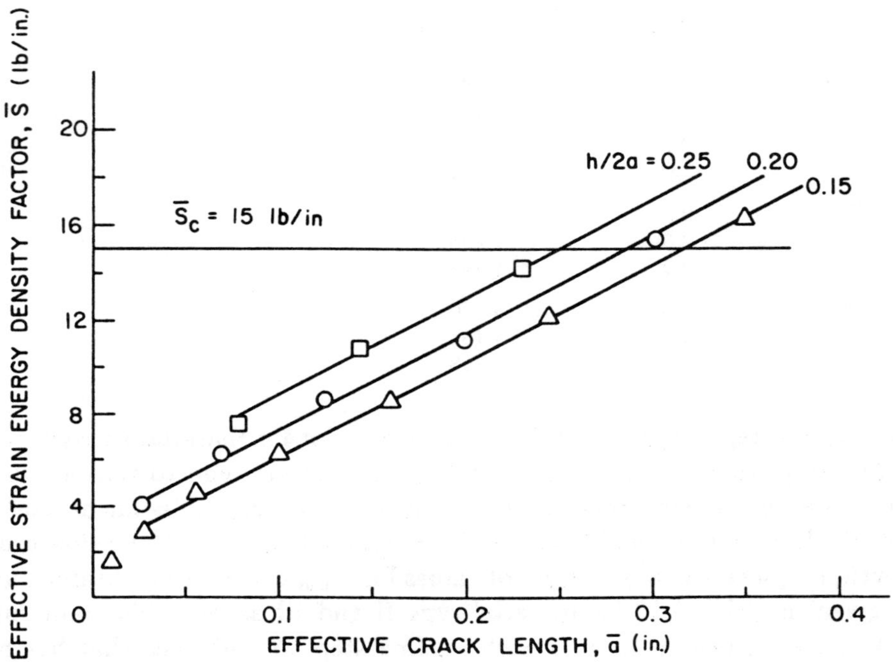

Figure 9.35. **R-curves for through thickness crack of different $h/2a$ ratios with $\Delta\sigma = 0.3$ ksi and $(dW/dV)_c = 0.15$ ksi.**

is not shown in Figure 9.35. Hence, specimen thickness effects become less influential in crack growth for higher loading steps. Similar trends have been obtained by Sih and Chen [12] for materials II and III.

By holding $h/2a = 0.15$ for material I, $\Delta\sigma$ can be varied to obtain an $\overline{S}$ versus $\bar{a}$ plot. The results are summarized in Figure 9.36 for three different values of $\Delta\sigma$. Again, straight lines are obtained but their slopes are now different and they increase with load step. All lines rotate in a counterclockwise manner as $\Delta\sigma$ is increased. This effect is not as pronounced for the thicker specimens.

The crack growth resistance curves for materials I, II and III with $h/2a = 0.15$ and $\Delta\sigma = 0.3$ ksi are given in Figure 9.37. Counterclockwise rotation of the lines is observed when the material toughness $(dW/dV)_c$ increases. Material III, being tougher, sustains more slow crack growth than materials I and II. This can be established by locating the interactions of the lines with $\overline{S}_c =$ const. for the three different materials I, II and III. The significant differences become more for larger loading steps.

The selection of a failure criterion for constructing R-curves has been problematic. A common procedure is to simply plot test data of the assumed parameter against crack growth. Among the common candidates are the J-integral, COD and others as discussed by Sih [14]. These empirical

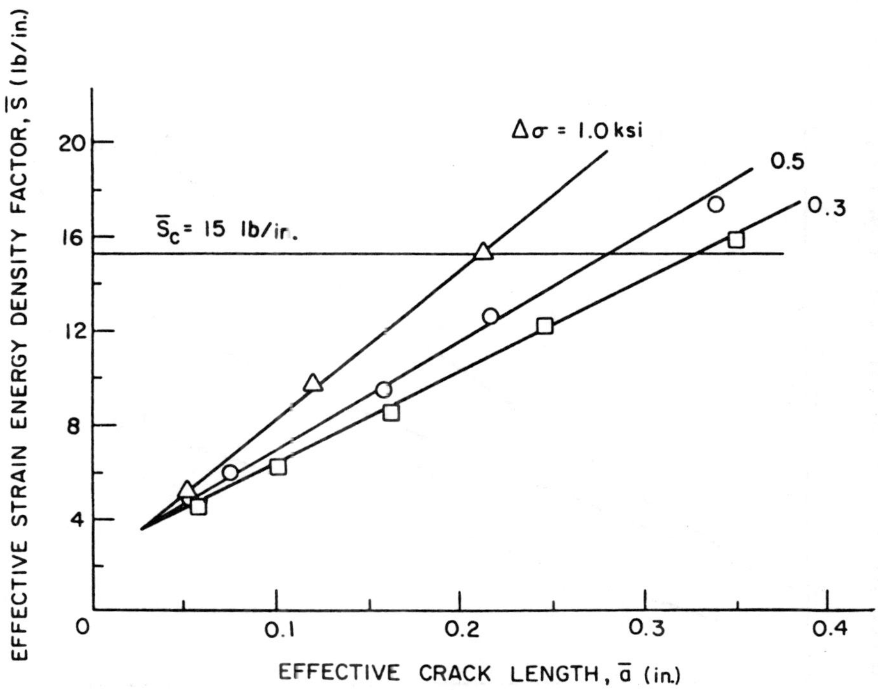

Figure 9.36. **R-curves for through thickness crack subjected to different load steps with $h/2a = 0.15$ and $(\mathrm{d}W/\mathrm{d}V)_c = 0.15$ ksi.**

approaches have no merits as they reproduce the nonlinear load–deflection relation in a nonlinear fashion. No advantage results as nonlinear data precludes the use of interpolation that simplifies the design rules. The straight line relation with $\mathrm{d}S/\mathrm{d}a =$ const. for constructing the R-curves is considered to be a fundamental feature of the strain energy density criterion. This criterion has been applied successfully to solve not just a few examples but a variety of crack problems with complex geometries and loading conditions [27]. Its versatility and consistency for explaining a wide range of physical phenomena makes it useful in application.

More specifically, the R-curves obtained earlier provide a quantitative assessment of the interaction of specimen size, loading step and material type. Their combined effect determines the failure mode that involves yielding and fracture. They are obtained for the three-dimensional elastic–plastic crack model and are qualitatively similar to those for the two-dimensional case. Of fundamental departure from the conventional approach is that K_{1c} and specimen thickness requirement are derived analytically from uniaxial data. Fracture testing is no longer required. Fracture resistance curves for fatigue cracks can be developed by the same approach using uniaxial data alone.

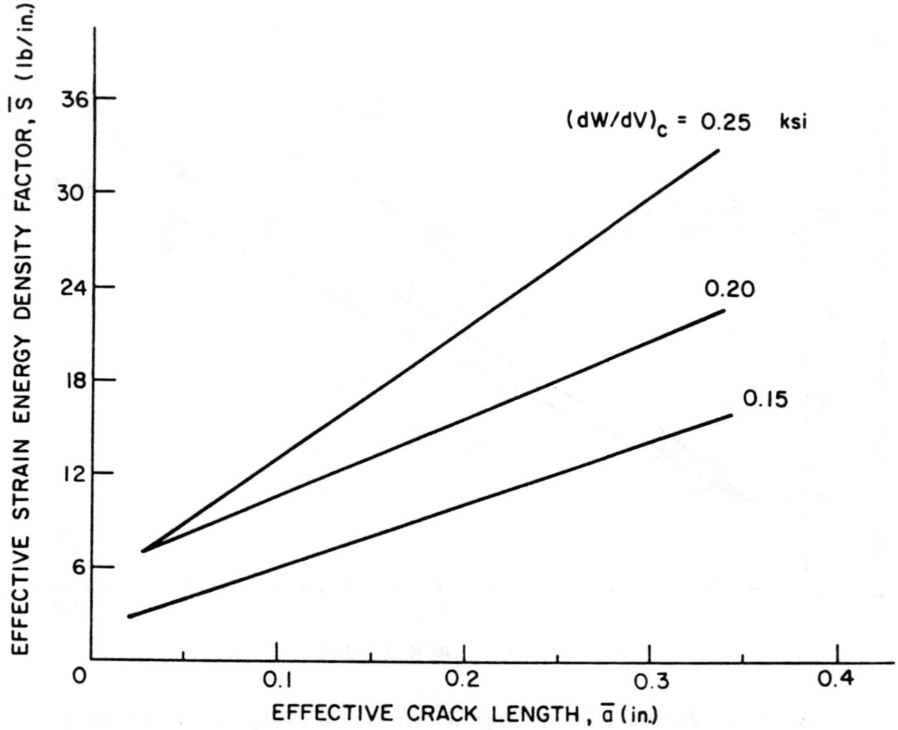

Figure 9.37. R-curves for through thickness crack of different materials with $h/2a = 0.15$ and $\Delta\sigma = 0.3$ ksi.

The versatility of the strain energy density method has thus been demonstrated in that the same criterion can be applied to all loadings, materials and specimen configurations.

Fatigue crack growth. The term 'fatigue' generally refers to material damage occurred through the load that has been cycled a number of times. The conventional approach of material testing does not account for the mode of failure or for the existence of initial flaws in the material and simply collects data on an uniaxial sinusoidally loaded specimen. This procedure results in a plot of the stress amplitude against the number of cycles at failure, commonly known as the SN-curves. Depending on the specimen type, material and nature of cyclic loading, the SN-curve data often contain large scatter that can lead to inaccurate prediction in the design life of structural components. The fracture mechanics approach to fatigue assumes a crack-like defect to exist in the material. Failure consists of an initial stage of subcritical crack growth until the crack size becomes critical and instability follows. Instability is defined as the sudden occurrence of rapid crack propagation to a point where the integrity of the specimen or structure has been seriously impaired and frequently to the point of complete material separation. Although attempts have been made to relate data collected on uncracked specimens to the cracked specimens, the results are either incomplete or restricted to special cases. There lacks a general framework within which the conventional material testing and crack growth data could be related.

A reliable estimate of the life expectancy of structural components necessitates the development of procedures for describing failure modes due to fatigue in terms of nominal parameters such as allowable loading (stress amplitude and number of life cycles), crack dimensions and material properties. Since flawed structures are not necessarily disqualified from service and may possess substantial remaining life, the capability for assessing the rate of crack growth is the prerequisite for assuring immunity of structures from unstable fracture. Although concepts and expressions for predicting fatigue crack growth are numerous in the literature, much of the existing work is concerned with specimen testing and data collection with little or no attempt made towards understanding the fundamental aspects of the fatigue process. Admittedly, fatigue is a complex phenomenon that cannot easily be translated into mathematical terms. The difficulties, however, are not being overcome in a consistent and systematic fashion. The mere collection of test data is not sufficient.

From the engineering viewpoint, the interest is in how an initial crack starts to grow in fatigue and how it propagates during the stage of subcritical growth. Under normal loading and environmental conditions, the design should be made such that the fatigue life is not so sensitive to changes in the microstructure of the material. Otherwise, cost of the product would be prohibitively high and considered to be impractical. To this end, it suffices to take the continuum mechanics approach which assumes that the fatigue properties of the uncracked material or continuum elements are obtainable from appropriate specimen testings. Recognizing that fatigue is basically a path-dependent process – i.e., the rate of crack growth is inherently governed by the history of loading – the irreversible damage of the material in the immediate vicinity of the crack tip region must obviously be recorded in some consistent fashion depending on the assumed criterion of failure. For each cycle of loading, the damage may be different and it is the accumulation of these damages over a number of cycles that leads to crack growth. Assumed is the accumulation of the strain energy density function, dW/dV, in an element ahead of the crack that leads to crack growth. The critical value of dW/dV is obtainable from an uniaxial specimen subjected to cyclic loading. For illustration, the incremental theory of plasticity will be used to calculate the strain energy density field ahead of the crack. The path-dependent nature of the fatigue crack growth process is discussed in connection with specifying either the increment of crack growth or the increment of number of load cycles. Because of the complexity involved in the nonlinear analysis, computation is carried out by the finite element method for the two-dimensional example of a central crack panel. There are no basic difficulties in the extension of the analysis to three dimensions. Once the nonlinear character of the fatigue crack growth process is understood, simpler models making use of the linear theory of elasticity can be developed and used within limited range of the operating parameters. Limitations on some of the current fatigue crack growth models are discussed in connection with their inability to account for changes in mean stress level and specimen geometry.

Damage cumulation. A material element can fail by stressing it repeatedly even if the stress or energy level is below the threshold. This process is known as fatigue where damage is accumulated according to the load cycle. Similar to the prediction of failure initiated under monotonically rising load, attention would be focused on an element at a finite distance ahead of the crack, as shown in Figure 9.15. The three hypotheses stated earlier in Section 9.3 remain valid because the strain energy density criterion applies to crack initiation when dW/dV reaches $(dW/dV)_c$ and to subcritical crack propagation when S_j/r_j equals $(dW/dV)_c$. The condition $(dW/dV)_c = S_c/r_c$, as given in equation (9.20), determines the onset of rapid fracture or global instability.

Sih and Moyer [41] applied the volume energy density theory to fatigue. An increment of crack growth Δr corresponding to ΔN cycles is assumed to occur when $(dW/dV)_{\min}$ in the leading element attains the critical value

$$\left(\frac{dW}{dV}\right)_c = \left(\frac{dW}{dV}\right)_0^{r=\Delta r} + \sum_{j=1}^{\Delta N} \Delta \left(\frac{dW}{dV}\right)_j^{r=\Delta r} \tag{9.32}$$

in which $(dW/dV)_0$ represents the static response due to the applied load and $\Delta(dW/dV)_j$ the amount of energy accumulated during the jth load cycle. Making use of equation (9.32), the accumulated energy becomes

$$\Delta \left(\frac{dW}{dV}\right)_j^{r=\Delta r} = \frac{(\Delta S)_j^{r=\Delta r}}{\Delta r} \tag{9.33}$$

with $(\Delta S)_j^{r=\Delta r}$ being defined as*

$$(\Delta S)_j^{r=\Delta r} = S(r = \Delta r;\ t = t_f^j) - S(r = \Delta r;\ t = t_0^j). \tag{9.34}$$

* Here, the symbol Δ stands for the change in a quantity from the beginning to the end of a cycle. The beginning and end do not necessarily correspond to the maximum and minimum values as in the case of linear theory.

In equation (9.34), t_f^j is the time at the completion of the jth cycle and t_0^j is the time at the start of the jth cycle. By means of equations (9.33) and (9.34), equation (9.19) can thus be applied to yield

$$\left(\frac{dW}{dV}\right)_c - \left(\frac{dW}{dV}\right)_0^{r=\Delta r} = \frac{1}{\Delta r}\sum_{j=1}^{\Delta N}(\Delta S)_j^{r=\Delta r}. \tag{9.35}$$

The above expression provides a relationship between the rate of crack growth, $\Delta r/\Delta N$, for a given crack configuration as a function of the local stress and strain response from which $(dW/dV)_0$ and ΔS_j can be determined. This rate can be written as*

$$\frac{\Delta r}{\Delta N} = \frac{dr}{dN} = \frac{\sum_{j=1}^{\Delta N}(\Delta S)_j^{r=\Delta r}}{\Delta N[(dW/dV)_c - (dW/dV)_0^{r=\Delta r}]} \tag{9.36}$$

which gives the crack growth rate relationship.

The evaluation of equation (9.36) is no small task. It involves many difficulties for an estimate of the accumulated energy that is needed for finding the summation. As a simplifying and conservative assumption, the damage accumulated during each cycle is taken to be constant and equal to the amount accumulated during the first cycle of the current increment of crack growth. Therefore, it suffices to calculate the energy accumulated for the first loading cycle of each crack length. Now, let ΔS be the change in S for the first cycle of energy accumulation, and equation (9.36) becomes

$$\frac{dr}{dN} = \frac{(\Delta S)^{r=\Delta r}}{(dW/dV)_c - (dW/dV)_0^{r=\Delta r}}. \tag{9.37}$$

The quantity $(dW/dV)_0^{r=\Delta r}$ may be defined by an average response of the fatigue process. For a sinusoidal or any other periodic and constant amplitude loading, $(dW/dV)_0^{r=\Delta r}$ can be calculated from

$$\left(\frac{dW}{dV}\right)_0^{r=\Delta r} = \frac{1}{2}\left\{\frac{1}{2}\left[\left(\frac{dW}{dV}\right)_{t_0}^{r=\Delta r} + \left(\frac{dW}{dV}\right)_{t_f}^{r=\Delta r}\right] + \left(\frac{dW}{dV}\right)_{\bar{t}}^{r=\Delta r}\right\}. \tag{9.38}$$

The time t_0 corresponds to the initiation of the load cycle; t_f to the completion of the load cycle; and $\bar{t}$ to the time at mid-cycle.

A fatigue life curve for a particular crack specimen can be obtained by integrating equation (9.37) in incremental steps to obtain the actual growth of the fatigue crack. The choice of these increments will affect the accuracy of the growth scheme in its ability to predict actual crack growth paths.

---

* If the load is applied normal to the crack plane, dr/dN may be replaced by da/dN where a is a crack dimension.

Crack growth model. There are two ways to pose the fatigue crack growth problem which depends inherently on the path history. It is possible either to select the increments of growth, Δr, or to specify the number of loading cycles, ΔN, between growth steps for the evaluation of equation (9.37). Experimental evidences [42] have shown that, at the macroscopic scale level, fatigue crack growth can be regarded as a discrete process with extremely small growth steps. This suggests that an accurate means of finding the r versus N relationship would be to integrate equation (9.37) continuously. Because ΔS corresponding to ΔN cycles has to be computed from a nonlinear stress analysis for the crack configuration, this process of continuous integration would involve an insurmountable amount of computation and is considered to be impractical at this time. Instead, finite increments of growth Δr will be chosen to determine the corresponding ΔN or finite numbers of loading cycles; ΔN will be assumed to obtain the corresponding crack growth increment Δr.

In order to be more specific, the center crack panel in Figure 9.22 is chosen to investigate the path-dependency nature of the crack growth process. The dimensions $2b$ and $2c$ in Figure 9.22 are, respectively, 20 in. and 10 in. The material and fracture properties of the panel are $E = 30 \times 10^3$ ksi, $\nu = 0.3$, $\sigma_{ys} = 75$ ksi, $S_c = 77$ lb/in. and $(dW/dV)_c = 7$ ksi. The critical stress intensity factor $k_{1c} = 94.2$ ksi$\sqrt{\text{in.}}$ which differs from the conventional fracture toughness K_{1c} by a factor of $\sqrt{\pi}$, i.e., $K_{1c} = \sqrt{\pi} k_{1c}$. More specifically, the nonlinear stress and strain behavior of the material can be described by equation (9.29) with $\phi = 0.02$ and $\psi = 5$, which are typical of many steels used in engineering application. A graphical representation of the uniaxial data for the material used in the future analysis is shown in Figure 9.38.

Because of symmetry, only the quadrant shaded in Figure 9.22 needs to be analyzed. An elastic–plastic finite element code is employed using the J_2 flow theory of plasticity and a singular crack tip element. The code is a modified version of the PAPST program [43]. The panel is assumed to be under the condition of plane strain and precracked with an initial flaw length of $2a = 2.0$ in. prior to any growth. A sinusoidal loading is taken

$$\Delta\sigma = \tfrac{1}{2}\sigma_{ys}[1 - 0.6\cos(2\pi t)] \tag{9.39}$$

where t is a nondimensional time parameter. For this material, the global instability of the panel occurs when the elastic strain energy density factor, S, at maximum load, i.e. $(\sigma_0)_{max} = 0.8\sigma_{ys}$, is equal to the critical value, S_c. Based on an elastic calculation for a center crack panel, the critical half crack length a_c for this problem is found to be 2.0 in. This establishes the growth range for the half crack length, 1.0 in. $\leq a \leq$ 2.0 in. The plastic flows around the crack tips are contained in that they do not intersect with the panel boundary. The sizes of the yielded zones, however, are significant

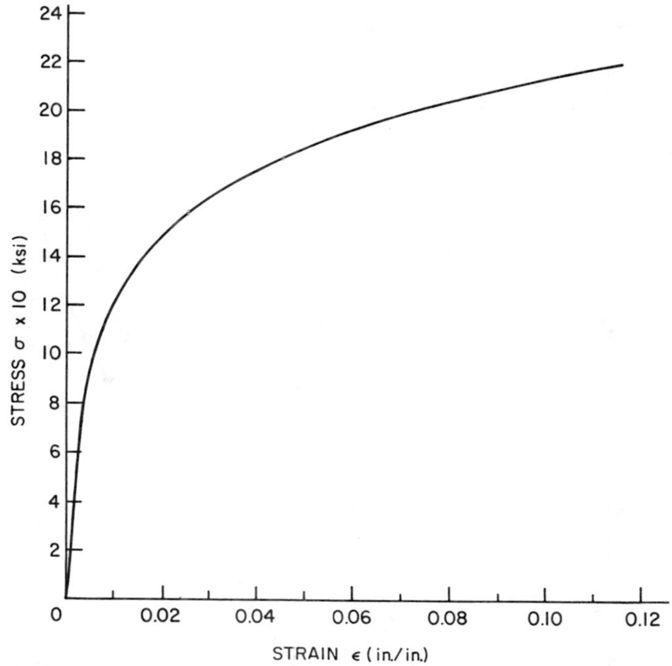

Figure 9.38. Uniaxial stress–strain curve for material used in fatigue.

as they are of the order of the half crack length.

Crack growth characteristics. Four crack growth processes will be chosen to study the path-dependency nature of fatigue crack propagation. They will be referred to as models I–IV and are devised to simulate either experimentally observed crack behavior of Bates and Clark [44] or to suggest additional experiments. The crack tip element and numerical interpolation scheme are developed for incremental crack growth in the range $0.02a \le \Delta a \le 0.08a$. The four models to be considered are:

Model I: The crack growth steps are assumed to be constant with $\Delta a =$ 0.05 in. and ΔN corresponding to each growth step is determined.

Model II: Crack growth step Δa is assumed to occur after a constant number of load cycles $\Delta N = 50$.

Model III: The same as model II except that each crack growth step Δa corresponds to $\Delta N = 100$.

Model IV: Each increment of crack growth Δa is assumed to be proportional to the square* of Δk_1, i.e.,

---

* The relationship in equation (9.40) is based on the observed striation spacing of metal alloys in fatigue which seems to correlate with $(\Delta k_1)^2$ as given by Sih and Barthelemy [45]. These striations tend to form macroscopic bands that presumably are related to the crack growth increments Δa.

Table 9.11. Number of cycles to failure for the four models.

Model	I	II	III	IV
N_f (cycles)	669	804	1261	1155

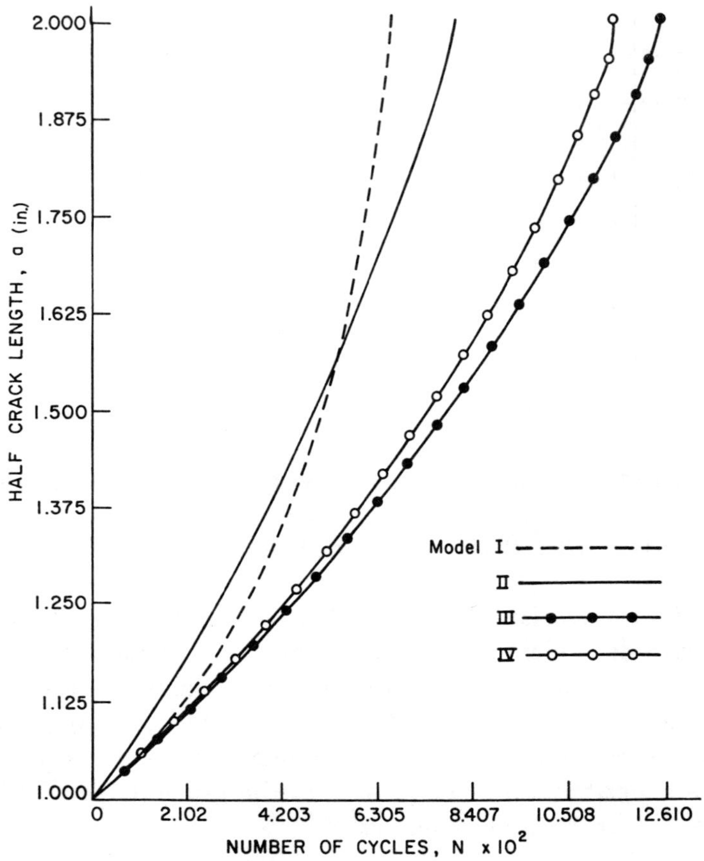

Figure 9.39. Crack length as a function of load cycle.

$$\Delta a = \Lambda(\Delta k_1)^2 \tag{9.40}$$

where Λ is chosen such that the first crack growth increment is 0.05 in.

A comparison of the results from these four models will be made to demonstrate the sensitivity of the analytical model and range of prediction. In view of the lack of existing experimental data on fatigue, it is not possible to draw quantitative conclusions on the physical assumptions leading to equation (9.37). Nevertheless, it is felt that the path-dependent nature of the nonlinear analysis is indicative of the true nature of the fatigue crack growth process. Moreover, it is worthwhile discussing the type of experi-

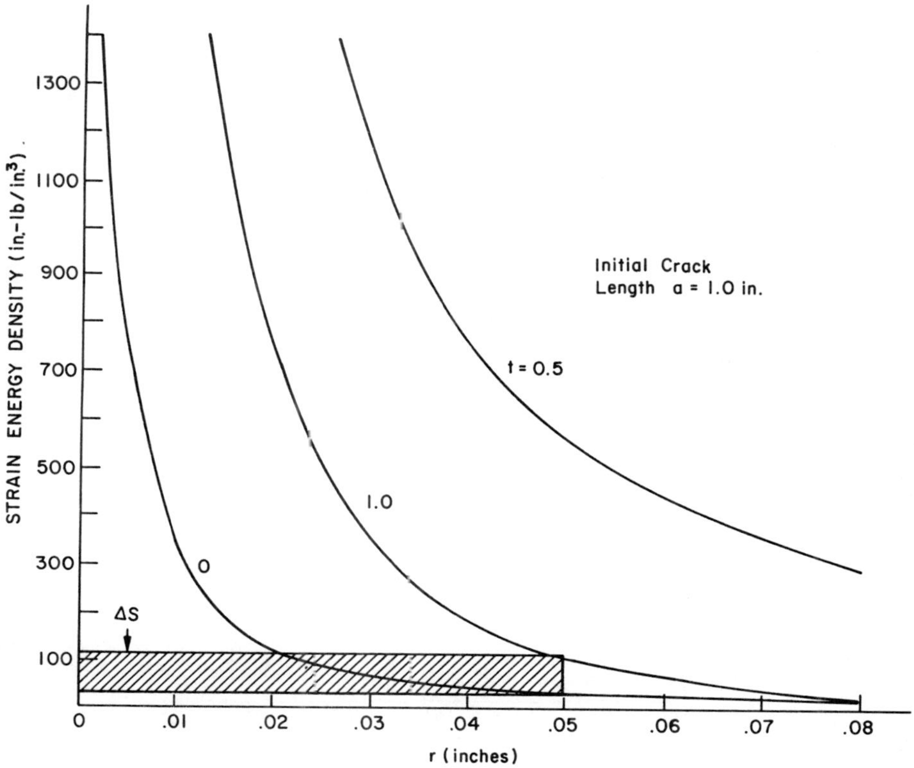

Figure 9.40. Variations of strain energy density with distance ahead of the crack.

ments that are needed to enhance the predictive capability of fatigue crack growth.

Performing the nonlinear stress analysis for each increment of crack growth, the a verus N curves for the four models are obtained from equation (9.37). Table 9.11 summarizes the results for N_f corresponding to the final number of cycles at crack instability. Figure 9.39 gives a plot of the fatigue life curves for the center crack panel and shows that the nonlinear analysis is very sensitive to the choice of crack growth steps. The procedure involves the calculation of the strain energy density as a function of the distance r ahead of the crack. The results for the first cycle are shown in Figure 9.40 for $t = 0.0$ (beginning of cycle), $t = 0.5$ (middle of cycle) and $t = 1.0$ (end of cycle). As an illustration, model I assumes a crack growth increment of 0.05 in. and hence δS corresponding to this growth step can be computed and is represented by the shaded area in Figure 9.40.

The a versus N curve predicted by model I is somewhat unrealistic when compared to available experimental data on steels. A plot of the crack growth rate da/dN against the half crack length a in Figure 9.41 indicates

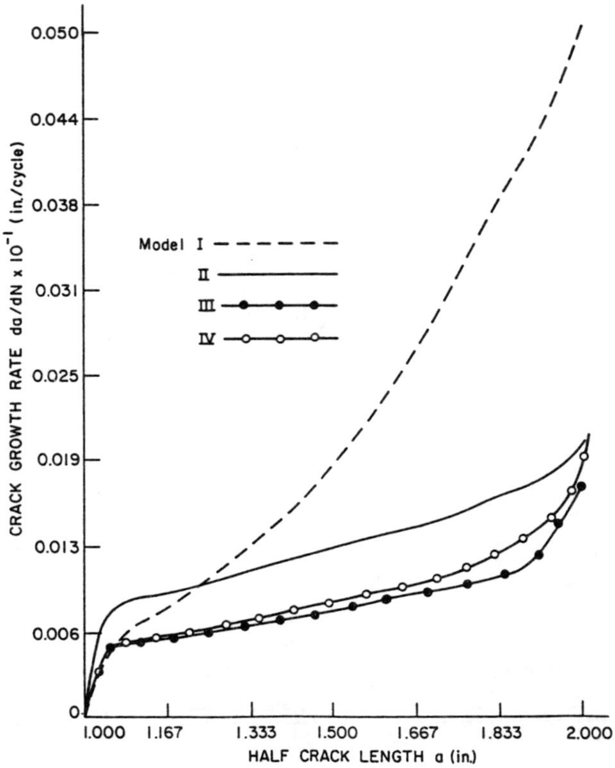

Figure 9.41. Crack growth rate versus crack length.

that the initial growth phase is extremely slow and then accelerates rapidly, which is in contrast to experimental observation. A more gradual acceleration is the normal pattern. The steep acceleration characteristic of the terminating stages of growth is not exhibited as the crack approaches critical length. Model I, however, is informative to demonstrate that the assumption of equally spaced crack growth is unrealistic and that the path-dependent nature of the fatigue process cannot be ignored.

Models II–IV all yield qualitatively acceptable da/dN versus a behavior, Figure 9.41. While the actual number of cycles to failure predicted by these three models is quite different, the shape and acceleration characteristics are qualitatively similar. All three models predict slow, steady acceleration with a sharp increase at the final stages of growth. This is similar to the trends observed experimentally. The fundamental difference between these three models is the number of cycles between growth steps and the predicted growth step sizes. Model II, due to shorter intervals between growth steps, is more strongly influenced by the static response of the material. This would be typical of a more brittle material whose crack profiles are experimentally observed to be closer together. In contrast, model III is typical of a ductile response. Growth steps are larger, occurring at less frequent intervals than in brittle materials.

The results predicted by models III and IV are extremely similar. Since the choice of the proportionality constant M used in model IV was arbitrary, it is reasonable to assume that a smaller value would produce results similar to those of model II. This suggests that while model IV is based on 'experimentally correlated' results, a model based on the range of stress intensity factors is not unique. One of the fundamental aspects of fatigue which has yet to be explored is the condition between the number of cycles for each crack growth step.

Plastic deformation. Under cyclic loading, the extent of plastic deformation will alter in accordance with the load history. The region that is stressed beyond yield is normally referred to as the 'plastic enclave' that prevails at both sides of the crack plane. Because of symmetry, it suffices to show only one-half. It is this process of yielding that causes material damage from cycle to cycle. The accumulation of this damage leads to crack growth. Figure 9.42 shows the yield zone for the upper half of the initial crack $2a = 2.0$ in. at the maximum applied load. Note that three different intensities of yielding are shown in terms of the ratio of $\sigma_{\text{eff}}/\sigma_{\text{ys}}$, where σ_{eff} is the effective stress. The most intense yielding occurs in the immediate vicinity of the crack tip. The height and width of the plastic enclave are of the order of one half of the crack length, i.e., 99 and 92.6% of a being equal to unity in this case. As the crack propagates, the plastic enclaves grow and spread until, for the final half crack length, $a_c = 2.0$ in., the height and width of the plastic enclave are now larger than the half crack length, i.e., 134.5 and 149.2% of a, respectively. Refer to Figure 9.43 for a pictorial view of the enlarged plastic enclave. The growth of the plastic zones produces a strong nonlinear response on the crack tip stress field and causes the nonuniform rate of crack growth.

Comparison with stress intensity factor model. Having solved the nonlinear fatigue crack growth problem, it is of interest to compare the data with those models based on the linear theory of elasticity. One of the most widely used relationships is that of Paris [16] or Paris and Erdogan [46]:

$$\frac{da}{dN} = C(\Delta K)^n \tag{9.41}$$

where ΔK is the stress intensity factor range and the parameters C and n are evaluated experimentally for a given material. Alternatively, da/dN may be associated with the range of strain energy density factor ΔS [45]:

$$\frac{da}{dN} = B(\Delta S)^m \tag{9.42}$$

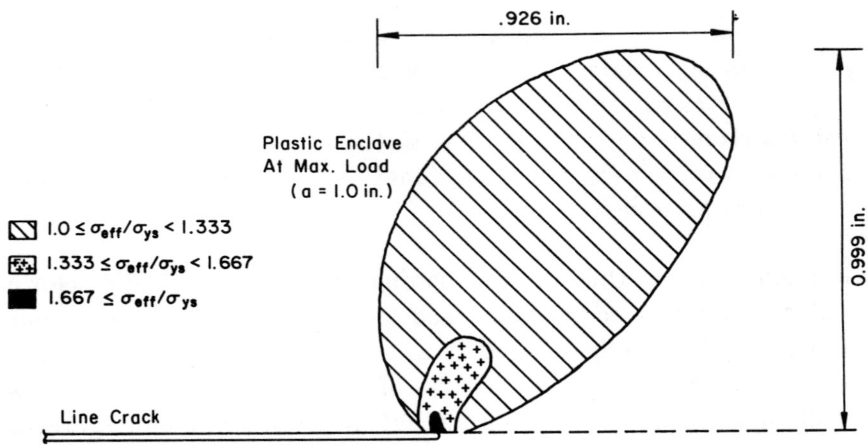

Figure 9.42. **Plastic enclave ahead of crack at maximum load.**

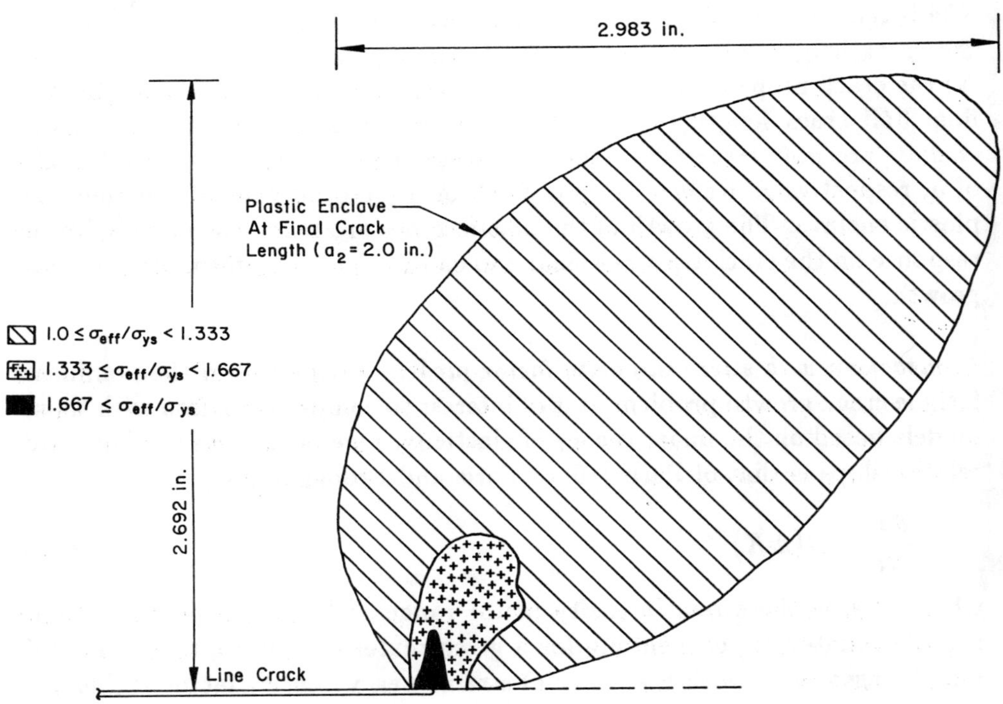

Figure 9.43. **Plastic enclave ahead of crack at final crack length.**

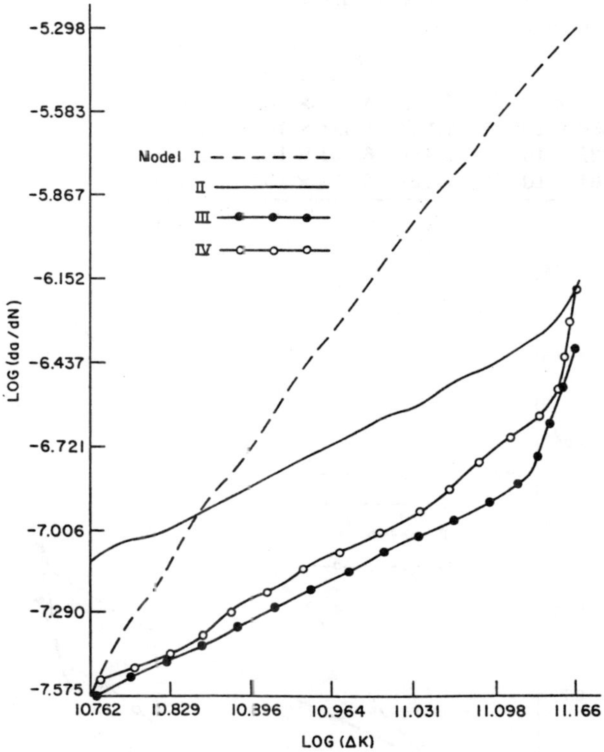

Figure 9.44. Crack growth rate as a function of stress intensity factor range.

in which B and m are a set of experimentally determined parameters. The quantity ΔS may, in general, be related to all three stress intensity factors if the crack is a three-dimensional stress field. In the special case of mode I loading, equation (9.42) may be expressed in terms of K alone since

$$\Delta S = \frac{(1-2\nu)}{2\pi\mu}(\Delta K)\overline{K} \tag{9.43}$$

with $\overline{K}$ being the mean stress intensity factor, which is not included in equation (9.41). The shear modulus of elasticity is denoted by μ. One common feature of equations (9.41) and (9.42) is that both ΔK and ΔS are computed from the linear theory of elasticity which is an oversimplified and nonphysical assumption.

The data obtained from the nonlinear theory in Figure 9.39 for the four models are reinterpreted in terms of equations (9.41) and (9.42). Figures 9.44 and 9.45 show the plots of $\log \mathrm{d}a/\mathrm{d}N$ versus $\log \Delta K$ and $\log \Delta S$, respectively, with ΔK and ΔS computed from the linear theory of elasticity. The parameters C, n, B and m from a least squares fit analysis are given in Table 9.12. Note that the parameters vary by several orders of magnitude

Table 9.12. Parameters in the linearized fatigue crack growth models.*

Model	c	n	B	m
I	1.084×10^{-29}	5.502	4.030×10^{-8}	2.751
II	7.356×10^{-14}	2.147	1.924×10^{-5}	1.074
III	1.632×10^{-15}	2.454	6.821×10^{-6}	1.227
IV	2.961×10^{-17}	2.828	3.619×10^{-6}	1.414

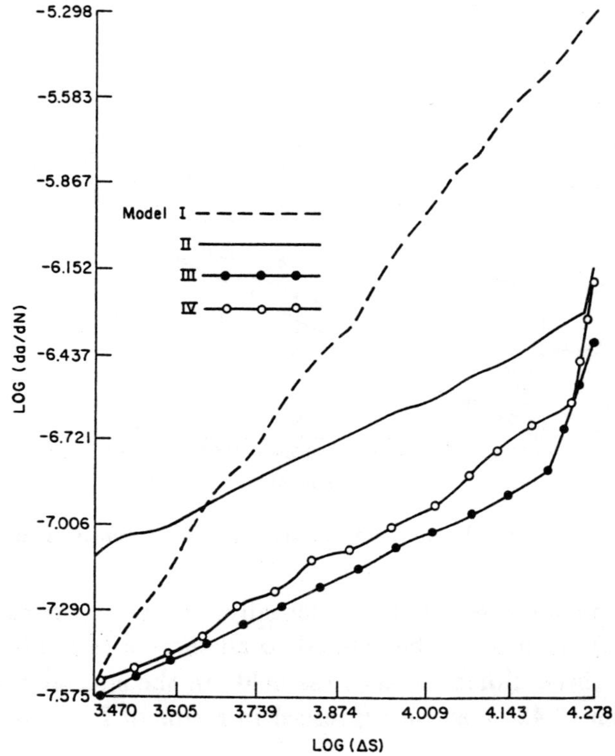

Figure 9.45. Crack growth rate versus strain energy density range.

and are extremely path dependent, i.e., they are sensitive to the history of the crack growth steps. This demonstrates the inadequacy of equations (9.41) and (9.42) to describe the fatigue growth process. It also implies that great error can be made on predictions of fatigue life from equations (9.41) and (9.42) if their limitations are not known from a nonlinear analysis which is an inherent character of the fatigue process.*

The nonlinear analyses, even though they are too complex for the ordinary designers to apply, do serve a useful purpose. They do not only yield physically realistic results but also provide the range of applicability on the

* The results in Table 9.12 are for ΔK being specified in units of psi$\sqrt{\text{in.}}$, ΔS in units of lb/in. and da/dN in in./cycle.

linear models such as those stated in equations (9.41) and (9.42). In general, equation (9.42) should be the preferred linear model since it accounts for the mean stress effect in addition to mixed mode crack propagation. For design purposes, a number of carefully selected pilot problems may be solved by varying the geometry of the crack and specimen, the material, and cyclic loading such that the validity of the parameters B and m for a given situation can be established.

The linear fatigue crack growth relations are inadequate not only because of their inability to describe the full range of observed crack growth behavior but also of their unrealistic basic assumption of excluding the mechanism of cumulative damage. The strain energy density criterion provides a consistent procedure for addressing the path-dependent nature of the fatigue process. Based on a nonlinear elastic–plastic stress analysis, fatigue data similar to those observed experimentally were produced by the finite element method. Other examples on fatigue crack growth prediction that accounts for damage accumulation can be found in [34, 35, 47].

Inhomogeneous material behavior. Modeling of the fatigue growth process would not be complete unless the effect of inhomogeneity as a result of change in material properties from one location to another is accounted for. Such an effect is particularly significant near crack tips and cannot be adequately described by a single constitutive relation as in the formulation of classical continuum theories. Local material properties can alter with strain rates and strain rate history; that is, the same functional form of the constitutive relation may not apply to all locations in the solid, particularly when the strain undergoes high gradients. These deficiencies are by no means obvious and will be brought into light by the isoenergy density theory developed by Sih [48, 49].

9.5. Interaction between surface and volume energy: nonhomogeneous energy dissipation

The freedom to choose different failure criteria and to use them in conjunction with continuum mechanics theories has led to diversified views and opinions as to the precise nature of this implementation when explaining the physical process of material damage. A case in point is the invokement of a fracture criterion on the stress solution obtained from the theory of plasticity. The end results are highly influenced by the assumptions made on the part of the investigator. There lacks a general theory that can treat stress and failure analysis in a unique fashion.

To begin with, there is a fundamental difference between the philosophy that underlines the continuum mechanics and material science approach for analyzing material damage behavior, even though both disciplines may use similar equations and perform similar experiments. Continuum mechanics assumes that the global behavior of any systems could be predicted by analyzing the interaction of an ensemble of material elements whose properties are homogeneous and obtainable from uniaxial specimen tests. Admittedly, great difficulties are encountered when the material response becomes highly nonlinear and nonhomogeneous on account of irreversible damage. Nevertheless, the objective of continuum mechanics is clear notwithstanding the shortcomings of applying nonlinear constitutive equations to model material behavior. The material science procedure, as being practiced today, focuses more attention on experimental observation and on devising models to describe material damage at disjointed scale levels. The unification of failure modes observed at the different scale levels is a challenge that must be overcome before failure prediction can be made.

One of the idealized concepts in formulating continuum mechanics theories is that a continuum can be divided and subdivided into smaller and smaller elements while its properties either remain constant or vary smoothly with space and time. This process, if continued indefinitely, will obviously lead to conceptual difficulties on account of the fact that nature is discrete. Any analyses or experiments will necessarily be restricted in their resolution. Differential calculus circumvents the issue by invoking the limit process. This is accomplished at the expense of size effect that classical continuum theories are unable to account for, unless some of the underlying basic ideas are altered. Solid mechanics theories such as elasticity, plasticity, viscoelasticity, viscoplasticity, etc., are developed to characterize the behavior of different materials. The results, however, are load-specific. A case in point is 'fatigue'. None of the existing theories can adequately explain the fatigue behavior of solids. What has changed after all is only the loading. Instead of increasing the load monotonically, it is now repeated many times at a lower amplitude. The restricted nature of the conventional approach arises from the *a priori* assumption that constitutive equations are material-specific while the interaction of material with the *local* stress and/or strain states has been neglected. That is, the same constitutive relation may not hold for all locations in a continuum and all loading steps. The procedure for assuming a known form* of the constitutive relation should be scrutinized.

Traditionally, forces distributed over the volume of a body (known as *body forces*) are distinguished from those that act on the surface (known as *surface forces* or *tractions*). Their independence follows from shrinking the volume element indefinitely, i.e., by letting the rate change of volume with respect

---

* Even the classical theory of elasticity that describes only the linear and reversible behavior of material has serious limitations when attempting to predict material behavior near a free surface or interface where the so-called modulus and Poisson's ratio cease to be independent. This is a first-order effect that is not negligible.

to surface, dV/dA, to zero in the condition of equilibrium. Similarly, bulk or ***volume energy***, being proportional to the volume of a solid, and ***surface energy***, being proportional to the surface, are also considered as independent quantities in classical mechanics and physics. Such an assumption when removed can lead to a general theory that explains many physical phenomena previously not understood.

Rate change of volume with surface area. Classical continuum mechanics defines an element as the smallest permissible division of a substance. The volume of this element must be large enough so that the average effects of the molecules upon its properties can be assumed to be constant or vary continuously with time and space coordinates. In particular, the mass per unit volume, M/V, of this element may be regarded as constant and approaches a definite limit as both M and V tend to zero. This definition of mass density omits the size effect. Predictions based on theories that avoid addressing element sizes, no doubt, become elusive as all experimental data are necessarily involved with size limitations. The resolution of experiments or numerical calculations must necessarily be defined in terms of some length parameter. The volume to surface relation plays a central role not only in describing material behavior but also in evaluating material damage.

Global consideration. It is difficult to compare the mechanical behavior of specimens with different sizes and shapes because geometric similitude does not hold. In the case of a circular cylinder and a rectangular plate which are geometrically dissimilar, the variations of the volume to surface ratio, V/A, with the radius or thickness aspect ratio can yield useful information on their relative mechanical behavior. For a cylinder of radius B and length $2H$, the V/A ratio can be written as

$$\frac{V}{A} = \frac{H}{1 + 2H/B}. \tag{9.44}$$

The same ratio for a rectangular plate with dimensions $2H \times 2B \times h$, as shown in Figure 9.33, is given by

$$\frac{V}{A} = \frac{H}{1 + H/B + 2H/h}. \tag{9.45}$$

A plot of V/A normalized with respect to H against $B/2H$ or $h/2H$ is shown in Figure 9.46. The ratio of H/B for the plate is taken as 2. The curve for the cylinder rises more quickly than that for the plate. This means that a cylinder would become more brittle as its radius increased. The rate at which this occurs is much faster than that of a plate with increasing thickness. The ratio V/A provides an indication of the brittle and/or ductile behavior of structural components and has been applied by Civallero, Mirabile and Sih [50].

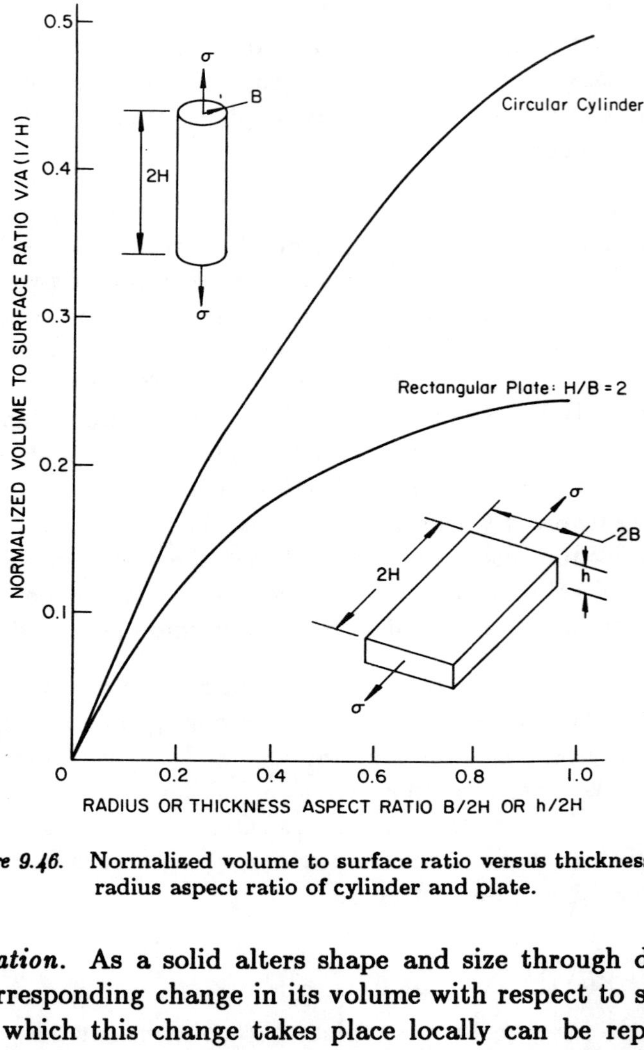

Figure 9.46. Normalized volume to surface ratio versus thickness or radius aspect ratio of cylinder and plate.

Local variation. As a solid alters shape and size through deformation, there is a corresponding change in its volume with respect to surface area. The rate at which this change takes place locally can be represented by dV/dA for a unit volume of material. Consider the unit element in Figure 9.47(a) as deformation changes its position and dimension according to the strain state, say ϵ_{ij}, referred to the coordinates (x, y, z) or $(\epsilon, \eta, \varsigma)$ in Figure 9.47(b). In general, dV/dA can be expressed in terms of the normal strains* ϵ_x, ϵ_y and ϵ_z or ϵ_ξ, ϵ_η and ϵ_ς. Consider a two-dimensional state of deformation with dV given by

$$dV = (n_\xi + \epsilon_\xi)(n_\eta + \epsilon_\eta)(n_\varsigma + \epsilon_\varsigma) - n_\xi n_\eta n_\varsigma \tag{9.46}$$

* The products of shear strain will also enter into dV/dA. They are left out in this work but should be included for situations where the strains are sufficiently large. Sih [49] has obtained the expression for dV with all higher order terms of deformation gradients.

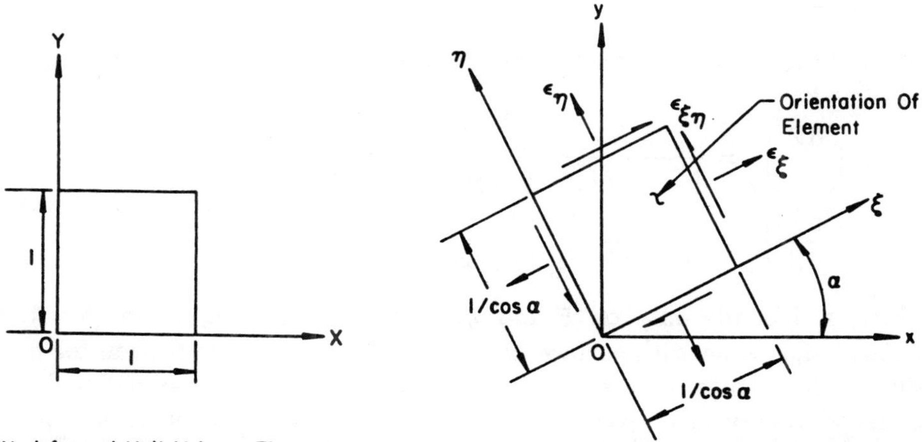

(a) Undeformed Unit Volume Element

(b) Deformed Volume Element

Figure 9.47. Orientation of volume element in undeformed and deformed state.

such that $n_\xi = n_\eta = 1/\cos\alpha$ and $n_\varsigma = 1$. By retaining only linear terms in strain, equation (9.46) reduces to

$$dV = \frac{(\epsilon_\xi + \epsilon_\eta)\cos\alpha + \epsilon_\varsigma}{\cos^2\alpha}. \tag{9.47}$$

The corresponding change in surface area $(dA)_i$ in a given direction can be obtained by referring to the respective normals $i = \xi, \eta, \varsigma$ as

$$dA_\xi = \frac{\epsilon_\varsigma + \epsilon_\eta \cos\alpha}{\cos\alpha}, \qquad dA_\eta = \frac{\epsilon_\xi + \epsilon_\varsigma}{\cos\alpha}, \qquad dA_\varsigma = \frac{\epsilon_\xi + \epsilon_\eta}{\cos\alpha}. \tag{9.48}$$

In view of equations (9.47) and (9.48), $(dV/dA)_i$ for $i = \xi, \eta, \varsigma$ follow immediately:

$$\begin{aligned} \left(\frac{dV}{dA}\right)_\xi &= \frac{(\epsilon_\xi + \epsilon_\eta)\cos\alpha + \epsilon_\varsigma}{(\epsilon_\eta \cos\alpha + \epsilon_\varsigma)\cos\alpha} \\ \left(\frac{dV}{dA}\right)_\eta &= \frac{(\epsilon_\xi + \epsilon_\eta)\cos\alpha + \epsilon_\varsigma}{(\epsilon_\xi \cos\alpha + \epsilon_\eta)\cos\alpha} \\ \left(\frac{dV}{dA}\right)_\varsigma &= \frac{(\epsilon_\xi + \epsilon_\eta)\cos\alpha + \epsilon_\varsigma}{(\epsilon_\xi + \epsilon_\eta)\cos\alpha}. \end{aligned} \tag{9.49}$$

Let $(dV/dA)_\varsigma$ be independent of ς, then

$$\left(\frac{dV}{dA}\right)_\varsigma = f(\xi, \eta) \tag{9.50}$$

and the last of equations (9.49) gives

$$\epsilon_\varsigma = (f-1)(\epsilon_\xi + \epsilon_\eta)\cos\alpha. \tag{9.51}$$

Eliminating ϵ_ς in the first two of equations (9.49), the following results are obtained:

$$\begin{aligned} \left(\frac{dV}{dA}\right)_\xi &= \frac{\epsilon_\xi + \epsilon_\eta}{[(1-1/f)\epsilon_\xi + \epsilon_\eta]\cos\alpha}, \\ \left(\frac{dV}{dA}\right)_\eta &= \frac{\epsilon_\xi + \epsilon_\eta}{[\epsilon_\xi + (1-1/f)\epsilon_\eta]\cos\alpha}. \end{aligned} \tag{9.52}$$

Since f depends only on ξ and η, as indicated in equation (9.50), the change of volume with surface area is constant in the ς-direction for a two-dimensional state of strain. In the case of plane strain, as defined in the classical theory of elasticity with $\epsilon_\varsigma \to 0$, it is easily seen from equation (9.51) that this condition corresponds to $f \to 1$.

The direction for which $(dV/dA)_\xi$ and $(dV/dA)_\eta$ are equal is unique in that it defines the ***isoenergy plane***, i.e.,

$$\left(\frac{dV}{dA}\right)_\xi = \left(\frac{dV}{dA}\right)_\eta \to \left(\frac{dV}{dA}\right)_0 \tag{9.53}$$

where $(dV/dA)_0$ can be identified with a known state of strain from experiment normally taken as the uniaxial test. For an initially isotropic and homogeneous material, equation (9.53) leads to*

$$\epsilon_\xi = \epsilon_\eta. \tag{9.54}$$

These are not the principal strains but those acting on the α-plane in Figure 9.47(b) with a dV/dA value that can be identified with the same energy transmitted across all surfaces of a volume element regardless of whether the strain state is uniaxial, biaxial or triaxial. More complicated relations of ϵ_ξ and ϵ_η will result if higher order terms in the strains ϵ_ξ, ϵ_η and ϵ_ς are retained in equations (9.49) for $(dV/dA)_j$ $(j = \xi, \eta, \varsigma)$.

Relation between surface and volume energy. As mentioned earlier, inconsistencies arise when continuum mechanics theories are used in conjunction with criteria based on dW/dA or dW/dV. This is partly because dW/dA and dW/dV are assumed to be independent by letting $dV/dA \to 0$ in the limit. Such a condition, of course, cannot be realized physically and can lead to inaccurate predictions near geometric discontinuities where fluctuations in dV/dA are an inherent part of the physical process.

---

* Expressions similar to equation (9.54) may be derived for anisotropic and nonhomogeneous materials in which case the expression between ϵ_ξ and ϵ_η will involve material constants.

It is well known that the crack edge stress singularity at a point near the free surface can differ from the familiar inverse square root behavior, a condition frequently encountered in practice. Although Benthem [51] predicted such a trend, large discrepancies between theory and experiment remain unexplained. Smith, Epstein and Olaosebikan [52] have shown that the stresses behave according to $1/r^{\omega}$ along a straight crack front, where r is the radial distance. The square root behavior corresponding to $\omega = \frac{1}{2}$ prevailed only at the intersection of the crack border with the mid-plane of the plate. The exponent ω increases *gradually* as the plate surface is approached, in contrast to the step-jump behavior predicted by the theory of elasticity where the $1/\sqrt{r}$ character changes suddenly to some other order across an infinitesimal surface boundary layer. What is questionable is the basic assumption in elasticity that Young's modulus and Poisson's ratio are everywhere independent. This may not hold in regions close to a free surface or interface.* It is physically unrealistic to assume that the same constitutive relation applies to all regions in a solid even in the absence of permanent damage. The equivalent uniaxial stress and strain behavior for elements in biaxial or triaxial stress and/or strain states can vary from one region to another. Even more serious errors can result when dealing with irreversible and nonlinear phenomena. The process becomes highly sensitive to strain rate and strain rate history.**

Isoenergy density theory. As a fundamental departure from the classical continuum mechanics approach, the surface and volume energy density are assumed to be related through a kinematic quantity that represents the rate change of volume with surface. Consequently, the body force and surface traction will no longer be independent of one another and there is no need to preassume the form of the constitutive relation. The stresses and strains in each element can be determined separately for each time increment. This is necessitated from the fact that the irreversible nature of any physical system is dictated by the load–time history.

The strain energy density theory in its most basic form can be formulated from the basic hypothesis that *the surface and volume energy density of each material element are related by the rate change of volume with surface*, i.e.,

$$\left(\frac{\mathrm{d}W}{\mathrm{d}A}\right)_i = \left(\frac{\mathrm{d}V}{\mathrm{d}A}\right)_i \left(\frac{\mathrm{d}W}{\mathrm{d}V}\right) \tag{9.55}$$

* Elasticity can yield peculiar results near the free surfaces ahead of a three-dimensional crack when the surface and volume interaction effects are not negligible, as indicated by Sih and Lee [53].

** Since material behavior depends directly on load–time history, the *a priori* assumption that the form of the constitutive relation can be stated independent of loading imposes serious restrictions on the resulting theory.

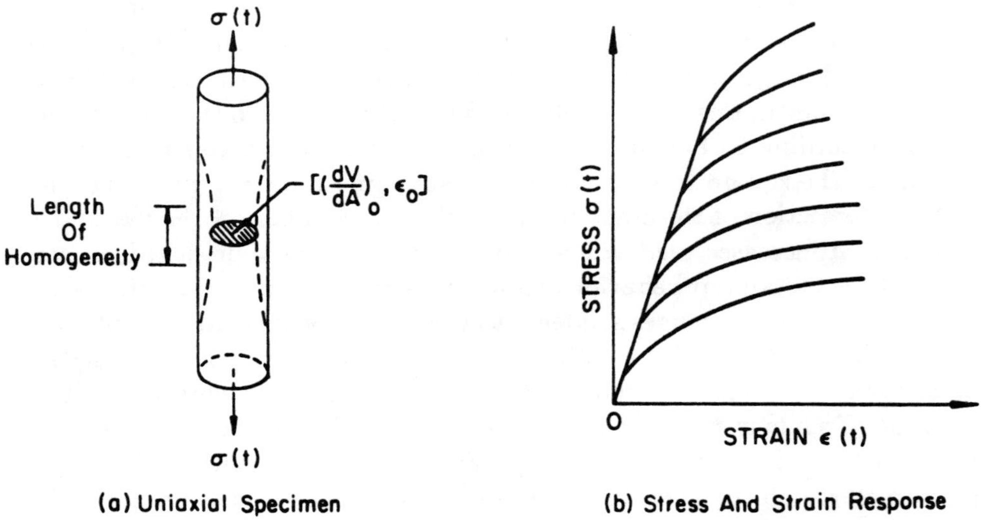

Figure 9.48. Uniaxial data bank for different combination of loading rate and/or specimen size.

in which the quantities $(\mathrm{d}V/\mathrm{d}A)_i$ $(i = \xi, \eta, \varsigma)$ are given by equation (9.49) and whose physical meanings have already been discussed. Equation (9.55) implies that both $(\mathrm{d}W/\mathrm{d}A)_i$ and $(\mathrm{d}V/\mathrm{d}A)_i$ are directionally dependent. For each material element, there prevails a plane of isoenergy such that in two dimensions

$$\left(\frac{\mathrm{d}W}{\mathrm{d}A}\right)_\xi = \left(\frac{\mathrm{d}W}{\mathrm{d}A}\right)_\eta \to \left(\frac{\mathrm{d}W}{\mathrm{d}A}\right)_0 \tag{9.56}$$

where $(\mathrm{d}W/\mathrm{d}A)_0$ corresponds to the known surface energy density obtained from a uniaxial test. Since $\mathrm{d}W/\mathrm{d}V$ in equation (9.55) is a scalar, equation (9.56) is equivalent to equation (9.53) and establishes an unique correspondence between the data of uniaxial tests and multiaxial stress or strain states. The proof of the existence of the isoenergy density plane in three dimensions can be found in Sih [49].

Referring to Figure 9.48(a), $(\mathrm{d}V/\mathrm{d}A)_0$ in equation (9.53) has the unit of length and may be regarded as a scaling parameter. A region may thus be defined within which the stress or energy state is uniform or homogeneous. As long as the uniaxial stress and strain response are linear, $\mathrm{d}V/\mathrm{d}A$ remains constant and the relation between $\mathrm{d}W/\mathrm{d}A$ and $\mathrm{d}W/\mathrm{d}V$ is fixed. The system is said to be reversible with reference to a given scale level of measurement. When the uniaxial stress $\sigma(t)$ changes nonlinearly with the uniaxial $\epsilon(t)$, $\mathrm{d}V/\mathrm{d}A$ will vary and can be uniquely identified with the slope of the stress–strain curve, Figure 9.48(b). Any point in the σ versus ϵ plane possesses a corresponding value of $[(dV/dA)_0, \epsilon_0]$. A uniaxial data bank may thus be established to include the entire range of local strain rates experienced by the material elements that may behave differently from one location to another. The homogeneity of any multiaxial stress or energy states described by $[(dV/dA)_\xi, \epsilon_\xi]$ or $[(dV/dA)_\eta, \epsilon_\eta]$ can thus be uniquely related to that in a uniaxial test. In this way, the actual stress and strain path of each element is derived as the system is loaded incrementally. The theory of isoenergy density, therefore, requires the knowledge of only a single parameter* $(dV/dA)_0$ to describe the mechanical properties of the material.

Boundary conditions. Let $\overset{n}{T}$ be the stress vector acting on an element of surface with the unit normal vector $\boldsymbol{n}$. Since dV/dA is finite in the energy density theory, the relation between the components $\overset{n}{T}_i$ and stresses** σ_{ij} are related as

$$\overset{n}{T}_i = \sigma_{ji} n_j + (\bar{\rho}\ddot{u}_i - F_i)\left(\frac{dV}{dA}\right)_n \tag{9.57}$$

in which $\bar{\rho}\ddot{u}_i$ is the inertia term with $\bar{\rho}$ being the average mass density of the element. The acceleration is given by twice time differential of the displacement vector $\ddot{\mathrm{u}}$. In equation (9.57), the F_i represent components of the body force.

The tractions $\overset{n}{T}_i$ specified on the boundary are connected with the stresses σ_{ij} and displacements u_i in the solid and depend on the rates at which the load and/or displacement is applied to the boundary. The boundary stresses or displacements can no longer be specified independently or arbitrarily. There prevails a surface layer within which equation (9.57) must be observed. This is in contrast to the classical continuum mechanics theory, which passes to the limit by letting $dV/dA \to 0$.

Local inhomogeneity. The material behavior is said to be inhomogeneous if the stress and strain response at each point is derived individually and differs from those at other locations. This is independent of whether inhomogeneity arising from the material microstructure is taken into account or not. While the uniaxial data bank in Figure 9.48(b) provides a range of different stress and strain curves, the actual response of an element is derived according to the load history and, in general, would not coincide with any of the curves in Figure 9.48(b). In other words, the stress and strain relation for an element is not ***preselected***. The procedure for constructing the stress and

---

* Poisson's ratio, Young's modulus, yield strength, etc., as used in the classical continuum mechanics theories are no longer needed as dV/dA, being a function of strain history, can uniquely describe the uniaxial data.

** The quantities σ_{ij} are taken as average values over an element of finite area.

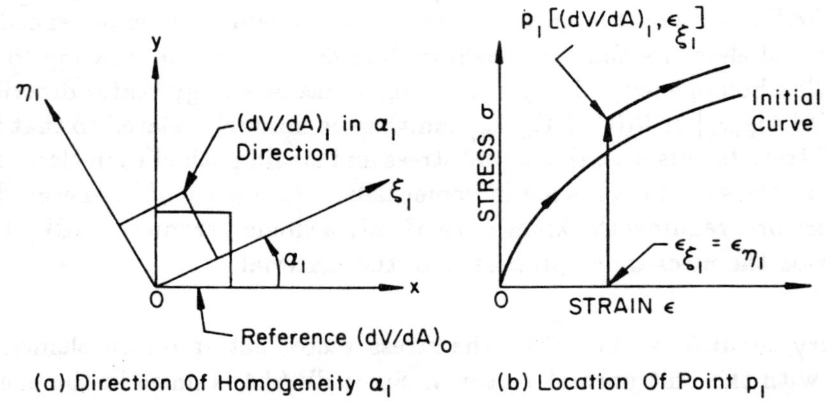

Figure 9.49. Searching of point p_1 in data bank for the first load increment.

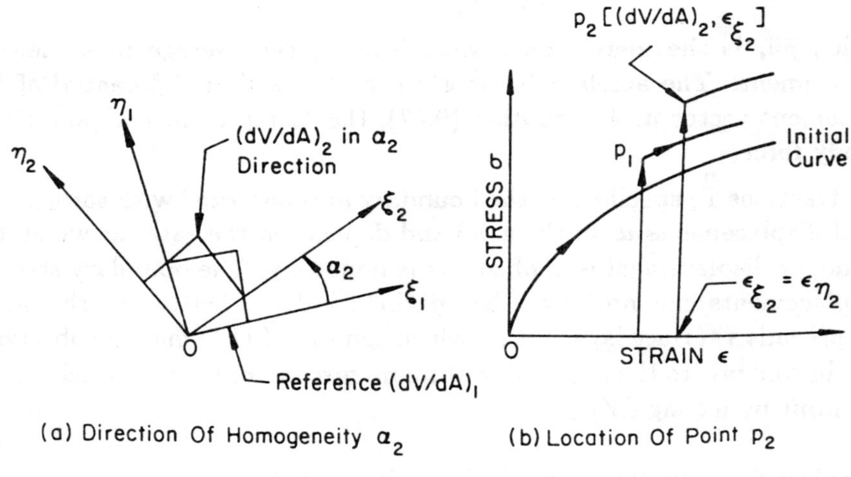

Figure 9.50. Searching of point p_2 in data bank for the second load increment.

strain history of a typical element is illustrated schematically in Figures 9.49 and 9.50.

An initial stress and strain of a given element is assumed for the first load increment. Equation (9.54) is then applied to find α_1, $(dV/dA)_1$, and $\epsilon_{\xi_1} = \epsilon_{\eta_1}$. The predicted damage state makes an angle α_1 with the reference state, Figure 9.49 (a). This locates the point p_1 in the data bank with coordinates $[(dV/dA)_1, \epsilon_{\xi_1}]$ as shown in Figure 9.49(b). The next increment of loading gives α_2, $(dV/dA)_2$ and $\epsilon_{\xi_2} = \epsilon_{\eta_2}$, and hence the point p_2 is obtained, Figures 9.50(a) and 9.50(b). A series of points p_1, p_2, etc., are found until the complete stress and strain history is derived. Unloading

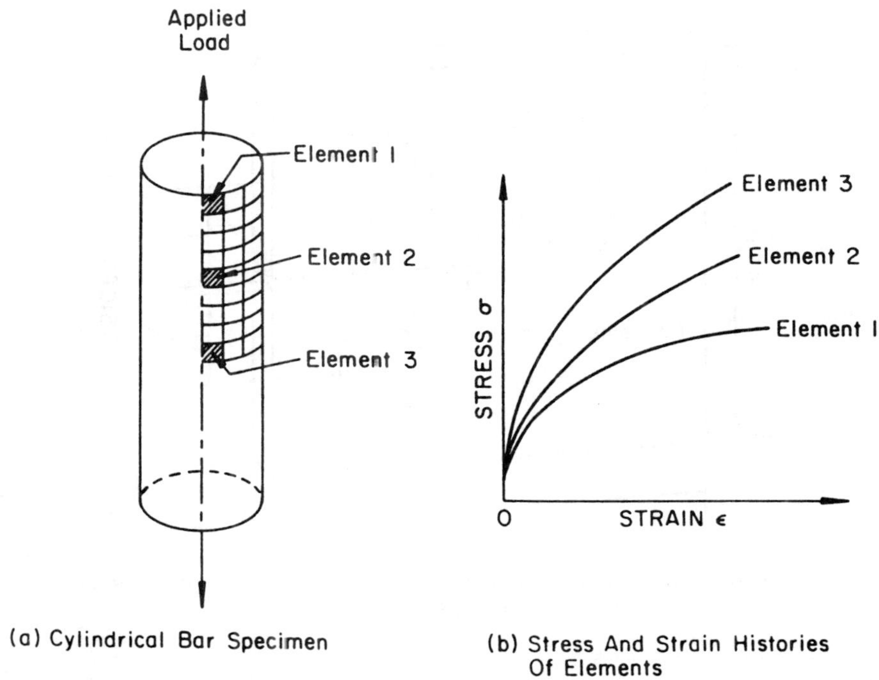

Figure 9.51. Schematic of stress and strain response for elements in uniaxial specimen.

can also be accomplished incrementally in the same way. Each element will follow its own path of loading or unloading. The energy dissipated in damaging each element can be assessed with the desired degree of accuracy depending on the refinement of the data bank. Reference can be made to the early version of the PEDDA computer program [54].

In general, elements near the specimen edge would behave differently from those near the center. This is illustrated schematically in Figures 9.51(a) and 9.51(b). Element 1 in Figure 9.51(a) would experience more distortion and the equivalent uniaxial stress and strain curve is more ductile. In contrast, element 3 is stiffer as it is near the specimen center which would dilate more. To be expected is the stress and strain curve of element 2 which lies between elements 1 and 3, Figure 9.51(b). The isoenergy density theory is particularly useful for analyzing problems that cover a wide range of strain rates.

Cooling/heating phenomenon. Thermal/mechanical interaction effects are seldom discussed in fracture mechanics because they require nonequilibrium considerations that are beyond the scope of ordinary mechanics theories such

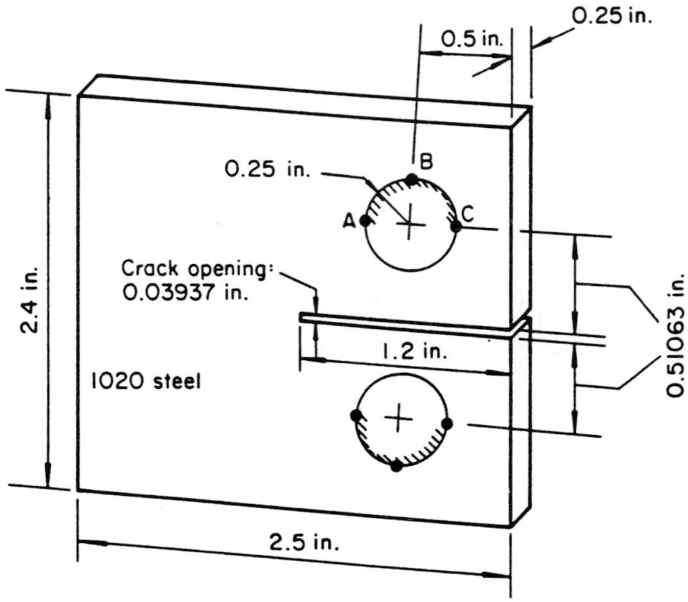

Figure 9.52. Compact tension specimen subjected to average displacement rate of 0.02 in./min.

as thermoelasticity, thermoviscoplasticity, etc. Equilibrium theories are not able to predict the cooling/heating phenomenon in solids when stretched under a rising load. The prediction of cooling and heating, contrived by Cernocky and Krempl [55], invalidates the theoretical development. This can be further evidenced from the wrong deduction made by the same theory of Cernocky and Krempl [56] that torsional loading produced no reversal of heat flow. Experiments by Lieu [57] at Lehigh University showed that the effect of cooling/heating is just as pronounced in torsion as in extension.

Although the phenomenon of solids undergoing cooling when stretched has been discussed since the time of Kelvin in the 1800's, it has not been quantified or understood. In contrast to previous notions that the material next to the crack tip undergoes heating immediately upon loading, the experimental data of Beghi, Bottani and Cagliotti [58] on a compact tension specimen made of SAFC-40R steel have shown that the local temperature dropped below that of ambient. Theoretical prediction of the crack tip cooling/heating behavior, however, was made by Sih and Tzou [59] for AISI 316 steel and by Sih, Tzou and Michopoulos [60] for 1020 steel.

Compact tension specimen. Consider the compact tension specimen in Figure 9.52 that is subjected to an average displacement rate of 0.02 in./min. Since the load is tensile in character, only the shaded portion of the holes *ABC* in Figure 9.52 will be in contact with the pins. The load distribution on the hole is assumed to vary in accordance with the expression

$$\dot{u}(\theta) = 0.01 \sin\theta, \tag{9.58}$$

such that the average

$$\bar{\dot{u}} = \int_0^{\pi} u(\theta)\, \mathrm{d}\theta = 0.02 \text{ in./min.} \tag{9.59}$$

corresponds to the specified displacement rate.

The specimen is made of 1020 steel whose stress-strain curve at normal strain rates would be sufficient to construct the material data bank. Let σ and ϵ denote the isostress and isostrain. They can correspond to any one of the nine pairs of stress and strain components on the isoenergy plane because any of these pairs would yield the same isoenergy density, as shown by Sih [49]:

$$\frac{\mathrm{d}W}{\mathrm{d}V} = \int \sigma\, \mathrm{d}\epsilon \tag{9.60}$$

where (σ, ϵ) could represent the pair $(\sigma_\xi, \epsilon_\xi)$. Defining λ as a proportionality parameter that relates $\mathrm{d}\sigma/\mathrm{d}\epsilon$ and $\mathrm{d}V/\mathrm{d}A$, the relation

$$\frac{\mathrm{d}\sigma}{\mathrm{d}\epsilon} = \lambda \frac{\mathrm{d}V}{\mathrm{d}A} \tag{9.61}$$

can be used to put equation (9.60) in the form

$$\frac{\mathrm{d}W}{\mathrm{d}V} = \iint \lambda \frac{\mathrm{d}V}{\mathrm{d}A}\, \mathrm{d}\epsilon\, \mathrm{d}\epsilon \tag{9.62}$$

where $\mathrm{d}W/\mathrm{d}V$ depends only on the strain ϵ. The isoenergy strain ϵ is in fact a deformation gradient. For the first time step, λ for the 1020 steel is taken as 8.37209 ksi/in. This corresponds to a slope of $\mathrm{d}\sigma/\mathrm{d}\epsilon = 2.88 \times 10^4$ ksi on the initial stress and strain curve that can be modelled by equation (9.29) with $\phi = 0.02$ and $\psi = 5$. A two-dimensional stress or strain state is assumed, with $(\mathrm{d}V/\mathrm{d}A)_\varsigma = 3.44$.

Energy fluctuation. In view of symmetry, only one-half of the specimen needs to be modelled by application of the 12 node isoparametric finite elements. Referring to Figure 9.53, 17 elements and 121 nodes are used. Element number 7, being directly ahead of the crack, is embedded with 64 Gaussian points of integration while the others contain only 16 Gaussian points.

The PEDDA program [54] is applied to solve for the displacements and strains from which all other quantities may be obtained. Thirteen time steps are taken as t is varied from 0 to 288 s. The stationary values of the volume energy density have been related to sites of possible failure initiation by Sih [27]. According to equation (9.55) the surface energy density also plays a role as both $\mathrm{d}W/\mathrm{d}A$ and $\mathrm{d}W/\mathrm{d}V$ are connected by $\mathrm{d}V/\mathrm{d}A$. The

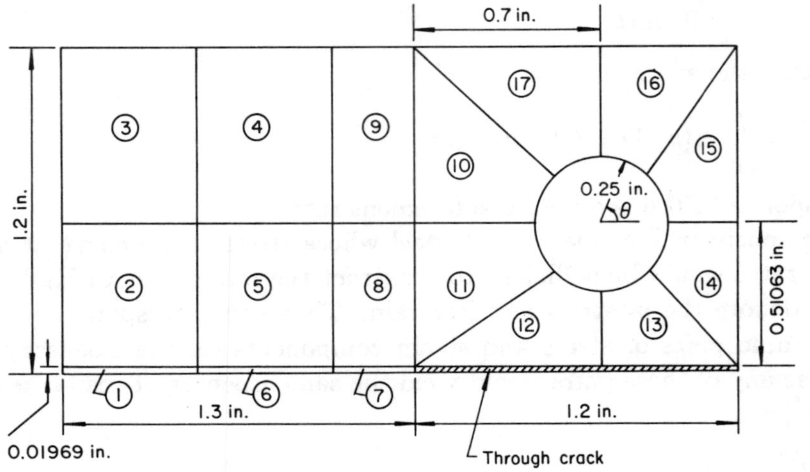

Figure 9.53. Finite element grid pattern of compact tension specimen with one-half symmetry.

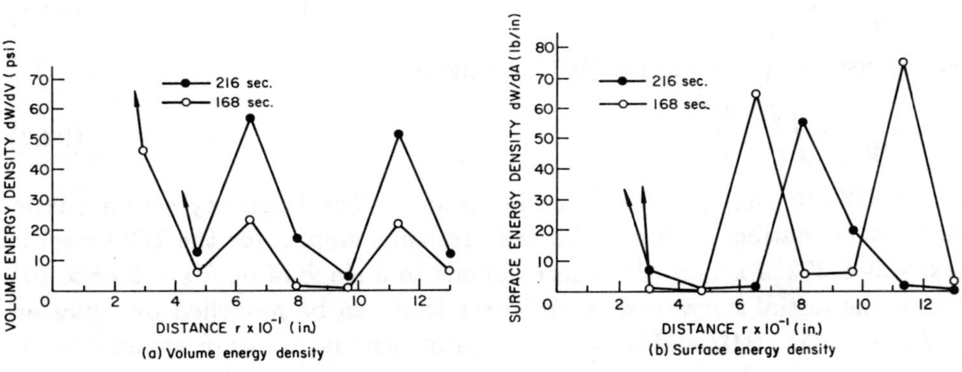

Figure 9.54. Volume and surface energy density distribution along the axis of symmetry at $t = 12$ s.

surface energy density tends to dominate locally when the load is initially transmitted across the boundary or surface while the volume energy density increases more readily as the disturbance occupies over a larger portion of the interior of the system. Their distribution as a function of time depends on the combined effects of geometry, material and load–time history. Figures 9.54(a) and 9.54(b) give the variations of dW/dV and dW/dA with the distance r along the prospective path of crack growth that coincides with the axis of symmetry for $t = 12$ s. They both decrease in magnitude very

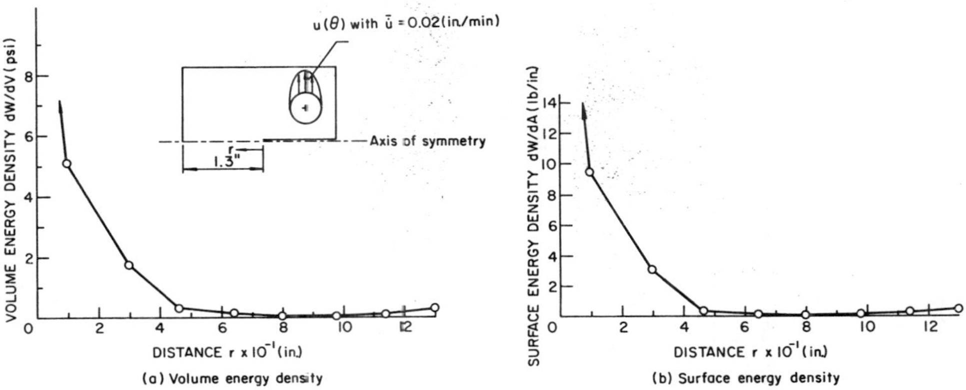

Figure 9.55. **Volume and surface energy density distribution along the axis of symmetry at $t = 168$ and 216 s.**

quickly as r increases, reaching a minimum at approximately $r = 0.8$ in. Similar plots are given in Figure 9.55(a) and 9.55(b) for $t = 168$ and 216 s. Oscillations in both the volume and surface energy densities are observed. The phase difference between dW/dV and dW/dA becomes significant as t increases from 168 to 216 s.

Nonequilibrium temperature. The incremental change in nonequilibrium temperature Θ can be determined from the incremental change in stress $\Delta\sigma_\xi$, change in strain $\Delta\epsilon_\xi$, and change in energy dissipation density $\Delta\mathcal{D}$, i.e.,

$$\frac{\Delta\Theta}{\Theta} = -\frac{\Delta\sigma_\xi \Delta\epsilon_\xi}{\Delta\mathcal{D}}, \tag{9.63}$$

in which $\mathcal{D}$ stands for $(dW/dV)_p$ in equations (9.17) and (9.18). The temperature Θ can be scaled in K. When energy dissipation takes place entirely in the form of heat Q, i.e., $\mathcal{D} = Q$, then Θ reduces to the classical temperature T. Equation (9.63) provides the means for determining heat transfer quantities directly from stress and strain data. Evaluation of the quantity $\mathcal{D}$ for a local element should be distinguished from the unloading process of a uniaxial specimen that represents the average response of all the elements. The tractions on a local element when isolated from its neighbors do not vanish even if the loads are removed from the system. The energy dissipation density $\mathcal{D}$ reflects the intrinsic recovery property* of the material. Without going into details, the values of $\mathcal{D}$ for each time step at a specified location will be given. The time history of $\mathcal{D}$ at a distance 0.03937

* The distinction between unloading and recovery behavior of material has been made by Sih and Chou [61].

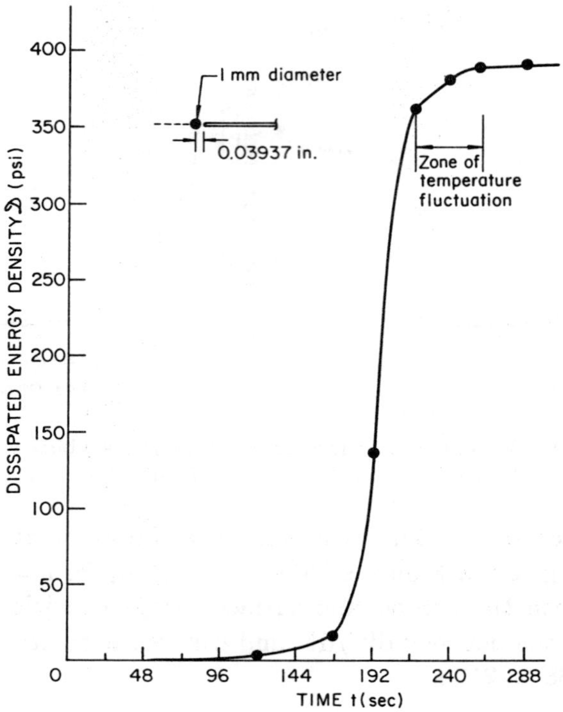

Figure 9.56. Time history of energy dissipation ahead of crack.

in. directly ahead of the crack tip is shown in Figure 9.56. Initially, the dissipated energy is negligibly small. It then starts to increase slowly until the time interval $t = 144$ and 192 s. is reached when $\mathcal{D}$ suddenly changes by an order of magnitude. A slight flattening of the $\mathcal{D}$ versus t curve is observed in the range $t = 216$ and 264 s. This corresponds to a fluctuation in the temperature history. Refer to the numerical data in Table 9.13 and Figure 9.57, where solid circles represent the theoretical prediction and the open circles are the experimental measurement.* Table 9.14 gives the measured values of Θ for 18 time steps starting from $t = 12$ s and ending at $t = 300$ s. The secondary fluctuation in temperature is found to occur between $t = 228$ and 240 s. This agreed sufficiently well with theory, as indicated in Figure 9.57. Such a phenomenon is not intuitively obvious and could have easily been dismissed as inaccuracy in measurement had it not been consistently predicted by the theory. What it does indicate is a change in the disorder of the system due to damage. A more elaborate interpretation of this can be found in [60].

---

* The temperature change is monitored over an area of 1 mm in diameter, being the size of the thermistor at a distance 0.03937 in. ahead of the crack tip.

Table 9.13. Dissipated energy density and temperature change ahead of crack.

Time t (s)	Dissipated energy D (psi)	Change of temperature $\Theta - \Theta_0$ (mK)
0	0	0
12	0.0102	− 0.8281
24	0.0114	− 5.6435
48	0.0126	−11.1994
72	0.0128	−20.1224
96	1.018	−25.8065
120	2.342	−29.4893
144	7.343	−29.5309
168	19.383	−23.7984
192	137.681	−14.2290
216	366.880	49.2377
240	380.303	37.8322
264	388.500	60.8983
288	390.256	91.896

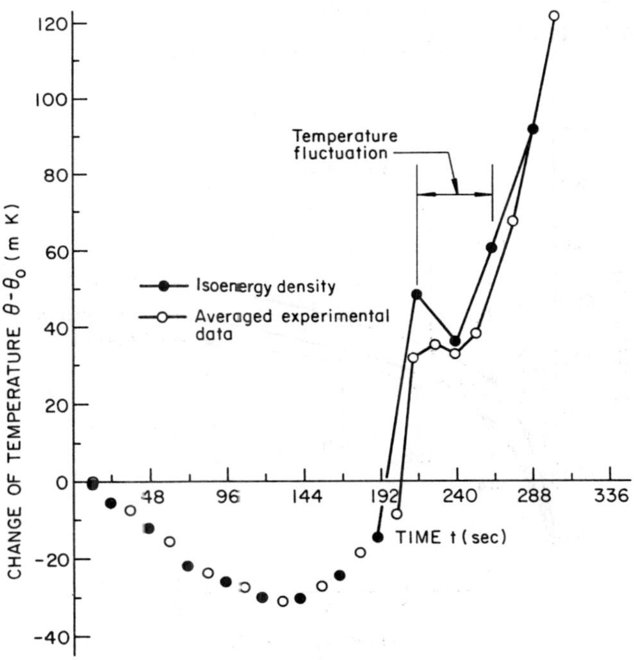

Figure 9.57. Time history of temperature ahead of crack.

Isostress and isostrain. It should be reiterated that stress and strain response is not sufficient to reveal the irreversible character of the physical process associated with change in the local temperature. In other words, the temperature disturbance for two seemingly similar stress and strain curves may be very different. Shown in Figure 9.58, for example, are the equiva-

Table 9.14. Measured temperature change $\Theta - \Theta_0$ in mK local to crack tip for three tests.

Time t (s)	Test 1	2	3	Average value
12	0	0	0	0
36	− 9.1	− 8.5	− 6.2	− 7.933
60	−18.2	−16.1	−13.1	−15.800
84	−24.5	−24.6	−21.0	−23.367
108	−27.2	−29.5	−26.9	−27.867
132	−29.1	−30.4	−31.1	−30.200
156	−25.0	−27.1	−29.9	−27.333
180	−15.9	−18.2	−23.8	−19.300
204	− 8.2	− 8.1	−10.1	− 8.800
210	− 5.4	− 5.3	− 9.5	− 6.733
216	− 4.1	− 4.1	− 5.2	− 4.467
222	− 0.4	− 0.5	− 1.8	− 0.900
228	36.4	35.1	33.2	34.900
234	31.9	32.1	26.5	30.167
240	31.0	31.2	35.4	32.533
246	35.5	35.6	35.7	35.600
252	43.3	36.2	38.5	39.333
276	72.9	65.4	65.6	67.967
300	146.0	115.6	105.1	122.233

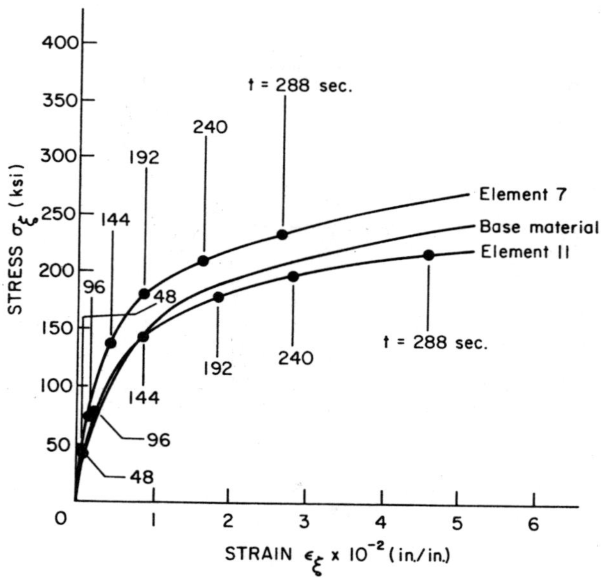

Figure 9.58. Stress and strain response on plane of isoenergy density.

lent uniaxial stress and strain response on the plane of isoenergy density for elements 7 and 11, as shown in Figure 9.53. The base material is displayed for the sake of reference. The curves rise smoothly with time and exhibit no change at locations where the local temperature increased monotonically or oscillated with time. In passing, it should be noted that the σ versus

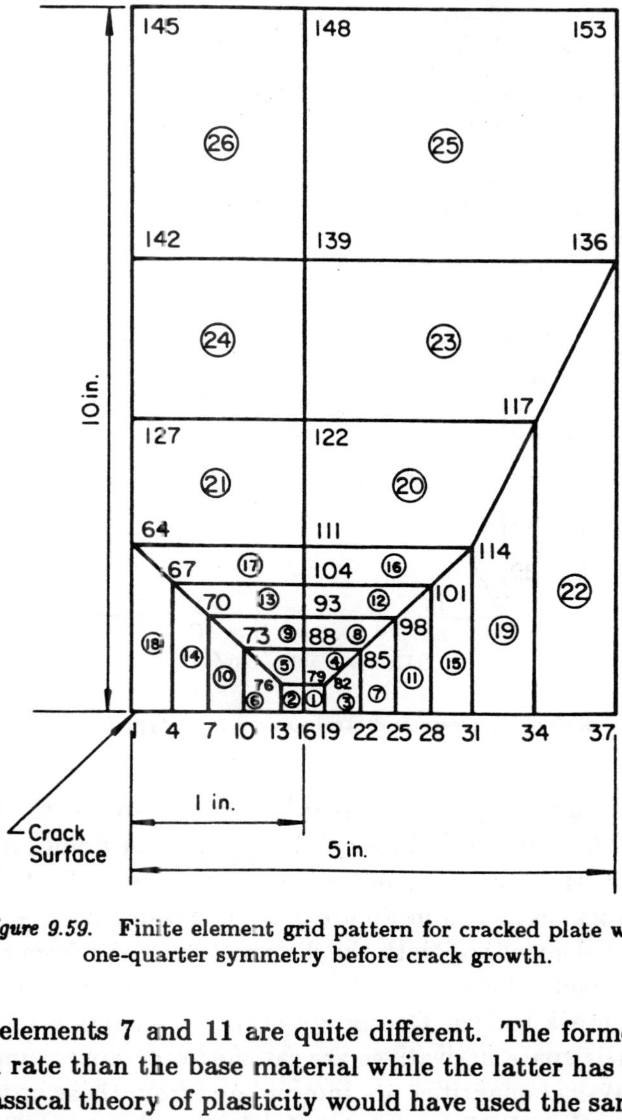

Figure 9.59. **Finite element grid pattern for cracked plate with one-quarter symmetry before crack growth.**

ϵ curves for elements 7 and 11 are quite different. The former possesses a higher strain rate than the base material while the latter has a lower strain rate. The classical theory of plasticity would have used the same base material curve for all elements in the cracked plate, an assumption that is overly idealized and can lead to large errors in estimating the state of affairs near the crack tip, as pointed out by Sih and Tzou [30].

Subcritical crack growth. Tzou and Sih [62] applied the isoenergy density theory of Sih [48, 49] to examine the subcritical crack growth of a center-cracked panel as shown in Figure 9.22, where $b = 10$ in., $c = 5$ in. and $a = 1$ in. A two-dimensional stress state is assumed as in Section 9.3 for the edge crack problem, i.e., $(dV/dA)_\varsigma$ is taken to be constant. Discretized by finite elements is one-quarter of the panel geometry, as shown in Figure 9.59,

Table 9.15. Applied stress increments on center-cracked panel.

Step	Time t (s)	Applied stress σ_0 (ksi)
1	0.0667	40
2	0.0833	50
3	0.1000	60
4	0.1167	70
5	0.1333	80
6	0.1500	90
7	0.1667	100
8	0.1833	110

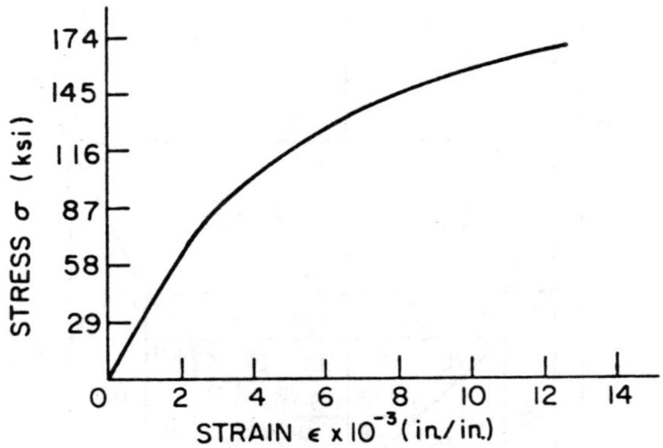

Figure 9.60. Reference stress and strain curve for a typical engineering steel.

where 26 standard isoparametric quadrilateral elements with 161 nodes are used. All elements are embedded with 16 Gaussian points of integration while a total of 64 Gaussian points is used for element 1 situated directly ahead of the crack, as shown in Figure 9.59. Increasing accuracy is achieved by reducing the element size with the distance measured from the crack tip. The lineal dimension of element 1 is 0.194 in., which corresponds to 1.5% of the initial half crack length. Continuity of displacements is required at the nodes while their space derivatives are satisfied only approximately. Nodal values of all physical quantities are interpolated from those at the Gaussian points. Refer to the PEDDA program [54] for details.

Material and loading steps. Equal and opposite uniform stress of magnitude σ_0 is applied with $\dot{\sigma}_0 = 600$ ksi/s. Eight time steps will be taken and their correspondence with the applied stress increments is summarized in Table 9.15.

Since the stress and stress response for each element will be determined, only the slope of the initial stress–strain curve needs to be specified. A structural steel with $d\sigma/d\epsilon = 3 \times 10^4$ ksi will be used, the stress–strain diagram of which is shown in Figure 9.60. The first step is taken to be sufficiently small, giving rise to a crack growth of the order of 10^{-2} to 10^{-1} cm of the initial half crack length such that the assumption of all elements possessing the same initial slope will hold. Loading is increased incrementally until the nearest element ahead of the crack breaks. This is assumed to occur when the volume energy density corresponding to the total area under the true stress–strain curve reaches the critical value $(dW/dV)_c$.

Growth analysis. Following the procedure outlined in Figures 9.49 and 9.50, the state of affairs in element 1 are determined on the plane of isoenergy density. Shown in Figure 9.61 are the isostress σ_ξ and isostrain ϵ_ξ for the four time increments leading up to the first crack growth increment. The shaded triangle represents $(dW/dV)_c^* = 0.329$ ksi, the amount of available energy density, while the unshaded portion under the σ_ξ versus ϵ_ξ plot corresponds to the amount dissipated prior to the onset of crack extension. A plot of $(dW/dV)^*$ as a function of the distance r along the prospective path of crack growth is shown in Figure 9.62. The intersection of $(dW/dV)_c^*$ with the curve gives $r_1^* = 0.1194$ cm for the first increment of crack growth. An energy density factor S_1^*, which represents the area of the cross-hatched rectangular $(dW/dV)_c^*$ by r_1^*, can thus be obtained:

$$S_1^* = r_1^* \left(\frac{dW}{dV}\right)_c^* = 15.873 \text{ lb/in.} \tag{9.64}$$

It has the same units as the energy release rate quantity in the conventional theory of linear elastic fracture mechanics although the physical and mathematical implications are quite different. Since volume and surface energy density interact as postulated in equation (9.55), it is also necessary to check whether dW/dA has surpassed its critical value or not. The quantity $(dW/dA)_\xi^*$, being equal to the product of $(dV/dA)_\xi$ and $(dW/dV)^*$ in element 1, can be found. It has also exceeded its critical value of $(dW/dA)_c^* = 4.3 \times 10^3$ lb/in. This would yield a crack growth segment of $r_1^* = 1.18 \times 10^{-3}$ in. which is negligible in comparison with $r_1^* = 4.70 \times 10^{-2}$ in. as predicted from $(dW/dV)_c^*$. More specifically, the damage predicted by $(dW/dA)_\xi^* \to (dW/dA)_c^*$ lies within the 'macrodamage free' zone as discussed by Sih and Tzou [59]. This reveals the necessity to specify the scale level of crack growth in formulating a predictive model of fracture. Similar plots of $(dW/dV)^*$ versus r may also be obtained for determining the subsequent crack growth increments according to the loading steps in Table 9.15.

Because the entire crack is engulfed in a uniform stress field, energy supplied to the crsck tip region increases continuously and this makes the crack growth process unstable. In other words, less and less energy is needed to drive the crack as $(dW/dV)_c^*$ tends to decrease with time. It is shown in Figure 9.63 that the maximum driving force is for initiating the crack growth.

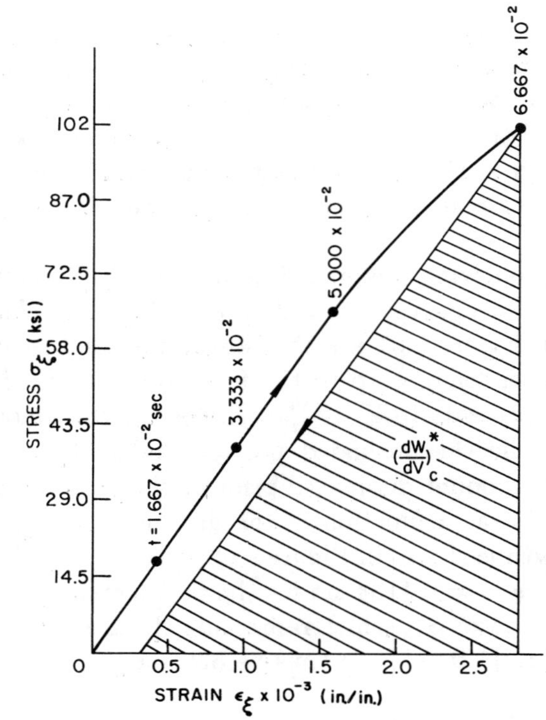

Figure 9.61. **Stress versus strain responses on the plane of isoenergy density in element 1 for time leading up to the first crack growth increment.**

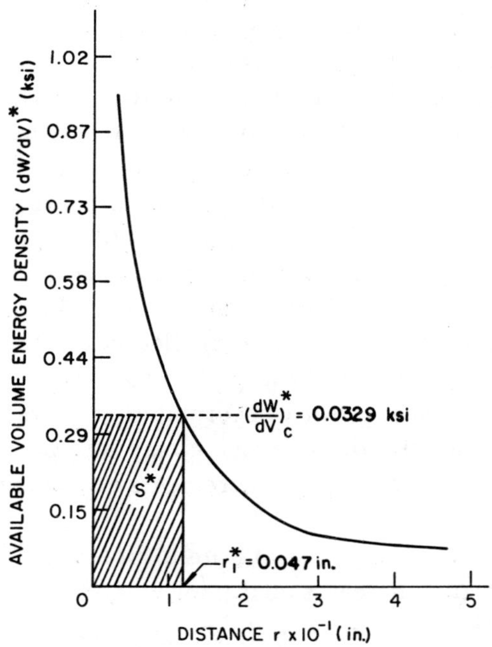

Figure 9.62. **Decay of available volume energy density for the first increment of crack growth at $t = 6.67 \times 10^{-2}$ s.**

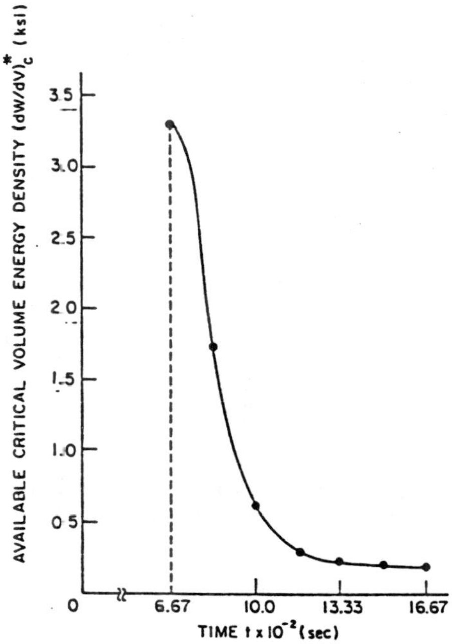

Figure 9.63. Variation of available critical energy density $(dW/dV)_c^*$ as a function of time for element 1.

The relation

$$\left(\frac{dW}{dV}\right)_c^* = \frac{S^*(t)}{r^*(t)} \tag{9.65}$$

will hold for each instance of crack growth* because $(dW/dV)_c^*$ changes as a new set of ϵ_ξ and $(dV/dA)_\xi$ values are obtained. Summarized in Figure 9.64 are the time histories of crack growth for the isoenergy density and plasticity theory. The former corresponds to the open circles and the latter to the solid circles. Large differences on the elapsed time are observed. This process, being loading rate dependent, is exhibited by the curve with a two-orders of magnitude higher loading rate of $\dot{\sigma} = 6 \times 10^4$ ksi/s. Referring to the isoenergy density prediction, the time for $a = 3.15$ in. is decreased from 15×10^{-2} s to 10×10^{-2} s when $\dot{\sigma}_0$ is increased from 6×10^2 to 6×10^4 ksi/s.

In view of equation (9.65), the crack driving force S^* in the isoenergy density theory is loading rate dependent. The variations of S^* with crack growth for the eight time steps in Table 9.15 are plotted in Figure 9.65.

The available strain energy density remained constant during crack

* The aforementioned feature will not occur if the same constitutive relation is assumed to hold as in the case of plasticity.

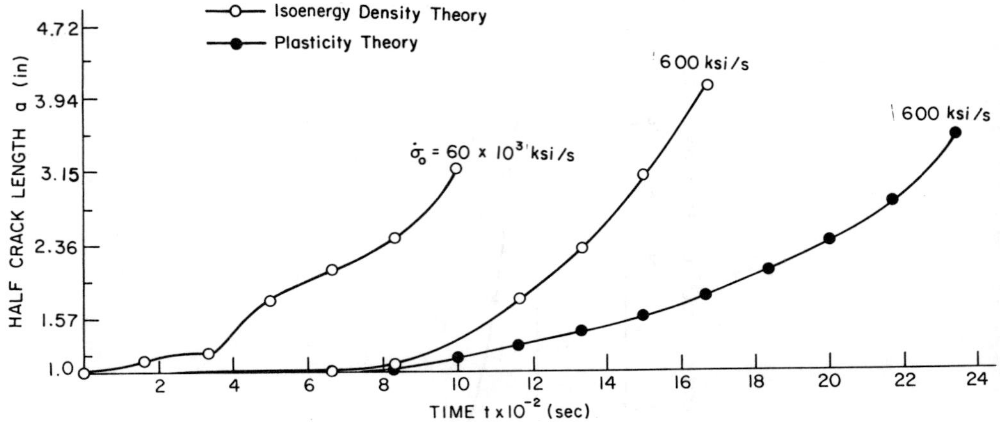

Figure 9.64. Time history of crack length: comparison of isoenergy density and plasticity theory.

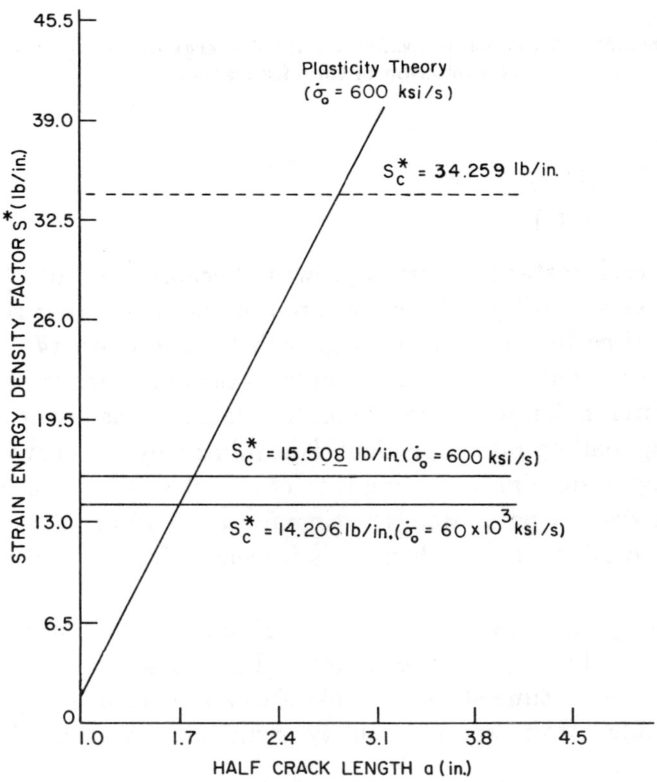

Figure 9.65. Variations of available critical strain energy density factor with crack growth.

growth. A lower value of $S^* = 14.206$ lb/in. is obtained when the loading rate is increased by two orders of magnitude to $\dot{\sigma} = 6 \times 10^4$ ksi/s. This is the crack growth resistance curve. The slanted line in Figure 9.65 represents data obtained from the incremental theory of plasticity for $\dot{\sigma} = 6 \times 10^2$ ksi/s, which satisfies the condition $dS^*/da =$ const. with a critical value of $S_c^* = 34.259$ lb/in. This differed substantially from that of the isoenergy density theory. To emphasize the critical value of S^* or S_c^* is indeed material-specific but loading rate dependent. This must be so because the creation of new crack surface depends on the rate at which energy is being supplied and dissipated.

Stress field inhomogeneity. Since the constitutive relation for each element is determined individually for each time step, the resulting local stresses and strains are said to be nonhomogeneous. More specifically, $(dV/dA)_\xi$ and ϵ_ξ may vary for each time increment of loading while the plane of isoenergy density located by the angle α also changes accordingly. If X is used to denote those physical quantities under consideration, then their weighted average over the entire isoparametric element consisting of 12 nodes can be obtained as

$$X = \frac{\sum_{j=1}^{12} x_j X_j}{\sum_{j=1}^{12} d_j} \tag{9.66}$$

where x_j $(j = 1, 2, \ldots, 12)$ is the distance measured from the jth node to the centroid of the element. Once the stress state on the plane of isoenergy density is known, coordinate transformation may be used to obtain the stress components referred to a rectangular Cartesian system with x and y aligned parallel and normal to the crack plane respectively. The z-axis is in the thickness direction of the panel. Referring to an element just outside a small region surrounding the crack tip with a radius equal to 1.5% of the half crack length, the stresses σ_{xx}, σ_{yy} and σ_{zz} are obtained for $t = 6.667 \times 10^{-2}$ s at which instance, $a = 1$ in. Figure 9.66 shows that the curves representing σ_{xx} for plasticity and energy density theory intersect at $r \simeq 0.6 \times 10^{-2}$ in. ahead of the crack. Plasticity solution yields smaller and smaller σ_{xx} as $r \to 0$. This difference arises because of the neglect of $(dV/dA)_i$ throughout the system as well as in the boundary conditions stated in equation (9.57). For a local element near the crack surface, even though $\bar{\rho}\ddot{u}_i$ in equation (9.57) may be small, the product $\bar{\rho}\ddot{u}_i(dV/dA)_n$ can be of the same order of magnitude as the stresses σ_{ij}. The situation for σ_{yy} in Figure 9.67 is reversed; that is, plasticity tends to yield higher values of σ_{yy} than that of the isoenergy density theory for $r < 0.6 \times 10^{-2}$ in. The degree of variance will change as the crack grows. As expected, the difference in σ_{zz} for the two theories is small because damage variations in the z-direction have been

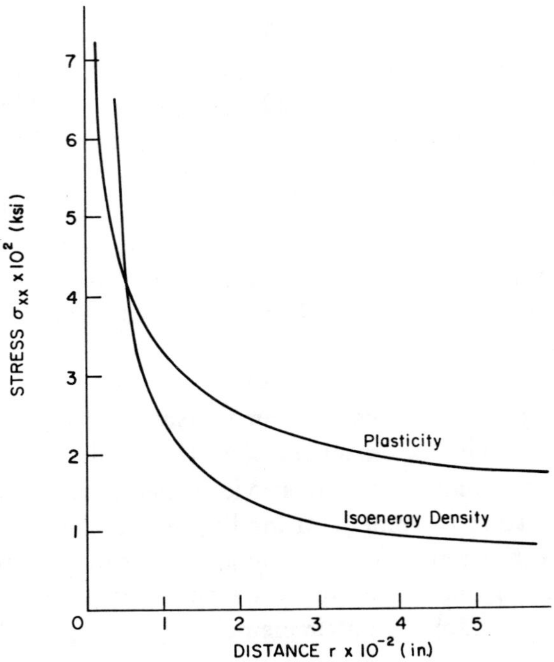

Figure 9.66. **Distribution of σ_{xx} ahead of the crack tip at $t = 6.667 \times 10^{-2}$ s for $a = 1.0$ in.**

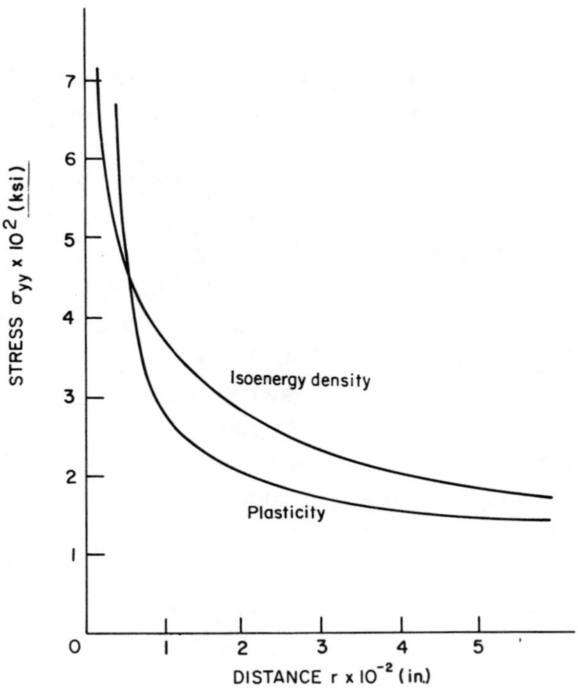

Figure 9.67. **Distribution of σ_{yy} ahead of the crack tip at $t = 6.667 \times 10^{-2}$ s for $a = 1.0$ in.**

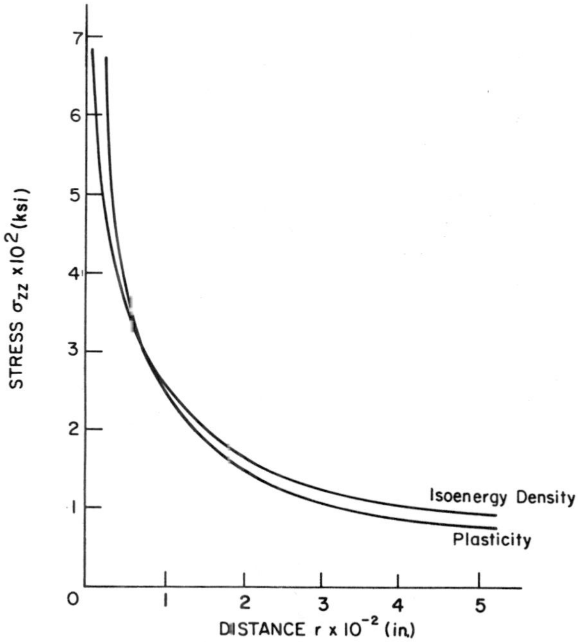

Figure 9.68. Distribution of σ_{zz} ahead of the crack tip at $t = 6.667 \times 10^{-2}$ s for $a = 1.0$ in.

neglected. This is illustrated in Figure 9.68.

The time history of stress and strain in element 1 as the crack advances is exhibited in Figure 9.69 in terms of σ_ξ and ϵ_ξ. While the grid pattern changes slightly with crack growth, the location and sizes of the local elements are not altered so as to preserve consistency in the numerical averaging process. Figure 9.69 shows that the curve for element 1 rises rapidly as an indication of the high strain rate. This effect becomes much more pronounced when σ_ξ and ϵ_ξ are referred to a small area of 3.93×10^{-3} in. diameter at a distance 0.02 in. from the moving crack front. Table 9.16 shows the difference for $(dV/dA)_\xi$ and σ_ξ in elements 1 and 4. Their magnitudes decreased quickly as the distance from the crack is increased slightly.

Plotted in Figure 9.70 are the effective stress σ_{eff} as a function of the effective strain ϵ_{eff} for the isoenergy density and plasticity theory, where

$$\sigma_{\text{eff}} = \frac{1}{\sqrt{2}}\sqrt{(\sigma_1 - \sigma_2)^2 + (\sigma_2 - \sigma_3)^2 + (\sigma_3 - \sigma_1)^2} \tag{9.67}$$

and

$$\epsilon_{\text{eff}} = \frac{1}{\sqrt{2}}\sqrt{(\epsilon_1 - \epsilon_2)^2 + (\epsilon_2 - \epsilon_3)^2 + (\epsilon_3 - \epsilon_1)^2} \tag{9.68}$$

In equations (9.67) and (9.68), σ_j and ϵ_j ($j = 1, 2, 3$) are, respectively,

Table 9.16. Quantities referred to the plane of isoenergy density in elements 1 and 4.

Time $\times 10^{-2}$ (s)	α (degrees)	$(dV/dA)_\xi$ (in.)	σ_ξ (ksi)
Element 1			
6.667	-12.26	4.484	103.052
10.000	-21.82	3.613	148.074
13.333	-17.03	2.840	200.303
18.333	-17.39	1.864	239.004
Element 4			
6.667	-13.77	0.939	24.810
10.000	-12.45	1.386	45.617
13.333	-11.40	1.512	80.243
18.333	-10.35	1.096	120.582

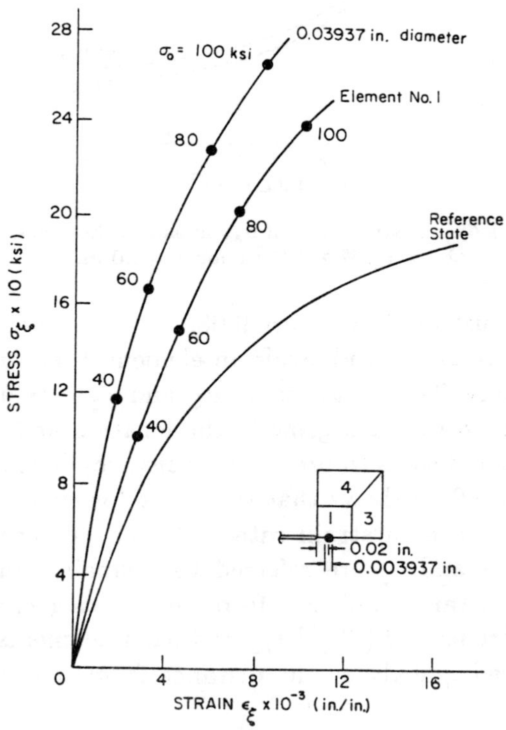

Figure 9.69. Stress and strain curves of element 1 and an area averaged over 3.39×10^{-3} in. ahead of the crack tip.

the principal stresses and strains. They should be distinguished from those referred to the plane of isoenergy density. Note that large differences are again seen for the two curves. This is largely due to the neglect of dilatation in plasticity, an assumption that is not justified ahead of the crack where the local stress state is hydrostatic in character. The fluctuation of volume with surface is responsible for changes in temperature.

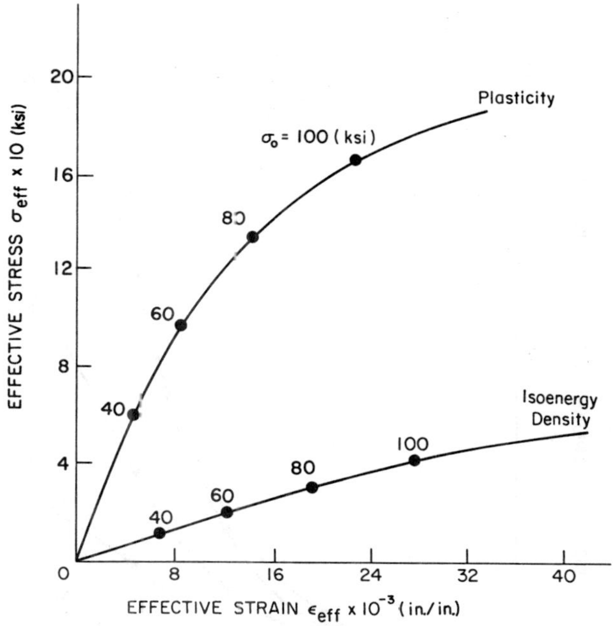

Figure 9.70. Comparison of effective stress and strain response in element 1 for the isoenergy density and plasticity theory.

Synchronization of temperature and deformation. Equation (9.63) shows that Θ and (σ, ϵ) are related via the dissipation energy density $\mathcal{D}$ or $(dW/dV)_p$. Applying the procedure devised by Sih and Chou [61] for determining $\mathcal{D}$ from the recovery property of the local material element, the results for some typical elements next to and away from the crack tip region are obtained in Figure 9.71. Prior to the first step of crack growth for $t < 6.67 \times 10^{-2}$ s, dissipation is negligibly small in all the elements. The curves begin to rise very sharply in the local elements, 1 and 7, that are situated along the path of prospective crack growth. Elements 21 and 23 are further from the site of fracture; the dissipation energy density $\mathcal{D}$ increased only gradually with time.

As a result of volume and surface change, the temperature will alter accordingly. With reference to a set of coordinate axes that move with the crack tip, Θ is computed numerically from equation (9.63) for a time increment of $t = 2 \times 10^{-2}$ s. A drop of temperature below that of ambient is detected at the initial stage of loading for all the elements as the crack grows. Constant reversal of heat transfer occurs between the element and its surrounding. This is reflected by the oscillation of Θ–Θ_0 about the reference Θ_0. Such a phenomenon is particularly pronounced in element 1, which is a manifestation of the dilatation and distortion of the local material as shown in Figure 9.72. The sign change in Θ–Θ_0 is not as pronounced for elements away from the crack tip. In general, a sudden jump in temperature

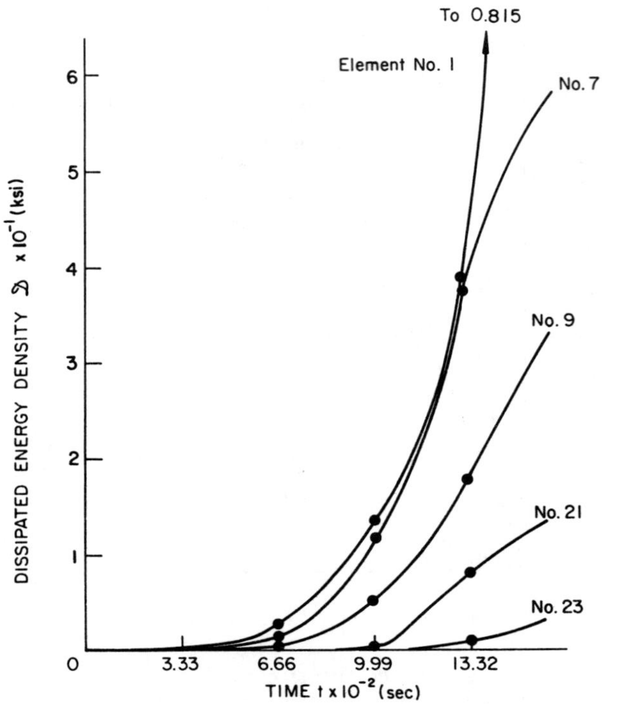

Figure 9.71. **Time variation of dissipated energy density in typical elements.**

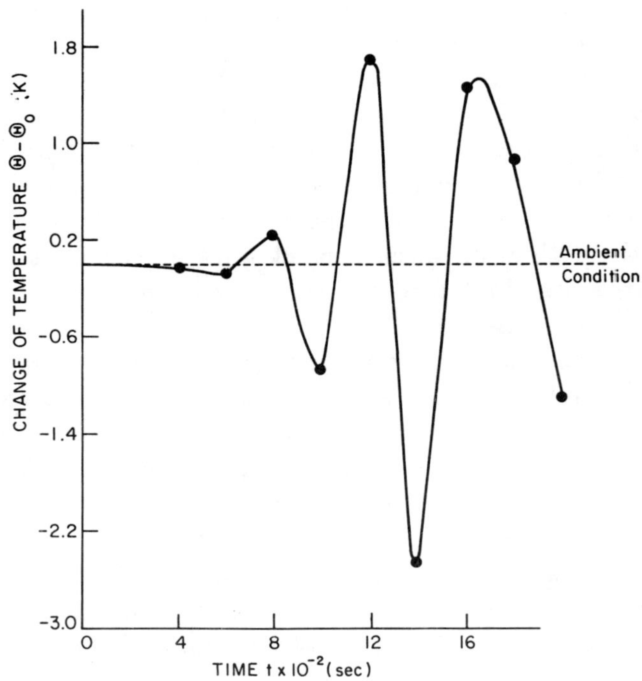

Figure 9.72. **Time variation of temperature change in element 1.**

Table 9.17. Distances d_i (in.) of concentric contours around crack tip.

d_1	d_2	d_3	d_4	d_5
0.60	0.40	0.20	0.10	0.025

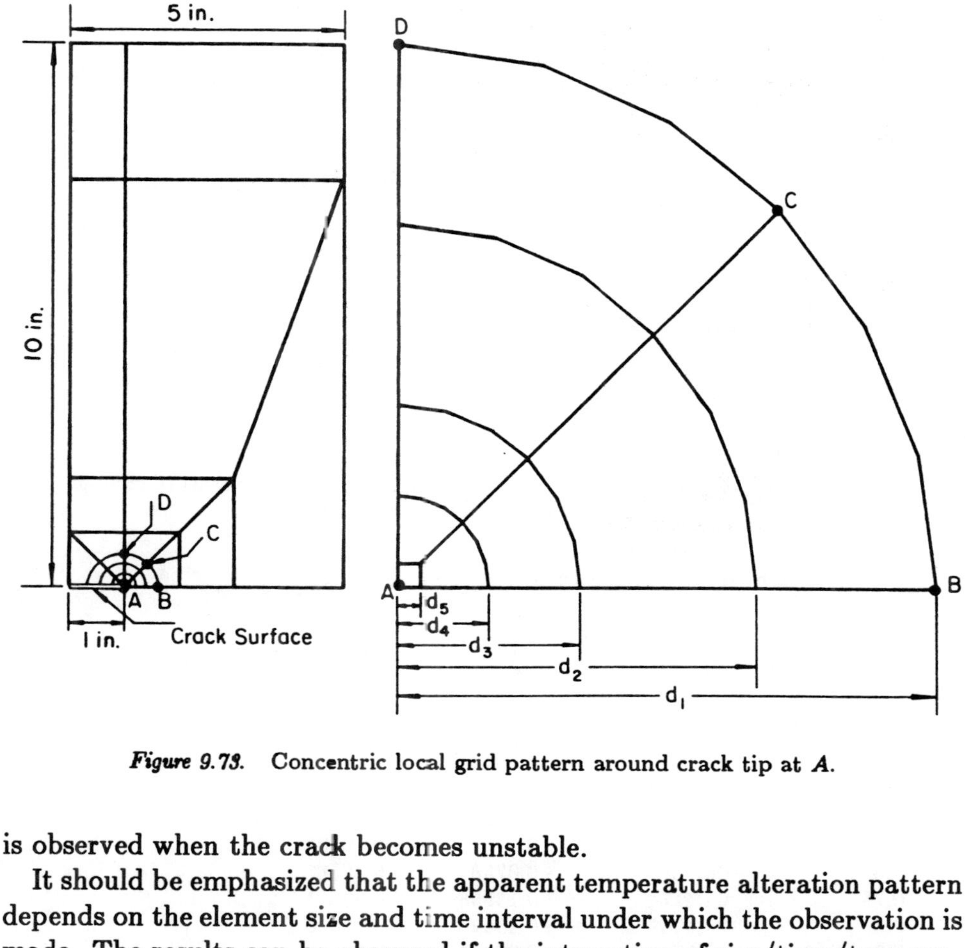

Figure 9.73. Concentric local grid pattern around crack tip at *A*.

is observed when the crack becomes unstable.

It should be emphasized that the apparent temperature alteration pattern depends on the element size and time interval under which the observation is made. The results can be obscured if the interaction of size/time/temperature, as discussed by Sih and Chao [63], is not taken into account. This implication can be illustrated by considering a typical element, say, at a distance 0.1 in. from the crack tip. Computation is referred to a system of concentric finite elements as shown in Figure 9.73. The outer peripheries of these contours are placed at distances d_i ($i = 1, 2$, etc.) from the crack front, Table 9.17. The smallest distance d_5 is 2.5% of the stationary half crack length $a = 1.0$ in. The change of temperature Θ–Θ_0 will be averaged over circular regions with diameters 2.92×10^{-3}, 3.54×10^{-3} and 2.92×10^{-4} in. All the curves in Figure 9.74 tend to oscillate with time and their behavior

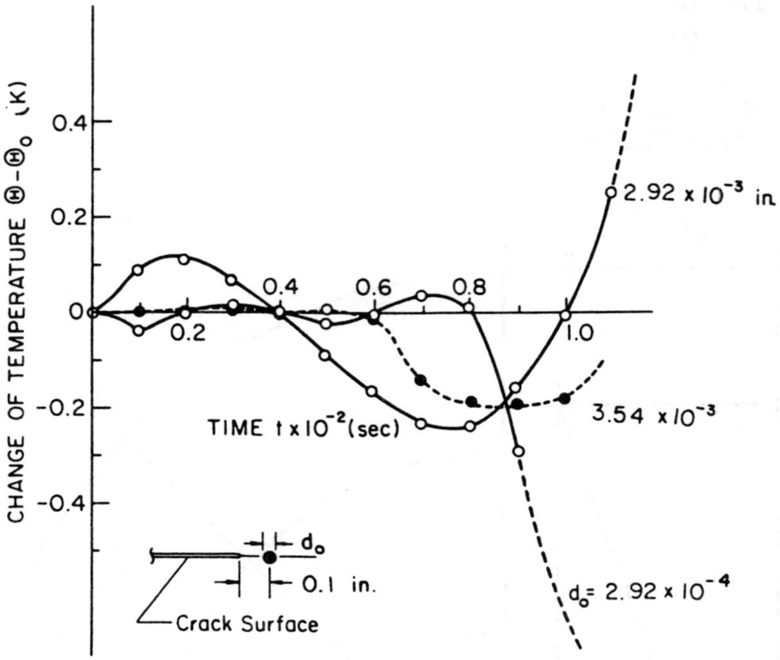

Figure 9.74. The effects of scale of observation made on temperature change at a distance 0.1 in. ahead of the crack tip.

Table 9.18. Change of temperature $\Theta - \Theta_0$ (K).

Time	Diameters (in.)		
$t \times 10^{-2}$ (s)	2.92×10^{-3}	3.54×10^{-3}	2.92×10^{-4}
0.0	0.0	0.0	0
0.1	−0.03820	0.0	0.09701
0.2	0	0.07045	0.11506
0.3	0.01649	0.09245	0.07558
0.4	0	0.06397	0
0.5	−0.01678	0	−0.09023
0.6	−0.00015	−0.0758	−0.17368
0.7	0.04174	−0.13958	−0.22891
0.8	0.01599	−0.18526	−0.23448
0.9	−0.29801	−0.19431	−0.16895
1.0	−0.85433	−0.17661	−0.01089
1.1	−0.92122	0	0.26114

is quite different. A reversal in the heat transfer occurs when the curve crosses the $\Theta = \Theta_0$ line, representing the ambient condition. For t smaller than 0.2×10^{-2} s, the temperature averaged over the region with diameters 2.92×10^{-4} in. and 2.92×10^{-3} in. is opposite in sign. This character is, by no means, obvious and cannot be detected by the classical continuum mechanics theories. The precise values of $\Theta - \Theta_0$ averaged over three different diameters can be found in Table 9.18. The interaction of size and temperature at a given time is highly nonlinear. Smaller regions may experience higher temperature change. What appears to be dilatation may switch to distortion as the element size of observation is altered. This flip-flop phenomenon can be associated with change in the $(dV/dA)_\xi$ values on the plane of isoenergy density. The scaling of size/time/temperature can be done selectively such that cooling/dilatation and heating/distortion are in phase for the time response and scale size at which the events are observed. Otherwise, the Θ–Θ_0 versus t domain could be filled with data that yield a disorderly pattern. A more fundamental discussion of size/time/temperature scaling can be found in the work of Sih and Chao [63].

Further comments. The development of the mechanics of fracture has not been consistent for obvious reasons. Too much arbitrariness prevails in the ways with which fracture criteria could be used in conjunction with solutions obtained from continuum mechanics theories. Approximations arising from stress analyses can be overlooked and are not compatible with assumptions made in the failure analyses. Some of the more common pitfalls will be pointed out in passing.

Traditionally, excessive shape change has been associated with yielding and volume change with fracture. Only in the linear theory of elasticity can these two effects be separated as a linear sum. The situation is no longer clear when permanent deformation or energy dissipation arises. Dilatational and distortional effects are interactive and inseparable; they are rate dependent and cannot be superimposed; nor can they be combined arbitrarily. Serious limitations on the outcome can result if a permanent deformation is assumed to be associated exclusively with the distortional component of the strain energy density function as reflected by the von Mises yield criterion in the classical theory of plasticity. This restricts the theory to *global* failure prediction of structures as in the case of limit analysis. Local effects – such as the initiation of crack growth, coalescence of voids, etc., that are dominated by dilatation – cannot be adequately explained. Although attempts [64–66] have been made to include the effects of hydrostatic stress in the yield condition, this does not remove the inadequacies of the theory if the coincidence of uniaxial data with the effective stress and effective strain is not removed. In other words, the effect of dilatation cannot be completely accounted for through the yield condition. Furthermore, the way with which such correction should be made appears to be empirical because the justification has always relied on *a priori* experimental evidence.

Incompleteness of continuum mechanics theories to account for the material damage process cannot be resolved by the piecemeal approach of imposing failure criteria on solutions for the undamaged states. Unless the intrinsic characteristics of energy dissipation in the *nonequilibrium* material damage process is included, the underlying controlling mechanisms would remain hidden. What has been modelled by the isoenergy density theory in this chapter is the *real time* crack growth behavior which is associated with highly nonhomogeneous deformation in segments of nonequilibrium states.

Synchronization of the thermal and mechanical effects was made possible by not allowing $\mathrm{d}V/\mathrm{d}A$ to vanish in the limit. The unique character of the isoenergy density theory can be summarized as follows:

1. Stress and strain behavior is not presumed but determined according to the combined effects of load/geometry/material. Only the initial or reference state needs to be specified.

2. Tractions on body or local inertia forces are related by the change of volume with surface, $(\mathrm{d}V/\mathrm{d}A)_n$, in the direction of the outward normal acting on the surface of the continuum. This effect is particularly significant near free surfaces, the region of load application and geometrical discontinuities such as cracks.

3. The concept of equivalent uniaxial stress and strain relation on the plane of isoenergy density establishes a one-to-one correspondence between the uniaxial and triaxial stress states without loss of generality.

4. Coupling of all energy forms – including thermal or mechanical forms – is retained. Temperature fluctuations are synchronized with mechanical deformation. This is accounted for by the rate change of volume with surface.

In particular, the failure of material elements involves the consideration of both $(\mathrm{d}W/\mathrm{d}A)_\xi$ and $\mathrm{d}W/\mathrm{d}V$ compared with their respective critical values which change for each load increment and are different from element to element. This is because the theory no longer assumes the same constitutive relation throughout the system. Material behavior depends on the rate of local energy transfer. As the accumulation of energy dissipation or damage can be assessed analytically regardless of loading type, the method applies equally well to cyclic loading, as in fatigue, and sustained loading, as in creep. The change in local strains and strain rates can be determined individually for each element of the specimen with or without an initial crack. Microdamage can also be included. When the loading rate is slowed down, as in the creep process, there will be ample time for the material microstructure to interact with the load. Remodelling or remeshing of the elements becomes necessary because the morphology of the microstructure would have undergone significant changes with time prior to global instability. This would be similar to the situation in phase transformation as discussed by Sih [67].

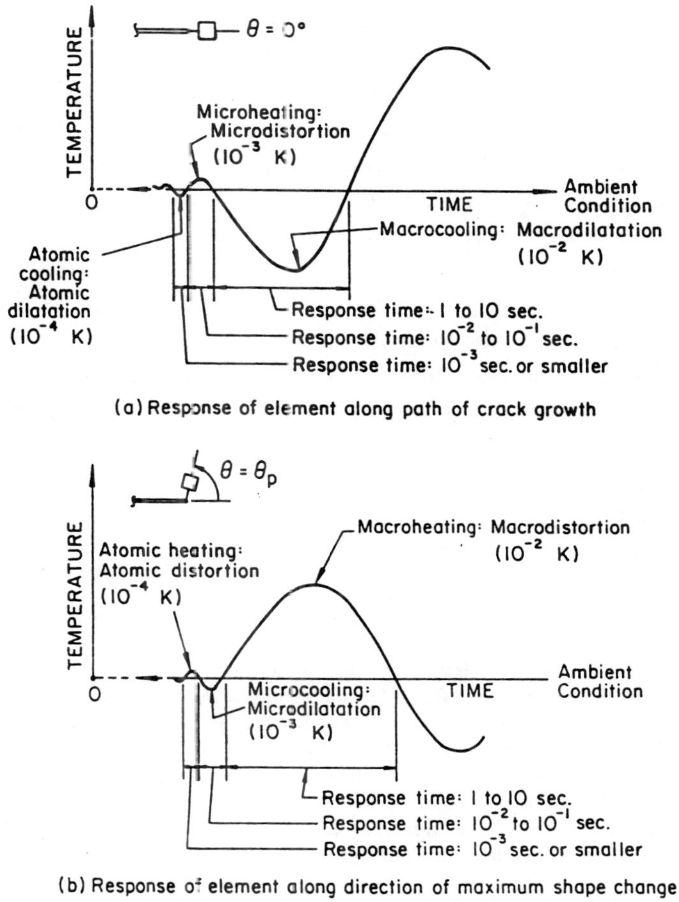

Figure 9.75. Size/time/temperature response of element ahead of crack at different scale level.

As the rate change of volume with surface area, dV/dA, undergoes high oscillation near the crack tip, its inclusion in the isoenergy density theory had made it possible to assess material damage at the atomic, microscopic and macroscopic level in a consistent fashion. This is accomplished by considering the transient character of size/time/temperature interaction. Failure modes at one scale level belonging to the same material damage process may have a different appearance at another scale level. Depending on the loading type and rate, Sih and Chao [63] have shown that dilatation/distortion and cooling/heating tend to flip-flop as the scale level of observation is altered; depending, of course, also on the selected time interval within which the events are observed. This alternating behavior is illustrated with reference to the state of affairs near the macrocrack tip, as in Figure 9.75(a) and 9.75(b). At an element along the path of possible crack growth at $\theta = 0°$ in

Figure 9.75(a), what appears to be macrocooling* and macrodilatation for response time of 1 to 10 s corresponds to microheating and microdistortion when the same event is viewed within the time interval of 10^{-2} to 10^{-1} s. Disturbances at the atomic scale refer to an even smaller response time and temperature fluctuation. The situation for an element in the direction of excessive distortion at $\theta = \theta_p$ in Figure 9.75(b) behaves opposite to that in Figure 9.75(a). What was macrocooling now becomes macroheating. The same applies to the micro- and atomic-scale together with the corresponding time response and temperature. Scaling of size/time/temperature is being assessed quantitatively such that seemingly different behavior of the same fracture process can be related uniquely when viewed at the atomic, microscopic and macroscopic levels. This interaction will become better understood as more actual examples and results are made available.

9.6. References

[1] Sih, G. C., Wei, R. P. and Erdogan, F. (eds.), *Linear Fracture Mechanics*, Envo Publishing Co., Inc., Bethlehem, Pa (1976).

[2] Pellini, W. S., *Principles of Structural Integrity Technology*, Office of Naval Research, Arlington, Va. (1976).

[3] Brown, W. F.,Jr. and Srawley, J. E., Plane strain crack toughness testing of high strength metallic materials, *ASTM Special Technical Publication* No. 410 (1966).

[4] Wells, A. A., Application of fracture mechanics at and beyond general yielding, *British Welding Journal* **10**, pp. 563–569 (1963).

[5] Begley, J. A. and Landes, J. E., The J-integral as a fracture criterion, *Fracture Toughness, ASTM Special Technical Publication* No. 514, pp. 1–20 (1972).

[6] Bucci, R. J., Paris, P. C., Landes, J. E. and Rice, J. R., J integral estimation procedure, *Fracture Toughness, ASTM Special Technical Publication* No. 514, pp. 40–69 (1972).

[7] Shih, C. F., DeLorenzi, H. G. and Andrews, W. R., Studies on crack initiation and stable crack growth, *Elastic–Plastic Fracture Mechanics, ASTM Special Technical Publication* No. 668, pp. 65–120 (1979).

[8] Bernstein, H. L., A study of J-integral method using polycarbonate, AFWAL-TR-82-4080, Air Force Wright Aeronautical Laboratories, Wright-Patterson Air Force Base, Ohio, August (1982).

[9] Sih, G. C. and Tzou, D. Y., Crack extension resistance of polycarbonate material, *Journal of Theoretical and Applied Fracture Mechanics* **2**, No. 2, pp. 220–234 (1984).

[10] Sih, G. C., Mechanics of subcritical crack growth, in *Fracture Mechanics Technology Applied to Material Evaluation and Structure Design*, edited by G. C. Sih, N. E. Ryan and R. Jones, Martinus Nijhoff Publishers, The Netherlands, pp. 3–18 (1982).

[11] Sih, G. C. and Tzou, D. Y., Mechanics of nonlinear crack growth: effects of specimen size and loading step, in *Modelling Problems in Crack Tip Mechanics*, J. T. Pindera, ed., Martinus Nijhoff Publishers, The Netherlands, pp. 155–169 (1984).

[12] Sih, G. C. and Chen, C., Non-self-similar crack growth in an elastic–plastic finite thickness plate, *Journal of Theoretical and Applied Fracture Mechanics* **3**, No. 2, pp. 125–139 (1985).

---

* Macro refers to dimensions of 10^{-3} to 10^{-2} in.; micro to 10^{-4} to 10^{-5} in.; and atomic to 10^{-6} to 10^{-8} in. The size/time/temperature scale can shift depending on the loading or local strain rate.

[13] Sih, G. C. and Tzou, D. Y., Dynamic fracture rate of Charpy V-notch specimen, *Journal of Theoretical and Applied Fracture Mechanics* **5**, No. 3, pp. 189–203 (1986).
[14] Sih, G. C., Fracture mechanics of engineering structural components, in *Fracture Mechanics Methodology*, edited by G. C. Sih and L. Faria, Martinus Nijhoff Publishers, The Netherlands, pp. 35–101 (1984).
[15] Sih, G. C., Outlook on fracture mechanics, in *The Mechanism of Fracture*, edited by V. S. Goel, *Proceedings of the Annual American Society of Metal Conference, Salt Lake City, Utah*, pp. 1–16, December 2–6 (1985).
[16] Paris, P. C., The growth of cracks due to variations in load, Ph.D. Dissertation, Department of Mechanics, Lehigh University (1962).
[17] Wei, R. P., Contribution of fracture mechanics to subcritical crack growth studies, in *Linear Fracture Mechanics*, edited by G. C. Sih, R. P. Wei and F. Erdogan, Envo Publishing Co., Bethlehem, Pa., pp. 287–302 (1976).
[18] Vecchio, R. S. and Hertzberg, R. W., A rationale for the apparent anomalous growth behavior of short fatigue cracks, *Journal of Engineering Fracture Mechanics* **22**, No. 6, pp. 1049–1060 (1985).
[19] Sih, G. C., Some basic problems in fracture mechanics and new concepts, *International Journal of Engineering Fracture Mechanics* **5**, No. 2, pp. 365–377 (1973).
[20] Sih, G. C., The mechanics aspects of ductile fracture, in *Proceedings on Continuum Models of Discrete Systems*, edited by J. W. Provan, University of Waterloo Press, Canada, pp. 361–386 (1977).
[21] Griffith, A. A., The theory of rupture, *First International Congress for Applied Mechanics, Delft*, pp. 55–63 (1924).
[22] Irwin, G. R., Analysis of stresses and strains near the end of a crack traversing a plate, *Journal of Applied Mechanics* **24**, pp. 361–364 (1957).
[23] Hilton, P. D. and Hutchinson, J. W., Plastic intensity factors for cracked plates, *Journal of Engineering Fracture Mechanics* **3**, pp. 435–451 (1971).
[24] Hutchinson, J. W., Plastic stress and strain fields at a crack tip, *Journal of Mechanics and Physics of Solids* **16**, pp. 337–342 (1968).
[25] Ohr, S. M., Horton, J. A. and Chung, S. J., Direct observations of crack tip dislocation behavior during tensile and cyclic deformation, in *Defects, Fracture and Fatigue*, edited by G. C. Sih and J. W. Provan, Martinus Nijhoff Publishers, The Netherlands, pp. 3–15 (1982).
[26] Orowan, E., Energy criteria of fracture, *Welding Research Supplement* **34**, pp. 157s-160s (1955).
[27] Sih, G. C., *Mechanics of Fracture*, Introductory Chapters, Vol. I to Vol. VII, edited by G. C. Sih, Martinus Nijhoff Publishing, The Hague (1972–1982).
[28] Sih, G. C. and Chu, R. C., Characterization of material inhomogeneity by stationary values of strain energy density, *Journal of Theoretical and Applied Fracture Mechanics* **5**, No. 3, pp. 151–161 (1986).
[29] Sih, G. C., Michopoulos, J. G. and Chou, S. C., *Hygrothermoelasticity*, Martinus Nijhoff Publishers, The Netherlands (1986).
[30] Sih, G. C. and Tzou, D. Y., Nonhomogeneous energy dissipation ahead of a slowly growing crack: Part I – Monotonic loading, Institute of Fracture and Solid Mechanics Technical Report IFSM-84-126 (1984).
[31] Barenblatt, G. I., On some basic ideas of theory of equilibrium cracks forming during brittle fracture, *Problems of Continuum Mechanics*, Society for Industrial and Applied Mathematics, Philadelphia, Pa., pp. 22–38 (1961).
[32] Barenblatt, G. I., The mathematical theory of equilibrium of cracks in brittle fracture, *Advances in Applied Mechanics* **7**, pp. 55–129 (1962).
[33] Gdoutos, E. E. and Sih, G. C., Crack growth characteristics influenced by load time record, *Journal of Theoretical and Applied Fracture Mechanics* **2**, No. 2, pp. 91–103 (1984).
[34] Sih, G. C. and Chao, C. K., Fatigue initiation in unnotched specimens subjected to monotonic and cyclic loading, *Journal of Theoretical and Applied Fracture Mechanics* **2**, No. 1, pp. 67–74 (1984).

[35] Sih, G. C. and Chao, C. K., Size effect of cylindrical specimens with fatigue cracks, *Journal of Theoretical and Applied Fracture Mechanics* **1**, No. 3, pp. 239–247 (1984).
[36] Ballarini, R., Shah, S. P. and Keer, L. M., Crack growth in cement-based composites, *Journal of Engineering Fracture Mechanics* **20**, No. 3, pp. 433–445 (1984).
[37] Weibull, W., A statistical theory of the strength of metals, *Proceedings of the Royal Swedish Institute of Engineering* **193**, No. 151 (1939).
[38] Sih, G. C. and Madenci, E., Fracture initiation under gross yielding: Strain energy density criterion, *Journal of Engineering Fracture Mechanics* **78**, No. 3, pp. 667–677 (1983).
[39] Carpinteri, A. and Sih, G. C., Damage accumulation and crack growth in bilinear materials with softening: application of strain energy density theory, *Journal of Theoretical and Applied Fracture Mechanics* **1**, No. 2, pp. 145–159 (1984).
[40] Gilbey, H. J., Murphy, G. and Bergman, E.O., *Material Testing*, McGraw-Hill, New York (1941).
[41] Sih, G. C. and Moyer, Jr., E. T., Path dependent nature of fatigue crack growth, *Journal of Engineering Fracture Mechanics* **17**, No. 3, pp. 269–280, (1983).
[42] Tompkins, B. and Biggs, W. D., Low endurance fatigue in metals and polymers. Part III: Mechanisms of failure, *Journal of Material Science* **4**, pp. 544–553 (1969).
[43] Gifford, L. N. and Hilton, P. D., Preliminary documentation of Papst–nonlinear fracture and stress analysis by finite elements, NSROC–Preliminary Documentation, Jan. (1981).
[44] Bates, R. C. and Clark, Jr., W. G., Fractography and fracture mechanics, *Trans. Quart. ASM* **62**, pp. 380–389 (1969).
[45] Sih, G. C. and Barthelemy, B. M., Mixed mode fatigue crack growth predictions, *Engineering Fracture Mechanics* No. 3, pp. 439–451 (1980).
[46] Paris, P. C. and Erdogan, F., A critical analysis of crack propagation laws, *Trans. ASME Journal of Basic Engineering* **85**, pp. 528–534 (1963).
[47] Moyer, Jr., E. T. and Sih, G. C., Fatigue analysis of an edge crack specimen hysteresis strain energy density, *Journal of Engineering Fracture Mechanics* **19**, No. 4, pp. 643–652 (1984).
[48] Sih, G. C., Mechanics and physics of energy density and rate of change of volume with surface, *Journal of Theoretical and Applied Fracture Mechanics* **4**, No. 3, pp. 157–173 (1985).
[49] Sih, G. C., Thermomechanics of solids: nonequilibrium and irreversibility, *Journal of Theoretical and Applied Fracture Mechanics* **9**, No. 3, pp. 175–198 (1988).
[50] Civallero, M., Mirabile, M. and Sih, G. C., Fracture mechanics in pipeline technology, in *Analytical and Experimental Fracture Mechanics*, edited by G. C. Sih and M. Mirabile, Sijthoff and Noordhoff Publishers, The Netherlands, pp. 157–174 (1981).
[51] Benthem, J. P., Three-dimensional state of stress at the vertex of a quarter–infinite crack in a half–space, Delft University of Technology Report No. 563, The Netherlands (1975).
[52] Smith, C. W., Epstein, J. E. and Olaosebikan, O., Experimental boundary layer studies in three-dimensional fracture problems, *Advances in Aerospace structures, Materials and Dynamics*, ASME–AO–06, pp. 199–209 (1983).
[53] Sih, G. C. and Lee, Y. D., Review of triaxial crack border stress and energy behavior, *Journal of Theoretical and Applied Fracture Mechanics* **12**, No. 1, pp. 1–17 (1989).
[54] Plane Energy Density Damage Analysis (PEDDA), Institute of Fracture and Solid Mechanics Technical Report IFSM–84–131, December (1984).
[55] Cernocky, E. P. and Krempl, E., A theory of thermoviscoplasticity for uniaxial mechanical and thermal loading, *Journal de Méchanique Appliquée* **5**, No. 3, pp. 293–321 (1981).
[56] Cernocky, E. P. and Krempl, E., A coupled isotropic theory of thermoplasticity based on total strain and overstress and its prediction in monotonic torsional loading, *Journal of Thermal Stresses* **4**, pp. 69–82 (1981).
[57] Lieu, F. L., Experiments on cooling/heating of aluminum tube under torsion, Institute of Fracture and Solid Mechanics, Lehigh University (1986).
[58] Beghi, M., Bottani, C. E. and Caglioti, G., Temperature variations around the

crack tip during a fracture test, ***Metallurgy and Science Technology*** **2(3)** pp. 102–104 (1984).

[59] Sih, G. C. and Tzou, D. Y., Heating preceded by cooling ahead of crack: macro-damage free zone, ***Journal of Theoretical and Applied Fracture Mechanics*** **6**, No. 2, pp. 1033-111 (1986).

[60] Sih, G. C., Tzou, D. Y. and Michopoulos, J. G., Secondary temperature fluctuation in cracked 1020 steel specimen loaded monotonically, ***Journal of Theoretical and Applied Fracture Mechanics*** **7**, No. 2, pp. 79–89 (1987).

[61] Sih, G. C. and Chou, D. M., ***Thermomechanics of Solids***, Elsevier Science Publishers, The Netherlands. (forthcoming)

[62] Tzou, D. Y. and Sih, G. C., Thermal/mechanical interaction of subcritical crack growth in tensile specimen, ***Journal of Theoretical and Applied Fracture Mechanics*** **10**, No. 1, pp. 59–72 (1988).

[63] Sih, G. C. and Chao, C. K., Scaling of size/time/temperature associated with damage of uniaxial tensile specimens. Part 1: Progressive damage in uniaxial tensile specimen, pp. 93–108 and Part 2: Progressive damage in uniaxial compressive specimen, pp. 109–119, ***Journal of Theoretical and Applied Fracture Mechanics*** **12**, No. 2 (1989).

[64] Yang, W. H., A generalized von Mises criterion for yield and fracture, ***Journal of Applied Mechanics*** **47**, pp. 297–300 (1980).

[65] Yang, W. H., A useful theorem for constructing convex yield functions, ***Journal of Applied Mechanics*** **47**, pp. 301–303 (1980).

[66] Casey, J. and Jahedmotlagh, H., The strength-differential effect in plasticity, ***International Journal of Solids and Structures*** **20**, pp. 377–393 (1984).

[67] Sih, G. C., Thermal/Mechanical Interaction Associated with the Micromechanisms of Material Behavior, Institute of Fracture and Solid Mechanics Monograph (1987).

Author index

Subject index

Engin.

TA409.S55 1991
Sih, G. C. (George C.).
Mechanics of fracture
initiation and propagation.

20127268

11-19-92

[redacted]

[redacted]

[redacted]

1855

[redacted]